EELS

EELS

Biology, Monitoring, Management, Culture and Exploitation

Proceedings of the First International Eel Science Symposium

Edited by Andy Don and Paul Coulson

5m Publishing

First published 2019

Published by
5M Publishing Ltd,
Benchmark House,
8 Smithy Wood Drive,
Sheffield, S35 1QN, UK
Tel: +44 (0) 1234 81 81 80
www.5mpublishing.com

A Catalogue record for this book is available from the British Library

ISBN 9781789180695

Book layout by Servis Filmsetting Ltd, Stockport, Cheshire
Printed and bound in Wales by Gomer Press Ltd
Photos and illustrations by the authors unless otherwise indicated

Front cover photo credits:
Top photo by Jason Burton
Bottom middle photo by Bruce Ellender of the South African Institute of Aquatic Biodiversity
Bottom right and left photos by Michael J Miller, Katsumi Tsukamoto

Back cover photo credits
Top photo by S.MARR/SAIAB
Middle photo by Jason Burton
Bottom photo by Darryl Clifton-Dey of the Environment Agency, UK

Contents

V Aquaculture

VI Exploitation

VII Appendices

About the editors

Andy has worked in Fisheries teams in the Environment Agency and its predecessors for almost 30 years. During that time both his area and national roles have found him observing, appreciating, protecting and trying to understand the confounding animals that are the freshwater eels. He is a Fellow of the Institute of Fisheries Management and uses the Institute as a forum to raise awareness of Anguillid research and management.

Paul has worked with fish and fisheries all of his professional life and is currently the Director of Operations at the Institute of Fisheries Management in the UK.

Paul grew up fishing the ponds, drains and rivers of East Yorkshire, all of which held a good, but declining, population of eels. This early introduction to the species sparked an interest that has never left him.

Andy and Paul have worked together on several eel focused initiatives including the European Eel Conference held in London in 2013, a specialist workshop on practical solutions to overcome barriers to migration in 2015 and the International Eel Science Symposium on which this book is based.

This book is kindly produced with the support of the Institute of Fisheries Management and Environment Agency.

Environment Agency

ifm
Institute of Fisheries Management

Contributors

A. Alonso Aguirre, George Mason University, Fairfax, VA, USA
Rebecca Andreini, Department of Veterinary Medical Sciences, University of Bologna, Italy
Kristen Anstead, The American Eel Technical Committee and Stock Assessment Subcommittee, Atlantic States Marine Fisheries Commission, Arlington, VA, USA
Claude Belpaire, INBO, Belgium
Jonathan D. Bolland, Hull International Fisheries Institute, University of Hull, Hull, UK
Jacques Boubée, Vaipuhi Freshwater Consulting Limited, Morrinsville, New Zealand
Cédric Briand, EPTB Vilaine, La-Roche Bernard, France
Kerry Brink, University of KwaZulu-Natal, School of Life Sciences, Pietermaritzburg, RSA; World Fish Migration Foundation, Groningen, the Netherlands
Michael Brown, Independent Author
Jon Bolland, Hull International Fisheries Institute, University of Hull, UK
Ian A.E. Butts, Technical University of Denmark, National Institute of Aquatic Resources. Lyngby, Denmark; Auburn University, School of Fisheries, Aquaculture and Aquatic Sciences, Auburn, AL, USA
Jo Cable, School of Biosciences, Cardiff University, UK
Antonio Casalini, Department of Veterinary Medical Sciences, University of Bologna, Italy
John Conneely, Zoology, Ryan Institute, National University of Ireland Galway, Ireland
Paul Coulson, Institute of Fisheries Management, UK
Karen Creed, Zoology, Ryan Institute, National University of Ireland Galway, Ireland
Gareth D. Davies, National Fisheries Laboratory, Environment Agency, UK
Willem Dekker, Sustainable Eel Group SEG, Brussells, Belgium; Department of Aquatic Resources, Institute of Freshwater Research, SLU-Aqua, Drottningholm, Sweden
Estibaliz Díaz, AZTI-Tecnalia, Sukarrieta, Spain
C. Andrew Dolloff, USDA Forest Service, Southern Research Station, Fish and Wildlife Conservation, Virginia Tech University, Blacksburg, VA, USA
Andy Don, National Fisheries Services, Environment Agency, Bridgwater, UK
Colleen Downs, University of KwaZulu-Natal, School of Life Sciences, Pietermaritzburg, RSA
Sylvie Dufour, Paris Natural History Museum, France
Pietro Emmanuele, Department of Veterinary Medical Sciences, University of Bologna, Italy
Derek Evans, Fisheries and Aquatic Ecosystems Branch, Agriculture Food and Environmental Science Division, Agri-Food and Biosciences Institute, Belfast, UK
Carlos Fernández-Delgado, Grupo de Investigación 'Aphanius' Departamento de Zoología, Universidad de Córdoba, Spain

Nobuto Fukuda, Yokosuka Branch, National Research Institute of Fisheries Science, Japan Fisheries Research and Education Agency, Yokohama, Japan
Martin de Graaf, Narrandera Fisheries Centre, Narrandera, NSW, Australia
A. Ben Griffioen, Wageningen University, the Netherlands
Olga Haenen, Wageningen Bioveterinary Research, National Reference Laboratory for Fish, Shellfish and Crustacean Diseases, Lelystad, the Netherlands
Tessa van der Hammen, Wageningen Marine Research, IJmuiden, the Netherlands
Céline Hanzen, University of KwaZulu-Natal, School of Life Sciences, Pietermaritzburg, RSA
Dave Hoole, Keele University, UK
Joanna James, National Fisheries Laboratory, Environment Agency, UK
Fernando Jiménez, Centro de Experimentación Pesquera, Escuela Náutico-Pesquera, Gijón, Spain
Kenzo Kaifu, Chuo University, Japan
Olvin A. van Keeken, Wageningen University, Wageningen, the Netherlands
Johanna S. Kottmann, Technical University of Denmark, National Institute of Aquatic Resources, Lyngby, Denmark
Frode Kroglund, County Governor Office, Agder, Norway
Björn Kullmann, University of Hamburg, Center of Natural History, Hamburg, Germany
Hiroaki Kurogi, Yokosuka Branch, National Research Institute of Fisheries Science, Japan Fisheries Research and Education Agency, Yokohama, Japan
Mari Kuroki, Graduate School of Agricultural and Life Sciences, The University of Tokyo, Japan
Colin Lawton, Zoology, School of Natural Sciences, Ryan Institute, National University of Ireland Galway, Ireland
Joep de Leeuw, Wageningen Marine Research, IJmuiden, the Netherlands
Eamonn S. Lenihan, Zoology, School of Natural Sciences, Ryan Institute, National University of Ireland Galway, Ireland
Neil C. Lewin, National Fisheries Laboratory, Environment Agency, UK
Martyn C. Lucas, University of Durham, Department of Biosciences, Durham, UK
T. Kieran McCarthy, Zoology, Ryan Institute, National University of Ireland Galway, Ireland
Oliviero Mordenti, Department of Veterinary Medical Sciences, University of Bologna, Italy
Michelle Morrissey, Zoology, Ryan Institute, National University of Ireland Galway, Ireland
Vincent Nijman, Oxford Brookes University, UK
Gordon O'Brien, University of Mpumalanga, School of Biology and Environmental Sciences, Nelspruit, RSA
Albamaria Parmeggiani, Department of Veterinary Medical Sciences, University of Bologna, Italy
Adam T. Piper, International Centre for Ecohydraulics Research, Faculty of Engineering and the Environment, University of Southampton, UK; Institute of Zoology, Zoological Society of London, UK
Sebastian N. Politis, Technical University of Denmark, National Institute of Aquatic Resources, Lyngby, Denmark
Amy J. Reading, National Fisheries Laboratory, Environment Agency, UK
Randolph Richards, Fish Pharm Supplies Ltd, Alloa, UK

Kirby Rootes-Murdy, The American Eel Technical Committee and Stock Assessment Subcommittee, Atlantic States Marine Fisheries Commission, Arlington, VA, USA
Pierre Sasal, PSL Research University, Université de Perpignan Via Domitia, France
Penthai Siriwat, Oxford Brookes University, UK
Sune Riis Sørensen, Technical University of Denmark, National Institute of Aquatic Resources, Lyngby, Denmark; Billund Aquaculture Service A/S, Billund, Denmark
Kristen Steele, University College of London, UK
Florian Stein, Technische Universität Braunschweig, Germany; Sustainable Eel Group (SEG), Belgium
Graeme Storey, Environment Agency, UK
Ralf Thiel, University of Hamburg, Center of Natural History, Hamburg, Germany
Jonna Tomkiewicz, Technical University of Denmark, National Institute of Aquatic Resources, Lyngby, Denmark
Katsumi Tsukamoto, Specially Appointed Professor, Professor Emeritus, Department of Aquatic Bioscience, Graduate School of Agricultural and Life Sciences, University of Tokyo, Japan
James F. Turnbull, University of Stirling, UK
Alan M. Walker, Centre for Environment Fisheries and Aquaculture Science, UK
Nick Walker, George Mason University, Fairfax, VA, USA
Erina Watene-Rawiri, National Institute of Water and Atmosphere, Hamilton, New Zealand
Olaf L.F. Weyl, DST/NRF Research Chair in Inland Fisheries and Freshwater Ecology, South African Institute for Aquatic Biodiversity, Grahamstown, RSA
Chris F. Williams, National Fisheries Laboratory, Environment Agency, UK
Erica Williams, National Institute of Water and Atmosphere, Wellington, New Zealand
Hendrik V. Winter, Wageningen University, Wageningen, the Netherlands
Karen E. van de Wolfshaar, Wageningen Marine Research, IJmuiden, the Netherlands
Rosalind M. Wright, Environment Agency, Feering, UK
Takashi Yada, Nikko Branch, National Research Institute of Fisheries Science, Nikko, Japan
Toshihiro Yamamoto, Yokosuka Branch, National Research Institute of Fisheries Science, Japan Fisheries Research and Education Agency, Yokohama, Japan
Kazuki Yokouchi, Yokosuka Branch, National Research Institute of Fisheries Science, Japan Fisheries Research and Education Agency, Yokohama, Japan
Annalisa Zaccaroni, Department of Veterinary Medical Sciences, University of Bologna, Italy
Lluís Zamora Hernández, Universitat de Girona, Institut d´Ecologia Aquàtica, Capmany, Spain
Anke Zernack, Department of Archaeology and Heritage Studies, Aarhus University, Denmark

Preface

Welcome to the Proceedings of the First International Eel Science Symposium.

We hope that this book captures the flavour of the Symposium which took place across three sunny days in June 2017 at the Zoological Society of London.

From the editors' perspective, the Symposium will be remembered as an instructive, happy and friendly forum where folk gathered to learn and to teach. The original vision was that the Symposium would, '... foster and develop international links between key academic, policy, industrial and other stakeholders to promote better understanding of the issues, parallels and conservation opportunities for these iconic species...'. At the Symposium close and now, many months after that event, those aspirations have certainly borne fruit.

We have made a conscious effort to ensure that these Proceedings include a mix of cultural, scientific and management information. The Symposium Organizing Committee, from the very start, were keen to include new research being carried out on Anguillids around the world, but wanted also to have sessions that included cultural and societal themes. We recognized that these enigmatic, contradictory and confounding species share similar challenges and pressures wherever they are found.

A common thread of the research we were seeing was the potential to inform management decisions and solutions, we used the theme of 'Science into Management' to garner interest in our call for papers and the Committee was overwhelmed by the quality and quantity of submissions. After much agonizing, we had our final programme, much of which is represented in the following pages. At the time, we imagined a forum where these compelling papers could be presented to a wide and diverse audience.

This turned out to be one of the Symposium's great strengths – its diversity of subjects and presenters. An early decision was made to try and avoid a purely euro-centric perspective and we strove to include as many wider global papers, presenters and species as possible. In the end, we had 25 countries represented, from as far afield as New Zealand, South Africa, Vanuatu, French Polynesia, Japan, Canada and the USA. This melting pot of people and ideas generated much lively debate – always a great indicator of people's engagement with the subject matter and the event.

Two of the legacies of the Symposium are this book and its companion, the dedicated Anguillid eel edition of the journal, *Fisheries Management and Ecology* (FME), where the reader will find more peer-reviewed research papers that were presented at the Symposium (https://onlinelibrary.wiley.com/toc/13652400/2019/26/1). Details on this journal and how to access it are included in Appendix 3.

The Symposium, as well as having oral presentations, had a poster evening at The Vaults within the RSA on The Strand, and a river cruise on the Thames. These extra mural activities were well-received and allowed delegates to socialize in less formal environments and enjoy London's sights. We have included the original Programme along with a selection of Symposium photographs in the appendices.

This event would never have happened at all were it not for the intellectual and practical support of fellow Organizing Committee members Dr Matt Gollock, Dr Alan Walker – who also acted as guest editors for the FME journal – and Mr Iain Turner. We had great financial support from the represented sponsors. Thanks must also go to the numerous helpers, facilitators and other Anguillophiles who ensured that the Symposium ran as smoothly as it did. You know who you are – thank you.

Hugh Fearnley-Whittingstall agreed to officially open the event for us. Hugh, with his background of environmental campaigning including 'Hugh's Fish Fight', gave a compelling speech on the threats to the wider environment and fisheries and on his involvement with eels in particular; in doing so he introduced our first session, 'Setting the Scene' perfectly.

A special Thank You must go to the Environment Agency, for their generous donation to enable publication of these Proceedings.

Clearly, we are biased, but all in all it was a great event and we would like to thank again all of the presenters, delegates, sponsors and our co-organizers for making it so.

A last and most important thank you, for tolerating unsocial hours, putting up with distracted conversations and for learning about eels (whether they wanted to or not), must go to our long-suffering partners, Jacqui and Natalie, without whose support the Symposium and these Proceedings would never have happened.

We hope you enjoy the fantastic diversity of papers presented and their astonishing images.

The Editors,
Andy Don FIFM
Paul Coulson MIFM

Foreword

There can be little doubt that across the world Anguillid eels are experiencing one of the most dramatic population declines in recent memory. The impacts are far reaching and transparent given the value societies place on eel as a source of food. There are likely to be less obvious impacts too, relating to the function eels have in our aquatic ecosystems.

We are continuing to build on our understanding of the role diverse pressures, such as a changing climate, disease and barriers to migration play in this decline. The unique life history demonstrated by Anguillid eels increases this challenge but, despite this, action is underway.

In 2007, the Council of the European Union introduced regulations to establish measures for the recovery of the stock of European eel. A key part of this regulation is the production of Eel Management Plans, which set out the measures each member state will employ. In England and Wales specific legislation has been developed to implement measures addressing two of the most significant pressures. Intake structures must be screened to prevent loss of eels across their age range and eel passes must be constructed to allow upstream passage.

These measures will not only help to improve the freshwater production of eel stocks, they will also contribute to our wider environmental goals. With an increasing focus on natural capital and the services provided by the natural environment, addressing the decline in eel populations will take on an increased significance worldwide.

It was heartening to witness the enthusiasm, expertise, ambition and innovation around this subject, demonstrated throughout the First International Eel Science Symposium and reflected in these proceedings. As scientists and fisheries managers it is essential that we come together to share our understanding and experiences. In doing so we can continue to identify and implement the most effective measures to help restore our eel stocks.

I would encourage us all to meet again soon.

Dr Graeme Storey
Fisheries Manager, Environment Agency

A note on nomenclature

The chapters use the following binomial names and their common equivalents drawn from the IUCN listings. Where the species is clear from the context plain eel/eels and other local names are used. The preferences of the individual chapter authors were followed in this regard.

Anguilla anguilla	European eel
Anguilla australis	Short-finned eel, shortfin eel
Anguilla bengalensis	African mottled eel
	Indian mottled eel
Anguilla bicolor (bicolar/pacifica)	shortfin eel (Indonesian/Indian)
Anguilla borneensis	Indonesian longfinned eel
Anguilla celebesensis	Celebes longfin eel
Anguilla dieffenbachii	New Zealand longfin eel
Anguilla interioris	highlands long-finned eel
	New Guinea eel
Anguilla japonica	Japanese eel
Anguilla luzonensis	Luzon mottled eel
	Philippine mottled eel
Anguilla marmorata	giant long-finned eel
	giant mottled eel
	marbled eel
Anguilla megastoma	Pacific long-finned eel
	Polynesian longfin eel
	Polynesian longfinned eel
Anguilla mossambica	African longfin eel
	longfin eel
Anguilla obscura	brown eel
	Pacific short-finned eel
	Pacific shortfin eel
	Pacific shortfinned eel
	Pacific shortfinned freshwater eel
	short-finned eel
	South Pacific eel
Anguilla reinhardtii	Australian long-finned eel, marbled eel
	speckled longfin eel
Anguilla rostrata	American eel

PART I

European eel, a local history

CHAPTER 1

Eel memories from the Somerset Levels

Michael Brown

Independent Author

Michael Brown has been, among other things, an eel fisherman, a commercial eel buyer, an eel smoker and an eel restaurateur. He is also the author of several books.

He has witnessed fundamental changes in the eel fishery, from its being a cottage industry to becoming a multi-million-pound operation spanning continents. Michael's passion and knowledge for the subject is unique.

The following are Michael's recollections from a working life on the Somerset Levels. More captivating tales from an eel catcher and trader can be found in Michael's 2011 book *Moonlighting: Tales and Misadventures of a Working Life with Eels.*

For thousands of years in Europe and around the world, eels – and elvers – have been a vital source of food for man. Wherever there were eels people did not starve for they were nutritious, high in protein, good to eat and not difficult to catch. Valuable in other ways too: their skin gave supple strength to implements, hinges, flails, connections; they were also a highly prized commodity that could be counter-traded and used as currency.

The eel was indeed so important to some communities like the Maori of New Zealand or the Aboriginal peoples of Australia, that it achieved almost spiritual status, figuring large in their art and culture. In nearly all the state museums 'down under' eel traps beautifully woven from dried grass and reed adorn the walls, thousands of years old. The Parramatta River, around which Sydney grew, means in the language of its indigenous people 'the place where the eels sit down.'

In South West England, in what is now the county of Somerset, an inland sea, a shallow watery waste, covered much of the northern area of what has since become the Somerset Levels. Six thousand years ago Stone Age people lived around its margins and on its islands of reed and willow, they built wooden trackways to move about. They too would not have gone hungry for these brackish waters that rose and fell with the tides were a perfect habitat for fish and fowl, and teemed with eel.

With the spread of the monastic orders in around AD 800 the Benedictine monks sought peace and contemplation in this rural fastness, establishing abbeys at Athelney, Muchelney (the suffix '-ney' in Anglo-Saxon denoted an island) and greatest of them all, Glastonbury. The monks prized eels and were ingenious at catching them. At Muchelney Abbey, just a mile from where I live, they developed an eel fishery, probably a banked area that could be drained and harvested, that is mentioned in the Domesday Book of 1086. It is recorded as yielding 5000 eels a year, 1400 of which were due to the abbot and his brethren, the 'whole worth three pounds'. Glastonbury too had a managed fishery that yielded eel and carp for the monks' table.

Meanwhile in East Anglia, on the other side of the country, the importance of eel as a trading currency was famously demonstrated in Ely – the island of eels – in Cambridgeshire. Here the stone for the great cathedral built there in the 12th century was quarried from the nearby abbey lands of Peterborough and paid for in kind – to the tune of 8000 eels a year.

Down through the centuries eels and elvers remained a vital source of food in South West England particularly in the counties bordering the Bristol Channel and the Severn Estuary. Travellers in the 1700s and 1800s passing through the town of Langport in Somerset observed with wonder, and mild distaste, that the locals caught and ate quantities of eel. Certainly, up until the Second World War eel was a vital source of food. When I first came to Somerset in the early 1970s and became involved with buying elvers to supply the restocking market in Europe, my mentor and helper at the time was a shepherd, Ernie Woods, who lived on the farm where our holding tanks were located.

During the Great Depression in the 1930s he had lived down on the Levels, 'on the moor' as he called it, on the banks of the River Parrett. It had been a hard, poor upbringing. His father worked as an agricultural labourer on the farms around, often being paid 'in kind', usually in cider for a day's work. Not much use for a growing family. As Ernie said, 'we lived on the moor and we lived off the moor', true hunter-gatherers. 'I've et everything that moved down on thet there moor, fish, eels, elvers, swan, rabbit, crow; you name 'em, I've ad en.' As a small boy he went eeling in the rhynes – the deep water ditches – around his home with a five pronged spear like a Greek trident, a common item that could be bought at any of the blacksmiths round the edge of the moor (Figure 1.1). What they didn't eat themselves he'd skin and sell around the village.

Figure 1.1a and b Traditional eel spears that would have been used in the Somerset Levels. The spear in b is from the 17th century. Photographs: Andy Don.

One of the highlights each year for his family was the spring migration of the elvers past his house on the big tides. The family would cut a notch in the bank so that they could be diverted straight into a trench dug in the garden. 'The ducks, cor they loved them. We did too. And mother used to cook 'em in a duck egg omelette … though I could never eat too many … they were rich mind … you could'n ave too many.'

Others have memories of river life in those days too: Dion Warner, a retired engineer, and now dedicated Somerset naturalist, recalls how as a child he was evacuated from London during the Second World War and joined his grandparents who also lived on the River Parrett, just below Ernie. His grandfather was a ditcher, working for the Drainage Board, paid by the chain (22 yards) of ditch cleared. The old man used to set night lines baited with chicken gizzards. Dion remembers him coming home on his bike often with an eel over the handle bars trailing on the road. 'Big ones like that, they'd bake, stuffed with parsley sauce.' Ernie also told the story of how shortly before the war the water authorities had drained a large lake about 10 miles (16 km) away to remove the silt. This was the Chard reservoir, an area of about 50 acres (20.2 ha), shallow, full of nutrients and warmed like a rich soup in the summer. Fish thrived in it. The size and spectacle of the eels they found in it had passed into local legend and history.

'I knew chap wot lived up Chard who were there at the time,' Ernie recalled, 'Ee said they were *massive*, six foot long, and they barked like dogs!', as they were being hauled away by the cartload.

Yet it is probably the arrival of the elvers each spring that will be remembered best. As a young man in the 1930s the migrations that Ernie witnessed were huge. Living beside the Parrett, untouched by commercial fishing, the river was so full of elvers, clogged by their sheer numbers, 'so thick that you felt you could walk across 'em.' For the locals, the coming of the elvers represented fresh food, a delicacy – and free. Part of nature's bounty, keenly anticipated after a long winter. They had always been part of the local diet, and it was only in the 1960s that their fishing became commercial in Somerset. Before decimal currency introduced in 1971, the local buyer paid 2/6 per pound (c. 27.5 pence per kilogram) and pay out was every Friday night in a pub called the Black Smock on the river bank. Not much fishing ever happened that night. In the early 1970s when I first arrived in Somerset it had risen to 50 pence per kilogram, most of the elvers were bought for restocking on the continent or for the Spanish 'tapas' market. At a rough guess, back then, the Parrett and its adjacent rivers on the Levels probably produced 10–15 tonnes per year.

Yet even when it had become fully commercial there were still local people who were not interested in selling because they were fishing only 'for a feed'. I remember one in particular, an ancient withy grower, who would rattle down to the river on his bike with net on his shoulder and bucket on the handle bars. When asked if he'd like to sell his catch, he'd refuse and explain, 'No my wife, she do love a feed of elvers, they do make er proper frisky!' And as they were both in their late 70s one could not fail to be impressed by the life-enhancing and aphrodisiacal properties of the elver.

When I first started buying elvers on the Parrett in the 1970s these were still vintage years: 1975 and 1981 were very good, but 1979 and 1982 were colossal, the stuff of legend. My wife and I used to fish for them ourselves then rush back early to weigh in the catchers who fished for us (Figures 1.2, 1.3 and 1.4).

I remember one warm spring night in April in 1979, seeing a band of elvers about 1 m wide straining upriver against the current like an undulating carpet, millions and millions of them;

Figure 1.2 A glass eel fisherman on the banks of the River Parrett in Somerset. Photograh: Rod Baillie-Grohman.

Figure 1.3 Freshly caught River Parrett glass eels. Photograph: Rod Baillie-Grohman.

Figure 1.4 Michael weighs in the catch. Photograph: Rod Baillie-Grohman.

putting my hand in the water, I could feel the points of their noses prinkle against my palm. And thinking, 'This is an elver migration, this has been happening for thousands of years.' It was both humbling and awe inspiring: the urgency of their swimming, their sheer numbers. And once the elvers came they seemed to keep on coming, tide after tide, as if nature were showing off, demonstrating her boundless supply. These were often some of the most challenging, stressful, times for us. We'd run out of markets, all orders fulfilled, our holding tanks overflowing (Figure 1.5). Finding new outlets became increasingly difficult; everyone had elvers. At the same time, we were too frightened to turn away our catchers for fear that we might never get them back. It was then, just as we were about to drown, that, often thanks to a friendly elver station, a lorry would appear bound for Spain or Eastern Europe just in time to clear us out and rescue us from going under. We'd have survived – just, for another day!

Back then it was common for a 'crew' of good fishermen to weigh in well over 1 tonne in a season. There were legendary catches. Back in the 1960s, Ernie, the shepherd, held the record for the biggest single net full caught in his ancient hand net, so heavy it took two strong men to haul it in: over 100 kg of elvers in a single dip and earning himself what was then a small

Figure 1.5 Michael's glass eel holding tanks. Photograph: Rod Baillie-Grohman.

fortune – nearly £15. It had been the talk of the river for years. Later, in the season of 1982, during a huge tide a muscular band of brothers netted nearly half a tonne in a single night.

In terms of catches however, those early years belonged to a character called George Cotty. He'd lived on the river and had been catching elvers all his life. In the 1950s, he'd sent them up by train to Bristol to the glue factories – it must have been an incredible smelly glue. Beside his cottage on the river he'd discovered a private goldmine, the perfect spot to fish: a freshwater outflow from the moor, controlled by a hinged clapper. On a good night when the elvers were running, George would wait until the river had dropped to its lowest then crack it open to release a flow of what he called 'that there sweet water' from the moor, which drew every elver downriver as if magnetized. Any fishermen who tried to poach his spot would be told to 'Bugger off' in no uncertain terms. So, for hour after hour, long after the others had gone home, he fished the river down to a thin trickle between the high muddy banks, mining it for its precious cargo. They were beautiful fish, perfectly caught, unstressed, and he stored them lovingly spread on trays in the back of his incredibly smelly van. Alone, each season he caught well over 1 tonne; night after night we airfreighted them to our main customer, the Deutsche Fischerei Verband in Hamburg who loved them; 'Viel, viel besser', we were told – better than those French elvers caught by trawlers in their boom nets.

Yet no two elver seasons were ever the same, bad seasons could easily follow boom. One thing became abundantly clear: it was not possible to rely on elvers alone to support a family. For that reason, we began to establish, alongside the elvers, an eel smokery to provide a steadier all-year-round income. At first, I caught my own eels, setting fyke nets by night down on the Levels but soon discovered that the brown eels I trapped were, as a Dutchman described them, 'Only good for kookink!' Then through a contact, I was lucky enough to go to Germany to learn the craft of smoking eel in a wood-fired smoker. Over the years as the business grew and flourished we needed increasingly larger quantities of quality eel. By chance I was put in touch with the river keepers on the chalk streams of southern England: rich in food, crustacea, snail and insect, these rivers, the Test, the Avon, the Itchen and the Stour, yielded superb eel, all caught on the racks. It was in fact with eel from the River Test that we won the Tesco Food Award for our smoked eel in 2005 – along with a very useful and sizeable cheque.

By the mid-1990s, it was plain that elver catches in the South West and on the Parrett were in serious decline despite a frenzied increase in catch effort, fuelled by the astronomic prices paid by the buyers supplying, in particular, the lucrative Chinese market. There was now no

Figure 1.6 Contemporary glass eel fishery on the River Parrett. Photograph: Environment Agency.

pause between tides, fishermen were on the river day and night, every day for 3 to 4 months (Figure 1.6). In Scotland where my wife and I had fished an average of 200 kg each May on the Laxford in Sutherland since the 1970s, that river was now yielding just a handful of elvers. And then nothing. Having started at 50 pence per kilo in 1974, by 1998 when I gave up my involvement in the elvers to concentrate fully on the smokery the price had reached £285 per kilo. Danger signs were flashing. With the adult eel population, by contrast, eel catches from the chalk streams remained remarkably stable, around 6–8 tonnes a year through the 1990s, only declining very gradually until 2007 when most of the estates decided to halt eel fishing on their rivers. The smokery had then to find other sources of supply. Yet the elver migrations could still surprise. In 2013 and 2014, when the Somerset Levels were hit by massive floods, there were very good elver catches, not quite like the old days but still significant and enough, after all markets had been satisfied, to allow a significant restocking of rivers and waters in both Somerset and Gloucester.

My main hope is that this wondrous, ancient and mysterious creature, so much part of the rural life and history of Somerset, will survive and thrive again. Instinct says that it will, and that it'll probably outlive us all.

CHAPTER 2

Smoked eel was not a delicacy! The history of smoked eel in Amsterdam and the rest of the world

Willem Dekker

Sustainable Eel Group SEG, Brussells, Belgium; Department of Aquatic Resources, Institute of Freshwater Research, SLU-Aqua, Drottningholm, Sweden

'Eels (...) that are smoked too long become hard, tough and less palatable – if they are not smoked enough, they easily become rancid and may spoil soon.' – What? Does Friedreich (1846) really describe one of the most expensive and tastiest delicacies in our modern fish shops? Yes, he does – but smoked eel was not expensive in 1846, and moreover, it was not a delicacy. The development of the modern, tasty smoked eel was a major step forward for the eel industry, which I would like to discuss in this chapter. A story smoking techniques, product development and promotion campaigns, trade and export, and, ultimately, the development of the entire European commercial eel fishery.

Hung in the chimney

This story starts with a recipe from 1772 – not because no eel was smoked before, but because this seems to be the oldest recipe that has been handed down. It is in Dutch – other recipes in German exist from the same period, but none in English or French; the first English recipe for smoked eel dates from 1871 – a direct translation from the French (Gouffé, 1869), which in turn was probably a translation from a German original. In 1772, Catharina Zierikhoven (pseudonym of J.A. de Chalmot, c. 1730–1801, from Kampen) wrote the *Volkoomen Neerlandsch Kookkundig Woordenboek* (*The Complete Dutch Cooking Vocabulary*), containing recipes for aalmoes (mashed eel), for white soup of eel, brown soup, stewed and boiled eel, fricassée of eel, fried and roasted eel, eel at sticks and in pie, and finally: smoked eel to be eaten with cabbage (Figure 2.1). A fairly simple recipe, of which the most prominent line reads: 'hang them in the chimney, until they are adequately dry'. Comparison with other recipes (in the Netherlands, Germany and Russia, among others) yields a number of variants, in which the eel was dried and/or salted for a shorter or longer period, sometimes the subcutaneous fat ('the second skin') was removed from the raw eel before smoking, and the duration of the smoking process varied from a few days, usually a few weeks, sometimes up to half a year. In doing so, one obtained a

Gerookte Aal om bij Kool te eeten.

Stroopt het vel van de *aal*, waar na men ze wel zuivert, wascht en ter deegen afdroogt; kerft vervolgens de geheele *aal* aan weerzijden met fijne kerven, en bewrijft het met *zout*, *nagelgruis*, *peper* en *ſalie*, trekt er als dan het vel weder over, na dat het wel gezuivert en afgewasſchen is, en hangt ze in de ſchoorſteen, tot dat ze bekwaamelijk droog zijn; wanneer men ze vervolgens tot het gebruik wil gereed maaken, trekt men het vel er weder af, en men braad ze een weinig op de rooster of in de pan, is een zeer goed eeten; men kan ze ook bij *witte* en andere *kool* ſtooven.

A 3 AAL.

Figure 2.1 The recipe for smoked eel, from the *Volkoomen Neerlandsch Kookkundig Woordenboek* by Catharina Zierikhoven (1772).

The text translates as 'Smoked eel to be eaten with cabbage. Strip the skin of the eel, after which it is cleansed, washed and dried; then carve the whole eel on both sides with fine notches, and rub it with salt, nail grit, pepper and sage, then pull on the skin again, after it has been purified and washed off. Hang them in the chimney, until they are adequately dry. If one then wishes to prepare them for use, remove the skin again, and roast them a little on the grid or in the pan, is very good to eat; they can also be stewed with white and other cabbage.'

product with a good shelf life, which could be easily stored (dry), and was nutritious (fat) – but actually not very edible. Before eating, the smoked eel first had to be soaked in water for a night, boiled for an hour, or cut in pieces and fried. In some places in Germany and further to the east, one wrapped individual eels in a piece of paper before being hung in the chimney, which obviously reduced the impact of the smoke, and thereby improved the taste; Reichskanzler Otto von Bismarck (1815–1898) regularly had smoked eel for breakfast. Since the smoking time was still in days/weeks, the paper-wrapped product must have remained as raw and rancid as before.

What was the big difference with the 'modern' smoked eel, our modern expensive delicacy? Actually, only one thing: the long smoking time, at a low temperature. Below 30°C, the product actually dries out more than it cooks, taking in a lot more smoke and losing far more moisture. At higher temperatures (in modern 'hot-smoking' practices, approx. 80°C) the fish will soon be cooked, and further smoking would make the fish fall apart – already within a few hours. A hard, dry, strongly smoked product (cold), compared to a soft, creamy and tasty fillet (warm) – that is about the same difference as between a thoroughly smoked bacon (cold) and a freshly baked bacon crack (hot). Most likely, the cold-smoked eel was truly hung between the bacon and sausages, in the smoking cupboard behind the hearth. In Germany, all smoked eel is still referred to as Spickaal, bacon-eel.

All over – none left

There are several historical sources describing smoked eel as an essential part of soldiers' rations (for example, Napoleon on his way to Moscow). Cheap, nutritious, unperishable and often rancid – the ideal soldiers' food. Furthermore, international trade of dried, salted or smoked eel occurred: from Holland to London, from Holland to St Petersburg, from Venice to Vienna, from Paris and Valencia to London, and so on. In all these cases, the eel was preserved, and traded over long distances – but the exported volumes remained small, much smaller than the local sales, and especially far less than the subsistence usage by the rural population, preserving their own smoked eel for consumption (or sometimes even as pig food in late winter). In fact, smoked, salted or dried eel was so well known, that nobody bothered to document the preparation method. In a Swedish cookbook from 1755, I found the instruction 'treat the eel just as you do with pike', without further details. Everyone knew how to do that, and everywhere in Europe, salted or smoked eel was produced specifically for home usage. Per kilogram, smoked eel was about equally expensive as fresh eel. However, recognizing that at least 2 kg of fresh eel were required to make 1 kg of smoked eel, it becomes clear that the smoked eel was not so appreciated at all. But eel was everywhere, so abundant, and so cheap and nutritious.

It is completely unclear how much eel was processed this way in total. There were countless farmers, watermill owners and local residents each catching and preserving some eel. Eels were present in all ditches and canals, rivers and streams, marshes and swamps – all over Europe. Countless numbers of people, each making a tiny small catch – we can only guess how much that has been in total.

Cold-smoked eel is no longer produced today. As far as I know, the production came to an end around 1900. In 1909, I found a last newspaper article that featured a question from a reader asking: 'Recently a reader asked me how to smoke eel' (*Nieuws van de Dag*, 24 November 1909). Whether that reader really sat down to make a 'nice cold-smoked eel', I doubt. Furthermore, there is a German theatre play of 1846, in which a thin character performed, wearing far too large clothes, nicknamed 'Spickaal' (Figure 2.2). That gives some impression of the looks of a cold-smoked eel, a Spickaal: shrivelled, in a far too large skin. For the rest, no single picture or drawing of a cold-smoked eel has been handed down, to my knowledge, nor is there anyone still producing it anymore. Cold-smoked eel, for centuries the most common product in large parts of Europe, disappeared without a trace.

A new delicacy

In 1908, Wilhelm Doose (1871–1937, Celle, near Hanover) wrote: 'The eel, a cheap folk food in the past, has become a table fish everywhere, and smoked, it is a delicacy, which can only be obtained for expensive money.' In less than two decades, the situation had changed completely: from cheap and simple, to pricey and exquisite. From that moment on, a rapid expansion of eel fisheries began, all over (northern) Europe (and later in the south), focusing on the market for hot-smoked eel, especially in Germany. Where did that development come from?

Figure 2.2 'Spickaal' (bacon-eel), the nickname for a skinny man in too large clothes. Illustration for a theatre play by Hopf, 1846 (background greyed-out).

The earliest record of hot-smoked eel, which I have been able to find anywhere in Europe, comes from Mr H.E. Sauvage, a French fishery inspector visiting the Netherlands on a study trip in 1881. In Harderwijk, he saw the local fishermen smoke their eel on the beach. In his words: 'On the ground, a fire is made under a barrel without a bottom. That fire makes a lot of smoke, so that the eel is not only smoked, but also cooked at the same time.' Without a doubt, a description of small-scale (non-commercial) hot-smoking, in which completely cooked fish is produced within short time. Although this is the first *certain* indication, I do not think this means that hot-smoking is a Harderwijk invention; Sauvage just happened to note it there and wrote it down. Moreover, decades earlier, there had already been quite clear indications of hot-smoking: the sale of 'Amsterdamsche Gerookte Paling' (Amsterdam smoked eel) in various cities in the Netherlands.

From 1847 onwards, advertisements appeared regularly in the Amsterdam newspapers, for 'fine smoked eel' (Figure 2.3 shows the first), explicitly mentioning 'dagelijks versch' (daily fresh) – that could not have been the eternally preserved, cold-smoked eel. From 1858, these campaigns expanded outside Amsterdam, and 'Amsterdam smoked eel' was sold in several other cities too (Figure 2.4). Note that all these cities were along the newly built railways (with the exception of Zwolle, which only got a rail connection in 1865, but advertised '*Amsterdam* smoked eel' since 1853); furthermore, note that the sale usually took place during the annual fair; and that the seller emphasized over and over again to come from Amsterdam, or to return there soon. Attempts to set up a smokehouse in Rotterdam, The Hague and Delft were more difficult (neighbours complaining about the smoke) – but even there they produced '*Amsterdam* smoked eel'.

Gerookte Paling-Berigt.

DE ONDERGETEEKENDE, genoegzaam bekend, berigt door deze, dat hij thans extra fijne GEROOKTE PALING heeft, dewelke dagelijks versch bij hem aan Huis te bekomen zijn.

J. C. F. KNAPP,

Korte Leidschedwarsstraat tusschen de Leidschegracht en Leidschestraat, N°. 87.

(8262)

Figure 2.3 Advertisement in the *Algemeen Handelsblad* of Saturday 22 May 1847 for smoked eel. The text translates as 'Smoked eel message. The undersigned, well known, reports that he now has extra fine smoked eel, which can be obtained daily fresh from him at his home address.'

In Amsterdam, it was mainly the brothers Johan C.F. Knapp (1828–1900) and Frederik H. Knapp (1832–1917) in the Korte Leidschedwarsstraat, in one of the slums, who took the lead. Later, a number of unnamed smokers (the Knapps themselves?) on the Visscherspad joined in (Figure 2.5, now Schollenbrugstraat; sale in Café Schollenburg, just behind the bridge). Although the eel was not caught in the stinking Amsterdam canals (there was probably no eel

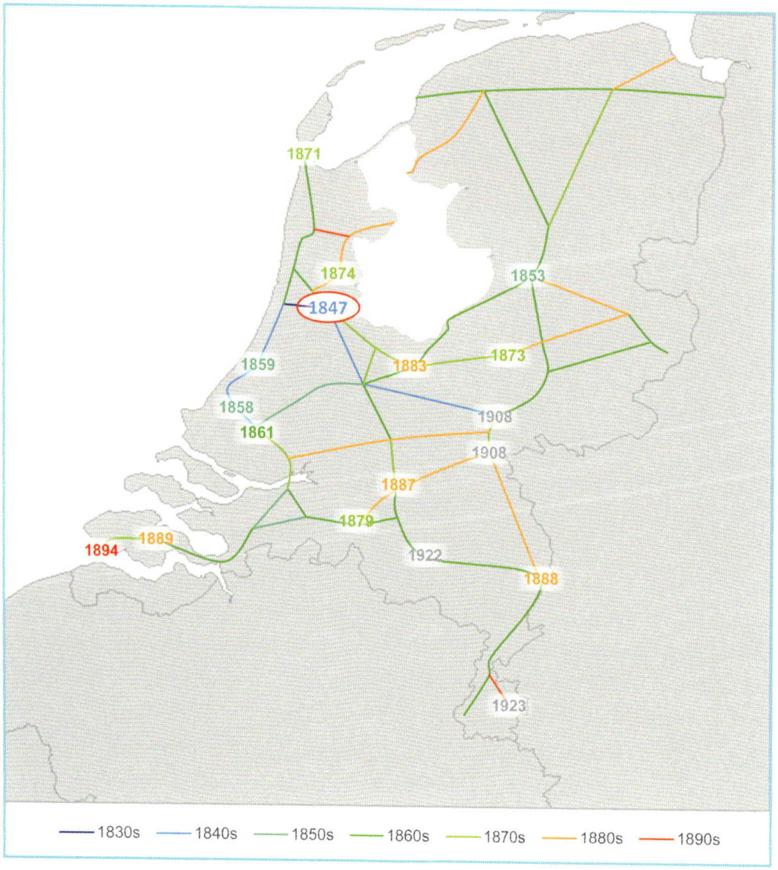

Figure 2.4 The year in which first advertisement appeared in various cities, offering 'Amsterdamsche Gerookte Paling' (Amsterdam smoked eel) in the Netherlands. The newly built railways are indicated, coloured according to their building years. Data: www.delpher.nl.

Figure 2.5 The Visscherspad in Amsterdam, where smoked eel was produced. The smoking took place in small, wooden sheds along the waterfront – it seems likely that the sheds in this picture actually are those. Unknown date, before 1931. Photo: www.vriendenvanwatergraafsmeer.nl.

there anymore), the smokehouses did bring eel into the city. In 1886, a violent revolt broke out in the slums of Amsterdam, which started with an eel-pulling competition (tearing a living eel from a rope it had been suspended on – a deliberate breach of animal-protection laws). Most likely, that eel was brought into the city by our smoke houses.

From slum to sleek

The whole development of eel-smoking in Amsterdam remained a bit ponderous until 1893 (Figure 2.6). 'Why are those low smoking lofts not a little higher? Because the eels are no lengthier than that, sir!' (*De Amsterdammer*, Sunday 2 June 1889). That changed completely in 1893, when fishmonger Saur started promoting the smoked eel.

Christiaan Saur (Saür, or Sauer, 1849–1928) was a relative of the Knapp brothers, a descendant of a Jewish family from central Germany. Grandfather Saur started a fish shop in Amsterdam at the beginning of the 1800s. (In French, 'saur' is the common word for smoked herring. Had the family purposely adopted that name? Or the other way around?) In 1879, the young Christiaan

Figure 2.6 A photo from 1907, of a professional eel smokehouse in Kortenhoef, 20 km southeast of Amsterdam. In the Amsterdam backyards, the smokehouses will not have been very different. Photo: *Op de Hoogte* 4 (9): 501–507.

took over his father's shop, financed a new building at Rokin 134 in Amsterdam (Figure 2.7, now a shop of ladies' underwear), and started an energetic campaign to renew and expand his commerce. In 1887, he actively participated in the 'Tentoonstelling van Voedingsmiddelen' (Exhibition of Foods) on what is now called Museumplein (the square behind the Rijksmuseum). He received a gold medal for 'smoked river fish' (he was the only participant) and he was mentioned frequently in the daily exhibition newspaper. Over the years, he sometimes advertised 'smoked eel', but on Sunday 20 August 1893 the text changed to 'Hollandsche Gerookte Paling' (Holland smoked eel). This cost Dfl 1.50/kg (€18/kg, 2000 price level), about twice the price of fresh eel. The next day, and all of the following days in the next 2 years, he advertised Holland smoked eel as being available in his shop, or even better, in his new and luxurious restaurant at the Reguliersbreestraat 11, in Amsterdam. Of the 234 advertisements for eel appearing in 1895 in the Netherlands as a whole, 148 were from Saur (Figure 2.8). After 1895, his campaign for smoked eel ended, and his other products were promoted.

Where Christiaan Saur got his Holland smoked eel from, I have not yet been able to figure out. But look at Figure 2.9, showing the founding years of the currently existing eel smokehouses in the Netherlands: apart from the modern smokehouses (after 1980, mostly aquaculture), the majority of the existing smokehouses mentions on their websites a foundation year ('Since …') between 1893 and 1900. The majority of those smokehouses originated in Saur's years! Whether

Figure 2.7 The interior of Christiaan Saur's fish shop in 1879, at the Rokin 134 in Amsterdam.

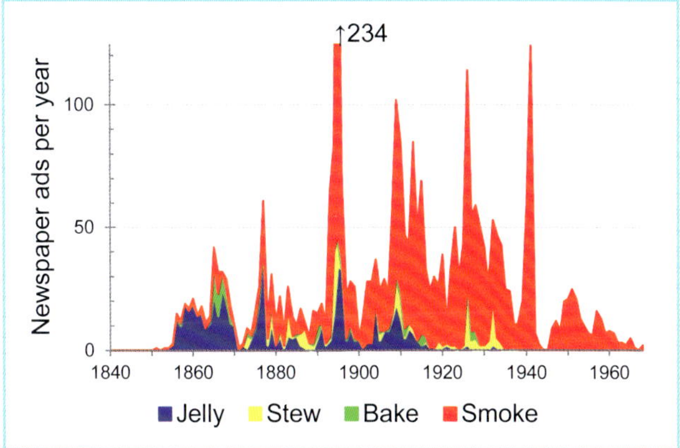

Figure 2.8 The number of newspaper advertisements per year for various eel products. Data: www.delpher.nl.

Saur visited them in person, or whether the news that money could be made from smoked eel spread like wildfire – in both cases it was Saur giving the impetus. Apparently, his smoked eel promotion campaign laid the foundation for a new business sector in 1893, and the smoking booths in the Amsterdam slum backyards were quickly replaced by real smokeries, mostly in fishing ports around the Zuiderzee (now IJsselmeer, the 1800 km^2 estuary northeast of Amsterdam). Soon after, smokeries were established in Germany too. The name 'Amsterdam smoked eel' remained in use until 1935.

From 1893 onwards, smoked eel was no longer a cheap countryside food, but a luxurious delicacy, offered in trendy shops in the city.

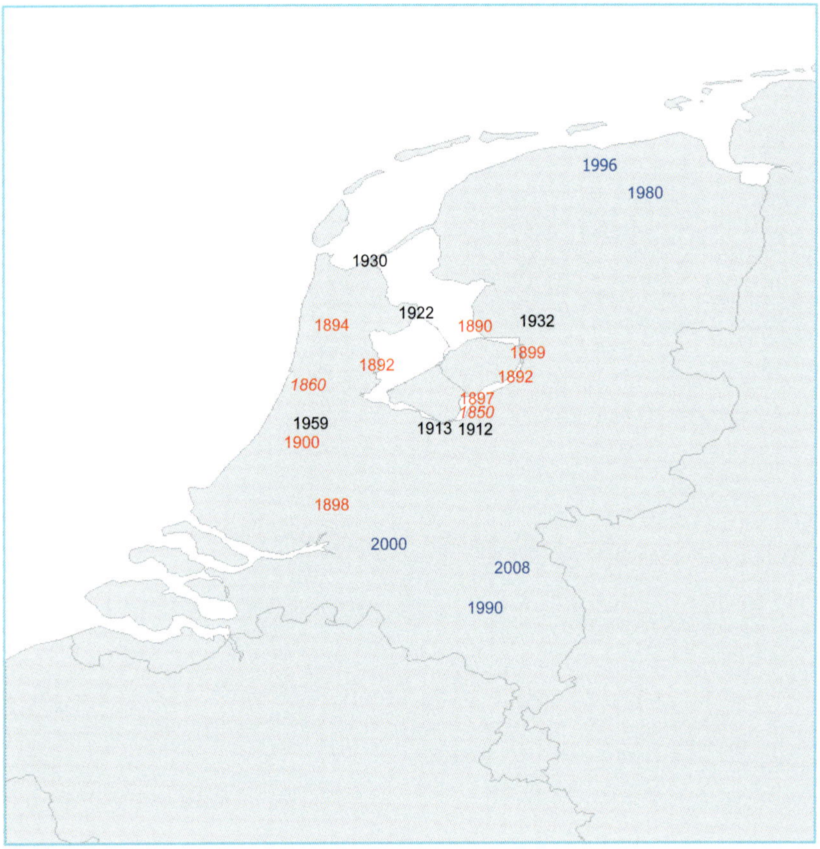

Figure 2.9 The founding year of the existing eel smokehouses in the Netherlands. Italicized: two herring-smokeries, converting to eel-smoking in the 1890s. Data: their own websites.

Igniting a new eel industry

The eel fishery in Europe developed from a centuries-old tradition of small-scale fishing, mostly for subsistence use. Such subsistence fishing had declined sharply in the course of the 19th century: damming, land reclamation, draining, 'normalization' of rivers and water pollution all had their effects. Many small subsistence-fisheries had disappeared early in the 19th century. In several countries, attempts were made to counteract this development, but none of those programmes survived for a long time – for the low price eel was worth, it did not pay off to take expensive measures. If nothing had changed, the eel would have died a relatively inglorious death, just like the salmon and sturgeon in so many rivers. That changed completely, with the development of a new market (Amsterdam and other cities), for a new product (smoked eel), at new prices – in 1893 and thereabouts. The rapidly increasing demand made it possible to expand the fishery significantly, to formerly unexploited, larger waters. The Germans, in particular, took the lead in the early 20th century, with a planned development of their own inland fisheries, and a strong stimulation of eel fishery in the neighbouring countries for import into Germany (Scandinavia, the Baltic states, the Netherlands, France, Italy and the rest of the Mediterranean Sea). However, shortly thereafter, catches began to decline, and with that, commercial fisheries eventually found the same destiny as the subsistence-fisheries had seen before – but that is another story, for another time.

The vast majority of all eel was and is processed by hot-smoking (stewed or jellied eel was a good runner-up for some time, inherited from the earlier subsistence-fisheries). Today, the largest volume of eel comes from aquaculture (that is, the on-growing of wild-caught young eels), mostly in the Netherlands. And as before, the major share of it is being processed into smoked eel, though no longer in Amsterdam.

The traditional eel culture in ponds and lagoons was an Italian specialty, for centuries. The development of 'modern' eel fishery over the past century was primarily a German investment. Danish technology development enabled modern indoor eel culture in recent decades. However, thinking of the Brothers Knapp and Christiaan Saur, we can now rightfully state: smoked eel is an original product of the Netherlands.

Bibliography

Dekker, W. (2019). The history of commercial fisheries for European eel commenced only a century ago. *Fisheries Management and Ecology*, *26*(1), 6–19.

Doose, W. (1908) Eine wenig bekannte Aalfangmethode [A less well known technique for catching eel]. *Allgemeine Fischerei-Zeitung* 33(18), 393–394.

Friedreich, J.B. (1846) *Handbuch der Gesundheitspolizei* [*Handbook of the Health Police*]. F.H. Gummi Verlag, Ansbach.

Hopf, A. (1846) *Spickaal auf der Berliner Kunst-Ausstellung im Jahre 1846*. Egbert Bauer Verlag, Charlottenburg.

Gouffé, J. (1869) *Le livre des conserves* [*The book of preserves*]. Hachette, Paris.

Sauvage, H.E. (1883) *Rapport sur la pêche en Hollande* [*Report on the fisheries in the Netherlands*]. Imprimerie nationale, Paris.

Zierikhoven, C. (=de Chalmot, J.A.) (1772) *Volkoomen Neerlandsch kookkundig woordenboek* [*The Complete Dutch Cooking Vocabulary*]. J.A. Chalmot, Campen.

PART II

Species and regions

CHAPTER 3

Distribution and conservation status of tropical eels

Vincent Nijman and Penthai Siriwat

Oxford Brookes University, UK

Species natural range

All anguillid eels, including the European, American and Japanese eel, spend at least part of their lifecycle in the tropics, that is, south of the Tropic of Cancer at 23°27'N and north of the Tropic of Capricorn at 23°27'S. The majority of anguillids in fact range completely in the tropics and, to a lesser extent, subtropical regions. The exception to the rule, that all anguillids range at least partially in the tropics and subtropics, is the New Zealand longfin eel (*Anguilla dieffenbachii*), a species confined to the temperate region of the southern hemisphere. In total, ten anguillid species can be classified as tropical eel, ranging in large parts of the Indian and Pacific Oceans, as well as parts of the Atlantic Ocean.

Tropical eels are found in a large number of countries and territories, including at least 12 in Oceania, 13 in South and Southeast Asia and 15 in Africa and the Indian Ocean. The countries with the most species of tropical eel are all in the Asia-Pacific region, with the Philippines having five species of tropical eel (in addition to the Japanese eel) and Indonesia and Papua New Guinea also having five species. Tropical eels can be differentiated into three species with large ranges and seven species with narrow, endemic or near-endemic ranges.

Large-range eel species

Shortfin eel (A. bicolor)

The shortfin eel (*A. bicolor*) has a range throughout the Indian Ocean, with genetically diverged populations in the Pacific Ocean (Indonesian shortfin eel [*A. bicolor pacifica*]) and the Indian Ocean (Indian shortfin eel [*A. bicolor bicolor*]) (Robinet and Feunteun, 2002; Robinet et al., 2003). The *A. bicolor pacifica* population is found on the coasts of Japan, China, Taiwan, Vietnam, Philippines, Indonesian Archipelago, including Borneo and Sulawesi, New Guinea and the Marianas (Jacoby et al., 2014a). Within the Indonesian Archipelago, the Indian Ocean population of *A. bicolor bicolor* occurs from the eastern African coastline up to the Arabian peninsula, and is widespread across the tropical Indian ocean (Seychelles, Madagascar, Mascarenes) across to India, Sri Lanka, Bangladesh, Myanmar, the Indomalaysian region and north-western Australia (within local streams in the Kimberley region) (Allen et al., 2002; Jacoby et al., 2014a). The Indian Ocean populations of *A. bicolor bicolor* can be found in the Indonesian Archipelago, from western Sulawesi island to Papua New Guinea (Sugeha and Suharti, 2008). The potential spawning grounds of *A. bicolor bicolor* are believed to be in the Indian Ocean near the Mentawai Trench for the Indian Ocean population (Arai, 2016). *A. bicolor* are morphologically similar to the Pacific shortfinned eel (*A. obscura*). The potential spawning locations for the Pacific Ocean population is associated with oceanic currents (Jacoby et al., 2014).

Indian mottled eel (A. bengalensis)

The Indian mottled eel (*A. bengalensis*) has wide geographic range occurring across the coasts of the Indian Ocean (Jacoby et al., 2014b; Robinet et al., 2008). It is found on the coasts of India, Sri Lanka, Bangladesh and Myanmar, and in major drainages to Sumatra. In some cases, *A. bengalensis* may be divided into two separate subspecies (Watanabe et al., 2013). The Indian Ocean population, *A. bengalensis bengalensis* occurs in Pakistan, India, Bangladesh, Nepal, Sri Lanka, Myanmar and in Southeast Asia. The Eastern African population, *A. bengalensis labiata* occurs along the east-African coast, as far north as Kenya and along the coast of South Africa, within local river systems as well as drainages, lakes and tributaries along the coast in Mozambique, Zimbabwe, Tanzania and Zambia, including Lakes Malawi and Kariba (Jacoby et al., 2014b; Seegers et al., 2003). This species is also found on various islands in the western Indian Ocean, such as Madagascar. In Ethiopia, *A. bengalensis* is the only anguillid species found in the country. A subpopulation of *A. bengalensis labiata* is also sporadically recorded in the Arabian peninsula (Jacoby et al., 2014b). As very few *A. bengalensis* leptocephali have been collected in the Indian Ocean, the geographic spawning location of *A. bengalensis* remains unclear and it has not been established whether the freshwater species have single or multiple spawning locations (Watanabe et al., 2013).

Marbled eel (A. marmorata)

The marbled eel (*A. marmorata*) is the second largest species of anguillids and is the most widely distributed across two ocean basins (the tropical and subtropical Pacific and Indian Oceans) (Ege, 1939; Jacoby and Gollock, 2014a; Watanabe et al., 2009a). The longitudinal geographic range extends from the east coast of Africa, India and Sri Lanka, to the Indo-Pacific region, including East Asia (Japan, Taiwan, China), Southeast Asia, the Pacific Islands, and as far east as the South Pacific islands (French Polynesia). The distribution in Africa is limited to South Africa. Populations at the edge of species range were recorded in the Galapagos Islands of the Pacific, although it is not known if it was due to natural migration, vagrant individuals or introduction (Jacoby and Gollock, 2014a). Introduced non-native species occurrences were recorded in North Pacific distributions in Hawaii (James and Suzumoto, 2006). It is the only anguillid species found in Laos. Multiple monophyletic populations of *A. marmorata* are reported in the separate populations in North Pacific (Japan to Sulawesi), Pacific (Papua New Guinea to Tahiti), Indian Ocean (Sumatra to Madagascar) and Guam (including Micronesia). *A. marmorata* spawn in same areas as other anguillids, such as the Japanese eel (*A. japonica*) (Kuroki et al., 2009). Spawning sites for the western Indian Ocean population are believed to be around Madagascar and the Mascarenes islands.

Endemic-range eel species

Indonesian longfinned eel (A. borneensis)

The Indonesian longfinned eel (*A. borneensis*) has the smallest range of all fresh water eels (Jacoby and Gollock, 2014b). It is endemic to Borneo Island, the Celebes Sea and the Makassar Strait region. Its native range country occurrences are Indonesia, Malaysia and the Philippines. This species has also been recorded in mainland China, however, it is not considered its natural range. Compared to other Anguilla species, the distance of migration for Indonesian longfinned eels are more restricted (Kuroki et al., 2006), where eels spawn close to their growth habitat. *A. borneensis* are morphologically similar to *A. japonica* (Jacoby and Gollock, 2014b).

Philippine mottled eel (A. luzonensis)

The Philippine mottled eel (*A. luzonensis*) have two narrow ranged populations in the northernmost and southernmost parts of the Philippines (Jacoby and Gollock, 2014c). Its endemic range is localized in Luzon Island, extending up the Pinacanauan River and Cagayan River tributary system (Watanabe et al., 2009b). The Cagayan tributary system is a very important site for this species as it is an isolated habitat site vital for the growth of glass eels. Furthermore, vagrant, unconfirmed and untested populations have also been reported in other islands in the Philippines and adjacent areas, for example, Taiwan (Jacoby and Gollock, 2014c). Their range requires further research.

Highlands longfin eel (A. interioris)

The Highlands longfin eel (*A. interioris*) are endemic to the northern half of New Guinea (Jacoby and Gollock, 2014d; Tsukamoto and Kuroki, 2014). Evidence of *A. interioris* leptocephali was recorded in the Indonesian archipelago around Central Sulawesi, Sumatra, the Western Central Pacific Ocean, and the eastern Indian Ocean (Wouthuyzen et al., 2009). This species may therefore have more distributed spawning sites than previously thought.

Celebes longfin eel (A. celebesensis)

The Celebes longfin eel (*A. celebesensis*) is found in the Western Pacific from Indonesia (Sulawesi, Borneo, surrounding islands) to the southern Philippines (Celebes Sea, Gulf of Tomini) (Arai et al., 2003; Jacoby and Gollock, 2014e). The full geographic range is still under consideration. Small vagrant populations of *A. celebesensis* are found outside of the known endemic range, for example in Papua New Guinea. *A. celebesensis* are morphologically similar to *A. interioris* and *A. luzonensisi* (Jacoby and Gollock, 2014e). Differing to temperate anguillids, *A. celebesensis* can occur in freshwater all year round (UNEP-WCMC 2015).

African longfin eel (A. mossambica)

The African longfin eel (*A. mossambica*) is found in the southwestern Indian Ocean, ranging along the African east coast, across to Madagascar and other western Indian Ocean inlands (Mascarene, Comoros and Seychelles islands) (Froese and Pauly, 2011; Jacoby and Gollock, 2014f). This species is commonly found inland, especially within east flowing rivers and drainage systems located east of Cape Agulhas (Robinet et al. 2008). Latitudinal, it is found from Kenya to northern South Africa (Froese and Pauly, 2011).

Pacific long-finned eel (A. megastoma)

The Pacific long-finned eel (*A. megastoma*) is endemic to the western and central Pacific islands from Pitcairn to New Guinea (Jacoby and Gollock, 2014i; Watanabe et al., 2011). It is found in French Polynesia, Fiji, the Cook Islands, Samoa, Salomon, Wallis and Futuna, New Caledonia and Papua New Guinea (Keith et al., 2013). *A. megastoma* was recorded in freshwater streams in Fiji (Keith et al., 2010).

Pacific shortfin eel (A. obscura)

The Pacific shortfin eel (*A. obscura*) has an endemic range throughout the Pacific Ocean from western New Guinea, Solomon Islands, Fiji, to French Polynesia (Jacoby and Gollock, 2014h; Silvergrip, 2009; UNEP-WCMC, 2015). The geographic range is similar to *A. megastoma*. The suspected spawning location of *A. obscura* is to the east of Tahiti but no clear evidence is yet available. *A. obscura* has been recorded in fresh water streams in Fiji and Tahiti.

IUCN Red List Category

For the primarily catadromous eels there is growing international concern for the population abundance and escapement trends of some of these species and yet incomplete knowledge of their life-histories hampers management and conservation. An overview of this is given by Jacoby et al. (2015). All species of eel faces a broad and dynamic range of threats, including habitat loss, habitat degradation, man-made or otherwise newly created barriers, pollution of both freshwater and marine environments, exploitation, and changing oceanic conditions. Combined these impacts will act in a synergistic manner, creating a patchwork of challenges and opportunities for the conservation and management of eels. The impact of these threats can be life-stage specific and regional (thus affecting not only species differently, but also populations of one species). Consequently, how these stressors combine to contribute to declines in abundance of particular life-stages is still poorly understood and as such the Red List assessments that are summarised below must be seen as an informed decision based on the available data at the time.

As noted by Jacoby et al. (2015), with trade and exploitation continually shifting comes the growing concern that protective measures banning or restricting the trade of temperate eel species has, and continues to adjust the focus to some of the lesser-studied tropical species in order to meet the high demand from East Asia (Crook and Nakamura, 2013; Nijman, 2015). Certainly compared to temperate eel species, far less is known about the conservation status of most of the tropical eels. A complicated factor is that within certain areas several eel species have broadly overlapping ranges and are morphologically difficult to distinguish making it difficult to accurately assess their global distribution ranges. To add to this, accurate knowledge of spawning grounds is lacking for most species, and almost completely absent for others. These gaps in our knowledge have to have taken into account when considering these species' conservation status and necessitate the need to improved monitoring and study of tropical freshwater eel recruitment, populations and escapement.

It is clear from the conservation and management of temperate eels that a broad and coordinated approach is needed, including involvement of stakeholders from often numerous countries and international bodies. This is currently lacking for most tropical eel species and more serious and effective commitment and political support is needed to ensure their sustainable use and conservation.

Least Concern

In terms of conservation status, ten of the tropical eel species have been assessed by the IUCN eel specialist group. Two species (*A. marmorata* and *A. mossambica*) have been assessed as Least Concern (LC), suggesting that their populations are large, stable and not of conservation concern.

Near Threatened

Four species (*A. bengalensis*, *A. bicolor*, *A. celebesensis* and *A. luzonensis*) have been assessed as Near Threatened (NT), meaning that while there is some concern regarding their conservation, this is not enough to be considered globally threatened.

Although *A. bicolor* have an extensively wide geographic range, the species is classified as NT as it is one of the main replacement species following the declined availability of European eel (*A. anguilla*) and *A. japonica*. Shortfin eel elvers are commonly traded on online platforms (Crook, 2014). The exploitation of the Pacific population of *A. bicolor pacifica* has been relatively greater when compared to the Indian ocean population *A. bicolor bicolor* (Jacoby et al., 2014a). However, overall, the global population of *A. bicolor* are in decline as a result of trade and under significant threat. Further impacts arising from the species extensive migration range, including changing oceanic currents, climate change and pollution is also not yet fully understood.

A. bengalensis is listed under NT due to the decrease in range and regional abundance. Natural habitats and migration paths are under threat due to dam building (Zimbabwe) and land conversion of fresh water bodies (India) (Jacoby et al., 2014b). In 2014, the Indian population experienced significant decline and has been completely absent in some regions, such as the Western Ghats.

A. celebesensis are classified as NT as a result of localized threats within its relatively narrow and already patchy distribution. Infrastructure development, such as dam building along key freshwater inlets, has directly impacted certain populations and caused a decline in the number of viable freshwater habitat sites (Indonesia). In addition to habitat modification, the threats of excessive and unsustainable trade *A. celebesensis* is also another reason the population are believed to be in decline (Jacoby and Gollock, 2014e). Further concerns arise from issues involving unclear identification of species due to similar morphologies to *A. interioris* and *A. luzonensis*, which may have led to inaccurate assessments of the threats imposed for this species.

The justification for the listing *A. luzonensis* as NT is the potential threats from overexploitation through indiscriminate or mixed eel fishing, habitat loss from mining and disruption from agricultural run-off in its narrow range in the Philippines (Jacoby and Gollock 2014c). The identification of *A. luzonensis* at the juvenile phase is difficult as it is morphologically identical to the other two species found in the same geographic region, *A. interioris* and *A. celebesensis*. An accurate assessment of true range and conservation status for this species still remains uncertain. Using knowledge of the species composition of glass eels in Cagayan River, it is estimated that up to 30 tonnes of *A. luzonensis* could have been potentially exported to East Asian countries in the first decade of the 21st century (Jacoby and Gollock 2014c).

Vulnerable

One species, the Indonesian longfinned eel (*A. borneensis*), is listed as Vulnerable (VU). This was based on observed, estimated, inferred or suspected population size reduction of ≥30% over the last three eel generations, based on a decline in area of occupancy, extent of occurrence and/or quality of habitat. The reduction in population size in somewhat limited geographic range has classified it as VU (Jacoby and Gollock, 2014b). Specifically, data from one river system in east Borneo suggested a marked population decline over a 35-year period following habitat degradation and disturbances of the riverside rainforests.

Data Deficient

Importantly, from a conservation perspective, three species (*A. interioris*, *A. megastoma* and *A. obscura*) are listed as Data Deficient (DD), meaning that the data available are inadequate to determine the degree of threat faced by the species, hence they could be threatened or not. For these three species especially, there is a need to obtain more information to determine the appropriate listing. While it is clear that at present most tropical eel species are not considered globally threatened, that there is a lack of data on their conservation and population status and it is telling that for none of the ten tropical eel species are population trend data (declining, stable, increasing) available.

The effects of potential threats (for example, freshwater habitat degradation, migration obstruction, river forest destruction, dam building, pollution and climate change) are generally more or less felt by all species of Anguilla eels. For example, *A. interioris* are found in the same Indonesian waters as the NT *A. bicolor* and *A. celebesensis*. Though reports suggest that *A. interioris* catchments are lower in volume most likely due to their lower abundance, it is evident that fishing of eels exists in the geographic distribution where it is found (Jacoby and Gollock, 2014d). Some eel species also face specific potential threats that still require more information. Eels with limited endemic island distributions, such as *A. megastoma*, for example, face potential threats from invasive species, such as the predatory freshwater turtle red-eared slider (*Trachemys scripta*), which prey on small fish and young eels. Overall, further research on the distribution, population status and trends, and biology of these species is required in order to better understand local and global threats to each. The combination of the declining abundance and stricter trade regulations of the three major eel species (*A. anguilla*, *A. japonica* and *A. rostrata*) along with the geographic proximity to major market areas may expose these DD species to rapid and large-scale exploitation, especially in areas where no clear regulation is established.

CITES listing

Currently none of the tropical eel species are listed on any of the appendices of CITES, but national governments may have restrictions in place regarding the domestic and international trade in eels.

Domestic regulations

The eel trade can be directly regulated by domestic regulations through respective domestic wildlife conservation and protection acts. For example, in Taiwan, *A. marmorata* are categorized as 'II Rare and Valuable', directly protected under the Wildlife Conservation Act (Article 4). The species is also considered a 'Natural monument' in Japan and is under protection (Jacoby and Gollock, 2014a). *Anguilla bengalensis*, listed as Near-Threatened on the IUCN Red List, is classified as 'Endangered' in the domestic legislation in India, however, harvesting of the species still takes place (Molur and Walker, 1998). Throughout Oceania, freshwater eels are strongly associated with mythical creatures. The eel (species unknown) forms part of the creation myth throughout Polynesia, and specifically the creation of the coconut tree (which all started with a local girl, Sina, and her pet eel) (Craig 2004). Traditionally, many islands impose local fishing restrictions and have existing taboos regarding fishing of eels (Jacoby and Gollock, 2014a).

The Philippines was recently highlighted as an important player in the export of Anguillid eels with eight native species found within the islands. Online market places, such as Alibaba, offered high volumes of glass eels (Crook and Nakamura, 2013). Following the increased trade and unregulated exploitation of glass eels, the Philippines issued a ban on juvenile eels or eel fry not exceeding 15 cm in length in 2012 (Fisheries Administrative Order No. 242). The ban covered at least five native tropical eel species including *A. bicolor*, *A. marmorata*, *A. japonica*, *A. celebesensis* and *A. luzonensis* (Crook, 2014), with penalties such as imprisonment for 8 years, catch confiscation, fines of double the export value or revocation of fishing or export permits. However, the effectiveness and enforcement of the ban is doubtful since the fishing, selling and export of juvenile eels smaller than 15 cm was still prevalent (Crook and Nakamura, 2013). Similarly, Taiwan imposed harvesting seasons limited to just 4 months for eels larger than 8 cm in most rivers (Jacoby and Gollock, 2014a).

The fishing of and trade in eels can also be regulated by weight set by each respective range country. For example, the trade of eels in Indonesia, which includes *A. bicolor*, *A. borneensis*,

A. celebesensis and *A. interioris*, is restricted in that all eels exported from the country must have a mass of 150 g or more (Nijman, 2015). In Japan, export of *A. marmorata* fry is set at equal or over 13 g by the Export Trade Control Order (Shiraishi and Crook, 2015). Similarly, in 2013, South Korea imposed specific custom codes in eel farming, further differentiating juvenile eels into glass and young by size (grams per eel), which may have led to clearer declarations in volumes of eel imports (Crook, 2014).

Bibliography

Affandi, R. (2005) Strategi pemanfaatan sumberdaya ikan sidat *Anguilla* spp. di Indonesia. *Jurnal lktiologi Indonesia* 5(2), 77–81.

Aida, K., Tsukamoto, K. and Yamauchi, K. (eds) (2003) *Eel biology*. Springer Verlag, Tokyo.

Allen, G.R., Midgley, S.H. and Allen, M. (2002) *Field guide to the Freshwater Fishes of Australia*. Western Australian Museum, Perth.

Aoyama, J., Wouthuyzen, S., Miller, M.J., Inagaki, T. and Tsukamoto, K. (2003) Short-distance spawning migration of tropical freshwater eels. *Biological Bulletin* 204, 104–108.

Arai, T. (2014) Do we protect freshwater eels or do we drive them to extinction? *SpringerPlus* 3(1), 534.

Arai, T. (2016) *Biology and Ecology of Anguillid Eels*, CRC Press, Abingdon, pp. 89–107.

Arai, T., Miller, M.J. and Tsukamoto, K. (2003) Larval duration of the tropical eel *Anguilla celebesensis* from Indonesian and Philippine coasts. *Marine Ecology Progress Series* 251, 255–261.

Baedah, M.A. (2010) *Strategi pengelolaan ikan sidat*. Marine Affairs and Fisheries Services, Province of Central Sulawesi, Poso.

Craig, R.D. (2004) *Handbook of Polynesian Mythology*. Santa Barbara, California.

Crook, V. (2014) *Slipping Away: International Anguilla Eel Trade and the Role of the Philippines*. TRAFFIC and ZSL, Cambridge, UK.

Crook, V. and Nakamura, M. (2013) Glass eels: assessing supply chain and market impacts of a CITES listing on Anguilla species. *TRAFFIC Bulletin* 25(1), 24–30.

Ege, V.A. (1939) A revision of the genus *Anguilla* Shaw. *DANA Report* 16, 8–256.

Froese, R. and Pauly, D. (eds) (2011) Fishbase. Available at: www.fishbase.org. (accessed 17 May 2012).

Ishikawa, S., Tsukamoto, K. and Nishida, M. (2004) Genetic evidence for multiple geographic populations of the giant mottled eel *Anguilla marmorata* in the Pacific and Indian oceans. *Ichthyological Research* 51, 343–353.

Jacoby, D. and Gollock, M. (2014a) *Anguilla marmorata. The IUCN Red List of Threatened Species 2014: e.T166189A45832585* (accessed 2 July 2018).

Jacoby, D. and Gollock, M. (2014b) *Anguilla borneensis. The IUCN Red List of Threatened Species 2014: e.T198973A2545681* (accessed 9 July 2018).

Jacoby, D. and Gollock, M. (2014c) *Anguilla luzonensis. The IUCN Red List of Threatened Species 2014: e.T18435966A45833133* (accessed 9 July 2018).

Jacoby, D. and Gollock, M. (2014d) *Anguilla interioris. The IUCN Red List of Threatened Species 2014: e.T198972A2545630* (accessed 2 July 2018).

Jacoby, D. and Gollock, M. (2014e) *Anguilla celebesensis. The IUCN Red List of Threatened Species 2014: e.T195503A2382716* (accessed 2 July 2018).

Jacoby, D. and Gollock, M. (2014f) *Anguilla mossambica. The IUCN Red List of Threatened Species 2014: e.T183155A1730550* (Accessed 20 July 2018).

Jacoby, D. and Gollock, M. (2014g) *Anguilla megastoma. The IUCN Red List of Threatened Species 2014: e.T196301A2443332* (accessed 9 July 2018).

Jacoby, D. and Gollock, M. (2014h) *Anguilla obscura. The IUCN Red List of Threatened Species 2014: e.T196302A2443435* (accessed 20 July 2018).

Jacoby, D., Harrison, I. and Gollock, M. (2014a) *Anguilla bicolor. The IUCN Red List of Threatened Species 2014: e.T166894A67015710* (accessed 20 July 2018).

Jacoby, D., Harrison, I. and Gollock, M. (2014b) *Anguilla bengalensis. The IUCN Red List of Threatened Species 2014: e.T61668607A15501445* (accessed 9 July 2018).

Jacoby, D.M., Casselman, J.M., Crook, V., DeLucia, M.B., Ahn, H., Kaifu, K., Kurwie, T., Sasal, P., Silfvergrip, A.M., Smith, K.G. and Uchida, K. (2015). Synergistic patterns of threat and the challenges facing global anguillid eel conservation. *Global Ecology and Conservation* 4, 321–333.

James, S.A. and Suzumoto, A.Y. (2006) Nonnative Occurrence of Anguilla marmorata in Hawai'i: Identification Using Morphological and Molecular Characters. *Bishop Museum Occasional Papers* 87, 55–61.

Keith P., Marquet, G., Gerbeaux, P., Lord, C. and Vigneux, E. (2013) *Polynesian Freshwater Fish and Crustaceans: Taxonomy, Ecology, Biology and Management*. Department of Conservation, Paris.

Keith, P., Marquet, G., Lord, C., Kalfatak, D. and Vigneux, E. (2010) *Poissons et crustaces d'eau douce du Vanuatu*. Societie Francaise d'Ichtyologie, Paris.

Kuroki, M., Aoyama, J., Miller, M.J., Wouthuyzen, S., Arai, T. and Tsukamoto, K. (2006) Contrasting patterns of growth and migration of tropical anguillid leptocephali in the western Pacific and Indonesian Seas. *Marine Ecology Progress Series* 309, 233–246.

Kuroki, M., Aoyama, J., Miller, M. J., Yoshinaga, T., Shinoda, A., Hagihara, S. and Tsukamoto, K. (2009) Sympatric spawning of *Anguilla marmorata* and *Anguilla japonica* in the western North Pacific Ocean. *Journal of Fish Biology* 74, 1853–1865.

Molur, S. and Walker, S. (1998) Freshwater fishes of India. Zoo Outreach Organisation, Tamil Nadu, India.

Ndobe, S. (2010) Struktur ukuran glass eel ikan sidat (*Anguilla marmorata*) di Muara Sungai Palu, kota Palu, Sulawesi Tengah. *Media Litbang Sulteng* 3, 144–150.

Nijman, V. (2015) CITES-listings, EU eel trade bans and the increase of export of tropical eels out of Indonesia. *Marine Policy* 58, 36–41.

Phelps, J., Webb, E.L., Bickford, D., Nijman, V. and Sodhi, N.S. (2010) Boosting CITES. *Science* 330, 1752–1753.

Ringuet, S., Muto, F. and Raymakers, C. (2002) Eels: their harvest and trade in Europe and Asia. *TRAFFIC Bulletin* 19, 1–26.

Robinet, T. and Feunteun, E. (2002) First observations of shortfinned *Anguilla bicolor bicolor* and longfinned *Anguilla marmorata* silver eels in the Reunion Island. *Bulletin francais de la peche et de la pisciculture* 364, 87–95.

Robinet, T., Lecomte-Finiger, R., Escoubeyrou, K. and Feunteun, E. (2003) Tropical eels *Anguilla* spp. recruiting to Réunion Island in the Indian Ocean: taxonomy, patterns of recruitment and early life histories. *Marine Ecology Progress Series* 259, 263–272.

Robinet T., Réveillac, E., Kuroki, M., Rabenevanana, M.W., Valade, P., Aoyama, J., Berrebi, P., Tsukamoto, K. and Feunteun, E. (2008) New clues for freshwater eels (*Anguilla* spp.) migration routes to eastern Madagascar and surrounding islands. *Marine Biology* 154, 453–463.

Sasongko, A., Purwanto, J., Mu'minah, S. and Arie, U. (2007) *Sidat: panduan agribisnis, penangkapan, pedederan, and pembesaran*. Penebar Swadaya, Depok.

Shiraishi, H. and Crook, V. (2015) Eel market dynamics: An analysis of Anguilla production, trade and consumption in East Asia. TRAFFIC. Tokyo, Japan.

Seegers, L., De Vos, L. and Okeyo, D.O. (2003) Annotated checklist of the freshwater fishes of Kenya (excluding the lacustrine haplochromines from Lake Victoria). *Journal of East African Natural History* 92(1), 11–47.

Silvergrip, A.M.C. (2009) CITES Identification Guide to the Freshwater eels (Anguillidae). Swedish Environmental Protection Agency, Stockholm, Sweden.

Siswomartono, W. (1998) Review of the policy and activities of wildlife utilization in Indonesia. *Mertensiella* 7, 37–44.

Sugeha, H.Y. and Suharti, S.R. (2008) Discrimination and distribution of two tropical short-finned eels (*Anguilla bicolor bicolor* and *Anguilla bicolor pacifica*) in the Indonesian Water. *The NaGISA Westpac Congress*, Jakarta, 26–28 October 2008, 1–14.

Tsukamoto, K. and Kuroki, M. (2014) *Eels and humans*. Springer, Tokyo.

UNEP-WCMC (2015) *Preliminary Overview of the Genus Anguilla*. UNEP-WCMC, Cambridge.

Watanabe, S., Miller, M.J., Aoyama, J. and Tsukamoto, K. (2009a) Morphological and meristic evaluation of the population structure of *Anguilla marmorata* across its range. *Journal of Fish Biology* 74: 2069–2093.

Watanabe, S., Aoyama, J. and Tsukamoto, K. (2009b) A new species of freshwater eel *Anguilla luzonensis* (Teleostei: Anguillidae) from Luzon Island of the Philippines. Fisheries Science 75, 387–392.

Watanabe, S., Miller, M.J., Aoyama, J. and Tsukamoto, K. (2011) Analysis of vertebral counts of the tropical anguillids, *Anguilla megastoma*, *A. obscura*, and *A. reinhardtii*, in the western South Pacific in relation to their possible population structure and phylogeny. *Environmental Biology of Fishes* 91(3), 353–360.

Watanabe, S., Miller, M.J., Aoyama, J. and Tsukamoto, K. (2013) Evaluation of the population structure of *Anguilla bicolor* and *A. bengalensis* using total number of vertebrae and consideration of the subspecies concept for the genus Anguilla. *Ecology of Freshwater Fish* 23(1), 77–85.

Wouthuyzen, S., Aoyama, J., Sugeha, H.Y., Miller, M.J., Kuroki, M., Minegishi, Y., Suharti, S.R. and Tsukamoto, K. (2009) Seasonality of spawning by tropical anguillid eels around Sulawesi Island, Indonesia. *Naturwissenschaften* 96, 153–15.

CHAPTER 4

Distribution, ecology and status of anguillid eels in East Africa and the Western Indian Ocean

Céline Hanzen[1], Olaf L.F. Weyl[2], Martyn C. Lucas[3], Kerry Brink[1,4], Colleen Downs[1] and Gordon O'Brien[5]

[1]University of KwaZulu-Natal, School of Life Sciences, Pietermaritzburg, RSA; [2]DST/NRF Research Chair in Inland Fisheries and Freshwater Ecology, South African Institute for Aquatic Biodiversity, Grahamstown, RSA; [3]University of Durham, Department of Biosciences, Durham, UK; [4]World Fish Migration Foundation, Groningen, the Netherlands; [5]University of Mpumalanga, School of Biology and Environmental Sciences, Nelspruit, RSA

Introduction

There are 19 anguillid eel species (hereafter, eels), found around the world (Aoyama, 2009). Five tropical species have been documented to occur along the east coast of Africa. The presence of *Anguilla obscura* Günther 1872 is, however, restricted to a single specimen (Jubb, 1957b) and is considered as incidental. The four others, longfin eel (*A. mossambica* Peters 1852), African mottled eel (*A. bengalensis* Gray 1831), giant mottled eel (*A. marmorata* Quoy Gaimard 1824) and shortfin eel (*A. bicolor* McClelland 1844), have been documented to migrate from the Indian Ocean into the east flowing rivers on the continent and associated islands. Here they are important components of freshwater ecosystems, contribute towards the food security and livelihoods of riparian communities, and are considered of spiritual significance by many people.

With less than 100 publications on African eels and few directed research programmes on them, this group is poorly studied. This is problematic as two of the species, *A. bengalensis* and *A. bicolor*, have already been categorized by the IUCN as near threatened (NT), and while *A. marmorata* and *A. mossambica* were evaluated as least concern (LC), there are few current data sets available with which to assess abundance or population status (Jacoby et al., 2015). This is of concern as globally eels are under pressure from degradation of inland habitats, chemical pollution, overfishing and the construction of dams and weirs, which are often insurmountable obstacles to migration (Baran et al., 2012; Dekker et al., 2003; Jacoby et al. 2015). Regionally, African eels are the only long-distance catadromous species, which makes them particularly vulnerable to river connectivity disruption. As a result, and owing to existing and planned dam development, they represent a particularly vulnerable component of the regional fish community. There is already evidence from the island of Réunion, that habitat loss, degradation of water quality and parasites are having an impact on eel abundance and recruitment (Valade et al., 2018). The lack of data and specialist knowledge from large parts of Africa is considered a severe constraint to the development of adequate assessments of the conservation status of eels by the IUCN-Species Survival Commission Anguillid Specialist Sub-Group (Jacoby et al., 2015).

The aim of this chapter is to review the current knowledge on the occurrence, abundance, ecology, behaviour, threats, values, policy status and management of eels on the East African coastline and connecting freshwater rivers, Madagascar and surrounding islands. It is our intention that this review will serve as a baseline for planning future research, to guide conservation efforts and, hopefully, to stimulate an interest in African eels and their vulnerability to the increasing pressures of global change.

Distribution

Basic global distribution maps (Figure 4.1) were redrawn based upon the IUCN shapefile data in the IUCN Red List summaries for the four African anguillids (Jacoby and Gollock, 2014a, 2014b; Jacoby et al., 2014a, 2014b). However, more detailed data used to infer distribution,

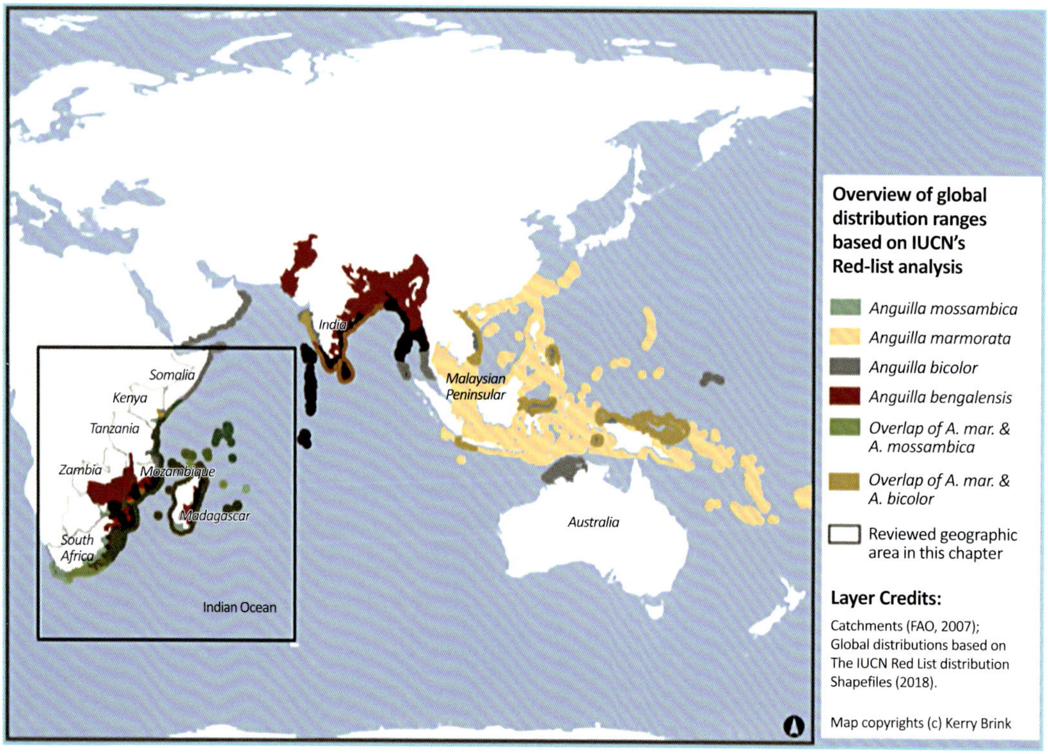

Figure 4.1 Geographic area covered in this review, including the global range of the four tropical eels discussed.

inland penetration and relative abundance (Figure 4.2 and text below) were based on available datasets from SAIAB (South African Institute for Aquatic Biodiversity), FishBase, IUCN, GBIF (Global Biodiversity Information Facility) and the River Health Programme (FishBase, 2018; GBIF, 2018; Kleynhans et al., 2007; Scott et al., 2006) with a focus on the east coast of Africa, Madagascar and surrounding islands. Overall the data availability was highest in southern Africa, with regions north of the Zambezi catchment in Mozambique, Kenya and Somalia being data deficient.

Anguilla bengalensis

According to van Someren (1959), *A. bengalensis* is the dominant eel species from the Limpopo River northward towards Kenya (Figure 4.2). More recently, Seegers et al. (2003) and Golubtsov et al. (2012) reported *A. bengalensis* from Kenya and Ethiopia, respectively. This species is rare in rivers south of the Limpopo River catchment (Bell-Cross and Minshull, 1988) and on the Indian Ocean islands (Valade et al., 2018). Robinet et al., (2007), for example, did not record *A. bengalensis* in a sample of more than 4000 eels from the Seychelles, Mayotte, Mauritius, Réunion and Madagascar, although the species is considered native to these regions (Jacoby et al., 2014a).

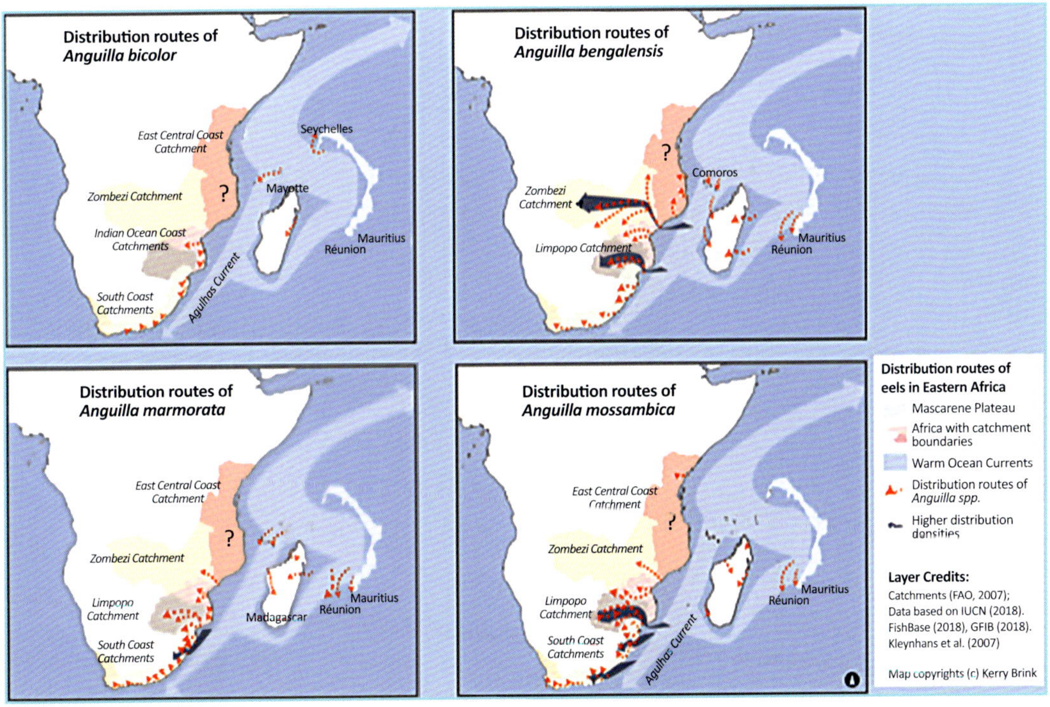

Figure 4.2 Distribution maps of *A. bicolor*, *A. bengalensis*, *A. marmorata* and *A. mossambica* in East Africa and the western Indian Ocean. Higher distribution density (grey) arrows in the key and on the maps refer only to southern Africa database information.

Anguilla bicolor

A. bicolor is widespread (Figures 4.1 and 4.2), occurring along the east coast of Africa, the Arabian Peninsula, tropical Indian Ocean islands, as well as eastward towards India, Sri Lanka, Bangladesh, Mayanmar, the Malaysian Peninsular and even in Australia (Jacoby et al., 2014b). Along the east coast of Africa, it is not abundant (Jubb, 1961); locality records extend from Tanzania to the eastern Cape region of South Africa (SAIAB collections, FishBase, 2018; GBIF: the Global Biodiversity Information Facility, 2018; Scott et al., 2006). In the Seychelles it is the only eel species present (Robinet et al., 2008). Distribution records indicate that in most instances this species tends to occupy the lower coastal regions of rivers.

Anguilla marmorata

A. marmorata (Figure 4.3) is the most widespread species (Figures 4.1 and 4.2) of the tropical African eels, with its distribution ranging from the Breede River in the southern Cape region in South Africa to the Pacific Islands (Jacoby and Gollock, 2014a) (and possibly to the Galapagos Islands [McCosker et al., 2003]). This species is also common in the western Indian Ocean Islands (including Réunion, Mauritius, Madagascar and Mayotte) and is found along the East African coast up to Somalia according to the IUCN (Jacoby and Gollock, 2014a).

Anguilla mossambica

The distribution of *A. mossambica* (Figure 4.4) is restricted to the African mainland and islands in the Agulhas Current region of the Indian Ocean (Figures 4.1 and 4.2). Its distribution extends from the southern Cape in South Africa to Kenya. This species is more common south of the Zambezi catchment (Jubb, 1967) and less common northward (Frost, 1957; van Someren, 1959). They occur through the altitudinal gradient up to 1700 m in Madagascar (Kiener, 1963) and migrate hundreds of kilometres inland from the African coastline. It is the most common species in South Africa (Jubb, 1961; Skelton, 2001) and is also fairly common in Mauritius (Valade et al., 2018). It is, however, less abundant than other eel species in Mayotte and Réunion islands (Valade et al., 2018).

Figure 4.3 A large specimen of the giant mottled eel (*A. marmorata*) (total length 158 cm) caught in South Africa in 2018. This specimen was under anesthesia at the time of the photograph and was released back in the river. © SAIAB/Sean Marr.

Figure 4.4 The longfin eel (*A. mossambica*). © Jeremy Shelton.

Migrations

East African eels exhibit the characteristic facultatively catadromous life cycle common to all anguillids, with ocean spawning, migration of leptocephali larvae to coastal regions, entrance of glass eels into estuaries, upstream migration as elvers and, after many years in the yellow eel stage, a return migration as silver eels to the spawning grounds (Jacoby et al., 2015).

East African riverine eel communities differ between geographic areas on the continent and across the islands in the region (Robinet et al., 2007). Robinet et al. (2008) hypothesized that this may be due to: (1) larvae of different species using different migration routes; (2) different species frequenting different breeding areas or the same area at different times; and (3) species-specific differences in behaviour. A short description on the current knowledge on these aspects follows.

Marine migration

Spawning grounds

For all the East African eels, spawning is hypothesized to occur over the Mascarene Plateau (Pous et al., 2010) (Figure 4.5). This was first suggested by Jespersen (1942) who found leptocephali in the area and delimited the location of the breeding grounds east of Madagascar as between 10°S and 20°S, and 60°E and 65°E.

For *A. bicolor*, two spawning grounds have been identified. One spawning ground was identified by Jespersen (1942) and confirmed by Arai et al. (1999) south west of Sumatra in the western Indian Ocean, while the other was placed by Robinet and Feunteun (2002) north-east of Madagascar. This latter spawning region was also supported by Robinet et al. (2008). Minegishi et al. (2005) found no differences in genetic structure between eastern and western Indian Ocean samples for *A. bicolor* suggesting that individuals in the African region could originate from either spawning ground depending upon prevailing currents (Robinet et al., 2008).

A. marmorata has a very wide distribution and has been shown to have more than one spawning area. The study by Ishikawa et al. (2004) showed at least five distinct geographic populations of *A. marmorata,* distinguished by mitochondrial DNA differences, originating from the North Pacific, Madagascar, Sumatra, Fiji and Tahiti. Some authors have suggested an influx of migrant *A. marmorata* from Sumatra to the southwestern Indian Ocean, which could be driven by the Equatorial Current (Gagnaire et al., 2009). This would also explain the documented high variability in larval life duration for *A. marmorata* (see Robinet et al., 2003c).

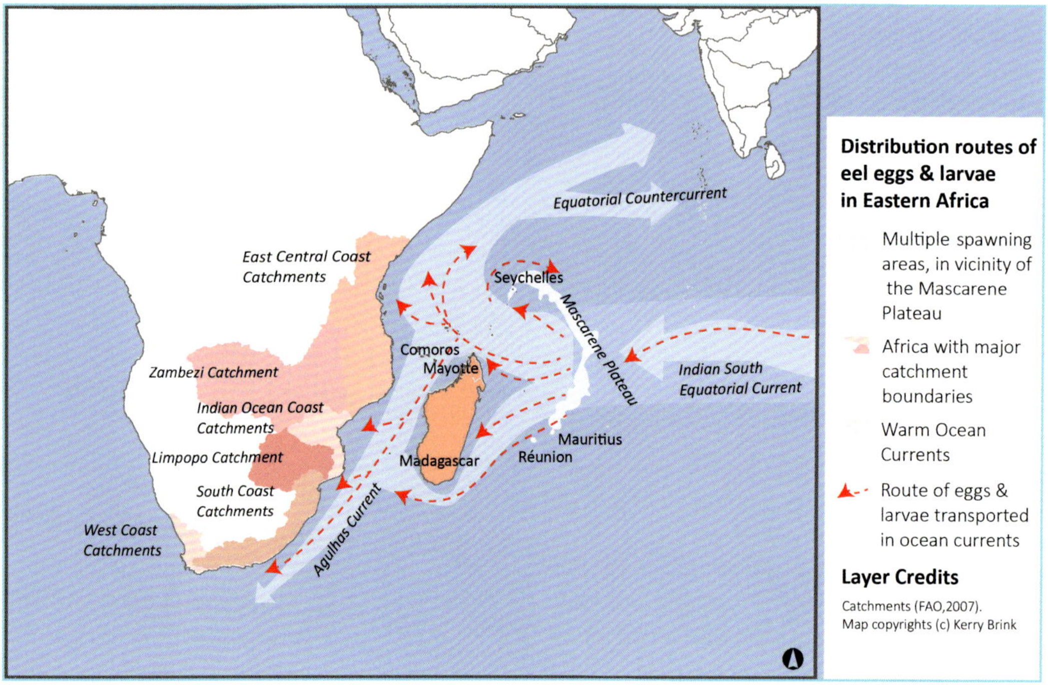

Figure 4.5 The suspected spawning area of East African tropical eels, east of Madagascar in the vicinity of the Mascarene Plateau. It is hypothesized the ocean currents carry eel larvae to Madagascar, the numerous islands around Madagascar and the east coast of Africa.

A. mossambica is endemic to the African Agulhas Current region and as such there is no doubt that the spawning area is in the southwestern Indian Ocean. The location of their breeding ground is between the east coast of Madagascar and the Mascarene Plateau (Robinet et al., 2008). Another study by Réveillac et al. (2009) observed a mean duration of the leptocephalus stage of 100 days and therefore estimated a migration length of 1000–2000 km, which would support the Jespersen (1942) and Robinet et al. (2008) hypothesis.

For *A. bengalensis,* including East African stocks, there is still uncertainty of its spawning ground in the Indian Ocean.

Ocean migration routes

Although the exact spawning areas of East African eels remain largely unknown, eggs and larvae are known to disperse with the main ocean currents (Figure 4.5). Jespersen's (1942) hypothesis that larvae are carried around the southern tip of Madagascar, is supported by Robinet et al. (2007) on the basis that eel communities in the islands of Mauritius and Réunion, which are on the southern migration route, are dominated by *A. mossambica* and *A. marmorata*. Robinet et al. (2007) suggested that *A. bicolor* and *A. bengalensis* follow a northern migration route around Madagascar.

Jubb (1961) suggested that young *A. mossambica* possibly reach the mainland coast between 21°S (at the border between Mozambique and Tanzania) and 31°S (southern coast of KwaZulu-Natal in South Africa) while *A. bengalensis* and *A. bicolor* reach the coast north of latitude 20°S based on his observations of size for glass eel, elvers and post-elvers, and those of Jespersen (1942). He also suggested that the post-elver stage can take a secondary migration outside of this primary range. Distribution data since then have shown a wider occurrence and more detailed evidence regarding ocean-coastal routes (Figure 4.5).

Larval life

Little is known about the oceanic larval life stages of the four East African eel species and the mechanisms triggering the metamorphosis of the leptocephali and the recruitment from the sea. Studies by Robinet et al., (2003b, 2003c, 2007) between 2002 and 2008 did, however, shed light on the recruitment of tropical eels and their larval stages around Réunion. The four East African eel species are believed to spend different times at sea before reaching the coast. For instance, *A. marmorata* elvers aged between 2 and 4 months were collected in rivers of Réunion (Robinet et al., 2003c).

It has also been shown that there is significant variation regarding the duration of the larval stage within a species, as observed for *A. marmorata* (Réveillac, 2008). This variability or flexibility is an interesting trait as this demonstrates that larval migration could happen at any time of the year and also be a key to East African eel life cycle success. Flexibility of duration of larval migration has also been observed in other parts of the world for the same species (Arai et al., 2002, 1999; Marui et al., 2001). This is also the case for *A. bicolor* that shows great variability for age at metamorphosis (Robinet et al., 2008). It has to be noted that differences in larval migration duration in different geographic areas of their distribution might also be explained by the fact they breed in different spawning areas.

Into freshwater

Recruitment

Once larvae have metamorphosed, glass eels start to appear along the coastal regions of East Africa, Madagascar and the other islands. One might expect recruitment times to differ depending on the regions and species. Although there are recruitment peaks, glass eels occur in this region all year long. On Réunion, glass eel migrations usually peak between November and April, which coincides with the wet season (Robinet et al., 2003a): *A. mossambica* recruitment occurs from November to December; *A. bicolor* in March; *A. bengalensis* between January and March and *A. marmorata* in January and February (Robinet et al., 2003b). The different species vary in size when they enter the rivers (Réveillac et al., 2009; Robinet et al., 2003b), which could be a result of breeding at different times or in different areas. River entry by glass eels is observed mostly at night (Bruton et al., 1987; Kiener, 1963).

On the continent, *A. mossambica* glass eels enter estuaries when they are about 50–60 mm long during spring and summer (September to February) (Harris and Cyrus, 1995). In South Africa, Wasserman et al. (2011) observed *A. marmorata* and *A. mossambica* in the glass eel stage during November and December migrating, at night, during periods coinciding with strong summer river flows (Bruton et al., 1987; Jubb, 1957c). Jackson (1976) also suggested that upstream migration in the rivers might vary across the years as seasonal rainfall in southern Africa can be erratic. A similar story has been reported in the islands that face cyclone seasons with variable rainfall (Robinet et al., 2003c).

Upstream migration

Recent otolith chemistry studies suggest that anguillids have more flexible life histories than previously believed (Arai and Chino, 2012; Chino and Arai, 2010; Shiao et al., 2003; Tsukamoto et al., 2002) and their migration up rivers is somewhat facultative. Some individuals have been known not to enter freshwater ecosystems, while others may shift repeatedly between river and estuarine habitats of multiple freshwater systems. Unfortunately, little is known of their abundance in estuaries and coastal waters in the East Africa region. It is also not known if, and for how long, they wait before they ascend the river as this probably varies with local conditions (Jackson, 1976). For instance, Whitfield (1998) suggested a limited residency in southern African estuaries as the records are sporadic. Moreover, the ratio between the migration phenotypes is not known. This knowledge could be of high significance to help decipher the importance of different river-estuary systems, their suitability for eels and, ultimately, to aid their conservation.

Lin et al. (2012) used otolith strontium : calcium ratios of *A. mossambica* and *A. marmorata* sampled from South African rivers to reconstruct their migratory life histories between freshwater and saltwater habitats. Interestingly, both species showed evidence of saltwater habitat experience after settlement in freshwater. This was primarily during their elver stage or early in the yellow eel stage. During the middle and late yellow eel stage, freshwater residency was preferred and only sporadic visits were made to saltwater habitats by many of the eels sampled. Lin et al. (2012) implied that elvers may reside in brackish estuaries, possibly waiting for favourable conditions for upstream migration, or move along the coast searching for suitable habitats. This hypothesis was also proposed, but could not, at the time, be demonstrated by Jackson (1976). Consequently, estuaries, river mouths or even coastal regions may be important habitat for both these eel species as a temporary shelter during periods when river flows are unfavourable for upstream migration.

Elvers that ascend the rivers will live most of their life in freshwater as yellow eels (sub-adult). On Réunion, Robinet et al. (2003c) observed that the upstream migration of elvers occurs from December to October (potentially all year round), with a peak in January to April. On the mainland, Skead (1959) observed strong elver migrations at night up the Laing Dam wall in South Africa, associated with rainfall and strong river flow, particularly between January and the end of April.

Settlement

One can observe an altitudinal gradient in yellow eel inland distribution and abundance that varies among the regions and the species. Kiener (1963) showed that in Madagascar *A. bengalensis* was restricted to high altitudes (over 900 m), while *A. bicolor* was only found in the lower catchment (below 300 m). *Anguilla marmorata* had a broader range up to 900 m, while *A. mossambica* was the most ubiquitous species and was found through the entire altitudinal gradient up to 1700 m (Kiener, 1963). Similarly van Someren (1959) reports that *A. bengalensis* is the dominant eel species above 950 m in altitude. On Mayotte Island, a small island with small rivers, *A. bicolor* is restricted to altitudes of less than 100 m, while *A. marmorata* is found at all available stream altitudes (up to 200 m) (Valade et al., 2018).

In the Limpopo, *A. mossambica* is considered the most common species and penetrates far up the catchment into Zimbabwean rivers (Figure 4.2). In the same river, little is known about the upstream penetration of *A. bicolor*, but it is suspected that this species stays in the lower river reaches (Marshall, 2011). *A. bengalensis* is the most common species in the Zambezi River catchment where it also travels far inland through Mozambique through to Zambia and Zimbabwe (Figure 4.2) (Marshall, 2011). *A. bicolor* is not known to penetrate far inland and the furthest it has been found in Zimbabwe was in the Sabi-Lundi River junction near the Mozambique border, about 290 km from the sea (Jubb, 1961). In South Africa, *A. marmorata* has been found at altitudes as high as 750 m and as far as 320 km inland (C. Hanzen, unpublished data), while *A. mossambica* has been recorded 1670 m above sea level in the Mzimkhulu River (Crass, 1964). *A. mossambica* has also been found as far as the Ncandu River, about 500 km from the sea in the Thukela catchment (Crass, 1964). This is still currently the case as the Thukela River remains mostly free flowing. The Thukela is also the only river where *A. bengalensis* has been sampled in recent surveys of the KwaZulu-Natal Province in South Africa (C. Hanzen, unpublished data).

The difference in altitudinal distribution between these eel species may be explained by differences in the size of glass eels reaching the continental coast and subsequent differentials in growth rates that might affect their tendency or ability to migrate. *A. mossambica* has been shown to be the shortest (51.2 ± 2.1 mm) at freshwater recruitment in the islands followed by *A. marmorata* (53.3 ± 2.5 mm) and *A. bicolor* (54.0 ± 2.1 mm) (Robinet et al., 2003b). Jubb (1961) pointed out that once elvers reach a critical size of 175 mm, large waterfalls or dams become very difficult to ascend. Skead (1959) observed larger eels (up to 335 mm) negotiating small sloping dams on the Buffalo River in South Africa, but he observed that the more vertical obstacles were a problem for these larger eels. Elvers can travel considerable distances in East African rivers. For example, an *A. mossambica* elver of 87 mm was found in the Sabi-Lundi River system in Zimbabwe 290 km from the sea (Jubb, 1961). Conversely, it has been observed that when *A. mossambica* reach about 25–30 cm they stop moving upstream and become quite sedentary until maturity (Skelton, 2001). A size gradient was evident in the Sundays River, South Africa, from the mouth to the upper river and all large individuals (>24 cm) occurred only at the top site sampled (Wasserman et al., 2011). Jubb (1961) proposed that when *A. mossambica* and *A. bengalensis* are found together in the same system their distributions relative to one another are not explained by water temperature or by any environmental variables, but rather by the size of individuals when they enter the freshwater system.

A sex ratio response associated with altitude and distance upstream has also been observed (Pienaar, 1978). Males of *A. marmorata* generally stay closer to the sea, grow less rapidly and do not live as long as the females (Pienaar, 1978). Similar trends have been observed for *A. mossambica* with males reaching a smaller size and living in freshwater up to 8–10 years compared to 10–15 years for the females (Jubb, 1961).

A matter of obstacles and flow

The ability of elvers to ascend natural and human-made obstacles is well known and numerous examples of ascent of large obstacles in East African rivers are available, although it has to be remembered that such observations fail to take account of the relative success rates in most cases. In the Sundays River, South Africa, *A. mossambica* has been found upstream of the Darlington Dam, the wall of which is 48 m high, and 104 km away from the sea (Weyl et al., 2009). In the Umgeni River, South Africa, *A. mossambica* has been recorded upstream of several major consecutive obstacles including the Inanda Dam (65 m), Nagle Dam (44 m), Albert Falls Dam (33 m) and, finally, the Howick Falls (111 m) (C. Hanzen, unpublished data). *A. bengalensis* has also been found upstream of impressive obstacles, such as Kariba Dam, with a wall height of 128 m and over 1000 km from the sea (Balon and Coche, 1974) and Cahora Bassa dams on the Zambezi River (Skelton, 2001). The same species in the Pungwe River was the only one found upstream of the Pungwe Falls in Zimbabwe, the total height of which is around 200 m (Jubb, 1961; Turnbull-Kemp, 1958). Historically elvers (104–161 mm) have also been reported far inland on the Tana River in Kenya, over 1165 m in altitude and over 500 km from the sea (Van Someren and Whitehead, 1959). Unfortunately, dams render the upper part of the catchments unavailable for larger eels or species that are already too big to climb up obstacles when they reach them. Jubb (1961) showed that in the Lundi River in Zimbabwe *A. mossambica* was the only species present upstream of the 27 m dam wall of the Umshandige Dam. However, just below the dam wall, while *A. mossambica* was the most abundant species, large specimens of *A. bengalensis* were also present. The author concluded that by the time *A. bengalensis* reached the dam wall individuals were already too large to negotiate the obstacle.

The spatial and temporal organization of eel populations in freshwater can be ruled by the local hydrological regime as well as by obstacles to their migration. Robinet et al. (2003c) suggested that eels of all sizes could be washed downstream when cyclones occur annually in the Indian Ocean islands. They also observed a general movement upstream for all size classes to colonize or re-colonize upper reaches after the annual cyclonic season. As stated earlier, recruitment of young eels into the freshwater system is also dependent on the hydrological regime. This means that their dispersal into a particular system could be impacted by the regulation of the flow. A flow-regulating dam, would therefore not only represent a physical obstacle to migration, but would also impede the natural stimulus for upstream migration.

With very limited data available, it is not known if East African eels currently still overcome dams and other physical obstacles in the upper parts of their former geographical range, especially with the growing numbers of barriers built on their migration paths up rivers (Zarfl et al., 2015). It is also unknown how regulated flow and/drought episodes impact recruitment and elver migration up East African rivers. Research is urgently needed to address these knowledge gaps.

Downstream migration of adults

The silvering metamorphosis is common to maturing, downstream-migrating anguillids (Aoyama and Miller, 2003; Tesch, 2003) including the East African eels (Pienaar, 1978). While little is known about this silvering process for tropical African eels, transformations have been observed during their downstream descent to their spawning grounds in the Indian Ocean. The first silver eel in East Africa was reported in the early 1950s by Barnard (1953). Jubb (1958) described *A. mossambica* silver eels migrating in October and November through to as late as February. More recently, Robinet and Feunteun (2002) reported silver eel observations for *A. marmorata* and *A. bicolor* on Réunion. Few observations were made for these species in the past (including Jubb, 1961) for both species in South Africa. The silver stage can be difficult to identify with certainty except when a good dataset is available for comparative purposes. For instance, existing data for *A. marmorata* from Réunion showed that the ocular index (OI) for yellow eel between 200–600 mm is below 6 while the suspected silver eels had an OI of 11.53–12.31 (Robinet and

Feunteun, 2002). These specimens caught on Réunion showed a more advanced maturation than described for most other eel species.

Even less is known about the downstream migrations of East African eels towards the sea. In Madagascar, migrating silver eels used to be caught in high numbers in traditional trap nets from the beginning of the warm season until the end of January (Kiener, 1963). While the migration upstream of formidable barriers does not seem to be too difficult for small elvers, their way downstream as adults is a mystery and the effects of dams and weirs on their downstream migration is undocumented. Once they reach the sea there is no information on their migration behaviour or routes.

A sedentary freshwater life between migrations

While being spawned at sea anguillids, including those from East Africa, can spend most of their life in freshwaters. This stage can last between 7 and 50 years depending on the species, sex and geographic location (McEwan and Hecht, 1984; Tesch et al., 2007). As riverine eel use the same freshwater systems, inter-specific competition might play a role in regulating their habitat use and population size (Jellyman and Sykes, 2003). To limit competition, sympatric eels may exhibit local differences in habitat preference (altitude, substrate, flow) or feeding habits for instance.

Sex, age and growth

Very few publications on these topics for East African eels are available. Few authors report the use of otoliths to age and determine growth rate for eels in the region (Balon, 1975; McEwan and Hecht, 1984; Robinet et al., 2003c). Robinet et al. (2003c) estimated a growth rate of 17.4 mm ± 3.4 mm per month for *A. marmorata* elvers in Réunion. It has been shown that it is difficult to use a length–age relationship for *A. mossambica* as for a given length, there is a great variation of up to 5 years, and more as eels grow older (McEwan and Hecht, 1984). Slower growing individuals may reach a length greater than 650 mm after 10 years. Harrison (1953) attempted to age individuals based on the use of scales but this technique was not successful.

It is considered that sexual differentiation occurs early in the yellow eel stage but the expression of sex in anguillid eels has been the subject of long debates (Davey and Jellyman, 2005). The length at differentiation for the African tropical eels is unknown. McEwan and Hecht (1984) reported the difficulty of sexing young individuals that are smaller than 200 mm by dissection.

Habitat

East African eels are known to occupy a great variety of habitats (Bell-Cross and Minshull, 1988). *A. mossambica* and *A. marmorata* are observed in fast-flowing areas of streams and rivers in the Indian Ocean islands, while *A. bicolor* seems to utilize slower-flowing habitat with very fine substrate (Valade et al., 2018). On the continent *A. mossambica* has also been reported to use shallow water habitat or even space among rocks in rapids or in pools just below rapids (Pienaar, 1978).

These preferences also vary with eel size. Larger *A. marmorata and A. mossambica* are known to occupy slower and deeper habitat than at smaller sizes (Pienaar, 1978; Valade et al., 2018). Large yellow eels might use exclusive micro-habitats, such as deep pools (Butler and Marshall, 1996). However, large *A. mossambica* have also been known to occupy small streams or small pools in seasonal streams as long as there is some feature to offer good cover (Pienaar, 1978). *A. mossambica* has also been reported to hide in crevices or even buried in soft sediments (Griffiths et al., 2015). Bell-Cross and Minshull (1988) also remarked that eels tend to prefer habitats that would give them shelter away from potential predators. *A. bicolor* seems to prefer rivers and swamps near the coast (Crass, 1964).

While representing barriers to their migration, impounded areas above dams have been observed to offer habitats for eels. Balon and Coche (1974) reported large eels in Lake Kariba at depths between 25 and 40 m but were unable to identify them. In Réunion, Robinet et al. (2003c) observed eels of all size classes in impounded areas. Balon and Coche (1974) suggested that after overcoming Kariba Dam on the Zambezi River, young eels would spend up to 7–9 years in small streams upstream of the dam and after some time would drift back to the lake where they would grow old and sedentary.

Unfortunately, very little literature is available describing habitat use and preference in the region for the African eel species and this also needs to be rectified for effective conservation.

Behaviour

Very little is known about the behaviour of the African tropical eels in the region. Searching in 2018, no publications referring to detailed behavioural studies were found. All previous studies agree that *A. marmorata* and *A. mossambica* are active at night (Jubb, 1961; Pienaar, 1978; Skelton, 2001). *A. bengalensis* is believed to be nocturnal but also active from early afternoon onward (Jubb, 1961; Pienaar, 1978). This species is also known to show minimal activity during the dry season (Butler and Marshall, 1996). Tweddle and Skelton (2016) observed bright yellow-orange elongated nostrils of *A. bicolor* and advanced the hypothesis that these could be used as a lure to attract prey. Their hypothesis needs to be tested experimentally.

Predator and prey

Eels are carnivorous predators and scavengers. In some parts of their freshwater distribution, like the eastern Cape in South Africa, they are the only large, indigenous, piscivorous fish making them trophically unique. Adults feed on various prey including crabs, frogs, worms, fish and small aquatic animals (Bruton et al., 1987; Pienaar, 1978). Their diet also depends on their habitat and their size. Young eels (<20 cm) will, for example, feed on small crustaceans and insects. Wasserman et al. (2012) showed that elvers mainly feed on small insect larvae, while copepods, fish and plant material were also found in the stomach contents. As eels grow, they feed on larger prey. In Zimbabwe, *A. bengalensis* has been found to prey mainly on river crabs (*Potamonautes perlatus*), mountain catfish (*Amphilius uranoscopus*) and dragonfly larvae (Butler and Marshall, 1996). Unexpected prey such as small snakes and birds have been reported in big eels (Kiener, 1963).

Freshwater eels have also been known to feed on introduced fish species, such as trout, in Zimbabwe (Bell-Cross and Minshull, 1988; Butler and Marshall, 1996; Turnbull-Kemp, 1958) or guppies (*Poecilia reticulata*) in Mayotte (Valade et al., 2018). Butler and Marshall (1996) found no evidence for trout predating on eels but suggested significant competition between small eels and trout, as well as an impact of otter predation on eels.

Cultural and spiritual values

Several legends and associated eel-like mythical creatures exist in the region, especially on the islands. In Madagascar, traditional living is regulated by *fady* or traditional law that forbids certain food, activities or gestures. These can vary between clans or even between families. For many clans, eels are taboo and eating them is forbidden (Poirier, 1978). There are many reasons for these interdictions, most of these being old tales of people who died after eating large eels. Some legends regard them as sacred creatures harbouring the souls of their ancestors. A legend from the Zafinatoandro clan tells the story of a fisherman who caught an enormous eel that immediately transformed into a beautiful woman. The fisherman married her, and they had many children. After a fight, she transformed back into her original form and returned to the river. Since then it is forbidden to eat eels as they and the Zafinatoandro clan all come from the same mother (Decary, 1950). In contrast, there exist many traditional recipes making use of smoked eel in Madagascar. This is also the case in Réunion, where large eels are highly valuable delicacies, and a traditional dish is 'Carry z'anguille' (which translates to eel curry) (Robinet et al., 2003c).

On the continent, the spiritual value of eels is less evident and largely unknown in areas in Kenya, Tanzania, Somalia and Mozambique. In South Africa, there is a fear factor associated with eels as they can be associated with snakes or with the *inkanyamba*, a mythical creature in Zulu and Xhosa culture that has been the source of awe and fear for many generations (Wood, 2000). *Inkanyamba* are river monsters living at the bottom of waterfalls and are believed to be migratory and able to move overland. They are also believed to be able to control the weather and have been associated with seasonal storms and drought. People fear them as they think that if they disturb the *inkanyamba*, it will lead to great disasters. Furthermore, in South Africa, the limited cultural value in eels is reflected in there is not being much recreational angling interest for them and Harrison (1950) even proposed that eels should be harvested as much as possible to reduce predation on 'more desirable' gamefish, such as trout.

A similar story exists in the Zambezi River region. In the legend of the river god Nyami Nyami, he is described as a 'fish-dragon' guardian and is a benevolent god to the Tonga people of the region, controlling the life in and around the river. The legend of Nyami Nyami became quite famous with the construction of the Kariba Dam in the 1950s (Siwila, 2015). During the dam wall construction, it is said he and his wife got separated while many of the locals were relocated mostly against their will. The following years, major flooding occurred, and the construction had to be halted. Locals said that the floods were Nyami Nyami's work, either because of his anger due to being separated from his wife, or because he was invoked by the Tongas. Elders claimed that the dam is still standing through their intervention but that to this day Nyami Nyami still wants to be reunited with his wife.

Many legends and stories have been inspired by freshwater eels around the world. In the East Africa region, they are not as vivid as some other parts of the world where they can be sacred and attract many tourists (as is the case in New Zealand and the Pacific islands). These legends tell the story of a fearful or benevolent god. These stories should not be allowed to fade away and leave space for indifference, as they can be relevant when it comes to conservation.

Conservation status

The global conservation status of most of the anguillid eels has been IUCN Red List evaluated (Jacoby and Gollock, 2014a, 2014b; Jacoby et al., 2014a, 2014b). In Table 4.1, we summarize the known classifications and some of the global conventions that include the protection of East African eel species and their habitats. *A. bicolor* and *A. bengalensis* are listed as near threatened, while *A. marmorata* and *A. mossambica* are least concern.

Until 2018, no comprehensive management and conservation strategies were available for East African eels at a local, national or international level. In 2018, Réunion produced a 9 year plan (2018–2027) for the conservation of eels within its territory (Valade et al., 2018). The plan covers three species (*A. mossambica, A. bicolor* and *A. marmorata)* and includes measures such as combating poaching, fisheries management, restoration of river connectivity and habitat quality, evolutionary ecology, and improved governance. It should be noted that as a French Department Réunion has to follow EU regulations, such as the EU Water Framework Directive, which supports ecologically sustainable management of water resources and requires free movement of fishes in rivers and estuaries.

Tropical anguillids in East Africa are absent on most national red lists (or lists of protected species) except for the Red List of Réunion (IUCN France et al., 2013). Although a slight increase in numbers of *A. marmorata* has been observed in Réunion by Valade et al. (2018), locally the species is considered as near threatened because the population size of mature individuals is small (Valade et al., 2018). The rarity of large, mature marbled eel in Réunion may be due to poaching, as they represent desirable fishing trophies. In Mayotte, where inland fishing is forbidden, *A. marmorata* appear to grow larger (Valade et al., 2018). In Réunion *A. mossambica* and *A. bicolor* are both considered as critically endangered and *A. bengalensis* is listed as data deficient.

Threats and opportunities for eel conservation in the region

Use and development of trade

Globally, anguillid eels are economically valuable (Kuroki et al., 2014). In the regions discussed in this review, the economic interest is rather low, but it was proposed in the 1970s that they probably have the highest economic value per unit weight of fish (Jackson, 1976). The lack of research in the area could possibly be linked to the absence of true commercial interest compared to Europe or Asia. In the European Union, there is a ban on export of anguillid eels to international markets. This, together with a constantly increasing demand from the Asian market, has resulted in a growing interest from international investors to farm the African species *A. mossambica.*

Table 4.1 Conservation status for East African eels and examples of the international conventions and treaties that include the protection these species.

Species	IUCN Red List (Jacoby et al., 2014a, 2014b; Jacoby and Gollock, 2014a, 2014b)	National protected species in Réunion (Valade et al., 2018)	International conventions & treaties protecting eel species (Brink et al., 2018)
A. mossambica	Least Concern	Critically endangered	African Convention on the Conservation of Nature and Natural Resources 1968
A. bicolor	Near Threatened	Critically endangered	
A. marmorata	Least Concern	Data deficient	
A. bengalensis	Near Threatened	Data deficient	Convention of Biodiversity 1992
			MARPOL
			OSPAR
			Paris Convention
			Ramsar Convention
			South African Development Community (SADC) treaties
			Stockholm Declaration 1979
			Sustainable Development Goals 2016
			UNCLOS III

Few reliable sources regarding fishing and aquaculture production in the region were found for this review. Commercial (or even subsistence and artisanal) eel fishing in continental East Africa is somewhat anecdotal and opportunistic. However, some countries (including Madagascar), have entered the global trade recently and could potentially play a growing part in the coming years. Madagascar represented 4% of eel imports into East Asia between 2011 and 2013 according to a TRAFFIC report (Wildlife Trade Monitoring Network) (Crook, 2014). The report showed that Madagascar, Mauritius and even South Africa have contributed to the global trade in recent years (Table 4.2). According to FAO (2019), the tropical eel trade from Africa started in 1998 with regular export from Madagascar and intermittent live export from Mozambique, Tanzania and South Africa. The Indian Ocean Commission (2016) reported a production of 10 tonnes per year through aquaculture for Madagascar, all of which was exported while commercial fishing generated 387 tonnes per year. Mossambica Angulas, a Malagasy company, claims to produce 440 tonnes per year (http://www.farm-africa.com/aquaculture.html, accessed on 30 July 2018). With these different sources, it is really difficult to have a global picture of the eel trade in Sub-Saharan Africa and its role on the species status and conservation.

In terms of small-scale fisheries, eels can contribute to food security and income for riparian communities, but data on such fisheries are lacking. On Réunion there is indeed a strong eel fishing tradition. Although fishing licenses allow the harvest of all life stages (Valade et al., 2018), eels are typically fished for by angling or a technique called 'à la tâte' involving a fisherman holding a baited stick and searching all possible eel spots. Harvests are regulated depending on the species, the time of the year and the classification of the rivers and/or the reach (Valade et al., 2018). In contrast on Mayotte Island (Comores Islands), fishing and trade is completely forbidden for any eel species. These regulations are planned to be revised (Valade et al., 2018). In Madagascar, the fishing can either be very traditional or completely forbidden depending on the area because of *fady* or traditional taboo. Historically, Kiener (1963) reported a flourishing eel fishery in the

Table 4.2 Export of live eels (kg) into East Asia from Sub-Saharan Africa, 2005–13. Adapted from Crook (2014).

Countries	2005	2006	2007	2008	2009	2010	2011	2012	2013
Madagascar (glass eel/elvers)	262	40	0	403	1370	22	1797	3783	5562
Madagascar (other sizes/stages)	0	0	0	0	0	100	900	600	2600
Mauritius (glass eel/elvers)	0	0	0	0	0	0	22	0	0
South Africa (glass eel/elvers)	0	0	0	0	0	0	0	0	10

Maningory region where over 15,000 eels are harvested per season and subsequently sold on local markets as live or smoked eels. In 2018, smoked eels could be bought online from Madagascar for around €24/kg, giving an indication of the potential economic value.

Parasites and disease

Madagascar, currently the only African producer of eels by aquaculture, has recently started exporting eels to Hong Kong, South Korea and Japan (Crook and Nakamura, 2013) and there has been interest in the culture of *A. mossambica* in the USA (Wyman-Grothem et al., 2018). It is well known that the global trade of live eels is a pathway for introductions of parasitic organisms into new ecosystems and hosts (Taraschewski, 2006). Some, such as the blood-feeding parasite of eel swimbladders *Anguillicoloides crassus* and the gill parasite *Pseudodactylogyrus anguillae*, have been implicated in the decline of *A. anguilla* stocks in Europe (Wielgoss et al., 2008; Wirth and Bernatchez, 2003). While little is known on the vulnerability of the East African species to parasites, Asian, European and American eels have been shown to be vulnerable to parasitic diseases (Taraschewski, 2006), which negatively affect their culture (Scholz, 1999) and have major effects on wild populations (Wielgoss et al., 2008).

Several helminth parasites including the African swimbladder parasite (*Anguillicolloides papernaii*), the gastro-intestinal nematodes (*Heliconema africana* and *Paraquimperia africana*) (Moravec et al., 2000; Taraschewski et al., 2005) and the introduced gill parasite (*Pseudodactylogyrus anguillae*), are widespread in *A. mossambica* populations in South Africa and Madagascar (Christison and Baker, 2007; McHugh et al., 2017; Parker et al., 2011; Weyl et al., 2014). Laetsch et al. (2012), based on a phylogenetic assessment of the subfamily Anguillicolidae Yamaguti, proposed that other *Anguillicoloides* species might transfer to and be pathogenic to native *Anguilla* species. This hypothesis is supported by the successful experimental infection of *A. anguilla* with the African *A. papernai* by Moravec and Taraschewski (Taraschewski et al., 2005). The gill parasite *P. anguillae* has also been sampled from eels in the Eastern Cape Province of South Africa (McHugh et al., 2017; Parker et al., 2011). Infected fish demonstrated considerable alterations including hyperplasia of gill tissue, fusion of the secondary lamellae and hyperanaemia. This has been linked to *P. anguillae* infestation in European eel (Kennedy, 2007) and has been previously documented to cause mortalities of both *A. rostrata* (American eel) and *A. anguilla* under aquaculture conditions (Kennedy, 2007; Køie, 1991). As a result, proactive approaches for the control of parasitic diseases will be required if African eels are to be used in aquaculture, or are to be traded globally (Weyl et al., 2014).

State of water resources influencing eel distribution

Throughout the distributional range of eels in eastern and southern Africa, human demands on freshwater ecosystems has affected the suitability of many rivers and other aquatic ecosystems for eels. Anthropogenic stressors on aquatic ecosystems include barrier formation, habitat alteration associated with water abstraction and flow alterations, deterioration of water quality, excessive harvesting of natural products, alien invasive species and direct disturbances from growing urban and peri-urban settlements (Abbott et al., 2015; King and Pienaar, 2011; Zarfl et al., 2015). Such stressors degrade the ecological wellbeing and status of rivers that eels use in eastern and southern Africa (Jacoby et al., 2015). Although limited, information indicative of the ecological well-being of rivers in the region, includes a general decline in well-being with regional losses of fisheries and other ecosystem services, and biodiversity of both species and habitats (Lorenzen et al., 2016; Thieme et al., 2007; Tweddle et al., 2015; UNEP, 2016). Within this region three entire ecoregions are now in a critical state in terms of current and future wellbeing of fauna and ecological and ecosystem processes, namely: the Madagascar Eastern Lowlands, Cape Fold and Drakensberg-Maloti Highlands (McClain et al., 2019; Thieme et al., 2007).

Although the upstream penetration ability of anguillid eels has been demonstrated (Jacoby et al., 2015), ongoing construction of dams for water storage and power generation on larger rivers in the region has strong potential to impede eel migrations, and resultant effects on eel distribution and recruitment in a manner seen for temperate eels (Jacoby et al., 2015). New dams are planned for many largely free-flowing rivers within the region including the Galana (Kenya), Pungwe, Save and Inkomati (Mozambique) and Thukela, Umkomas, Mzimkulu Tina and Tsitsa Rivers (South Africa). Other impacts, including water quality pollution, sedimentation and alien plant and animal invasions, have affected the well-being of small- and moderate-sized rivers throughout the region, including impacts to the many estuaries that have been identified to be in unsustainable conditions in the region (Beilfuss and Brown, 2010; Okuku et al., 2013; Shaghude, 2016).

The management of water resources, and the multiple stressors affecting ecosystem well-being throughout the region, includes policies and procedures developed to balance use and protection to ensure sustainability (Malzbender and Earle, 2007; King and Pienaar, 2011). This includes the maintenance of the biodiversity of the rivers in the region and associated ecosystem processes, which, although poorly implemented, includes river connectivity maintenance and as such the migrations of eels in the region (King and Pienaar, 2011). Regional policies and procedures are dominated by principles and requirements of the South African National Water Act (108 of 1998). This includes the concept of the ecological reserve in water management – the amount of water and its quality required to sustain viable functioning ecosystems, as well as to maintain human livelihoods (King and Pienaar, 2011). From the South African foundations, the ecological reserve has now been integrated into the Zambian (Water Resources Management Act of 2011), Kenyan (Act 43 of 2016) and Tanzanian (Act 11 of 2009) legislation with international trans-boundary agreements between South Africa and Mozambique being established on the principles of the policies (SADC regional water policy of 2005). In Madagascar, the National Water Code of 1999 has established principles of integrated water resource management that address sustain-ability and minimization of water resource use (Bidaud et al., 2013; Smakhtin et al., 2004). Each country within the region also has complementary conservation legislation, that along with the water resources management legislation, cooperatively provides a legal framework for the conservation of water resources and their biodiversity that includes habitats and the species themselves.

Complementary to the ecological reserve, regional water resource protection and conservation legislation has adopted principles of establishing environmental flows for rivers within eastern and southern Africa, Madagascar and other islands in the region (Tharme, 2003; Smakhtin et al., 2004). Environmental flows describe the quantity, timing, and quality of freshwater flows and levels necessary to sustain aquatic ecosystems that, in turn, support human cultures, economies, livelihoods and well-being (Arthington et al., 2018). This includes maintaining the biodiversity of habitats and species within the region and river connectivity for, among others, the anguillids that are one of the migratory fish taxa of the region. In the region, although many thousands of ecological reserve and/or environmental flow assessment studies have been undertaken, with consideration of river connectivity and fish migrations (King and Pienaar, 2011; O'Brien et al., 2018), many dams, and other barriers that hamper migrations, still exist and many new dams are being planned that do not adequately address eel migrations.

Fishways are measures to mitigate instream barrier impacts (Silva et al., 2018). There are less than 100 fishways in Africa and many of the designs are non-functional as they are based on American migratory fish species (Bok et al., 2007). Few have researched the design of fishways in the region with anguillid eel migrations in mind (Bok et al., 2007; Ross, 2015). No significant research in East Africa has evaluated the design of fishways or other instream measures to improve the migrations of eels in coastal and freshwater habitats in the region.

For the conservation of our charismatic migratory eels of the East African rivers we need a better understanding of their populations, biology, ecology and resilience to multiple stressors so that we can effectively manage them and the rivers that they occur in to maintain sustainable populations.

Conclusion: identifying knowledge gaps, trends and the way forward

This review is part of urgently needed action to develop the knowledge base required to direct future research and conservation measures for Africa's eels. We have seen that despite their large geographical range very little is known about the elusive East African species, particularly on the main continent. Their migration strategies and lifecycles are broadly similar, although evidence suggests that there are differences in their ecology and migration behaviour that could affect conservation planning given dam developments under consideration (Zarfl et al., 2015). The cumulative impacts on the eel populations from thousands of instream barriers, together with flow modifications, water quality changes and habitat changes remain unknown. This issue requires urgent consideration, over and above the expected responses of these fishes to climate change and changing ocean habitats.

Some conservationists have already started to recognize that both the range and abundance of the species are decreasing (Jacoby et al., 2015). IUCN classified *A. bengalensis* and *A. bicolor* as near threatened, while the French government on Réunion has considered *A. mossambica* and *A. bicolor* critically endangered on the island. It is vital that further information is collected since there are concerns that the situation is deteriorating further. As most information is outdated, we recommend research on all aspects of the biology and ecology of these species, or other work relevant to their conservation including:

- eel recruitment into freshwater systems of the Agulhas Current region and the associated drivers, habitat use and behaviour in freshwater systems, as well as the contribution of marine and brackish-water eels to the larger population
- developing monitoring baselines of abundance and distribution of eels in the Agulhas region to provide the basis for assessment of conservation status and impacts
- impacts of habitat change and river fragmentation on eel distributions in the region
- socio-economic importance of eels in East Africa and associated islands
- detailed knowledge about the passage performance and migration behaviour of eels that could help develop fishway designs suitable for the African continent and surrounding islands
- regional transboundary environmental flow policies, requirements and frameworks that specifically address river connectivity and the requirements of indicator migratory fishes
- water resource management and conservation policies in the region for the specific management of river connectivity for migratory fishes and the socio-ecological services they provide to the region.

Bibliography

Abbott, J.G., Hay, C.J., Naesje, T.F., Tweddle, D. and van der Waal, B.C. (2015) Rain and copper: the evolution of a fish marketing channel in a rapidly changing region of southern Africa. *Journal of Southern African Studies* 41(1), 29–45.

Aoyama, J. (2009) Life history and evolution of migration in catadromous eels (genus *Anguilla*) *Aqua-BioScience Monographs* 2(1).

Aoyama, J. and Miller, M.J. (2003) The silver eel. In: Aida, K., Tsukamoto, K. and Yamauchi, K. (eds) *Eel Biology*. Springer, Tokyo, pp. 107–117.

Arai, T. and Chino, N. (2012) Diverse migration strategy between freshwater and seawater habitats in the freshwater eel genus *Anguilla*. *Journal of Fish Biology* 81(2), 442–455.

Arai, T., Limbong, D., Otake, T. and Tsukamoto, K. (1999) Metamorphosis and inshore migration of tropical eels *Anguilla* spp. in the Indo-Pacific. *Marine Ecology Progress Series* 182, 283–293.

Arai, T., Marui, M., Otake, T. and Tsukamoto, K. (2002) Inshore migration of a tropical eel, *Anguilla marmorata*, from Taiwanese and Japanese coasts. *Fisheries Science* 68(1), 152–157.

Arthington, A.H., Bhaduri, A., Bunn, S.E., Jackson, S.E., Tharme, R.E., Tickner, D., Young, B., Acreman, M., Baker, N., Capon, S., Horne, A.C., Kendy, E., McLain, M.E., Le Roy Poff, N., Richter, B.D. and Ward, S. (2018) The Brisbane Declaration and global action agenda on environmental flows. *Frontiers in Environmental Science* 6, 45.

Balon, E.K. (1975) The eels of Lake Kariba: distribution, taxonomic status, age, growth and density. *Journal of Fish Biology* 7(6), 797–815.

Balon, E.K. and Coche, A.G. (1974) The eels. In: Balon, E.K. and Coche, A.G. (eds) *Lake Kariba: A Man-Made Tropical Ecosystem in Central Africa*. *Monographiae Biologicae*, Vol. 24. Springer, Dordrecht, pp. 446–458.

Baran, P., Basilico, L., Larinier, M., Rigaud, C. and Travade, F. (2012) Management plan to save the eel. Optimizing the design and management of installations. Meeting Recap. ONEMA, Paris, France.

Barnard, K.H. (1953) Migration of freshwater eels. *Piscator* 25, 8–9.

Beilfuss, R. and Brown, C. (2010) Assessing environmental flow requirements and trade-offs for the Lower Zambezi River and Delta, Mozambique. *International. Journal of River Basin Management* 8(2), 127–138.

Bell-Cross, G. and Minshull, J.L. (1988) *The Fishes of Zimbabwe*. National Museums and Monuments of Zimbabwe, Harare, Zimbabwe.

Bok, A., Kotze, P., Heath, R. and Rossouw, J., (2007) Guidelines for the planning, design and operation of fishways in South Africa. Pretoria: Water Research Commission. Report No. TT287/07.

Bidaud, C., Méral, P., Andriamahefazafy, F., Serpantié, G., Cahen-Fourot, L. and Toillier, A. (2013) Institutional and historical analysis of payments for ecosystem services in Madagascar. In: Muradian, R. and Rival, L. (eds) *Governing the Provision of Ecosystem Services*. Dordrecht, Springer, pp. 207–233.

Brand, M., Maina, J., Mander, M. and O'Brien, G. (2009) Characterization of the social and economic value of the use and associated conservation of the yellowfishes in the Vaal River. Water Research Commission Report, Gezina, South Africa.

Brink, K., Gough, P., Royte, J., Schollema, P.P. and Wanningen, H. (2018) *From Sea to Source 2.0. Protection and Restoration of Fish Migration in Rivers Worldwide*. World Fish Migration Foundation, Groningen.

Bruton, M.N., Bok, A.H. and Davies, M.T. (1987) Life history styles of diadromous fishes in inland waters of southern Africa. *American Fisheries Society Symposium* 1, 104–121.

Butler, J.R.A. and Marshall, B.E. (1996) Resource use within the crab-eating guild of the upper Kairezi River, Zimbabwe. *Journal of Tropical Ecology* 12(4), 475–490.

Chino, N. and Arai, T. (2010) Migratory history of the giant mottled eel (*Anguilla marmorata*) in the Bonin Islands of Japan. *Ecology of Freshwater Fish* 19(1), 19–25.

Christison, K.W. and Baker, G.C. (2007) First record of *Pseudodactylogyrus anguillae* (Monogenea) from South Africa. *African Zoology* 42(2), 279–285.

Crass, R.S. (1964) *Freshwater fishes of Natal*. Schuter & Shooter, Pietermaritzburg.

Crook, V. (2014) Slipping away: international *Anguilla* eel trade and the role of the Philippines. London, TRAFFIC International & ZSL.

Crook, Vicki and Nakamura, M. (2013) Glass eels: assessing supply chain and market impacts of a CITES listing on *Anguilla* species. *TRAFFIC Bulletin* 24(1), 24–30.

Davey, A.J. and Jellyman, D.J. (2005) Sex determination in freshwater eels and management options for manipulation of sex. *Reviews in Fish Biology and Fisheries* 15(1–2), 37–52.

Decary, R. (1950) *La faune malgache: Son rôle dans les croyances et les usages indigènes*. Payot, Lausanne.

Dekker, W., Casselman, J.M., Cairns, D.K., Tsukamoto, K., Jellyman, D. and Lickers, H. (2003) Worldwide decline of eel resources necessitates immediate action. Quebec Declaration of Concern. *Fisheries* 28, 28–30.

Eel farming in Madagascar – Mossambica Angulus LTD (n.d.) Available at http://www.farm-africa.com/aquaculture.html (accessed 30 July 2008).

FAO AQUASTAT (2007) Dams Africa – 070524: Dams and Agriculture in Africa., s.l.: Prepared by the AQUASTAT Programme: Water Development and Management Unit (NRLW) Land and Water division (NRL) Food and Agriculture Organization of the United Nations (FAO).

FAO (2019) Capture and Aquaculture Production (1950–2017) and Fisheries Commodities Production and Trade (1976–2016) data. (Accessed: May, 2019).

FISHBASE (2018) Available at www.fishbase.org (accessed 1 June 2018).

Frost, W.E. (1957) First record of the elver of the African Eel *Anguilla nebulosa labiata*. *Nature* 179(4559), 594–594.

Gagnaire, P.A., Minegishi, Y., Aoyama, J., Réveillac, E., Robinet, T., Bosc, P., Tsukamoto, K., Feunteun, E. and Berrebi, P. (2009) Ocean currents drive secondary contact between *Anguilla marmorata* populations in the Indian Ocean. *Marine Ecology Progress Series* 379, 267–278.

GBIF (2018) GBIF Occurrence Download. Available at https://doi.org/10.15468/dl.bcoxmo (accessed 12 July 2018).

Golubtsov, A.S., Tefera, F. and Darkov, A.A. (2012) Marine elements in fish fauna of southeastern Ethiopia. *Journal of Ichthyology* 52(11), 813–816.

Griffiths, C., Picker, M. and Day, J. (2015) *Freshwater Life: A Field Guide to the Plants and Animals of Southern Africa*. Random House Struik, Cape Town.

Harris, S.A. and Cyrus, D.P. (1995) Occurrence of fish larvae in the St Lucia Estuary, KwaZulu-Natal, South Africa. *South African Journal of Marine Science* 16(1), 333–350.

Harrison, A.C. (1950) The South African freshwater eel *Anguilla mossambica*. *Piscator* 15, 89–91.

Harrison, A.C. (1953) The age and scales of freshwater eels. *Piscator* 25, 9–12.

Ishikawa, S., Tsukamoto, K. and Nishida, M. (2004) Genetic evidence for multiple geographic populations of the giant mottled eel *Anguilla marmorata* in the Pacific and Indian oceans. *Ichthyological Research* 51(4), 343–353.

IUCN France, MNHN, SEOR, ARDA, Insectarium de La Réunion, GLOBICE and Kélonia (2013) La Liste rouge des espèces menacées en France – Faune de La Réunion. Paris, France.

IUCN (2018) The IUCN Red List of Threatened Species. Version 2018–1. Available at http://www.iucnredlist.org (accessed 1 July 2018).

Jackson, P.B.N. (1976) The distribution and growth of eels of the genus *Anguilla* Shaw, 1830, in the Western Indian Ocean. *Travaux et Documents de l'Office de la Recherche Scientifique et Technique d'Outre-Mer* 47, 189–192.

Jacoby, D. and Gollock, M. (2014a) *Anguilla marmorata*. The IUCN Red List of Threatened Species 2014: e.T166189A45832585. http://dx.doi.org/10.2305/IUCN.UK.2014–1.RLTS.T166189A45832585.en (accessed 30 July 2018).

Jacoby, D. and Gollock, M. (2014b) *Anguilla mossambica*. The IUCN Red List of Threatened Species 2014: e.T183155A1730550. http://dx.doi.org/10.2305/IUCN.UK.2014–1.RLTS.T183155A1730550.en. (accessed 30 July 2018).

Jacoby, D., Harrison, I.J. and Gollock, M. 2014a. *Anguilla bengalensis*. The IUCN Red List of Threatened Species 2014: e.T61668607A15501445. http://dx.doi.org/10.2305/IUCN.UK.2014–1.RLTS.T61668607A15501445.en (accessed 30 July 2018).

Jacoby, D., Harrison, I.J. and Gollock, M. 2014b. *Anguilla bicolor*. The IUCN Red List of Threatened Species 2014: e.T166894A67015710. http://dx.doi.org/10.2305/IUCN.UK.2014–1.RLTS.T166894A67015710.en (accessed 30 July 2018).

Jacoby, D.M.P., Casselman, J.M., Crook, V., DeLucia, M.-B., Ahn, H., Kaifu, K., Tagried, K., Sasal, P., Silvergrip, A.M.C., Smith, K., Uchida, K., Walker, A.M. and Gollock, M. J. (2015) Synergistic patterns of threat and the challenges facing global anguillid eel conservation. *Global Ecology and Conservation* 4, 321–333.

Jellyman, D.J. and Sykes, J.R.E. (2003) Diel and seasonal movements of radio-tagged freshwater eels, *Anguilla* spp., in two New Zealand streams. *Environmental Biology of Fishes* 66(2), 143–154.

Jespersen, P. (1942) *Indo-Pacific Leptocephalids of the genus Anguilla: systematic and biological studies. Dana Report n°22, RETZELS C.A. (ed.), Copenhagen, 13–103.* Dana Report No. 22.

Jubb, R.A. (1957a) Eels in Southern Africa. *Nature* 180(4575), 47–48.

Jubb, R.A. (1957b) The freshwater eel *Anguilla obscura* Gunther, 1871, in Southern Africa. *Nature* 180(4596), 1216–1216.

Jubb, R.A. (1957c) The mottled eels of King William's Town District. *Piscator* 39, 115–117.

Jubb, R. A. (1958). Notes on silver eels of the species *Anguilla mossambica* Peter 1852. *Piscator, 41*, 204–208.

Jubb, R.A. (1961) *An Illustrated Guide to the Freshwater Fishes of the Zambezi River, Lake Kariba, Pungwe, Sabi, Lundi and Limpopo Rivers*. Stuart Manning, Cape Town.

Jubb, R.A. (1967) *Freshwater Fishes of Southern Africa*. Balkema, Cape Town.

Kennedy, C.R. (2007) The pathogenic helminth parasites of eels. *Journal of Fish Diseases* 30(6), 319–334.

Kiener, A. (1963) *Poissons, Pêche et Pisciculture à Madagascar*. Centre Technique Forestier Tropical, Nogent-sur-Marne.

King, J. and Pienaar, H. (2011) Sustainable use of South Africa's inland waters: a situation assessment of resource directed measures 12 years after the 1998 National Water Act. *Water Research Commission Report No. TT* 491(11), 1–245.

Kleynhans, C.J., Louw, M.D. and Moolman, J. (2007) Reference frequency of occurrence of fish species in South Africa. Report Produced for the Department of Water Affairs and Forestry (Resource Quality Services) and the Water Research Commission.

Køie, M. (1991) Swimbladder nematodes (*Anguillicola spp.*) and gill monogeneans (*Pseudodactylogyrus* spp.) parasitic on the European eel (*Anguilla anguilla*). *ICES Journal of Marine Science* 47(3), 391–398.

Kuroki, M., Righton, D. and Walker, A. (2014) The importance of anguillids: a cultural and historical perspective introducing papers from the World Fisheries Congress. *Ecology of Freshwater Fish* 23(1), 2–6.

Laetsch, D.R., Heitlinger, E.G., Taraschewski, H., Nadler, S.A. and Blaxter, M.L. (2012) The phylogenetics of Anguillicolidae (Nematoda: Anguillicoloidea), swimbladder parasites of eels. *BMC Evolutionary Biology* 12(1), 60.

Lin, Y.-J., Jessop, B.M., Weyl, O.L.F., Iizuka, Y., Lin, S.-H., Tzeng, W.-N. and Sun, C.-L. (2012) Regional variation in otolith Sr:Ca ratios of African longfinned eel *Anguilla mossambica* and mottled eel *Anguilla marmorata*: a challenge to the classic tool for reconstructing migratory histories of fishes. *Journal of Fish Biology* 81(2), 427–441.

Lorenzen, K., Cowx, I.G., Entsua-Mensah, R.E.M., Lester, N.P., Koehn, J.D., Randall, R.G., So, N., Bonar, S.A., Bunnell, D.B., Venturelli, P., Bower, S.D. and Cooke, S.J. (2016) Stock assessment in inland fisheries: a foundation for sustainable use and conservation. *Reviews in Fish Biology and Fisheries* 26(3), 405–440.

Malzbender, D. and Earle, A. (2009) Water resources of SADC: Demands, dependencies and governance responses. In Pressend, M. and Othieno, T. (eds) *Rethinking Natural Resources in Southern Africa*. Midrand, Institute for Global Dialogue, pp. 85–106.

Marshall, B. (2011) The fishes of Zimbabwe and their biology. *Smithiana* Monograph 3. Publications in Aquatic Biodiversity, SAIAB, Grahamstown.

Marui, M., Arai, T., Miller, M.J., Jellyman, D.J. and Tsukamoto, K. (2001) Comparison of early life history between New Zealand temperate eels and Pacific tropical eels revealed by otolith microstructure and microchemistry. *Marine Ecology Progress Series* 213, 273–284.

McClain, M.E., Fouchy, K., O'Brien, G.C. and Conallin, J. (2019) Multiple stressors in Africa: insight from the developing world. In: Sabater, S., Elosegi, A. and Ludwig, R. (eds) *Multiple Stressors in River Ecosystems: Status, Impacts and Prospects for the Future*. Elsevier, Amsterdam, pp. 179–191.

McCosker, J.E., Bustamante, R.H. and Wellington, G.M. (2003) The freshwater eel, *Anguilla marmorata*, discovered in Galápagos. *Noticias de Galapagos* 62, 2–6.

McEwan, T. and Hecht, A. (1984) Age and growth of the longfin eel, *Anguilla mossambica* Peters, 1852 (Pisces: Anguillidae) in Transkei rivers. *African Zoology* 19(4), 280–285.

McHugh, K.J., Weyl, O.L.F. and Smit, N.J. (2017) Parasite diversity of African longfin eel *Anguilla mossambica Peters* with comments on host response to the monogenean *Pseudodactylogyrus anguillae* (Yin and Sproston). Journal of Fish Diseases 40(7), 959–961.

Minegishi, Y., Aoyama, J., Inoue, J.G., Miya, M., Nishida, M. and Tsukamoto, K. (2005) Molecular phylogeny and evolution of the freshwater eels genus *Anguilla* based on the whole mitochondrial genome sequences. *Molecular Phylogenetics and Evolution* 34(1), 134–146.

Moravec, F., Boomker, J., & Taraschewski, H. (2000) *Paraquimperia africana* n. sp.(Nematoda: Quimperiidae), a new intestinal parasite of the eel *Anguilla mossambica* Peters, in South Africa. Journal of Parasitology, 86(1), 113–118.

O'Brien, G.C., Dickens, C., Hines, E., Wepener, V., Stassen, R., Quayle, L., Fouchy, K., MacKenzie, J., Graham, P.M and Landis, W. G. (2018) A regional-scale ecological risk framework for environmental flow evaluations. *Hydrology and Earth System Sciences* 22(2), 957–975.

Okuku, E.O., Ohowa, B., Kiteresi, L., Wanjeri, V.O., Okumu, S., Ochola, O. and Ongore, C.O. (2013) Screening of potential ecological risk of metal contamination in some Kenyan estuaries. *Research Journal of Physical and Applied Science* 2(4), 52–63.

Parker, D., Weyl, O.L.F. and Taraschewski, H. (2011) Invasion of a South African *Anguilla mossambica* (Anguillidae) population by the alien gill worm *Pseudodactylogyrus anguillae* (Monogenea). *African Zoology* 46(2), 371–377.

Pienaar, U. de V. (1978) *The freshwater Fishes of the Kruger National Park.* V & R.

Poirier, J. (1978) Les fady dans la société malgache. *Collection IDERIC* 9(1), 395–411.

Pous, S., Feunteun, E. and Ellien, C. (2010) Investigation of tropical eel spawning area in the South-Western Indian Ocean: influence of the oceanic circulation. *Progress in Oceanography* 86(3), 396–413.

Réveillac, E. (2008) Histoires de vie larvaire et dispersion des Anguillidés: vers une approche bio-évolutive, 249. PhD thesis. Université de La Rochelle, La Rochelle.

Réveillac, É., Robinet, T., Rabenevanana, M.-W., Valade, P., Feunteun, É., (2009) Clues to the location of the spawning area and larval migration characteristics of *Anguilla mossambica* as inferred from otolith microstructural analyses. *Journal of Fish Biology* 74, 1866–1877.

Robinet, T. and Feunteun, E. (2002) First observations of shortfinned *Anguilla bicolor bicolor* and longfinned *Anguilla marmorata* silver eels in the Reunion island. *Bulletin Français de La Pêche et de La Pisciculture* (364), 87–95.

Robinet, T., Sbaihi, M., Guyet, S., Mounaix, B., Dufour, S. and Feunteun, E. (2003a) Advanced sexual maturation before marine migration of *Anguilla bicolor bicolor* and *Anguilla marmorata* at Réunion Island. *Journal of Fish Biology* 63(2), 538–542.

Robinet, Tony, Guyet, S., Marquet, G., Mounaix, B., Olivier, J.-M., Tsukamoto, K., Valade, P., Feunteun, E. (2003c) Elver Invasion, population structure and growth of marbled eels *Anguilla marmorata* in a tropical river on Réunion Island in the Indian Ocean. *Environmental Biology of Fishes* 68(4), 339–348.

Robinet, T., Lecomte-Finiger, R., Escoubeyrou, K. and Feunteun, E. (2003b) Tropical eels *Anguilla* spp. recruiting to Réunion Island in the Indian Ocean: taxonomy, patterns of recruitment and early life histories. *Marine Ecology Progress Series* 259, 263–272.

Robinet, T., Feunteun, E., Keith, P., Marquet, G., Olivier, J.-M., Réveillac, E. and Valade, P. (2007) Eel community structure, fluvial recruitment of *Anguilla marmorata* and indication for a weak local production of spawners from rivers of Réunion and Mauritius islands. *Environmental Biology of Fishes* 78(2), 93–105.

Robinet, T., Réveillac, E., Kuroki, M., Aoyama, J., Tsukamoto, K., Rabenevanana, M. W., Valade, P., Gagnaire P.A., Berrebi, P. and Feunteun, E. (2008) New clues for freshwater eels (*Anguilla* spp.) migration routes to eastern Madagascar and surrounding islands. *Marine Biology* 154(3), 453–463.

Ross, M.J. (2015) Determining the biological requirements of selected important migratory fish species to aid in the design of fishways in South Africa. PhD thesis. University of Johannesburg, Johannesburg.

Scholz, T. (1999) Parasites in cultured and feral fish. *Veterinary Parasitology* 84(3–4), 317–335.

Scott, L.E.P., Skelton, P.H., Booth, A.J., Verheust, L., Harris, R. and Dooley, J. (2006) *Atlas of Southern African Freshwater Fishes*. South African Institute for Aquatic Biodiversity, Grahamstown.

Seegers, L., De Vos, L. and Okeyo, D. O. (2003) Annotated checklist of the freshwater fishes of Kenya (excluding the lacustrine haplochromines from Lake Victoria) *Journal of East African Natural History* 92(1), 11–47.

Shaghude, Y.W. (2016) Estuarine environmental and socio-economic impacts associated with upland agricultural irrigation and hydropower developments: the case of Rufiji and Pangani Estuaries, Tanzania. In: Diop, S., Scheren, P. and Ferdinand Machiwa, J. (eds) *Estuaries: A Lifeline of Ecosystem Services in the Western Indian Ocean*. Springer, pp. 169–182.

Shiao, J.C., Iizuka, Y., Chang, C.W. and Tzeng, W.N. (2003) Disparities in habitat use and migratory behavior between tropical eel *Anguilla marmorata* and temperate eel *A. japonica* in four Taiwanese rivers. *Marine Ecology Progress Series* 261, 233–242.

Silva, A.T., Lucas, M.C., Castro-Santos, T., Katopodis, C., Baumgartner, L.J., Theim, J.D., Aarestrup, K., Pompeu, P.S., O'Brien, G.C., Braun, D.C., Burnett, N.J., Zhu, D.Z., Fjeldstad, H.P., Forseth, T., Rajaratnam, N., Williams, J.G. and Cooke, S.J. (2018) The future of fish passage science, engineering and practice. *Fish and Fisheries* 19, 340–363.

Siwila, L.C. (2015) An encroachment of ecological sacred sites and its threat to the interconnectedness of sacred rituals: A case study of the Tonga people in the Gwembe valley. *Journal for the Study of Religion* 28(2), 138–153.

Skead, C.J. (1959) The climbing of juvenile eels. *Piscator* 46, 74–87.

Skelton, P.H. (2001) *A Complete Guide to the Freshwater Fishes of Southern Africa*. Struik Nature, Cape Town.

Smakhtin, V., Revenga, C. and Döll, P. (2004) A pilot global assessment of environmental water requirements and scarcity. *Water International* 29(3), 307–317.

Taraschewski, H. (2006) Hosts and parasites as aliens. *Journal of Helminthology* 80(2), 99–128.

Taraschewski, H., Boomker, J., Knopf, K. and Moravec, F. (2005) *Anguillicola papernai* (Nematoda: Anguillicolidae) and other helminths parasitizing the African longfin eel *Anguilla mossambica*. *Diseases of Aquatic Organisms* 63(2–3), 185–195.

Tesch, F.-W. (2003) The Eel, 3rd edn. Blackwell Science, Oxford.

Tharme, R.E. (2003) A global perspective on environmental flow assessment: emerging trends in the development and application of environmental flow methodologies for rivers. *River Research and Applications* 19(5–6), 397–441.

Thieme, M., Lehner, B., Abell, R., Hamilton, S.K., Kellndorfer, J., Powell, G. and Riveros, J.C. (2007) Freshwater conservation planning in data-poor areas: an example from a remote Amazonian basin (Madre de Dios River, Peru and Bolivia) *Biological Conservation* 135(4), 484–501.

Tsukamoto, K., Aoyama, J. and Miller, M.J. (2002) Migration, speciation and the evolution of diadromy in anguillid eels. *Canadian Journal of Fisheries and Aquatic Sciences* 59(12), 1989–1998.

Turnbull-Kemp, P.S.J. (1958) Eels in Rhodesian trout streams. *Piscator* 41, 210–212.

Tweddle, D., Cowx, I.G., Peel, R.A. and Weyl, O.L.F. (2015) Challenges in fisheries management in the Zambezi, one of the great rivers of Africa. *Fisheries Management and Ecology* 22(1), 99–111.

Tweddle, D. and Skelton, P.H. (2016) Could the elongate yellow-orange nostrils of *Anguilla bicolor* McClelland, 1844 function as fishing lures? *African Journal of Aquatic Science* 41(4), 495–497.

UNEP (2016) *GEO-6 Regional Assessment for Africa*. United Nations Environment Programme, Nairobi.

Valade, P., Hoarau, P. and Bonnefoy, A. (2018) Plan Directeur de Conservation en faveur des Anguillidae à l'île de La Réunion 2018–2027. DEAL Réunion (Direction de l'environnment, de l'aménagement et du logement. Available at: www.reunion.developpment-durable.gouv.fr.

Van Someren, V.D. (1959) A study of a small basket-trap river fishery in Kenya. *The East African Agricultural Journal* 24(4), 257–267.

Van Someren, V.D. and Whitehead, P.J.P. (1959) Records of young eels in Kenya rivers. *Nature* 183 (4666), 950–971.

Wasserman, R.J., Weyl, O.L.F. and Strydom, N.A. (2011) The effects of instream barriers on the distribution of migratory marine-spawned fishes in the lower reaches of the Sundays River, South Africa. *Water SA* 37(4), 495–504.

Wasserman, R., Pereira-da-Conceicoa, L., Strydom, N. and Weyl, O. (2012) Diet of *Anguilla mossambica* (Teleostei, Anguillidae) elvers in the Sundays River, Eastern Cape, South Africa. *African Journal of Aquatic Science* 37(3), 347–349.

Weyl, O.L.F., Booth, A.J., Winker, H., McCafferty, J., Wartenberg, R., Peel, R. and Traas, G.R.L. (2009) Task 3: Assessment of the fishery potential for alien species on Darlington Dam. Sundays River Freshwater Fish Research Project. Report No. 5. 57 pp.

Weyl, O.L.F., Taraschewski, H. and Moravec, F. (2014) Are African longfin eel imports a potential pathway for parasite invasions? *Journal of Fish Diseases* 37(9), 843–845.

Whitfield, A.K. (1998) *Biology and Ecology of Fishes in Southern African Estuaries*. Ichthyological Monographs of the J.L.B. Smith Institute of Ichthyology, 2, http://dx.doi.10.1023/A:1008968200973.

Wielgoss, S., Taraschewski, H., Meyer, A. and Wirth, T. (2008) Population structure of the parasitic nematode *Anguillicola crassus*, an invader of declining North Atlantic eel stocks. *Molecular Ecology* 17(15), 3478–3495.

Wirth, T. and Bernatchez, L. (2003) Decline of North Atlantic eels: a fatal synergy? *Proceedings of the Royal Society of London B: Biological Sciences* 270(1516), 681–688.

Wood, F. (2000) The snake in the sky: tornadoes in clay and local narrative in the Hogsback-Alice area. *Critical Arts* 14(2), 79–95.

Wyman-Grothem, K.E., Popoff, N., Hoff, M. and Herbst, S. (2018) Evaluating risk of African longfin eel (*Anguilla mossambica*) aquaculture in Michigan, USA, using a Bayesian belief network of freshwater fish invasion. *Management of Biological Invasions* 9, 395–403.

Zarfl, C., Lumsdon, A.E., Berlekamp, J., Tydecks, L. and Tockner, K. (2015) A global boom in hydropower dam construction. *Aquatic Sciences* 77(1), 161–170.

CHAPTER 5

The management, fishery, and stock status of American eel along the US Atlantic coast

Kirby Rootes-Murdy and Dr. Kristen Anstead

The American Eel Technical Committee and Stock Assessment Subcommittee, Atlantic States Marine Fisheries Commission, Arlington, VA, USA

Introduction

American eel (*Anguilla rostrata*) is a unique species managed by the Atlantic States Marine Fisheries Commission (ASMFC) along the US Atlantic coast. While fisheries targeting American eel have spanned centuries, there is still much unknown about the species' life history throughout its range. Management, science, and the fisheries focused on American eel are complex and continually evolving as more information becomes available. This chapter outlines the history of management through ASMFC; descriptions of the different fisheries along the Atlantic coast targeting American eel; the latest stock assessment information on the resource; and research needs to further understanding the species.

Management of American eel

American eel is a challenging species to conserve and manage for a number of reasons. During its lifespan, American eel navigate through and reside in a wide range of habitats, from the oceanic waters of the Sargasso Sea to the brackish waters of coastal estuaries and the inland freshwater river systems. Throughout this journey, American eel inhabit areas under a myriad of management authorities, from international to multiple federal, state, and local governments. Information is limited about their abundance, status at all life stages, and habitat requirements. Additionally, while much is still unknown about the species from biological perspective, it is understood that American eel experience a late age of maturity and a tendency to aggregate during certain life stages. These life history characteristics further confound conservation efforts. American eel are an important resource from both a biodiversity and human-use perspective. They serve as an important prey species for many fish, aquatic mammals, and fish-eating birds. Although fisheries are a fraction of what they were historically, eel support valuable commercial, recreational and subsistence fisheries. For these reasons, it is critical to manage, assess, and conserve this important species.

The ASMFC is a deliberative body made up of representatives from the Atlantic coastal states that coordinates the conservation and management of 27 nearshore fish species, including American eel. The management unit for American eel under the jurisdiction of ASMFC includes that portion of the American eel population occurring in the territorial seas and inland waters along the Atlantic coast from Maine to Florida. Each state is represented on the Commission by three commissioners: the director of the state's marine fisheries agency, a state legislator, and an individual appointed by the state's governor to represent stakeholder interests. These commissioners participate in deliberations in the Commission's main policy arenas: interstate fisheries management, fisheries science, habitat conservation, and law enforcement.

There are several collaborative committees involved in the management and assessment of American eel for ASMFC including the American Eel Management Board (Board), Advisory Panel (AP), Technical Committee (TC), and Stock Assessment Subcommittee (SAS). The Board is responsible for approving, developing, and implementing any management decisions and is comprised of representatives from all member states/jurisdictions of ASMFC. The AP is comprised of stakeholders from a range of interests and provides guidance and input about the American eel fishery and its management. The TC addresses specific technical or scientific needs as posed by the Board and is made up of representatives from state agencies, federal fisheries agencies, the ASMFC, and academia. TC members with expertise in stock assessment and fish population dynamics make up the SAS, which is responsible for assessing the stock and determining stock status. Together these committees and the Board work together to address scientific and management needs of the American eel along the Atlantic coast through stock assessment reports and the fishery management plan.

The Interstate Fishery Management Plan (FMP) for American Eel was approved in November 1999 with the goals of conserving and protecting the American eel resource to ensure ecological stability while providing for sustainable fisheries. The FMP has several requirements that have been in place since its inception. Beginning in 2001, all states and jurisdictions were required to implement an annual young-of-year (YOY) abundance survey to monitor recruitment. In addition, states must implement a minimum recreational size and possession limit, as well as a state license for recreational fishermen to sell eels. States and jurisdictions must maintain existing or more conservative commercial fishery regulations for all life stages, including minimum size limits. Each state is responsible for implementing management measures within its jurisdiction to ensure the sustainability of its American eel population.

Since the establishment of the FMP, there have been five Addenda developed and approved by the Board (ASMFC, 2006a, 2008, 2013, 2014, 2018). Addendum I (ASMFC, 2006a) established a mandatory catch and effort monitoring program. Addendum II (ASMFC, 2008) placed increased emphasis on improving the upstream and downstream passage of American eel. Addendum III (ASMFC, 2013) sought to reduce mortality on all life stages of American eel through a reduced yellow eel recreational possession limit, an increased minimum commercial and recreational size limit, and additional gear restrictions. In October 2014, the Board approved Addendum IV, which established a coastwide quota (known as the coastwide cap) of 907,671 lb (411,713 kg) of yellow eel, reduced Maine's glass eel quota to 9688 lb (4394 kg) (based on 2014 landings), and allowed for the continuation of New York's silver eel weir fishery in the Delaware River. For yellow eel fisheries, the coastwide cap was implemented starting in the 2015 fishing year and established two management triggers: (1) if the cap is exceeded by more than 10% in a given year; or (2) the coastwide quota is exceeded for two consecutive years regardless of the percent overage. If either one of the triggers are met then states would implement state-specific quotas based landings information from 2010–2013. Addendum III and IV were initiated in response to the findings of the 2012 American Eel Benchmark Stock Assessment (ASMFC, 2012), which declared American eel stock along the US Atlantic coast as depleted.

Since the implementation of Addendum IV, states have raised several issues about the current management program. In particular, states are concerned about the second management trigger, which would require a change to state-specific quotas if there is a cap overage of any amount in the second year. This concern is largely based on the inherent uncertainty of the landings data. While the FMP requires states to report commercial landings by life stage, gear type, month, and region, not all states were able to provide this level of information for either the 2012 benchmark stock assessment or the 2017 stock assessment update. In addition to not always having a complete data set to distinguish landings by life stage, there are other potential biases present in the commercial yellow eel data. At least a portion of commercial American eel landings are from non-marine waters. Even with mandatory reporting, requirements do not always extend beyond marine districts. Further, misreporting between conger eel, hagfish, slime eel, and American eel has been known to occur. Despite these uncertainties, the commercial landings do represent the best data available and are indicative of the trend of total landings over time.

In August 2018, the Board approved Addendum V. The Addendum increased the yellow eel coastwide cap starting in 2019 to 916,473 lb (415,705 kg). This modest increase in the cap (less than 1%) reflects a correction in the historical harvest. Further, the Addendum adjusted the method (management trigger) to reduce total landings to the coastwide cap when the cap has been exceeded and removes the implementation of state-by-state allocations if the management trigger is met. If the management trigger is met, only those states accounting for more than 1% of the total yellow eel landings will be responsible for adjusting their measures. Lastly, the Addendum slightly modified the glass eel aquaculture provisions, maintaining the 200 lb (91 kg) limit for glass eel harvest but modifying the criteria for evaluating the proposed harvest area's contribution to the overall population.

American eel fishery

In the USA, many life stages are subject to fishing pressure and the degree of fishing varies. American eel undergo six distinct life stages. The life cycle begins when the eggs hatch and leptocephali (larvae) are carried by ocean currents from the spawning grounds in the Sargasso Sea. The prevailing currents along coastal areas disperse the leptocephali, which metamorphose into glass eels on the continental shelf. Glass eels move toward inland areas and become pigmented elvers before or during their entry into coastal estuaries. Elvers and yellow eels settle in habitats ranging from estuaries to far upstream freshwater reaches. Upon maturity, eels reach the silver stage and return to the Sargasso Sea, where they spawn and die.

Glass eel fisheries are prohibited in all states except for Maine and South Carolina, although Maine is the only state reporting substantial glass eel harvest. Over the past 7 years, there has been an increase in the demand for glass eel due to concerns over population levels of European and Japanese eels (*A. anguilla* and *A. japonica*), as well as tighter restrictions on the export of European eel. Harvest, by dip net or fyke net, has increased as the average market price has risen to over US$1000/lb (US$454/kg). The highest annual value reported in Maine in the past 5 years was US$38.78 million in 2012 for 21,610 lb (9802 kg) (Table 5.1). Since the implementation of Addendum IV (ASMFC, 2014), Maine's glass eel quota has been set at 9688 lb (4394 kg) (a 17.5% reduction from the 2014 quota). In 2017, preliminary landings indicate 9343 lb (4238 kg) of glass eels were sold for a value of US$12.16 million. Given the value of the fishery and demand from Asian markets, there has been an increase in illegal harvest of glass eels throughout the US Eastern seaboard in recent years. The US Fish and Wildlife Service (USFWS) in conjunction with state law enforcement agents have worked to investigate and eliminate illegal harvest through coordinated efforts (known as 'Operation Broken Glass') that are ongoing. Illegal harvest of glass eels has been found in many states from Maine through South Carolina. In May 2018, Maine closed its glass eel/elver fishery early due to the discovery of illegal harvest occurring outside of the state's swipe card system.

Historically and currently, the majority of commercial landings come from the yellow eel fishery, which occurs in all Atlantic coast states with the exception of Pennsylvania. While accounts of eel harvest date back to colonial times, with some commercial fishery harvest records available beginning in the late 1880s, consistent record keeping began in 1950. After an initial decline in the 1950s, commercial yellow eel landings increased to a peak of 3.67 million lb (1.66 million kg) in 1979, declined again in the 2000s, and exceeded 1 million lb (453,592 kg) three times since 2004 (Figure 5.1). Addendum IV implemented a coastwide cap of 907,671 lb (411,713 kg). Landings in 2016 are estimated at 946,116 lb (429,151 kg), which is above the cap. Eel pots are the typical gear used in the commercial yellow eel fishery; however, weirs, fyke nets, and other fishing methods are also employed. Although yellow eel were harvested for food historically, today's fishery sells yellow eel primarily for food in international Asian markets, as well as bait domestically for recreational fisheries such as striped bass, cobia, and catfish fisheries.

There is a small recreational fishery for American eel. The primary source of recreational harvest catch and effort data is NOAA Fisheries' Marine Recreational Information Program (MRIP). Atlantic coast recreational harvest (fish are kept or released dead) ranged from 3062 to 220,596

Table 5.1 Maine's glass/elver eel landings and value, 2007–17.

Year	Landings (lb)	Value (US$)
2007	3714	1,287,479
2008	6951	1,486,353
2009	5199	514,629
2010	3158	592,405
2011	8585	7,656,345
2012	21,610	38,791,627
2013	18,081	32,926,991
2014	9688	8,440,333
2015	5260	11,389,891
2016	9399	13,446,828
2017*	9343	12,155,672

Note: *Preliminary landings.

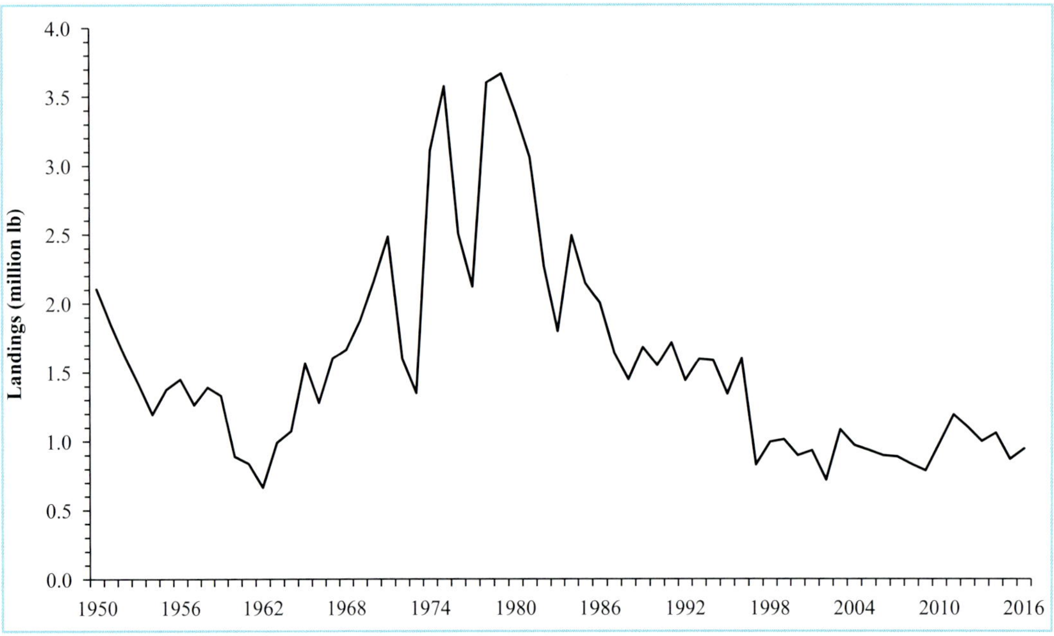

Figure 5.1 Commercial landings of American eel along the US Atlantic coast, 1950–2016.

eels per year during 1981 through 2016. Recreational harvest demonstrated an overall decline over the available time series, with some large peaks in the mid-1980s, early 1990s, and 2010 (Figure 5.2). The number of American eels released alive by recreational anglers ranged from a low of 26,707 eels in 1997 to a high of 157,189 eels in 2003. Live releases generally declined from the late 1980s through the late 1990s to early 2000s. Numbers of live releases increased again from 2002 to 2014, but have declined since then. The precision of these estimates is low due to the limited numbers of American eels that have been encountered during recreational surveys.

Since the approval of Addendum IV, silver eel fisheries are only permitted on a limited basis on the Delaware River in New York. The Delaware River eel weir fishery is restricted to nine annual permits which were initially limited to those who fished and reported landings from 2010 to 2013.

In recent years, a facility in North Carolina has been granted permission by the Board to harvest up to 200 lb (91 kg) of glass eels for domestic aquaculture purposes. In 2016 and 2017 combined, less than 1 lb (454 g) of glass eels were harvested as part of their efforts and no yellow eels have been produced for domestic markets from this harvest. In June 2018, Maine submitted a proposal to harvest up to 200 lb (91 kg) of glass eels for domestic aquaculture purposes. The Board will consider this proposal in August 2018.

Stock assessment and status of American eel

The ASMFC produces two types of stock assessment reports: a benchmark stock assessment and a stock assessment update. A benchmark stock assessment refers to either a new stock assessment or a stock assessment for which existing data inputs and model structure are modified and must

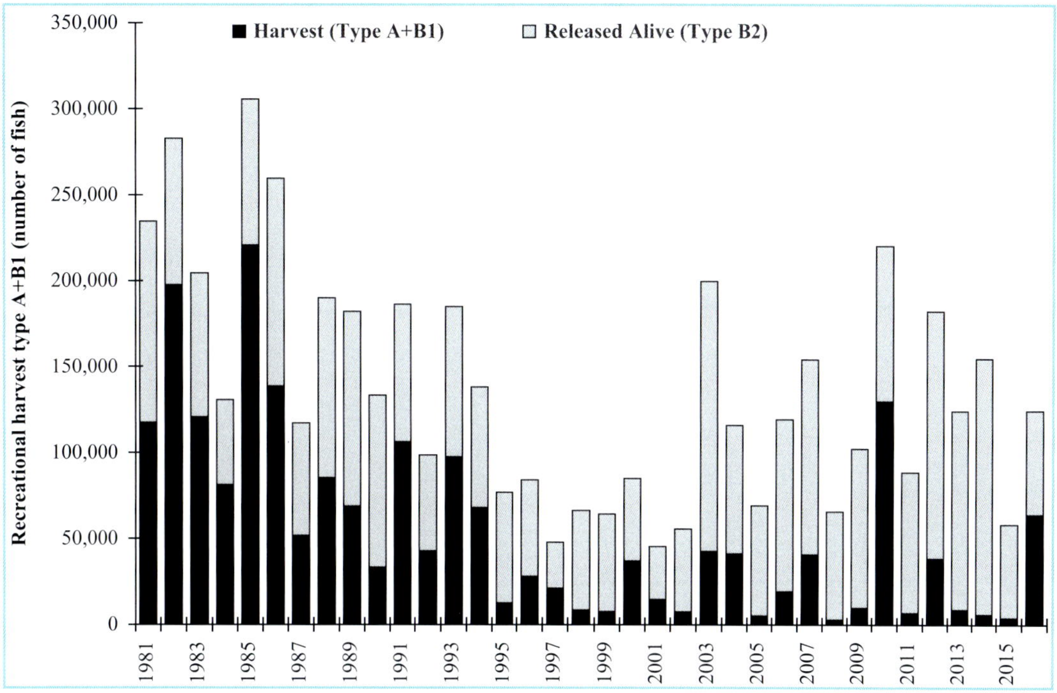

Figure 5.2 Recreational harvest and releases for American eel, 1981–2016, as estimated by the National Marine Fisheries Service's MRIP, formerly the Marine Recreational Fishery Statistics Survey (MRFSS) program. Harvest types reported by MRIP are those caught and brought back to the dock (Type A), caught and reported dead (Type B1), and caught and released alive (Type B2).

therefore be subject to an external peer review. Benchmark changes to data, parameterization, and model type or structure are often made in response to previous peer review recommendations. A stock assessment update consists of adding the most recent years of data to an existing, peer-reviewed, and board-accepted stock assessment model without changing the model type or structure. A stock assessment report is developed by the SAS and approved by the TC and is a product of both committees. Once completed, it is presented to the Board for consideration of any management actions that result from the findings.

In 2005, the first American eel benchmark stock assessment was conducted and reviewed by a panel of independent experts (ASMFC, 2005). The peer review panel recognized sufficient shortcomings with the assessment to warrant additional action prior to its use for future technical and management purposes. The 2005 stock assessment was not accepted by the Board; therefore, the stock status of American eel was deemed unknown by the ASMFC. In 2006, the Board tasked the American Eel SAS and TC with reviewing the recommendations from the peer review advisory report and recommending a follow-up plan. Subsequently, a report was issued in October 2006 containing updated datasets and the short-term analyses suggested by the review panel (ASMFC, 2006b).

In 2012, ASMFC conducted another benchmark assessment. This assessment evaluated American eel coastwide (Maine–Florida) and by watershed-based geographic regions: Gulf of Maine, Southern New England, Hudson River, Delaware Bay/Mid-Atlantic Coast Bays, Chesapeake Bay, and the South Atlantic (Figure 5.3). For this assessment, the SAS evaluated nearly 100 fishery-dependent and independent US data sources representing several life stages and geographical and temporal scales. Very few fishery-independent surveys target American eels (with the exception of the state-mandated YOY surveys and a few surveys in Maryland). Fifty-two fishery-dependent and independent data sources were selected for use in the assessment because they were considered adequate for describing life history characteristics and abundance trends of eels on either a coastwide or regional basis. Trends in fishery-dependent catch-per-unit effort were used to describe the fisheries but were not included in analyses because they were thought to not represent trends in eel abundance over time due to either poor participation in the fishery (that is, few fishers), major unquantified changes in the fishery over time, or insufficient time series.

Despite the large number of surveys and studies available for use in the 2012 assessment, the American eel stock is still considered data-poor because very few surveys target eels and collect information on length, age, and sex of the animals caught. Also, eels have an extremely complex life history that is difficult to describe using traditional stock assessment models. Trend analyses of abundance indices provided evidence of declining or, at least, neutral abundance of American eels in the USA in recent decades. All three trend analysis methods used (Mann–Kendall, Manly, and ARIMA) detected significant downward trends in numerous indices over the time period examined. Also, there was consensus for a decline for both YOY and yellow eel life stages through time. Both the ARIMA and Mann–Kendall analyses indicated decreasing trends in the Hudson River and South Atlantic regions. In contrast, survey indices from the Chesapeake Bay and Delaware Bay/Mid-Atlantic Coastal Bays regions showed no consistent increasing or decreasing trends. Overall, however, the prevalence of significant downward trends in multiple surveys across the coast were cause for concern.

Analyses and results indicated the American eel stock had declined and there were significant downward trends in multiple surveys across the coast. It was determined the stock was depleted but no overfishing determination could be made based on the analyses performed. The assessment concluded the seeds of the depletion lay in part in a 'fishing up/fishing down' episode that occurred on American eel in the 1970s into the 1980s as export demand rose. Roughly during the same period, river damming intensified and hydroelectric facilities on dams caused additional mortality. A suite of stressors including habitat loss from dams or urbanization, turbine mortality, the nonnative swim-bladder parasite, toxic pollutants, and climate change are all factors that act in concert with fishing mortality on American eel.

A stock assessment update was completed in 2017, adding 5 years of data to the landings, indices, and models used in the 2012 benchmark. The 2017 stock assessment update included 22 YOY surveys and 15 yellow eel surveys for use as indices of abundance based on the number of years surveyed, survey design, appropriateness of gear used for catching eel, and frequency of eel catches as determined by the 2012 benchmark stock assessment. Survey data were statistically standardized to account for factors that affect catchability of eels (for example, moon phase, season, flow rates). Survey data were analyzed separately and then combined to create YOY and yellow eel indices at the regional and coastwide level to look for trends at different spatial scales.

Trend analyses found evidence of declining or, at least, neutral abundance of American eels along the US Atlantic coast in recent decades. All three trend analysis methods used

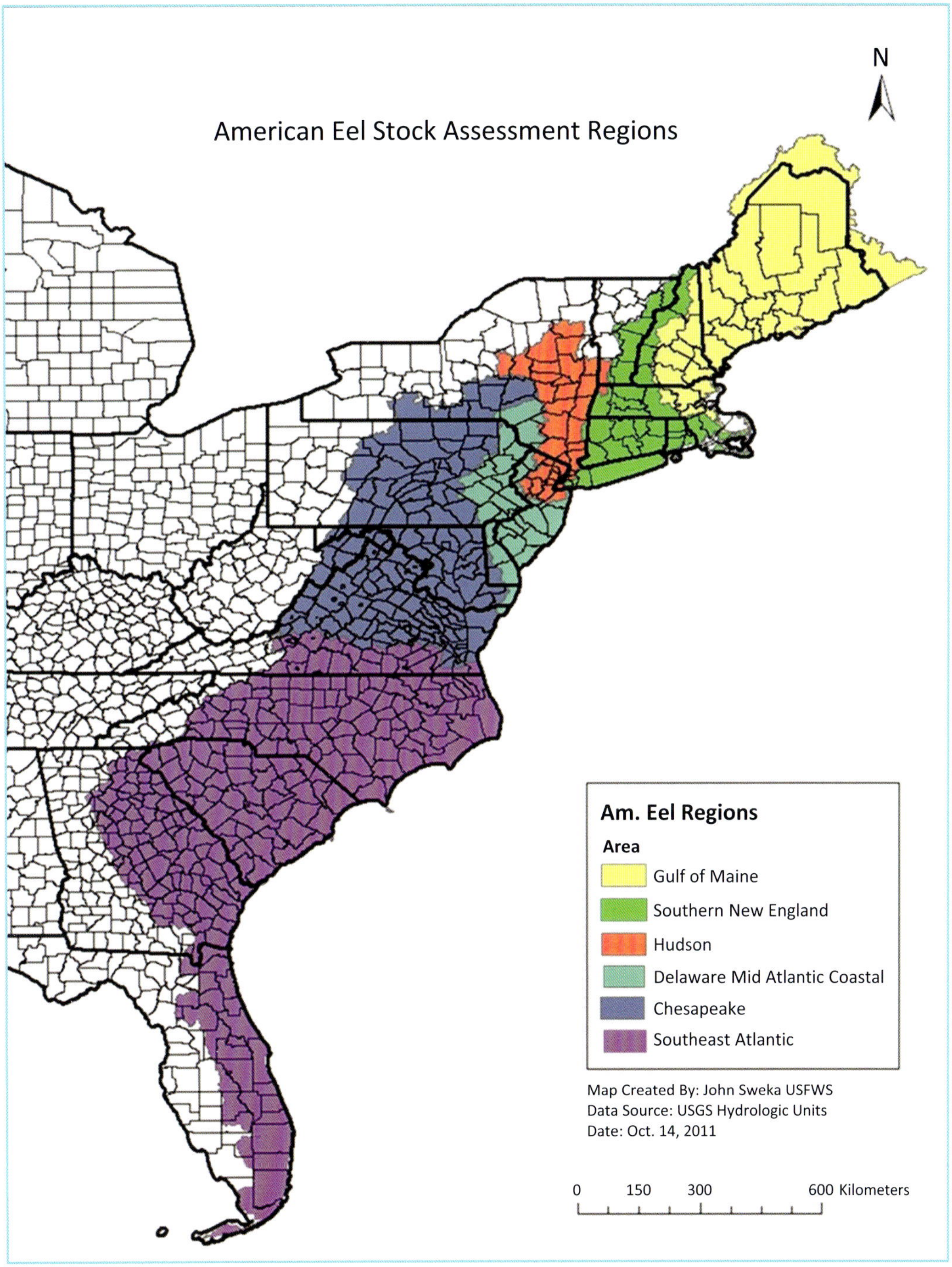

Figure 5.3 Watershed-based geographic regions used in the 2012 benchmark and 2017 update stock assessments.

Table 5.2 Results of the Mann–Kendall trend analysis applied to regional and coastwide indices of American eel abundance by YOY and yellow eel life stages. The arrows indicate the direction of the trend if a statistically significant trend was detected (p-value < α; $\alpha = 0.05$). NS = no significant trend detected. A dash (–) = indices that were not updated.

Region	Life stage	Time period	2012 trend	2017 trend
Gulf of Maine	YOY	2001–2016	NS	NS
Southern New England	YOY	2000–2016	NS	NS
	Yellow	2001–2010	NS	–
Hudson River	YOY	1974–2009	↓	–
	Yellow	1980–2016	↓	↓
Delaware Bay/Mid-Atlantic Coastal Bays	YOY	2000–2016	NS	NS
	Yellow	1999–2016	NS	NS
Chesapeake Bay	YOY	2000–2016	NS	NS
	Yellow	1990–2009	↑	↑
South Atlantic	YOY	2001–2015	NS	↓
	Yellow	2001–2016	↓	↓
Atlantic coast	YOY (short-term)	2000–2016	NS	NS
	YOY (long-term)	1987–2013	NS	NS
	Yellow (40+ year)	1974–2016	NS	↓
	Yellow (30-year)	1987–2016	↓	↓
	Yellow (20-year)	1997–2016	NS	NS

(Mann–Kendall, Manly, and ARIMA) detected significant downward trends in some indices over the time period examined. The Manly meta-analysis found consensus for a decline in YOY and yellow eels through time. The Mann–Kendall test detected a significant downward trend in six of the 22 YOY indices, 5 of the 15 yellow eel indices, 3 of the 9 regional abundance indices, and the 30-year and 40-year yellow eel coastwide abundance indices. The remaining surveys tested had no trend, except for two which had positive trends. In total, 3 of the 14 surveys included in the ARIMA modeling had greater than a 0.50 probability of being less than the 25th percentile reference point in the terminal year of the survey. While results varied among individual abundance indices and analyses, both the ARIMA and Mann–Kendall analyses identified decreasing trends in the Hudson River and South Atlantic regions for yellow eel (Figures 5.4 and 5.5).

The decline in yellow eels can also be seen in the 40-year index developed from multiple surveys (Figure 5.6). While not a comprehensive list of indices evaluated or analyses performed, Table 5.2 compares the regional trends between 2012 and the 2017 update.

Several other data sources were used to characterize length-, age-, and sex-structure of the population. Significant variation in length at age and a broad overlap in lengths across multiple age groups were observed in the data even within a regional analysis. Pooled data for all regions amplified these variations in length at age. These analyses confirm the relationship between age and length for American eel is not well defined and that age is a poor predictor of length for American eel. Aging error and uncertainty around aging estimates may also play an additional role in the weak relationship of length and age.

The trend analysis results in the 2017 stock assessment update are consistent with the 2012 results, with few exceptions. Despite downward trends in the indices, commercial yellow eel landings have been stable in the recent decades along the Atlantic coast although landings still remain much lower than historical landings. Compared to 2012, there are more significantly downward trends in indices as indicated by the Mann–Kendall test and similar results for the ARIMA. This trend analysis and stable low landings support the update conclusion that the American eel population in the assessment range remains depleted. Therefore, the stock status is depleted and no overfishing determination can be made based on the trend analyses performed.

Throughout the development of a stock assessment and attempts to determine stock status by the ASMFC, there have been several petitions for Endangered Species Act (ESA) listing for the species. In response to the extreme declines in American eel abundance in the Saint Lawrence River-Lake Ontario portion of the species' range, the ASMFC requested the USFWS and NOAA Fisheries conduct a status review of American eel in 2004. The ASMFC also requested an evaluation of a distinct population segment (DPS) listing under the ESA for the Saint Lawrence River/Lake Ontario and Lake Champlain/Richelieu River portion of the species range, as well as an evaluation of the entire Atlantic coast American eel population. A preliminary status review conducted by USFWS determined American eel was not likely to meet the requirements of DPS determinations. However, the USFWS initiated a coastwide status review of the American eel in coordination with NOAA Fisheries and ASMFC. At the same time, two private citizens submitted a petition to the USFWS and NOAA Fisheries to list American eel under the ESA.

In February 2007, the USFWS announced the completion of a Status Review for American Eel (50 CFR Part 17; USFWS, 2007). The report concluded protecting eels as an endangered or threatened species was not warranted. The USFWS did note that while the species' overall population was not in danger of extinction or likely to become so in the foreseeable future, the eel

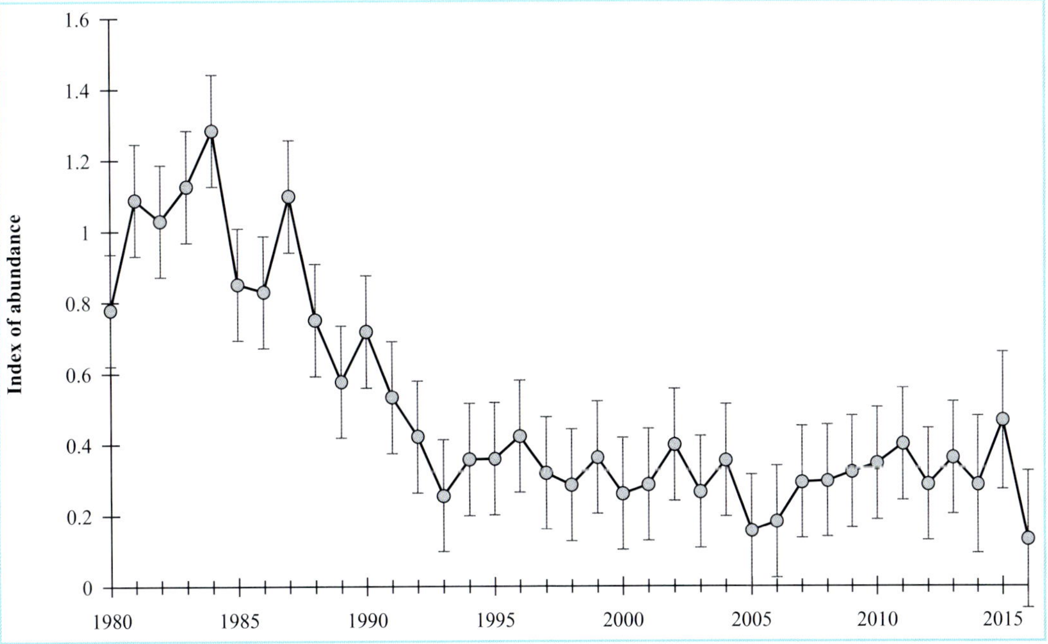

Figure 5.4 Index of abundance for yellow-stage American eel in the Hudson region, 1980–2016. The error bars represent the standard errors about the estimates.

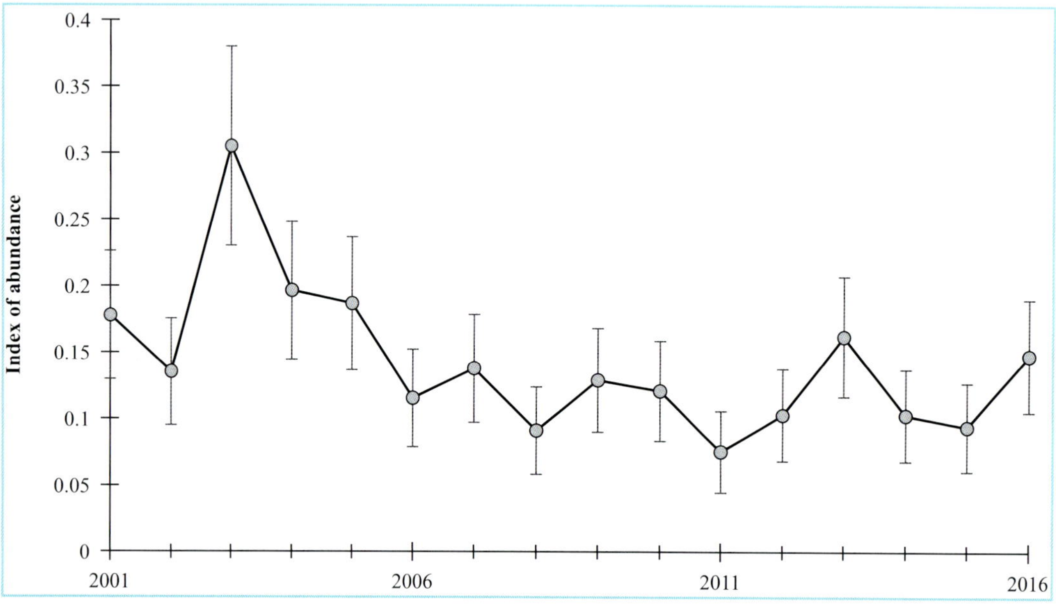

Figure 5.5 Index of abundance for yellow-stage American eel in the South Atlantic region, 2001–2016. The error bars represent the standard errors about the estimates.

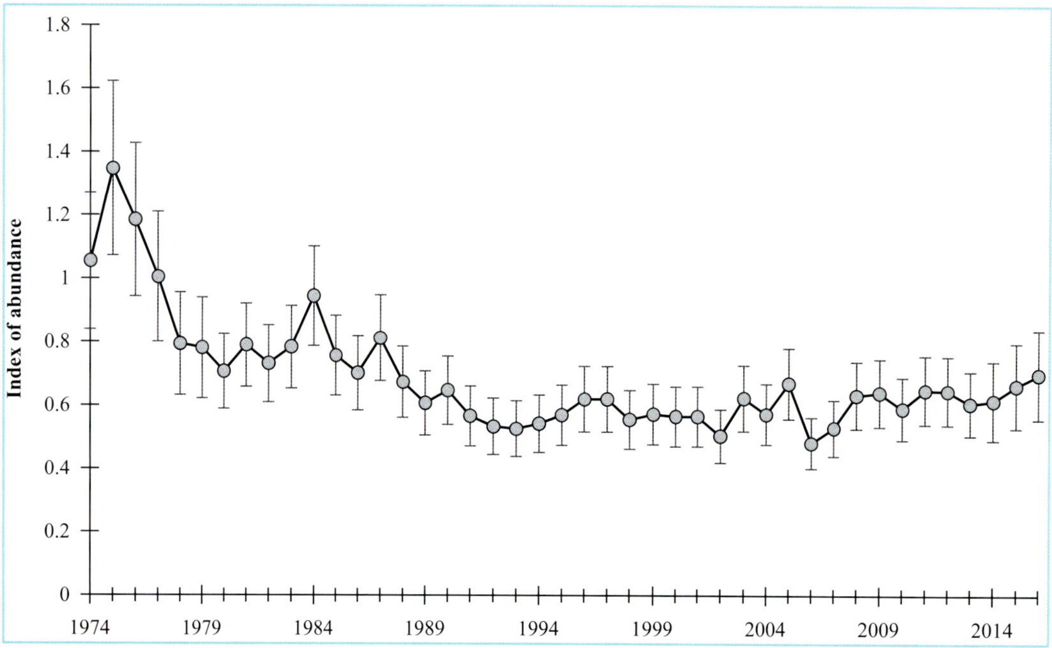

Figure 5.6 GLM-standardized index of abundance for yellow-phase American eel along the Atlantic coast, 1974–2016 (40-plus-year index). The error bars represent the standard errors about the estimates.

population has 'been extirpated from some portions of its historical freshwater habitat over the last 100 years… [and the species abundance has declined] likely as a result of harvest or turbine mortality, or a combination of factors'.

In 2010, the Center for Environmental Science Accuracy and Reliability filed a petition to the USFWS to consider listing the American eel on the endangered species list. The proposal was based on new information that had become available since the last status review. In September 2011, the USFWS published a positive 90-Day Finding, which stated the petition contained enough information to warrant conducting a status review (USFWS, 2011).

In 2015, USFWS announced that the American eel population is stable and protection under ESA was not warranted although the agency did recommend continuing efforts to maintain healthy habitats, monitor harvest levels, and improve river passage (USFWS, 2015). Conversely, the International Union for the Conservation of Nature (IUCN) listed American eel as 'Endangered' on the Red List in 2014 (Jacoby and Gollock, 2014). While this has no legal implications, it is an important metric and the ASMFC remains committed to closely monitoring this species and making management adjustments as necessary.

Data and research needs

Monitoring landings and stock status of American eel relies on the quality of data available. Direct and detailed monitoring of the population and fisheries trends is needed to improve the stock assessment. Accuracy of commercial catch and effort data should be improved through better compliance with landings and effort reporting requirements as outlined in the FMP. Targeted fishery-independent surveys for yellow and silver eels would greatly improve the assessment. In the absence of such surveys, the collection and processing of length, age, and sex information for yellow and silver eels caught in already existing surveys would be useful, especially in the South Atlantic where few surveys are conducted that catch American eel. The collection of age structure data outside the range of the fishery would provide much needed information about animals not typically caught by the fishery that may be contributing to stock productivity. Quantification of mortality rates due to dam turbines, disease, and other environmental stressors would be helpful. Also, a comprehensive map of coastwide habitat loss would aid in quantifying the potential magnitude of reduced productivity. Because of the panmictic nature of American eel, fisheries outside US jurisdiction are relevant to ASMFC management efforts, although they are not subject to management regulations implemented though the ASMFC. Given the intercontinental range of the American eel stock, the American eel SAS, TC, and Board encourage that the next assessment be a cooperative effort between North and South American countries if possible.

Bibliography

ASMFC (1999) Interstate Fishery Management Plan for American Eel (*Anguilla rostrata*). ASMFC, Fishery Management Report No. 36, Washington, DC.

ASMFC (2005) American Eel Stock Assessment Report for Peer Review. ASMFC, Washington, DC.

ASMFC (2006a) Addendum I to the Interstate Fishery Management Plan for American Eel. ASMFC, Washington, DC.
ASMFC (2006b) Update of the American Eel Stock Assessment Report. ASMFC, Washington, DC.
ASMFC (2008) Addendum II to the Interstate Fishery Management Plan for American Eel. ASMFC, Washington, DC.
ASMFC (2012) American Eel Benchmark Stock Assessment. ASMFC, Arlington, VA.
ASMFC (2013) Addendum III to the Interstate Fishery Management Plan for American Eel. ASMFC, Arlington, VA.
ASMFC (2014) Addendum IV to the Interstate Fishery Management Plan for American Eel. ASMFC, Arlington, VA.
ASMFC (2017) American Eel Stock Assessment Update. ASMFC, Arlington, VA.
Jacoby, D. and Gollock, M. (2014) *Anguilla anguilla*. The IUCN red list of threatened species. Version 2014.2.
USFWS (2007) Endangered and threatened wildlife and plants – 12-month finding on a petition to list the American eel as threatened or endangered. Notice of 12-month petition finding. Federal Register 72:22 (2 February 2007):4967–4997.
USFWS (2011) Endangered and threatened wildlife and plants – 90-day finding on a petition to list the American eel as threatened. Proposed rules. Federal Register 76:189 (29 September 2011):60431–60444.
USFWS (2015) Endangered and threatened wildlife and plants – 12-month finding on a petition to list the American eel as threatened or endangered. Notice of 12-month petition finding. Federal Register. US Government Printing Office, Washington, DC (8 October 2015). Docket Number FWS-HQ-ES-2015–0143.

PART III

Research and management

CHAPTER 6

Excerpts from 'Eels On The Move – Mysterious Creatures over Millions of Years'

By kind permission of Mari Kuroki, Katsumi Tsukamoto and Tokai University Press

Information re-worked by Katsumi Tsukamoto[1], Mari Kuroki[1] and Andy Don[2]

[1]Graduate School of Agricultural and Life Sciences, The University of Tokyo, Tokyo, Japan;
[2]National Fisheries Services, Environment Agency, Bridgwater, Somerset, UK

Hatching

Eggs

Eels hatch from eggs. A simple fact – yet it took more than 2000 years from Aristotle's day for this to be proved beyond doubt. For direct evidence was not obtained until 1973, when eels were first artificially hatched, and 2009, when natural eel eggs were first collected (Figure 6.1) – comparatively recent events in the long history of science.

So why did it take so long for the reproductive cycle of eels to be observed? The answer is simple: because they spawn their eggs far out in the open seas. If they had spawned in rivers or lakes, or even near the coast, we would have seen parent eels with their bellies full of eggs long ago. Then we would have guessed something of the truth.

The problem was that we had no idea where eels laid their eggs. To fill this gap in our knowledge, the search for eel spawning areas in open seas has continued for more than a century. Even now we have only managed to find the spawning areas of a few of the world's 19 species and subspecies of anguillid eels. Where the other spawning areas are located is still unknown. For research on the spawning ecology of eels is still very much at an embryonic stage.

Research on eel spawning in the Atlantic

Scientific research on eel spawning areas started in Europe. After collecting a European eel leptocephalus off the Faeroe Islands in 1904, Danish biologist Johannes Schmidt (1877–1933) (Figure 6.2) had an idea: could the spawning area of eels actually be in the Atlantic Ocean

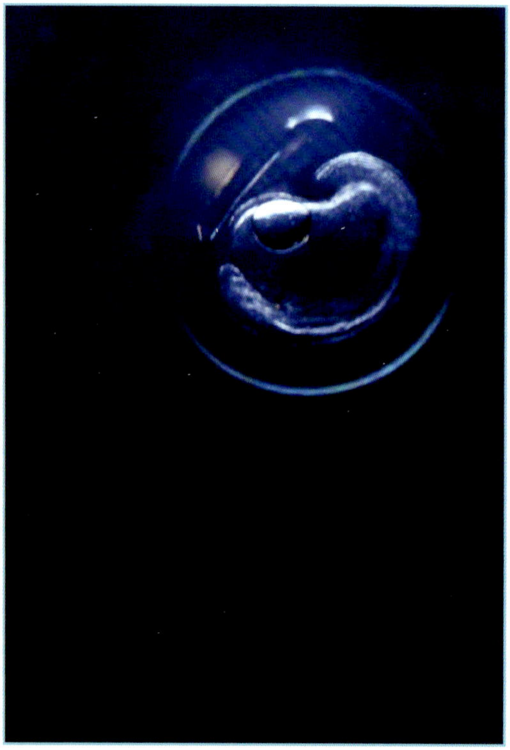

Figure 6.1 Natural egg of a Japanese eel (*Anguilla japonica*) collected by the research vessel *Hakuho Maru*. Photo: Katsumi Tsukamoto.

Figure 6.2 Johannes Schmidt, (1877–1933). Photo: DTU-Aqua.

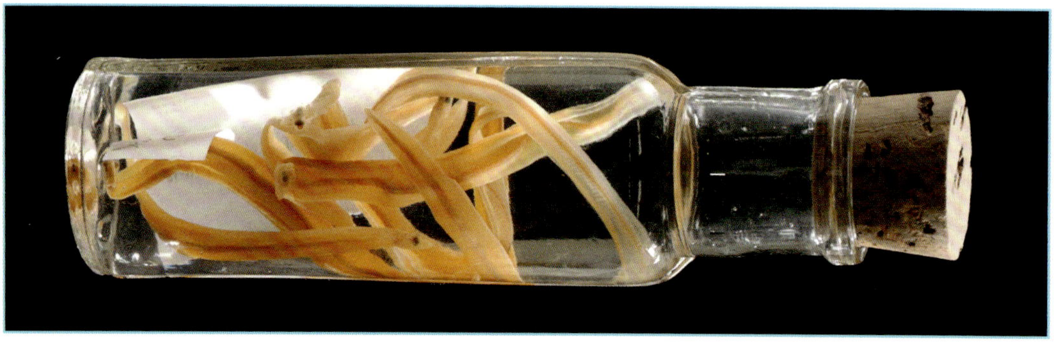

Figure 6.3 European eel leptocephali at their maximum growth stage, collected during the *Thor*'s Mediterranean expedition (1905–1906). Photo: Zoological Museum, Natural History Museum of Denmark, University of Copenhagen.

and not in the Mediterranean, as had previously been supposed? Schmidt then embarked on an exhaustive search for their spawning area, using four research ships in all (*Thor*, *Margrethe*, *Dana* and *Dana II*). He was quite single-minded in his quest; he even asked merchant vessels that regularly plied the Atlantic between North America and Europe to collect plankton samples for him.

On comparing the body length and distribution of eel leptocephali collected during his journeys across the Atlantic, Schmidt noticed something interesting. The distribution ranges of similarly sized leptocephali (Figure 6.3) formed a pattern of concentric circles growing larger with size. And the centre of those concentric circles was in the Sargasso Sea south of Bermuda.

The larvae of European eels (*Anguilla anguilla*) born there were dispersed by ocean currents, which transported them all the way to Europe. Leptocephali of the American eel (*A. rostrate*), which mostly lives in the eastern part of North America, showed the same concentricity centring on the Sargasso Sea. Although American eels appear to spawn slightly further west than European eels in the Sargasso Sea, the fact that two species of eels both spawn in the same place was an astonishing discovery (Schmidt, 1922). What is still unknown, though, is how the two species can tell each other apart when they spawn in the same area.

The round-the-world expedition by Dana II

After finding the spawning areas of Atlantic eels, Johannes Schmidt conducted a round-the-world voyage in search of other anguillid spawning areas, gathering leptocephali in seas all over the world as he went (Schmidt, 1935). This expedition on the *Dana II* in 1928–1930 spurred dramatic advances in eel biology. In fact, the expedition also gathered a wealth of knowledge on marine biology in general, from plankton to deep-sea fishes, besides the issue of eel spawning areas. The findings were compiled in 16 volumes and a total of 91 monographs in the 'Dana Expedition Report', published by the Carlsberg Foundation.

Schmidt and his colleagues gathered eels at each port they visited, laying the foundations for eel biology worldwide. After Schmidt's death, two of his younger colleagues – Vilhelm Ege (1887–1962) and Poul Jespersen (1891–1951) – wrote scholarly works on the taxonomy of eels and the distribution of leptocephali (Ege, 1939; Jespersen, 1942). Today, seven decades later, these are still cited as useful reference materials by researchers all over the world.

Back then, the round-the-world expedition on the *Dana II* captured the attention not only of researchers but also of the general public, and its progress was eagerly followed by the newspapers. One reporter asked Schmidt, as head of the research effort, to cite the biggest outcome of the expedition. Schmidt replied, 'That the ecology of eels has been revealed.' In his lifelong devotion to research on eels Schmidt's aim had not been to breed eels or conserve resources. It had merely been to unravel the mysteries of eel ecology.

Research expeditions in the Pacific

The discovery of Atlantic eel spawning areas triggered a desire to locate anguillid spawning areas in the Pacific Ocean – particularly those of the Japanese eel (*A. japonica*) distributed in East Asia.

In 1973, the Ocean Research Institute of the University of Tokyo (later renamed the Atmosphere and Ocean Research Institute) started searching for the Japanese eel spawning area using the research vessel *Hakuho Maru,* which now belongs to JAMSTEC. The second 'Eel cruise' of the *Hakuho Maru* led to the collection of 52 larvae with a body length of about 50 mm in waters east of Taiwan in 1973 (Tanaka, 1975), followed by another 21 larvae measuring about 40 mm in waters east of Luzon Island in the Philippines in 1986 (Kajihara et al., 1988). Then in 1988 and 1990 the training vessel *Keiten Maru* of Kagoshima University collected 7 and 21 specimens of larvae respectively measuring around 20–30 mm in waters even further to the east (Ozawa et al., 1989, 1991). In 1991, finally, the *Hakuho Maru* collected about 1000 larvae measuring about 10 mm to the west of the Mariana Islands. With this, the spawning area of Japanese eels was narrowed down to the area west of the Marianas (Tsukamoto, 1992).

Essentially, historical conjecture on the location of the Japanese eel spawning area followed the route travelled by the larvae, but in reverse: first, southwards from waters south of Okinawa to waters east of Taiwan in the Kuroshio current area, and then from east of Luzon Island in the Philippines, advancing eastwards against the North Equatorial Current, until finally reaching the seas west of the Marianas. As this 'journey' continued, the size of the leptocephali collected gradually grew smaller. The history of research on eel spawning areas could be described as an ongoing search for ever smaller leptocephali in the vastness of the ocean.

The seamount hypothesis

The distribution of Japanese eel leptocephali collected in the North Equatorial Current up to 1994 was concentrated around the latitude of 15°N in the western North Pacific. In terms of longitude, catches were recorded west of 142°E, but no leptocephali were found east of 143°E. The body size of leptocephali decreased towards the east within the North Equatorial Current, which flows slowly westward. Putting all this information together, it was apparent that the spawning point must be around 15°N latitude and between 142° and 143°E longitude.

A seabed topographical chart of this area revealed the presence of three seamounts (from the south: Suruga (Figure 6.4), Arakane and Pathfinder) These are peaks in the ocean similar in height to Mount Fuji in Japan, soaring from the seabed at depths of 3000–4000 m to just below the ocean surface. The hypothesis was proposed that Japanese eels could spawn in the vicinity of these seamounts (Tsukamoto et al., 2003).

The new moon hypothesis

Inside their inner ears, fish have three types of hard tissue called 'otoliths', mainly consisting of calcium carbonate crystals. Each of the three types (sagitta, lapillus, asteriscus) comes in a pair, one on the right and one on the left. The largest of them is the sagitta. When observed under a microscope, the sagitta shows concentric rings that accumulate day by day. By counting these rings (otolith increments) from the centre of the otolith outwards, the age of a fish can be determined, that is, the number of days since it hatched. This enabled the hatching date (birth date) of individual leptocephali collected near the spawning area to be calculated by first working out their age in days and then subtracting this from the collection date. The results suggested that

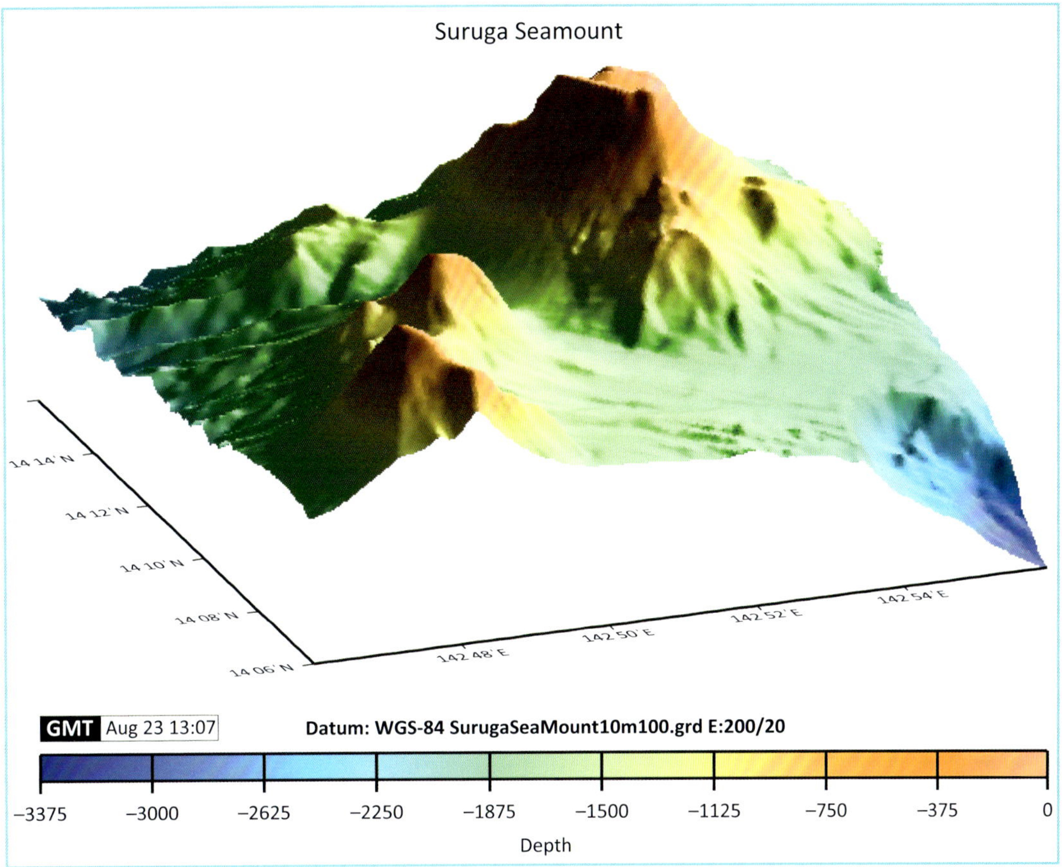

Figure 6.4 3D bird's-eye view of the Suruga Seamount (Bank), 14°16′N, 142°55′E, with relative elevations of 2000–3000 m, their peaks reach surface waters at depths of between several metres and about 40 m. Image: Atmosphere and Ocean Research Institute, The University of Tokyo; JAMSTEC.

leptocephali hatch around the time of the new moon each month during the spawning season (Tsukamoto et al., 2003).

This would mean that, during the long spawning season of Japanese eels, they would not spawn haphazardly or continuously, they would spawn in accordance with a regular rhythm, governed by the monthly lunar cycle. They would all spawn at once, with the new moon each month.

Fourteen years of stagnation

Surveys proceeded in line with these two hypotheses, that is, the new moon hypothesis on the timing of spawning and the seamount hypothesis on its location. In 1998, the small submersible *JAGO* was brought to the Pacific from the Max Planck Institute in Germany to conduct deep-sea searches along the seamounts. Despite a massive investment of man-power, however, neither spawning eels nor eel eggs were found. As a result, the achievement

Figure 6.5 Handling the big fish plankton net on the *Hakuho Maru*. Photograph: Katsumi Tsukamoto.

in collecting large quantities of leptocephali in 1991 was not surpassed for another fourteen years (Figures 6.5 and 6.6). The spawning locations would have to be narrowed down further still.

Next, the locations where leptocephali had been collected and the physical conditions of their marine environments were investigated. Many of these leptocephali were distributed south of salinity fronts that formed in an east–west direction across the spawning area (Kimura et al., 1994; Tsukamoto, 1992). These salinity fronts were thought to be a factor affecting the formation of spawning locations by adult eels. Continuous observations were then made during each research cruise to track the location of these salinity fronts, which are constantly changing. There are similar temperature fronts within the spawning areas of the two Atlantic eel species in the Sargasso Sea; these are considered important as environmental conditions that determine the locations of their spawning areas (Kleckner and McCleave, 1988).

Figure 6.6 Collected plankton samples being transferred to buckets on the afterdeck of the *Hakuho Maru*. Photograph: Katsumi Tsukamoto.

Discovery of preleptocephali

On the day of the new moon in June 2005 a total of 130 preleptocephali (early stage larvae) (Figure 6.7) that had hatched the previous day were collected to the west of the Suruga Seamount, in the southern part of the West Mariana Ridge. As expected, they were found immediately to the south of a salinity front. Judging from their body length of about 5 mm, the number of days after spawning and the speed of ocean currents, it was calculated that they had spawned near the Suruga Seamount (Tsukamoto, 2006). This was the first time anywhere in the world that the iotation of eel spawning activity had been pinpointed so clearly.

So why did it take 14 years to discover these preleptocephali? And why has it been so difficult to collect preleptocephali or eggs? Eels produce a huge number of eggs in their spawning areas, and many of them must hatch into preleptocephali. Yet they have been difficult to collect in the past, because both eggs and preleptocephali are concentrated in narrow areas and are distributed in dense patches. With time, the patches gradually increase in size, but are still trivial in scale compared to the vastness of the ocean. Even towing a huge plankton net three metres in diameter gives us nothing more than 'a scoop in the ocean'.

To make matters more difficult, eel eggs hatch only 1.5 days after fertilization (Yamamoto and Yamauchi, 1974). After that they are only preleptocephali for only about a week (Figure 6.8).

Figure 6.7
Preleptochephali collected in 2005. Photograph: Katsumi Tsukamoto.

Figure 6.8
Preleptochephali of Japanese eels. Preleptocephalus on the second day after hatching (top). It is still undeveloped; its eyes are not yet pigmented and it has no mouth (7 June 2005). Preleptocephalus on the fifth day after hatching (bottom). Its eyes are pigmented and its mouth is open, with fine teeth already formed (10 June 2005). Photographs: Katsumi Tsukamoto.

Unless our nets hit upon these patches within 1.5 days after spawning, the eggs will no longer be eggs but preleptochephali. And once they have reached that stage there is only a week in which to find them. These constraints of time and space make the probability of collecting eel eggs and preleptochephali astronomically small.

Discovery of eggs

In the early hours of 22 May 2009 (2 days before the new moon), 31 fertilized Japanese eel eggs were collected near the southern end of the seamount chain of the West Mariana Ridge (Tsukamoto et al., 2011). This was the first time humans had ever set eyes on eel eggs spawned in the wild. The first of these eggs were immediately subjected to DNA analysis (Watanabe et al., 2004) on board, and were identified as Japanese eels. The remaining eggs were confirmed using DNA in the laboratory (Yoshinaga et al., 2011). Both the seamount hypothesis and the new moon hypothesis had been correct.

The collection point was at the intersection between the row of seamounts on the West Mariana Ridge and a salinity front, formed by contact between water masses with higher and lower salinity. This also corroborated the salinity front hypothesis, that is, that spawning eels appear to use salinity fronts formed within the spawning area when deciding exactly where to spawn. The eggs were distributed in an extremely narrow range of about 10 × 10 km in the vastness of the ocean (Tsukamoto et al., 2011). They had an average diameter of 1.6 mm, and were at a stage of embryonic development characterized by a large perivitelline space (Yoshinaga et al., 2011). They were thought to have been spawned three nights before the new moon, about 30 hours after fertilization.

The moment of birth

Japanese eel eggs are epipelagic, meaning that they become separated and float in seawater. They are relatively large for pelagic eggs, having a large yolk and a single oil globule. As cleavage progresses and the embryo is formed, the basic forms of the eyes and inner ears appear. The heart starts to beat before hatching, which occurs at a temperature of 25°C about 24 hours after fertilization. Eel larvae hatch while still barely developed (Figures 6.9 and 6.10); there is no pigment in their eyes and their mouths have not yet formed. They may wriggle vigorously when moving, but have hardly any ability to escape from predators.

The egg membrane hardens after fertilization and forms a tough protective shield. Development proceeds inside the egg, thus protected from various external stimuli. This tough egg membrane could actually be a hindrance when hatching from the egg. So how do eels emerge from their eggs?

The answer is that they have a special mechanism to help them hatch smoothly. As they are about to hatch, numerous cells called hatching gland cells appear at the rostral end of the embryo and the front end of the yolk sac. Proteolytic enzymes called hatching enzymes that act specifically on the egg membrane are secreted from these cells, and these enzymes dissolve the egg membrane from the inside (Hiroi et al., 2004). Eel embryos can easily break through this weakened egg membrane to enter the world.

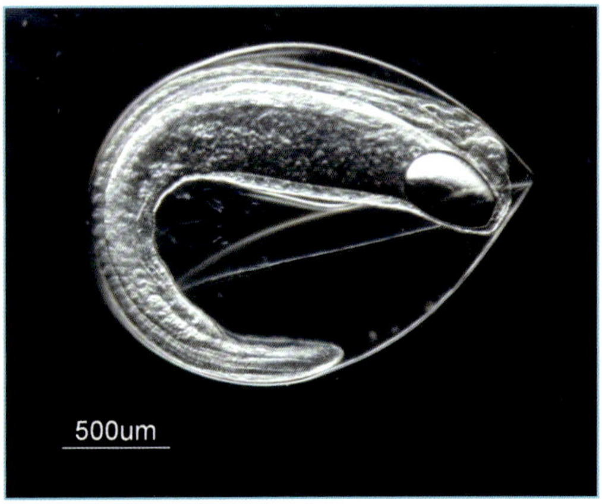

Figure 6.9 Natural Japanese eel egg just before hatching. Photograph: Katsumi Tsukamoto.

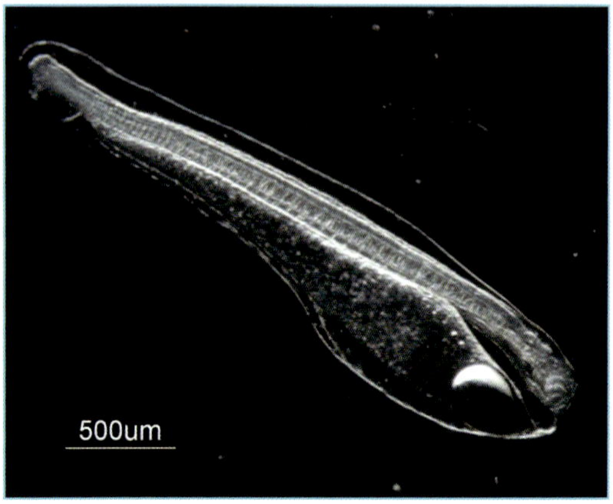

Figure 6.10 A natural Japanese eel immediately after hatching on board. Photograph: Katsumi Tsukamoto.

Depth at birth

Eel eggs hatch at surprisingly shallow depths, thought to be between 150 and 200 m (Tsukamoto et al., 2011). Although the eggs were discovered in seamount areas including some relatively low seamounts, these were deep sea areas with average depths of 3000–4000 m. Eels are born near the surface of these deep sea areas. The depth corresponds exactly to the top of a thermocline where the water temperature suddenly changes. It is also directly below the layer (150 m) where phytoplankton, which multiplies in the light-rich euphotic zone, accumulates together with its decaying remains. This is where the chlorophyll concentration is at its greatest. Probably spawned at depths of around 200 m, the eggs slowly float upwards as they develop. This is because they are slightly lighter than the seawater at those depths. Preleptocephali hatched from the eggs accumulate in the thermocline where the density of seawater changes significantly (Tsukamoto et al., 2011). The density of seawater in the upper layer of the thermocline is low, while the larvae are heavier, and this prevents them from floating further upwards. Therefore, eel eggs and preleptocephali rise up from deeper waters, while 'marine snow' formed from decaying plankton, thought to be a food source for the eel larvae, precipitates downwards from shallower depths. The two meet at the top of the thermocline, in what appears to be a brilliantly effective strategy for the eels' early survival.

Drifting

Leptocephali

Leptocephalus means 'slim head' in Greek, and is a general term for a specific type of fish larvae that have flat, transparent bodies resembling willow leaves (Figure 6.11). As well as the many eels of the order Anguilliformes, other fish in the traditional orders Elopiformes, Albuliformes,

Figure 6.11 Anguillid leptocephali. Photograph: Katsumi Tsukamoto, Michael J Miller.

Notacanthiformes and Saccopharyngiformes also have a leptocephalus larval stage. All of these orders are closely related to eels in phylogenetic terms, and the fact that they grow through a leptocephalus phase before reaching adulthood is a shared taxonomic trait.

A characteristic of leptocephalus larvae is that they have a long larval stage, during which they grow to quite a considerable size. Another noteworthy fact is that their growth rate is faster than that of many other fish larvae. About a week after hatching from their eggs as preleptocephali, the eel larvae have already consumed the yolk matter received from their mothers and have grown to a total body length of about 6 mm. Then they start looking for their own food.

As their bodies increase in depth and they start to resemble willow leaves, the larvae enter the growth phase when they are known as leptocephali. They remain in this form for quite a long time, continuing for about 4 months or so in the case of the Japanese eel (Figure 6.12). During this time, they grow about 0.5 mm every day. Then, when they reach a body length of around 60 mm, they start their first metamorphosis.

While still in the leptocephalus stage, eels are carried by ocean currents that take them on a long journey – a journey from their spawning area to freshwater habitats inland. For an eel, this is the outward journey of a lifelong round-trip.

Defining 'leptocephalus'

When a plankton net is towed through the sea, it collects a variety of different planktonic organisms. Among them are some strikingly conspicuous creatures called leptocephali. As well as eels, all other fish in the order Elopiformes (including conger eels, moray eels, pike eels and sea snakes) spend their larval stage in the form of leptocephali. Leptocephali are entirely different in body shape from their parent eels. That's why, when they were first recorded in 1763, they were thought to be a different species of fish altogether; they were even classified under the separate

Figure 6.12 Japanese eel leptocephali collected in the ocean to the west of the Mariana Islands (1991). Body lengths (from the top) 9.8, 21.6 and 33.5 mm. Photograph: Katsumi Tsukamoto, Noritaka Mochioka.

genus of 'Leptocephalus'. Then, in the second half of the 19th century, the idea was first mooted that leptocephali might not be a different type of fish at all, but merely the larval forms of fish that were already well known. Gill (1864) suggested that larvae classified as *Leptocephalus morrisii* could in fact be the young of *Conger conger*, a type of conger eel. Delage (1886) then proved Gill's hypothesis by monitoring the metamorphosis of leptocephali in a water tank. A particularly famous discovery in relation to eels was made by the Italian zoologists Giovanni Battista Grassi (1854–1925) and Salvatore Calandruccio in 1892. When they collected leptocephali then known as *Leptocephalus brevirostris* in the Strait of Messina and reared them in a water tank, the body depth of the larvae gradually decreased until they metamorphosed into juvenile European eels (Grassi, 1896). This historic discovery proved that these creatures, treated as a completely independent genus unrelated to eels until then, were in fact the larvae of eels. And with this, the parent–child relationship of eels and leptocephali was revealed.

Classifying leptocephali

Most leptocephali belong to eels of the order Anguilliformes. This order is currently recognized as comprising 15 families, 141 genera and 751 species (Nelson, 2006). The body forms of leptocephali vary widely from species to species (Miller and Tsukamoto, 2004). When identifying species of leptocephali, key distinguishing traits include body shape, body length, morphology of the digestive tract, pigmentation, head (Figure 6.13), snout, eyes, morphology of the tail, position of the end of gut and dorsal fin, position of vertical vessels, and myomere (muscle segment) counts. Leptocephali in the family Anguillidae (freshwater eels) have myomere counts of 101–120, a somewhat pointed snout and poor pigmentation. The digestive tract is simple and straight, and the anus is positioned at a point 75–90% along the body, depending on the size of the larva. These characteristic morphological traits distinguish anguillid eel leptocephali very clearly from those of other species (Miller and Tsukamoto, 2004). Different types of anguillid eel larvae are more difficult to distinguish from each other, however. The only useful taxonomic traits are the total myomere count and the myomere count between the anus and the start of the

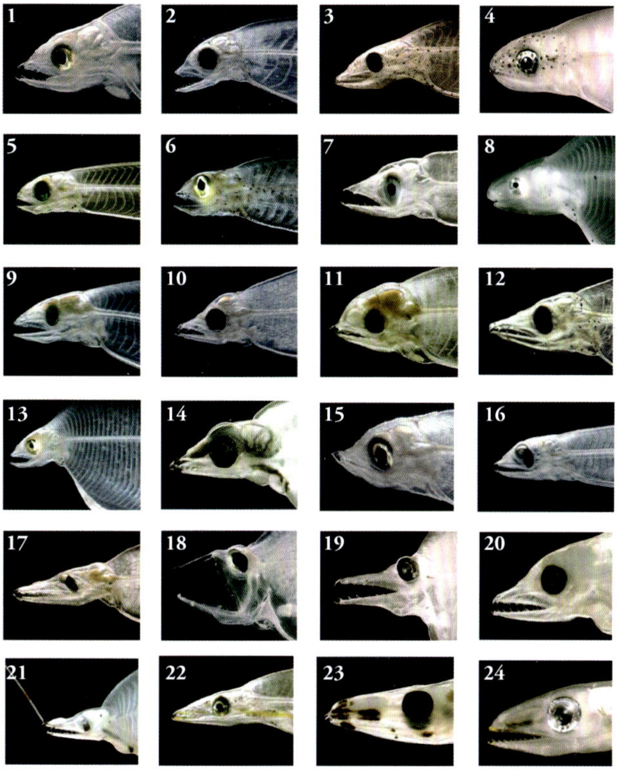

Figure 6.13 Heads of leptocephali in the superorder Elopomorpha. 1: Anguilla marmorata, 2: Ariosoma (normal gut type), 3: Heteroconger (garden eel), 4: Muraenidae, 5: Conger, 6: Bathycongrus, 7: Gnathophis, 8: Chlopsidae (metamorphic phase larva), 9: Muraenesox cinereus, 10: Serrivomeridae, 11: Muraenidae, 12: Gnathophis, 13: Chlopsidae, 14: Nettenchelys, 15: Anguilla japonica, 16: Synaphobranchinae, 17: Ilyophinae, 18: Eurypharynx pelecanoides, 19: Cyema atrum, 20: Ariosoma (exterillium gut type), 21: Ilyophinae (with rostral filament), 22: Neenchelys, 23: Nettastomatidae, 24: Nettastomatidae. Photographs: Michael J. Miller, Katsumi Tsukamoto.

dorsal fin. Leptocephali of European and American eels in the North Atlantic can be identified to some degree of certainty by the total myomere count (Boetius and Harding, 1985). However, in tropical waters where many different species share the same area of distribution, the values of these traits tend to overlap among species. To solve this problem, species are identified using DNA sequence information, in addition to preliminary information from morphological traits (Aoyama et al., 1999; Kuroki et al., 2006, 2008a).

Needle-like teeth

Inside their mouths, eel preleptocephali have needle-like teeth that are disproportionately long compared to the rest of their bodies. At first these threatening teeth led to the assumption that eel larvae were fiercely carnivorous, but now the teeth are thought unsuited to capturing fast-moving prey. In sharks and other piscivorous fishes, the front teeth certainly do not protrude forwards: they point backwards, and slant towards the inside of the mouth, thus preventing captured prey from escaping. This is not limited to fish alone, but is also evident in the teeth of carnivorous land animals. The relative size of the teeth of leptocephali gets smaller as the larvae get bigger. One early theory was that the teeth work as a calcium reservoir rather than to capture prey, and are reabsorbed in the process of growth into leptocephali. A recent idea is that the teeth appear

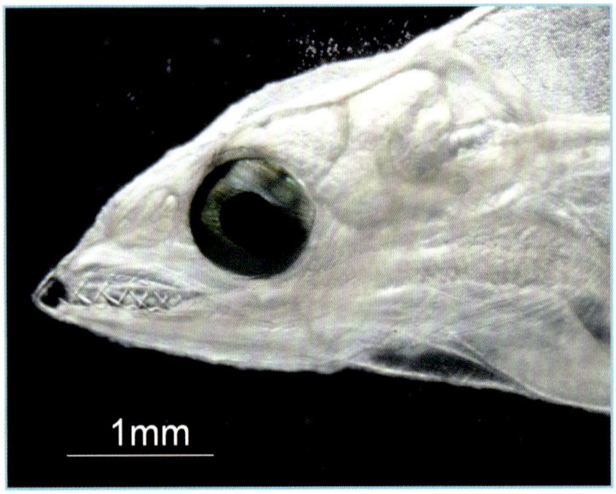

Figure 6.14 Head of Anguillid leptocephalus. Photograph: Michael J. Miller, Katsumi Tsukamoto.

to be useful for grasping food particles, such as faecal pellets of copepods and larvacean houses (Miller, 2009; Mochioka and Iwamizu, 1996) (Figure 6.14).

Invisible food

Compared to other fish, eel leptocephali go through a longer larval stage and reach a larger size in the process. So, what do they eat during this stage? Investigation of their gut contents reveals nothing but a mass of slimy amorphous matter. For this reason, it was hypothesized by some scientists that leptocephali did not ingest food through their mouths but instead absorbed dissolved organic matter directly through their large body surface. Recently, however, zooplankton faecal pellets and larvacean houses were found inside leptocephalus guts, proving that leptocephali do ingest through their mouths after all (Mochioka and Iwamizu, 1996; Otake, 1996). Analysis of isotopic and fatty acid compositions shows the trophic levels of eel leptocephali to be among the lowest of all marine organisms (Miyazaki et al., 2011; Otake et al., 1993). Marine snow including larvacean houses and faecal pellets is now thought to be the diet of leptocephali.

Body plan for planktonic life

The majority of marine organisms live near the ocean surface, at least in the larval stages (Figure 6.15). This is because they are dispersed by shallow ocean currents or winds and can most efficiently find food there. If we actually measure the specific gravity of marine organisms, however, we find that many species are heavier than seawater, and would normally be expected to sink down to the abyssal ocean bed (Tsukamoto et al., 2009a). Sea creatures have developed a variety of ingenious mechanisms to prevent this from happening and enable them to remain in surface waters. Jellyfish, salps and heteropods have transparent, gelatinous bodies that are low in specific gravity. As they are only slightly heavier than seawater, they can easily maintain neutral buoyancy by contracting their bodies or swimming. Pteropods (sea butterflies), copepods, shrimp and crabs, with their heavy shells of chitin and calcium carbonate, avoid sinking by vigorously moving wings or pleopods (swimming legs). Squid, octopus, juvenile fish and others

Figure 6.15 Growth stages from egg to leptocephalus. Photographs: Katsumi Tsukamoto, Mari Kuroki, Michael J. Miller.

without heavy shells can more easily remain in surface waters by using smaller swimming movements. Some creatures, in particular, prevent themselves from sinking by developing extremely flat bodies in the larval stage, or extending fins or spines, thereby increasing the frictional resistance of their body surface. Similar mechanisms to prevent sinking can be seen across organisms separated by huge taxonomic differences. These mechanisms have likely played an important role in the survival and evolution of marine plankton and small larval fish.

Weird shapes

Why do leptocephali come in such weird shapes? One thing for certain is that the design of leptocephali is related to planktonic adaptation. Their large, flat body surface area increases friction resistance in the water, making it more difficult for the body to sink. In fact, the body surface area of a leptocephalus before metamorphosis is more than three times that of a glass eel after metamorphosis. The high water content of leptocephali bodies also reduces their specific gravity and helps them to keep from sinking. These bodily characteristics are what enable leptocephali to live in this buoyant state for so long (Tsukamoto et al., 2009a). The large skin area of leptocephali also assists the respiratory functions of their undeveloped gills and seems to promote cutaneous respiration. Another adaptive feature of leptocephalus larvae is the fact that the high transparency of their bodies makes them more difficult for predators to see. When swimming calmly in a glass dish, leptocephali are difficult to see even if we stare really hard.

Planktonic adaptation

Leptocephali have large quantities of jelly-like polysaccharide materials called glycosaminoglycans (GAGs) stored inside their bodies (Bishop et al., 2000; Pfeiler et al., 1991, 2002). This is an extracellular matrix, which creates numerous intermolecular spaces inside the leptocephalus body. These spaces are filled with water, increasing the water content of the body as a whole. The body surface area is also extensively covered with chloride cells for osmoregulation (Figure 6.16). These keep the osmotic pressure of the water retained in GAG crevices in a more hypotonic state than external seawater, further reducing the specific gravity of the body (Kaneko et al., 2003; Tsukamoto et al., 2009a). GAGs continue to accumulate as the leptocephali grow, making their specific gravity even lower. When their total length exceeds around 30 mm, they are lighter than the specific gravity of seawater, and eventually, they attain their lightest point just before metamorphosis.

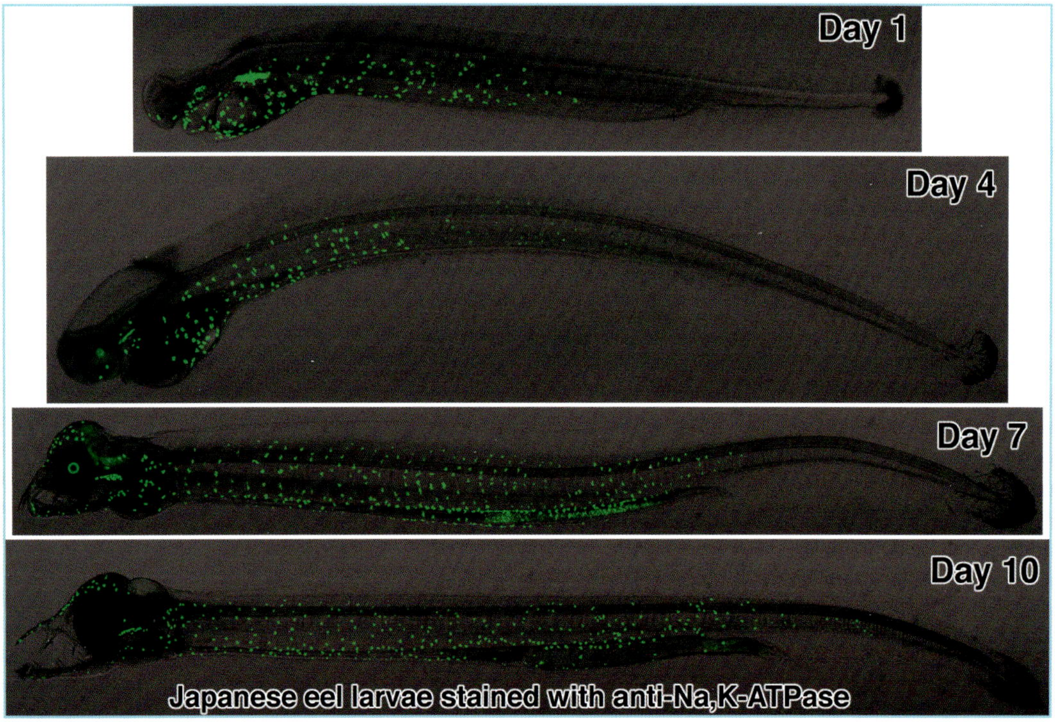

Figure 6.16 Chloride cells of artificially spawned Japanese eel leptocephali on the 1st, 4th, 7th and 10th day after hatching. The chloride cells, highlighted in green, are found all over the body surface. Photographs: Junya Hiroi, Toyoji Kaneko.

Larval transport

Japanese eel leptocephali are first carried westward towards the Philippines by the North Equatorial Current. They then transfer into the Kuroshio travelling northwards in seas off the Philippines, eventually reaching Taiwan, China, South Korea and Japan in East Asia (Kuroki et al., 2009). While they are being carried by the North Equatorial Current, the leptocephali start diurnal vertical migration as they grow. They rise to waters at depths shallower than 100 m at night, then move to deeper depths of about 200 m in the daytime (Otake et al., 1998). They repeat this pattern every day. This behaviour is thought to reduce the chance of being caught by predators and to increase the likelihood of encountering their own food sources. Rising to near the surface at night also makes them susceptible to the influence of Ekman transport, assisting their transfer from the North Equatorial Current to the Kuroshio (Kimura et al., 1994). One study suggests that, if this transfer between ocean currents were to fail and the eels were to remain in the southern part of the North Equatorial Current, they would be taken into the Mindanao Current that flows southward near the Philippines and would be carried to tropical waters where Japanese eels are not distributed. As a result, the numbers of eels recruiting to East Asia would decrease (Kim et al., 2007).

Transforming

Metamorphosing larvae

Caterpillars turn into butterflies, tadpoles into frogs. This is called metamorphosis – that stage of development when a creature's form changes dramatically. Eels are a great example of this. They go through a metamorphosis lasting about 2 or 3 weeks, during which they change from the larval stage of leptocephali to the juvenile stage of glass eels (Figure 6.17). In the process the

Figure 6.17 Metamorphic phase larvae of artificially hatched Japanese eels. Photograph: Yoshiaki Yamada.

flat, leaf-like leptocephali are transformed cylindrical, cord-shaped glass eels resembling their parents.

Metamorphosis reduces an eel's total length by several millimetres. Metamorphosis not only changes an eel's shape and size; its physiological functions and ecological characteristics also change dramatically at this time. The metamorphic phase could be seen as a period of transition in which the larval physiology and ecology change to the same type as the adult fish. This is a major turning point, from the larval stage mainly consisting of survival and development, to the juvenile and adult stages designed for growth and reproduction.

Otolith increments

If an otolith from a glass eel is removed and observed under an electron microscope, a deeply etched hole with a diameter of around 10 μm in the centre can be seen (Tzeng, 1990). The outside edge of this hole is thought to be a hatching check formed in response to sudden changes in the environment when the larva hatches from the egg (Shinoda et al., 2004). On the outside of the hatching check are about ten slightly unclear otolith increments. These increments are thought to be formed while the early larva grows, with the yolk matter and oil globule as its main sources of nutrients. Therefore, the otoliths appear in the preleptocephalus phase. Between the ten or so unclear otolith increments and the clear otolith increments outside them is a pronounced discontinuous ring (check) called the first feeding check. This is thought to be formed when the last of the yolk matter is absorbed and the larva switches to external feeding; the leptocephalus phase starts outside this ring. In Japanese eels, diurnal otolith increments suddenly grow wider on reaching the 90th–150th ring from the centre of the otolith. This corresponds to the metamorphic phase from the leptocephalus to the glass eel stage (Kuroki et al., 2005). As this shows, otoliths hold the key to revealing the early history of an eel's life.

Metamorphosis and otolith microchemistry

Otoliths contain small amounts of the element strontium. The relationship between metamorphosis and otolith increment width has been discovered by analysing this trace element. In particular, a comparative study has been made of otolith increment width and changes in strontium between leptocephali at their maximum body size immediately before metamorphosis and glass eels immediately after metamorphosis (Figure 6.18) (Kuroki et al. 2005). Although otoliths consist almost completely of calcium carbonate crystals, a tiny trace of strontium with the same divalent ions is incorporated in the crystals in place of calcium. Glycoaminoglycans (GAGs) in the leptocephalus body have strong affinity with strontium, and eels in the leptocephalus phase incorporate strontium in seawater into the GAGs, increasing the concentration of strontium in the body. With metamorphosis, strontium is released from the body as the GAGs are metabolized. The sharp fall in strontium concentration inside the body triggers a similarly sharp fall in the strontium concentration in otoliths (Otake et al., 1994). The sudden increase in ring width coincides with this sharp fall in strontium, signalling the start of metamorphosis. When the ring width reaches its maximum, metamorphosis may be almost complete.

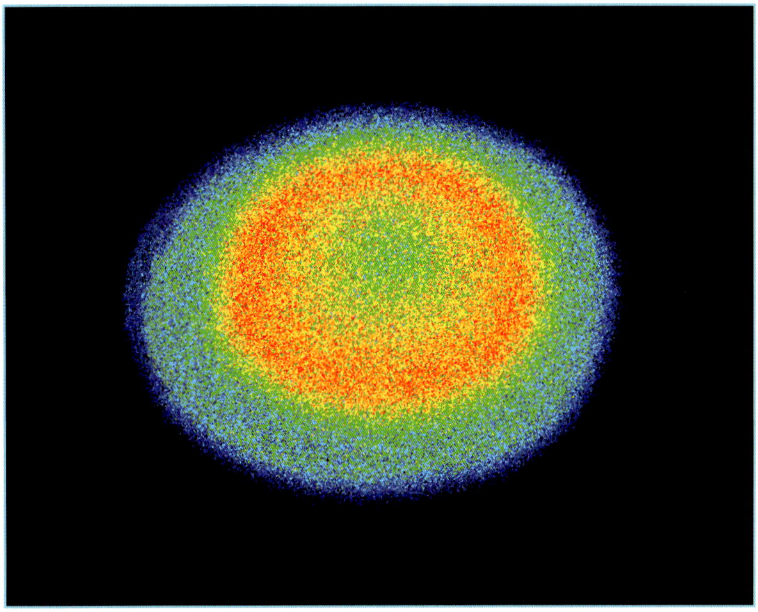

Figure 6.18 Distribution of strontium in a glass eel otolith. The high concentration of strontium in the leptocephalus phase (red) decreases sharply in the metamorphic phase (green-blue) reaching its lowest value (almost zero – blue) as the glass eel enters freshwater. Photograph: Mari Kuroki.

What triggers metamorphosis?

Leptocephali can start to metamorphose when they reach their maximum body length, but little is known about what triggers this metamorphosis. The minimum size at which they can begin metamorphosis is more or less fixed, depending on the species, but some larvae can grow slightly larger, although their growth appears to slow down (Kuroki et al., 2006). Sometimes, however, metamorphosis is not triggered even after reaching the maximum body size. In this case, growth sometimes seems to continue until a state of overgrowth, as apparently seen in some species of *Ariosoma* leptocephali. *Leptocephalus giganteus* of the order Notacanthiformes have been reported to grow to a total length of 893 mm (Castle, 1959). The trigger for metamorphosis has been investigated using artificially produced leptocephali (Kuroki et al., 2010). When premetamorphic leptocephali raised at a temperature of 22°C were transferred to water tanks at different temperatures, metamorphosis most often started in the 25°C group. But, although water temperature is thought to have a big impact on metamorphosis, the possibility that metamorphosis is triggered by olfactory cues originating from land has also been suggested. Recent experiments suggest that starvation may be able to trigger metamorphosis in eels (Okamura et al., 2012).

Metamorphosis and functions

The outward appearance of an eel is utterly transformed with metamorphosis (Figure 6.19). The central nervous system, sensory organs and internal organs also change dramatically, and significant changes occur in their various functions. Leptocephali have no erythrocytes (red blood cells) and very few leukocytes (white blood cells). But when metamorphosis is complete and they become glass eels, erythrocytes appear in the blood. Leptocephali have no erythrocytes because relative to their body size, the muscle, which is a major oxygen consuming tissue, only forms a thin single layer beneath the skin; oxygen appears to be adequately supplied by cutaneous respiration (Suzuki and Otake, 2001). Even in 10 mm leptocephali, the basic structure of the brain, including the olfactory bulb, telencephalon, diencephalon, mesencephalon, metencephalon, myelencephalon, and so on, is already in place (Tomoda and Uematsu, 1996). In leptocephali, the optic tectum of the visual centre accounts for about 30% of overall brain capacity, but this falls to just over 15% in glass eels and 10% in yellow eels. The olfactory and mechanosensory regions in the brain continue to grow from the leptocephalus to the glass eel and into the yellow eel stage. This shows that, while the behaviour of leptocephali depends largely on the visual senses, the emphasis shifts from vision to smell, mechanosensory and posture control information from the postmetamorphic glass eel onwards.

While leptocephali spend their time in relatively shallow depths, after metamorphosing into glass eels and recruiting to coasts, they become benthic (bottom dwellers). They conceal themselves in mud or holes in riverbanks and lead a nocturnal life. This must be why the functions of their visual senses diminish while chemical senses vital to feeding at night and lateral line sensitivity related to positive rheotaxis necessary for benthic activity develop.

The end of larval migration

Only six metarmorphosing larva of Japanese eels have ever been collected in the ocean (Shinoda et al., 2011). This is in complete contrast to the thousands of leptocephali collected so far. One

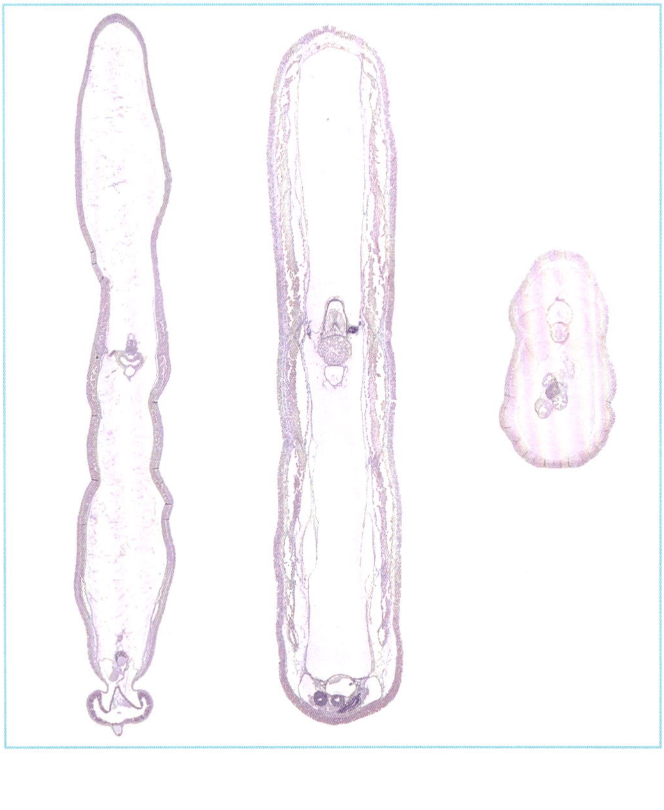

Figure 6.19 Body cross sections, from left to right, of leptocephalus, metamorphosing larvae and glass eel. Photograph: Mari Kuroki and Toyoji Kaneko.

reason for their elusive behaviour must be that, compared to the long leptocephalus stage, the metamorphic larval stage only lasts three weeks; moreover, their distribution and behaviour change with metamorphosis. Collection sites of leptocephali in mid-metamorphosis have been limited to eddy areas formed to the east of the Kuroshio in waters east of Taiwan; all eels collected west of the Kuroshio were postmetamorphic glass eels (Otake et al., 2006; Shinoda et al., 2011). This would suggest, then, that metamorphosis starts just before entering the Kuroshio, progresses while drifting in the Kuroshio, and is complete when metamorphosed glass eels leave the Kuroshio. Completion of metamorphosis determines the eels' detrainment from ocean currents, as well as the location and start time of active recruitment migration (Cheng and Tzeng, 1996; Kuroki et al., 2008b). As a result, the species range and the number of glass eels recruiting in each region is thought to be determined by the progression of metamorphosis and the flow regime of the Kuroshio.

Producing

Artificial seedlings

The eel farming industry today relies 100% on natural glass eels for its seedlings. Owing to massive fluctuations in glass eel catches every year, seedling prices alternate between inflation and decline, making business unstable. Research on technology for producing artificial glass eels started around 1960. The aim was to establish a stable supply of aquaculture seedlings.

In 1961, Hideo Satoh and Takashi Hibiya of the University of Tokyo successfully induced spermiation in male eels using gonadotropic hormones (Satoh et al., 1962). In 1973, Kiichiro Yamamoto and Kohei Yamauchi of Hokkaido University succeeded in artificially hatching eels for the first time in the world by injecting female eels with salmon pituitary (Yamamoto and Yamauchi, 1974). In 1976, Kohei Yamauchi's group successfully reared artificially hatched eel larvae for 14 days, paving the way for artificial seedling production (Yamauchi et al., 1976). In 2003, Hideki Tanaka and colleagues at the National Research Institute of Aquaculture reared leptocephali with a diet based on shark egg yolk until they metamorphosed into glass eels (Tanaka et al., 2003). And in 2010, finally, the Fisheries Research Agency reared artificial glass eels up to spawning adult eels, then succeeded in obtaining second generation artificial glass eels from these. As a result, the whole life cycle of eels has been brought under human management, making 'complete aquaculture' possible (Figure 6.20).

It took about half a century from the start of research until complete aquaculture was achieved, yet the original target of supplying artificial seedlings to the eel farming industry has still not been met. The eel farming industry probably requires hundreds of millions of seedlings every year, but current production capacity is only around a thousand per year. Therefore, some kind of epoch-making breakthrough is needed as soon as possible. Commercial mass production of artificial glass eels will reduce pressure on catches of natural glass eels and help to protect the species. Researchers are engaged in ongoing, painstaking research with the aim of making this a reality.

Eel farming currently involves obtaining natural glass eels and rearing them in aquaculture ponds until they reach commercial size. But despite the term 'aquaculture': the fish used as

Figure 6.20 Anatomical specimen of a female European eel with its abdomen opened. The eel had been artificially matured for the first time by Hideo Satoh and Takashi Hibiya at the University of Tokyo. At the start of the experiment (14 January, 1986), the total length was 91.5 cm and the body weight 1805 g. On completion (2 March 1986), the total length was 91.4 cm and the body weight 2228.5 g. Hormones were injected 12 times. Photograph: Mari Kuroki and Katsumi Tsukamoto.

seedlings are originally natural, so the result should perhaps be called 'quasi-natural'. Other species produced through aquaculture, such as sea bream and flounder, are now easy to raise artificially from eggs. But with eels, even inducing maturation in adult fish is difficult. To obtain eggs from a matured eel and then hatch and raise these up to glass eels as seedlings for aquaculture was always going to be harder. Eels could be described as the most difficult fish humans have ever tried to produce as seedlings. Nevertheless, it has now become possible, albeit only at laboratory level, to collect eggs from artificially matured adults and raise the hatched larvae to the glass eel stage (Figure 6.21) (Tanaka et al., 2003). Of all the world's eels, Japanese eels are the only ones to have gone through the whole process of seedling production from adult fish to eggs, then from leptocephali to glass eels.

Inducing maturation

Unlike ayu and sea bream, it is difficult to artificially induce maturation in adult eels by manipulating photo-periods and water temperature. A more direct method is to administer exogenous hormones to the adult fish, thereby bringing them to maturity. In 1936, Maurice Fontaine of France succeeded in using the urine of pregnant women to induce sperm formation in male European eels (Fontaine, 1936). Today, male eels are brought to maturity and their semen obtained by injecting them several times with human chorionic gonadotropin, which is a constituent of that type of urine. In females, conversely, eggs are obtained by first injecting them about ten times with salmon pituitary extract, then with an ovulation inducing agent called DHP (Figure 6.22). When the female's belly is squeezed, 1 to 3 million eggs can be collected

Figure 6.21 Kreizel tanks for artificial leptocephali (IRAGO Institute, Japan). Research on producing artificial seedlings is now in continuous progress 24 hours a day. Photograph: Katsumi Tsukamoto.

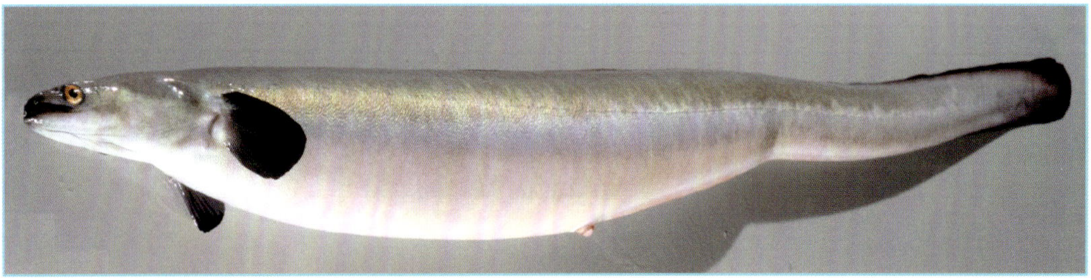

Figure 6.22 An artificially matured female Japanese eel (IRAGO Institute, Japan). Photograph: Yoshiaki Yamada.

(Figure 6.23). These are mixed with the semen to artificially induce fertilization. Another more natural way of obtaining fertilized eggs is to place artificially matured male and female eels in a tank and spontaneously induce spawning behaviour.

Barriers to perfect aquaculture

There are three problems that need to be solved. The first is the problem of egg quality. If the quality is poor, the larvae have low rates of survival and growth (Figures 6.24 and 6.25).

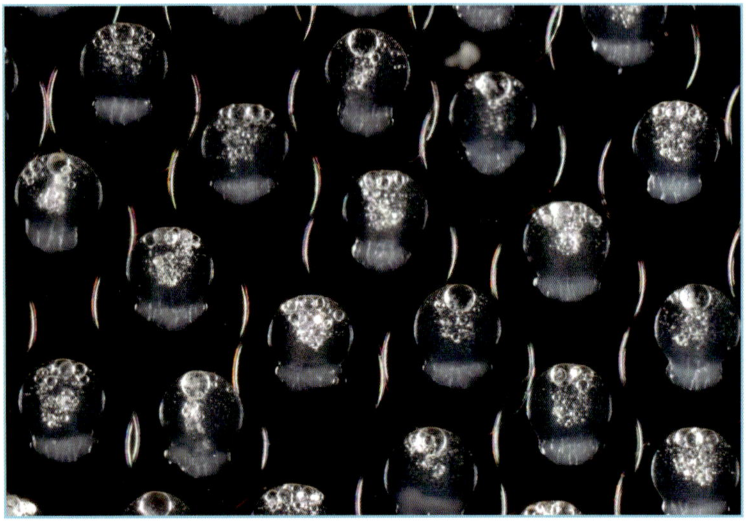

Figure 6.23 Artificially fertilized Japanese eel egg. Photograph: Yoshiaki Yamada.

Figure 6.24 Hatching Japanese eel larvae. Photograph: Yoshiaki Yamada.

Deformity of larvae is common at present. A method of inducing maturation in adult eels just by manipulating environmental conditions and without using hormones needs to be developed, unlike the hormone-intensive methods currently practised. Second is the problem of food for the larvae. The existing food, consisting mainly of shark egg yolk, is a mushy, semi-liquid material, (Figures 6.26 and 6.27) which easily dissolves in water and tends to pollute the rearing tanks. An alternative food that will achieve high growth without polluting water tanks needs to be developed. There is also a danger that the supply of shark eggs will be exhausted in future. And the third problem is a rearing method that will allow leptocephali to be reared in high density. Large numbers of glass eels (for example, more than 100 million in Japan) are required for commercial use, so it is important that we develop a rearing system to facilitate intensive, large-scale production.

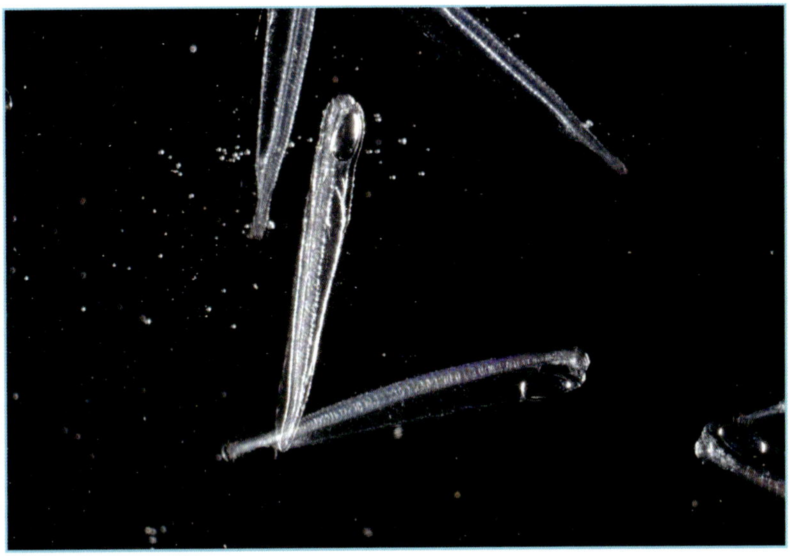

Figure 6.25 Newly hatched Japanese eel larvae. Photograph: Yoshiaki Yamada.

Figure 6.26 Artificial eel leptocephali feed, in easily swallowed paste form. Photograph: Yoshiyaki Yamada.

The outlook for aquaculture

In future, we will need to breed eels repeatedly using selective breeding techniques in order to create new strains distinct from natural resources, as has been done with livestock. Eventually, this will produce bloodlines that are easier to rear and have good growth and survival rates.

We could also create variants with a good taste and strong disease resistance. Then eels would become an even more familiar animal food resource, like chickens, cows and pigs.

Most importantly, the pressure from catches of natural eels would be vastly reduced, and eel populations now in decline could be expected to recover.

Figure 6.27 Eel leptocephali feeding on food paste scattered at the bottom of a rearing tank (IRAGO Institute, Japan). Photograph: Yoshiaki Yamada.

Rearing

Aquaculture

According to Aristotle eel farming was already being practised in Italy more than 2000 years ago. As if to reflect this historical background, Italy was the centre of European eel farming until the mid-1980s. Italian eel farm output in 1984 was 2600 tonnes, accounting for 95% of Europe's total. Italy's output has not grown significantly since then, but those of the Netherlands and Denmark have increased sharply, reaching 3800 tonnes and 2700 tonnes, respectively, in 2000. European eel farm production peaked at 10,900 tonnes in 2000, but has fallen back since then, reflecting a sharp decline in glass eel resources.

Eel farming in Asia started in Japan at the end of the 19th century. As rearing methods and eel feed improved, aquaculture technology made dramatic advances. At its peak in 1989, production stood at around 40,000 tonnes. Eel farming also started to thrive in Taiwan, China and Korea. Due to a shortage of Japanese glass eel seedlings for aquaculture in East Asia, European glass eels were airlifted from France and Spain for aquaculture in China. As a result of these developments, eels joined the ranks of internationally traded commodities.

Early days of eel farming

Japanese eel farming is said to have started in 1879, when Kurajiro Hattori established a 2 ha aquaculture pond in Fukagawa, Tokyo, and attempted to farm eels in it. At the time, elvers were used as seedlings, and it was not until around 1920 that glass eels started to be used. Eel farming spread, particularly in the Tokai region, and aquaculture in open ponds became a thriving business (Figure 6.28). In around 1930, the production volume passed 3000 tonnes, outstripping natural eel catches for the first time.

Development of eel farming

Eel farming temporarily declined during the Second World War in Japan, but enjoyed a revival in around 1960, when production surpassed pre-war levels (Figure 6.29). In 1955, waterwheels were introduced to replenish oxygen in eel farm ponds, and manufactured eel feed started to be

Figure 6.28 One of the few remaining open ponds in Yoshida-cho, Shizuoka, Japan. Photograph: Mari Kuroki.

Figure 6.29 Aerial photo of Yoshida-cho, Shizuoka, Japan, at the peak of eel farming in the 1960s. Photograph: Maruhai Yoshida Eel Fisheries Cooperative.

developed in 1964. Until then, the practice had been to boil up masses of sardines, sand lance and other cheaply available fish, which would then be dropped into the open ponds in bulk. The development of formula feed promoted eel growth and led to efficient, more systematic production. By 1972, attention had shifted from the rather crude eel farming using open ponds, the norm until then, towards 'greenhouse eel farming' that is the trend now (Figure 6.30). Concrete

Figure 6.30 Large greenhouse eel farm in Osaki-cho, Kagoshima, Japan. Photograph: Mari Kuroki.

tanks were housed inside greenhouse-like structures and eel growth was promoted by warming the water. With these techniques, eel farming spread beyond the Kansai region to Shikoku and Kyushu in the west of Japan.

In Taiwan, eel farming on a commercial scale started in the late 1960s. Triggered by an outbreak of disease in Japanese eel farms in 1968, Taiwanese eel farming has experienced rapid growth because of higher water temperatures, overtaking Japan's production volume in around 1990. In the 1990s, a new eel farming industry using cheap labour and inexpensive feed emerged in China. Its output passed 40,000 tonnes in 1994 eclipsing both Japan and Taiwan. This success can partly be attributed to imports of European eel glass eels, which enabled Chinese production to increase. The Chinese also developed a method of producing commercial products using a slower rearing method called the 'flow-through system' in inland regions, where water temperatures are lower.

In 2002, however, prohibited antibacterial agents were found in eels produced in China, and exports were subjected to regulation. This, coupled with a shortage of seedlings for aquaculture resulting from the decline in European eel resources and new restrictions on exporting glass eels, brought a downturn in China's eel farming industry. Eel farming is vulnerable to cycles of rampant inflation and sudden drops in prices, caused by three main factors: good or bad catches of glass eels for seedlings, outbreaks of disease and international trading factors.

The present status of eel farming

In Japanese eel farming, glass eels with a body weight of 0.2 g are reared up to around 200 g in around 6 months before being harvested and sold. This means that they grow by about 1 g every day, eventually reaching a thousand times their initial body weight. But as there is huge variation in the growth of eels, they are repeatedly sorted until they are ready for harvesting (Figure 6.31).

The letter 'p' (short for 'piece') is used as a size standard to show approximately how many eels make 1 kg. For example, if 5 p or 5 eels equals 1 kg, the body weight of each individual will be 200 g. Previously, small eels in 5 p and 6 p categories were called 'yota', those in the 4 p class were called 'boku' and large 2 p and 3 p eels were called 'oboku'.

Most aquacultured eels end up as males. Rearing at high temperatures and high density is said to promote male sex differentiation, but this has not been satisfactorily studied.

The future of eel farming

'Eel farming is all in the water', they say. That shows how important it is to manage water quality in this industry. In the old days, water bloom of the blue-green microalgae *Microcystis* sp. would be mixed into open ponds to prevent a sudden deterioration in water quality. The function of the water bloom was to promote photosynthesis and thus increase the concentration of oxygen in the water. Even today, when greenhouse eel farming is the norm, water quality management in the ponds is still important, and the skill involved requires a craftsman's touch (Figure 6.32).

In Denmark, indoor recirculating purification systems are used in eel farming, making intensive production possible. Because these are closed systems, the amount of water used is minimal, and this in turn places a low burden on the environment. In Taiwan, meanwhile, attempts at safe

Figure 6.31 Eels being sorted for processing and distribution. Photograph: Mari Kuroki.

Figure 6.32 Japanese eels reared on an aquaculture farm. Photograph: Mari Kuroki.

'organic eel farming' avoiding dense cultivation and without the use of pharmaceutical products are being made. At the first sign of disease, the amount of feed is reduced and salt or other natural substances are used to prevent it from spreading. If consumers were to attach special value to healthy eels raised in this way, this method of aquaculture could be expected to grow.

Bibliography

Aarestrup, K., Øikland, F., Hansen, M.M., Righton, D., Gargan, P., Castonguay, M., Bernatchez, M., Howey, P., Sparholt, H., Pedersen, M.I. and McKinley, R.S. (2009) Oceanic spawning migration of the European eel (*Anguilla anguilla*). *Science* 325, 1660.

Albert, V., Jonsson, B. and Bernatchez, L. (2006) Natural hybrids in Atlantic eels (*Anguilla anguilla., A. rostrata*): evidence for successful reproduction and fluctuating abundance in space and time. *Molecular Ecology* 15, 1903–1916.

Aoyama, J., Mochioka, N., Otake, T., Ishikawa, S., Kawakami, Y., Castle, P.H.J., Nishida, M. and Tsukamoto, K. (1999) Distribution and dispersal of *anguillid* leptocephali in the western Pacific Ocean revealed by molecular analysis. *Marine Ecology Progress Series* 188, 193–200.

Avise, J.C., Nelson, W.S., Arnold, J. and Koehn, R.K. (1990) The evolutionary genetic status of Icelandic eels. *Evolution* 44, 1254–1262.

Barbin, G.P. (1998) The role of olfaction in homing and estuarine migratory behavior of yellow-phase American eels. *Canadian Journal of Fisheries and Aquatic Sciences* 55, 564–575.

Bertin, L. (1956*) Eels, A Biological Study.* Cleaver-Hume Press Ltd., London 1–192.

Bishop, R.E., Torres, J.J. and Crabtree, R.E. (2000) Chemical composition and growth indices in leptocephalus larvae. *Marine Biology* 137, 205–214.

Boeseman, M. (1947) Revision of the fishes collected by Burger and Von Siebold in Japan. *Zoologische Mededelingen* 28, 1–242.

Boetius, I. and Boetius, J. (1980) Experimental maturation of female silver eels., *Anguilla anguilla* - Estimates of fecundity and energy reserves for migration and spawning. *Dana* 1, 1–28.

Boetius, J. and Harding, E.F. (1985) A re-examination of Johannes Schmidt's Atlantic eel investigations. *Dana* 4, 129–162.

Bonhommeau, S., Castonguay, M., Rivot, E., Sabatie, R. and Pape, O.L. (2010) The duration of migration of Atlantic Anguilla larvae. *Fish and Fisheries* 11, 289–306.

Casellato, S. (2002) European eel: a history which must be rewritten. *Italian Journal of Zoology* 69, 321–324.

Castle, P.H.J. (1959) A large leptocephalid (Teleostei., Apodes) from off South Westland., New Zealand. *Transactions of the Royal Society of New Zealand* 87, 179–184.

Cheng, P.W. and Tzeng, W.N. (1996) Timing of metamorphosis and estuarine arrival across the dispersal range of the Japanese eel *Anguilla japonica*. *Marine Ecology Progress Series* 131, 87–96.

Chiba, H., Iwatsuki, K., Hayami, K. and Yamauchi, K. (1993) Effect of dietary estradiol-17β on feminization, growth and body composition in the Japanese eel (*Anguilla japonica*). *Comparative Biochemistry and Physiology* 106A, 367–372.

Chow, S., Kurogi, H., Mochioka, N., Kaji, S., Okazaki, M. and Tsukamoto, K. (2009) Discovery of mature freshwater eels in the open ocean. *Fisheries Science* 75, 257–259.

Colombo, G. and Grandi, G. (1996) Histological study of the development and sex differentiation of the sex organ in the European eel. *Journal of Fish Biology* 48, 493 512.

Dadswell, M., Cunjak, R., Cooper, J., Beal, K. and Avery, T. (eds) (2009) Challenges for diadromous fishes in a dynamic global environment. *American Fisheries Society Symposium* 69 Bethesda, MD, pp. 887–889.

Deelder, C.L. and Tesch, F.W. (1970) HeimfindevermÖgen von Aalen (*Anguilla anguilla*) die Über grosse Entfernungen verpflanzt worden waren. *Marine Biology* 6, 81–92.

Dekker, W. (2003) On the distribution of the European eel (*Anguilla anguilla*) and its fisheries. *Canadian Journal of Fisheries and Aquatic Sciences* 60, 787–799 .

Dekker, W., Pawson, M. and Wickstrom, H. (2007) Is there more to eels than slime? An introduction to papers presented at the ICES Theme Session in September 2006. *ICES Journal of Marine Science* 64, 1366–1367.

Delage, Y. (1886) Sur les relations de parente du Congre et du Leptocephale. *Comptes rendus de l'Academie des Sciences* 103, 698–699.

Dou, S., Yamada, Y., Okamura, A., Tanaka, S., Shinoda, A. and Tsukamoto, K. (2007) Observations on the spawning behavior of artificially matured Japanese eels *Anguilla japonica* in captivity. *Aquaculture* 266, 117–129.

Ege, V. (1939) A revision of the genus *Anguilla* Shaw: a systematic, phylogenetic and geographical study. *Dana Report* 16, 1–256.

Elie, P., Lecomte-Finiger, R., Cantrelle, I. and Charlon, N.V. (1982) Definition des limites des differents stades pigmentaires durant la phase civelle d' *Anguilla anguilla*. L. Vie et Milieu 32, 149–157.

Fontaine, M. (1936) Sur la maturation complete des organes genitaux de l'anguille male et l'emission spontanee de ses produits sexuels. [On the complete maturation of male eel gonads and spontaneous emission of sex products.] *Comptes Rendus de l'Academie des Sciences* 202, 1312–1314 (in French).

Freud, S. (1877) Beobachtungen iiber Gestaltung und feineren Bau der als Hoden beschriebenen Lappenorgane des Aals. *Sitzungsber Akad Wiss Wien* (Math-Naturwiss Kl), 1(75), 419–431.

Friedland, K.D., Miller, M.J. and Knights, B. (2007) Oceanic changes in the Sargasso Sea and declines in recruitment of the European eel. *ICES Journal of Marine Science* 64, 519–530.

Fukuda, N. (2010) Inshore and upstream migration of Japanese eels in the Hamana Lake system, Japan. Doctoral dissertation, the University of Tokyo, Tokyo.

Gill, T.N. (1864) On the affinities of several doubtful British fishes. *Proceedings of the Academy of Natural Sciences of Philadelphia* 16, 199–208.

Grassi, G.B. (1896) The reproduction and metamorphosis of the common eel (*Anguilla vulgaris*). *Proceedings of the Royal Society of London* 60, 260–271.

Hayata, H. ed. and Sakurai, K., ill. (1984) *Mie Ken Suisan Zukai* (*Encyclopaedia of Mie Prefecture Fisheries*), Collection in 1 volume. Tokai Suisan Kagaku Kyokai, Toba Sea Folk Museum ed. (in Japanese).

Hirase, T. (ed.) and Hasegawa, M. (ill.) (1979) *Nippon Sankai Meibutsu Zue* (*Encyclopaedia of Famous Japanese Products*). Meicho Kankokai (in Japanese).

Hiroi, J., Maruyama, K., Kawazu, K., Kaneko, T., Ohtani-Kaneko, R. and Yasumasu, S. (2004) Structure and developmental expression of hatching enzyme genes of the Japanese eel *Anguilla japonica*: an aspect of the evolution of fish hatching enzyme gene. *Development Genes and Evolution* 214, 176–184.

Ida, T. (2007) *Unagi - Chikyū kankyō-wo kataru sakana* (*Eels — Fish that reflect the global environment*). Iwanami Shinsho (in Japanese).

Ide, C., De Schepper, N., Christiaens, J., Van Liefferinge, C., Herrel, A., Goemans, G., Meire, P., Belpaire, C., Geeraerts, C. and Adriaens, D. (2011) Bimodality in head shape in European eel. *Journal of Zoology* 285, 230–238.

Ikenami, S. (1998) *Onihei Hankachō*. Kōdansha (in Japanese).

Inoue, J.G., Miya, M., Miller MJ., Sado T Hanel, R., Lopez, J.A., Hatooka, K., Aoyama, J., Minegishi, Y., Nishida, M. and Tsukamow, K. (2010) Deep ocean origin of the freshwater eel. *Biology Letters* 6, 363–366.

Inoue, J.G., Miya, M., Venkatesh, B. and Nishida, M. (2005) The mitochondrial genome of Indonesian coelacanth *Latimeria menadoensis* (Sarcopterygii, Coelacanthiformes) and divergence time estimation between the two coelacanths. *Gene* 349, 227–235.

Jansen, H.M., Winter, H.V., Bruijs, M.C.M. and Polman, H.J.G. (2007) Just go with the flow? Route selection and mortality during downstream migration of silver eels in relation to river discharge. *ICES Journal of Marine Science* 64, 1437–1443.

Jellyman, D.J. and Sykes, J.R.E. (2003) Diet and seasonal movements of radio-tagged freshwater eels., *Anguilla* spp., in two New Zealand streams. *Environmental Biology of Fishes* 66, 143–154.

Jellyman, D.J. and Tsukamoto, K. (2002) First use of archival transmitters to track migrating freshwater eels *Anguilla dieffenbachii* at sea. *Marine Ecology Progress Series* 233, 207–215.

Jellyman, D.J. and Tsukamoto, K. (2005) Swimming depths of offshore migrating longfin eels *Anguilla dieffenbachii. Marine Ecologyrogress Series* 286, 261–267.

Jellyman, D.J. and Tsukamoto K. (2010) Vertical migrations may control maturation in migrating female *Anguilla diefenbachii. Marine Ecology Progress Series* 404, 241–247.

Jespersen, P. (1942) Indo-Pacific leptocephalids of the genus *Anguilla*, Systematic and biological studies. *Dana Report* 22, 1–128.

Jessop, B.M., Cairns, D.K., Thibault, I. and Tzeng, W.N. (2008) Life history of American eel *Anguilla rostrata*, new insights from otolith microchemistry. *Aquatic Biology* 1, 205–216.

Johnson, D., Ida, H., Sakaue, J., Sado, T., Asahida, T. and Miya, M. (in press) A 'living fossil' eel (Anguilliformes, Protoanguillidae., fam nov) from an undersea cave in Palau. *Proceedings of the Royal Society of London Series B.*

Kaifu, K., Tamura, M., Aoyama, J. and Tsukamoto, K. (in press-a) Dispersal of yellow phase Japanese eels *Anguilla japonica* after recruitment in the Kojima Bay-Asahi River system, Japan. *Environmental Biology of Fishes* DOI 10.1007/s10641–010–9640-y.

Kaifu, K., Yokouchi, K., Aoyama, J. and Tsukamoto, K. (in press-b) Head shape polymorphism in Japanese eels *Anguilla japonica* in relation to somatic growth in the Kojima Bay-Asahi River system, Japan. *Journal of Fish Biology.*

Kajihara, T., Tsukamoto, K., Otake, T., Mochioka, N., Hasumoto, H., Oya, M. and Tabeta, O. (1988) Sampling leptocephali with reference to the diel vertical migration and gears. *Bulletin of the Japanese Society for the Science of Fish* 54, 941–946.

Kamei, T. (2006) Mandan, Ao-unagi wo kurau (Thoughts on eels, eating blue eels). *Otemon Gakuin "Hakuru"* 57, 1–39 (in Japanese).

Kaneko, T., Hasegawa, S. and Sasai, S. (2003) Chloride cells in Japanese eels during their early life stages and downstream migration. In: Aida, K., Tsukamoto, K. and Yamauchi, K. (eds) *Eel Biology.* Springer, Tokyo, pp. 457–468.

Kimura, S., Shinoda, A., Kitagawa, T., Sasai, Y. and Sasaki, H. (2007) Effect of El Niño on migration and larval transport of the Japanese eel, *Anguilla japonica. ICES Journal of Marine Science* 64 1387–1395.

Kimura, S., Tsukamoto, K. and Sugimoto T (1994) A model for the larval migration of the Japanese eel, Roles of the trade winds and salinity front. *Marine Biology* 119, 185–190.

Kleckner, R.C. (1980) Swim bladder volume maintenance related to initial oceanic migratory depth in silver-phase *Anguilla rostrata. Science* 208, 1481–1482.

Kleckner, R.C. and McCleave, J.D. (1988) The northern limit of spawning by Atlantic eels (*Anguilla* spp.) in the Sargasso Sea in relation to thermal fronts and surface water masses. *Journal of Marine Research* 46, 647–667.

Kotake, A., Okamura, A., Yamada, Y., Uto, T., Arai, T., Miller, M.J., Oka, H.P. and Tsukamoto, K. (2005) Seasonal variation in migratory history of the Japanese eel., *Anguilla japonica.*, in Mikawa Bay, Japan. *Marine Ecology Progress Series* 293, 213–221.

Kuroki, M., Aoyama, J., Miller, M.J., Arai, T., Sugeha, H.Y., Minagawa, G., Wouthuyzen, S. and Tsukamoto, K. (2005) Correspondence between otolith microstructural changes and early life history events in *Anguilla marmorata* leptocephali and glass eels. *Coastal Marine Science* 29, 154–161.

Kuroki, M., Aoyama, J., Miller, M.J., Wouthuyzen, S., Arai, T. and Tsukamoto, K. (2006) Contrasting patterns of growth and migration of tropical anguillid leptocephali in the western Pacific and Indonesian Seas. *Marine Ecology Progress Series* 309, 233–246.

Kuroki, M., Aoyama, J., Miller, M.J., Watanabe, S., Shinoda, A., Jellyman, D.J. and Tsukamoto, K. (2008a) Distribution and early life history characteristics of anguillid leptocephali in the western South Pacific Ocean. *Marine and Freshwater Research* 59, 1035–1047.

Kuroki, M., Kawai, M., Jonsson, B., Aoyama, J., Miller, M.J., Noakes, D.L.G. and Tsukamoto, K. (2008b) Inshore migration and otolith microstructure/microchemistry of anguillid glass eels recruited to Iceland. *Environmental Biology of Fishes* 83, 309–325.

Kuroki, M., Aoyama, J., Miller MJ., Yoshinaga, T., Shinoda, S., Hagihara, S. and Tsukamoto, K. (2009) Sympatric spawning of *Anguilla marmorata* and *Anguilla japonica* in the western North Pacific Ocean. *Journal of Fish Biology* 74, 1853–1865.

Kuroki, M., Fukuda, N., Yamada, Y., Okamura, A., Aoyama, J. and Tsukamoto, K. (2010) Morphological changes and otolith growth during metamorphosis of Japanese eel leptocephali in captivity. *Coastal Marine Science* 34, 31–38.

LaBar, G.W. and Facey, D.E. (1983) Local movements and inshore population sizes of American eels in Lake Champlain. *Vermont Transactions of the American Fisheries Society* 11, 111–116.

LaBar, G.W., Hernando, C. and Delgado, D.F. (1987) Local movements and population size of European eels, *Anguilla anguilla*, in a small lake in southwestern Spain. *Environmental Biology of Fishes* 19, 111–117.

Lammens, E.H. and Visser, J.T. (1989) Variability of mouth width in European eel., Anguilla anguilla., in relation to varying feeding conditions in three Dutch lakes. *Environmental Biology of Fishes* 26 63–75.

Leander, N.J., Shen, K.N., Chen, R.T. and Tzeng, W.N. (2012) Species composition and seasonal occurrence of recruiting glass eels (Anguilla spp.) in the Hsiukuluan River., eastern Taiwan. *Zoological Studies* 51, 59–71.

Lecomte-Finiger, R. (1992) History and age at recruitment of European glass eels (*Anguilla anguilla*) as revealed by otolith microstructure. *Marine Biology* 114, 205–210.

Lopez, J.A., Westneat, M.W. and Hanel, R. (2007) The phylogenctic affinities of the mysterious Anguilliform genera *Coloconger* and *Thalassenchelys* as supported by mtDNA sequences. *Copeia* 959–966.

Manabe, R., Aoyama, J., Watanabe, K., Kawai, M., Miller, M.J. and Tsukamoto, K. (2011) First observations of the oceanic migration of the Japanese eel using pop-up archival transmitting tags. *Marine Ecology Progress Series* 437, 229–240.

Marui, M., Arai, T., Miller, M.J., Jellyman, D.J. and Tsukamoto, K. (2001) Comparison of early life history between New Zealand temperate eels and Pacific tropical eels revealed by otolith microstructure and microchemistry. *Marine Ecology Progress Series* 213, 273–284.

McCleave, J.D. and Kleckner, R.C. (1982) Selective tidal stream transport in the estuarine migration of glass eels of the American eel (*Anguilla rostrata*). *Journal du Conseil, Conseil International pour l' Exploration de la Mer* 40, 262–271.

Miller, M.J. (2009) Ecology of anguilliform leptocephali, remarkable transparent fish larvae of the ocean surface layer. *Aqua-BioScience Monographs* 2(4), 1–94.

Miller, M.J. and Tsukamoto, K. (2004) An introduction to leptocephali, biology and identification. Ocean Research Institute, The University of Tokyo, Tokyo.

Miyagawa, M. (1981) *Zokuji Hyakkō Kigen (Origins of Mundane Affairs., Crafts and Arts)*. Gendai Shichōsha (in Japanese).

Miyazaki, S., Kim, H.Y., Zenimoto, K., Kitagawa, T., Miller, M.J. and Kimura, S. (2011) Stable isotope analysis of two species of anguilliform leptocephali (*Anguilla japonica* and *Ariosoma major*) relative to their feeding depth in the North Equatorial Current region. *Marine Biology* 158, 2555–2564.

Mochioka, N. and Iwamizu, M. (1996) Diet of anguillid larvae, Leptocephali feed selectively on larvacean houses and fecal pellets. *Marine Biology* 125, 447–452.

Nakagawa Hozand, O. (orig. ed.) and Hanasaki, K. (ed.) (1972) *Edo Kaimono Hitori Annai* (*Traveller's Guide to Edo*). Watanabe Shoten (in Japanese).

Nelson, J.S. (2006) *Fishes of the World*, 4th edn. John Wiley and Sons, New York.

Okamura, A., Yamada, Y., Mikawa, N., Horie, N. and Tsukamoto, K. (December 2012) Effect of starvation., body size., and temperature on the onset of metamorphosis in Japanese eel (*Anguilla japonica*). *Canadian Journal of Zoology* 90(12), 1378–1385.

Okamura, A., Yamada, Y., Yokouchi, K., Horie, N., Mikawa, N., Utoh, T., Tanaka, S. and Tsukamoto, K. (2007) A silvering index for the Japanese eel *Anguilla japonica. Environmental Biology of Fishes* 80, 77–89.

Oliveira, K. (1999) Life history characteristics and strategies of the American eel., *Anguilla rostrata. Canadian Journal of Fisheries and Aquatic Sciences* 56, 795–802.

Otake, T. (1996) Fine structure and function of the alimentary canal in leptocephali of the Japanese eel *Anguilla japonica. Fisheries Science* 62, 28–34.

Otake, T. (2003) Metamorphosis. In: Aida, K., Tsukamoto, K. and Yamauchi, K. (eds) *Eel Biology*. Springer, Tokyo, pp. 61–74.

Otake, T., Nogami, K. and Maruyama, K. (1993) Dissolved and particulate organic matter as possible food sources for eel leptocephali. *Marine Ecology Progress Series* 92, 27–34.

Otake, T., Ishii, T., Nakahara, M. and Nakamura, R. (1994) Drastic changes in otolith strontium/calcium ratios in leptocephali and glass eels of Japanese eel *Anguilla japonica. Marine Ecology Progress Series* 112, 189–193.

Otake, T., Inagaki, T., Hasumoto, H., Mochioka, N. and Tsukamoto, K. (1998) Diel vertical distribution of *Anguilla japonica* leptocephali. *Ichthyological Research* 45, 208–211.

Otake, T., Miller, M.J., Inagaki, T., Minagawa, G., Shinoda, A., Kimura, Y., Sasai, S., Oya, M., Tasumi, S., Suzuki, Y., Uchida, M. and Tsukamoto, K. (2006) Evidence for migration of metamorphosing larvae of *Anguilla japonica* in the Kuroshio. *Coastal Marine Science* 30, 453–458.

Ozawa, T., Kakizoe, F., Tabeta 0, Maeda, T. and Yuwaki, Y. (1991) Japanese eel leptocephali from three expeditions in the western North Pacific. *Nippon Suisan Gakkaishi* 57, 1877–1881.

Ozawa, T., Tabeta, O. and Mochioka, N. (1989) Anguillid leptocephali from the western North Pacific east of Luzon, in 1988. *Nippon Suisan Gakkaishi* 55, 627–632.

Palstra, A.P., Heppener, D.F.M., van Ginneken, V.J.T., Szekely, C. and van den Thillart, G. (2007) Swimming performance of silver eels is severely impaired by the swim bladder parasite *Anguillicola crassus. Journal of Experimental Marine Biology and Ecology* 352, 244–256.

Palstra, A.P., van Ginneken, V.J.T., Murk, A.J. and van den Thillart, G. (2006) Are dioxin-like contaminants responsible for the eel (*Anguilla anguilla*) drama? *Naturwissenschaften* 93, 145–148.

Parker, S.J. (1995) Homing ability and home range of yellow-phase American eels in a tidally dominated estuary. *Journal of the Marine Biological Association of the United Kingdom* 75, 127–140.

Parker, S.J. and McCleave, J.D. (1997) Selective tidal stream transport by American eels during homing movements and estuarine migration. *Journal of the Marine Biological Association of the United Kingdom* 77, 871–889.

Pfeiler, E. (1991) Glycosaminoglycan composition of anguilliform and elopiform leptocephali. *Journal of Fish Biology* 38, 533–540.

Pfeiler, E., Toyoda, H., Williams, M.D. and Nieman, R.A. (2002) Identification, structural analysis and function of hyaluronan in developing fish larvae (leptocephali). *Comparative Biochemistry and Physiology B Comparative Biochemistry* 132, 443–451.

Poole, W.R. and Reynolds, J.D. (1996) Growth rate and age at migration of *Anguilla anguilla*. *Journal of Fish Biology* 48, 633–642.

Proman, J.M. and Reynolds, J.D. (2000) Differences in head shape of the European eel, *Anguilla anguilla*. *Fisheries Management and Ecology* 7, 349–354.

Santo, K. (1787) *Tsugen Somagaki* (in Japanese).

Sasai, S., Aoyama, J., Watanabe, S., Kaneko, T., Miller, M.J. and Tsukamoto, K. (2001) Occurrence of migrating silver eels *Anguilla japonica* in the East China Sea. *Marine Ecology Progress Series* 212, 305–310.

Satoh, H., Nakamura, N. and Hibiya, T. (1962) Studies on the sexual maturation of the eel. -I. On the sex differentiation and the maturing process of the sex organs. *Bulletin of the Japanese Society for the Science of Fish* 28, 579–584.

Schmidt, J. (1906) Contributions to the life-history of the eel (*Anguilla vulgaris*, Flem.). *Rapports et Proces-Verbaux des Reunions du Conseil International pour l'Exploration de la Mer* 5, 137–264.

Schmidt, J. (1922) The breeding places of the eel. *Philosophical Transactions of the Royal Society B* 211, 179–208.

Schmidt, J. (1935) *Danish eel investigations during 25 years (1905–1930)*. The Carlsberg Foundation, Copenhagen.

Shinoda, A., Aoyama, J., Miller, M.J., Otake, T., Mochioka, N., Watanabe, S., Minegishi, Y., Kuroki, M., Yoshinaga, T., Yokouchi, K., Fukuda, N., Sudo, R., Hagihara, S., Zenimoto, K., Suzuki, Y., Oya, M., Inagaki, T., Kimura, S., Fukui, A., Lee, T.W. and Tsukamoto, K. (2011) Evaluation of the larval distribution and migration of the Japanese eel in the western North Pacific. *Reviews in Fish Biology and Fisheries* 21, 591–611.

Shinoda, A., Tanaka, H., Kagawa, H., Ohta, H. and Tsukamoto, K. (2004) Otolith microstructural analysis of reared larvae of the Japanese eel *Anguilla japonica*. *Fisheries Science* 70, 339–341.

Shōtōken, C. ed. and Hasegawa, S. ill. (1893) *Edo meisho zue* (*Guide to Famous Sights of Edo*). Hakubunkan (in Japanese).

Strubberg, A. (1913) The metamorphosis of elvers as influenced by outward conditions. Meddelester fra Kommissionen for Havunderogesler, *Serie Fiskeri Copenhagen* 4, 1–11.

Sudo, R. (2011) The initiation mechanisms of spawning migration of the Japanese eel, *Anguilla japonica* Doctoral dissertation, University of Tokyo, Tokyo.

Suzuki, Y. and Otake, T. (2001) Skin lectin and lymphoid tissues in leptocephalus larvae of the Japanese eel *Anguilla japonica*. *Fisheries Science* 66, 636–643.

Tanaka, H., Kagawa, H., Ohta, H., Unuma, T. and Nomura, K. (2003) The first production of glass eel in captivity, fish reproductive physiology facilitates great progress in aquaculture. Fish *Physiology and Biochemistry* 28, 493–497.

Tanaka, S. (1975) Collection of leptocephali of the Japanese eel in waters south of the Okinawa Islands. *Bulletin of the Japanese Society for the Science of Fish* 41, 129–136.

Thibault, I., Dodson, J.J. and Caron, F. (2007) Yellow stage American eel movements determined by microtagging and acoustic telemetry in the St. Jean River watershed, Gaspe (Quebec, Canada). *Journal of Fish Biology* 71, 1095–1112.

Thurow, F. (1957) Beiträge zur Biologie des Fluβaales Anguilla vulgaris Turt Phil Diss Kieler Forde. *Archiv fuer Fischereiwissenschaft* 8, 79–93.

Todd (1980) Size and age of migrating New Zealand freshwater eels (*Anguilla* spp.). *New Zealand Journal of Marine and Freshwater Research* 14, 283–293.

Todd (1981) Timing and periodicity of migrating New Zealand freshwater eels (*Anguilla* spp.). *New Zealand Journal of Marine and Freshwater Research* 15, 225–235.

Tōin, K. (1937) *Shūgaisho* (*Compendium of Fragments*). Koten Hozonkai (in Japanese).

Tomoda, H. and Uematsu, K. (1996) Morphogenesis of the brain in larval and juvenile Japanese eels, *Anguilla japonica. Brain, Behavior and Evolution* 47, 33–41.

Torlitz, M. (1922) Anatomische und entwicklungsgeschichtliche Beiträge zur Artfrage unseres Fluβaales. *Z Fisch* 21, 1–48.

Tsukamoto, K. (1990) Recruitment mechanism of the eel, *Anguilla japonica,* to the Japanese coast. *Journal of Fish Biology* 36, 659–671.

Tsukamoto, K. (1992) Discovery of the spawning area for the Japanese eel. *Nature* 356, 789–791.

Tsukamoto, K. (2006) Spawning of eels near a seamount. Nature 439, 929.

Tsukamoto, K., Aoyama, J. (1998) Evolution of freshwater eels of the genus *Anguilla*, a probable scenario. *Environmental Biology of Fishes* 52, 139–148.

Tsukamoto, K., Nakai, I. and Tesch, F.W. (1998) Do all freshwater eels migrate? Nature 396, 635–636.

Tsukamoto, K., Aoyama, J. and Miller, M.J. (2002) Migration, speciation and the evolution of diadromy in anguillid eels. *Canadian Journal of Fisheries and Aquatic Sciences* 59, 1989–1998.

Tsukamoto, K., Otake, T., Mochioka, N., Lee, T.W., Fricke, H., Inagaki, T., Aoyama, J., Ishikawa, S., Kimura, S., Miller, M.J., Hasumoto, H., Oya, M. and Suzuki, Y. (2003) Seamounts, new moon and eel spawning, The search for the spawning site of the Japanese eel. *Environmental Biology of Fishes* 66, 221–229.

Tsukamoto, K., Yamada, Y., Okamura, A., Tanaka, H., Miller, M.J., Kaneko, T., Horie, N., Utoh, T., Mikawa, N. and Tanaka, S. (2009a) Positive buoyancy in eel leptocephali, An adaptation for life in the ocean surface layer. *Marine Biology* 156, 835–846.

Tsukamoto, K., Miller M,J., Kotake, A., Aoyama, J. and Uchida K, (2009b) The origin of fish migration, The random escape hypothesis. In: Haro, A., Smith, K., Rulifson, R., Moffitt, C., Klauda, R., Dadswell, M., Cuniak, R., Cooper, J., Beal, K. and Avery, T. (eds) Challenges for diadromous fishes in a dynamic global environment. *American Fisheries Society Symposium* 69 Bethesda, MD, pp. 45–61.

Tsukamoto, K., Chow, S., Otake, T., Kurogi, H., Mochioka, N., Miller M,J., Aoyama, J., Kimura, S., Watanabe, S., Yoshinaga, T., Shinoda, A., Kuroki, M., Oya, M., Watanabe, T., Hata, K., Ijiri, S., Kazeto, Y., Nomura, K. and Tanaka H, (2011) Oceanic spawning ecology of freshwater eels in the western North Pacific. *Nature Communications* 2, 179.

Tzeng, W.N. (1990) Relationship between growth rates and age at recruitment of *Anguilla japonica* elvers in a Taiwan estuary as inferred from otolith growth increments. *Marine Biology* 107 75–81.

Tzeng, W.N., Lin, H.R., Wang, C.H. and Xu, S.N. (2000) Differences in size and growth rates of male and female migrating Japanese eels in Pearl River, China. *Journal of Fish Biology* 57, 1245–1253.

Van den Thillart, G., Palstra, A.P. and van Ginneken, V. (2007) Simulated migration of European silver eel, Swim capacity and cost of transport. *Journal of Marine Science and Technology* 15, 1–16.

Van Ginekken, V.J.T. and van den Thillart, G.E.E.J.M. (2000) Eel fat stores are enough to reach the Sargasso. *Nature* 403, 156–157.

Vollestad, L.A. (1992) Geographic variation in age and length at metamorphosis of maturing European eel: environmental effects and phenotypic plasticity. *Journal of Fish Biology* 61, 41–48.

Watanabe, S., Minegishi, Y., Yoshinaga, T., Aoyama, J. and Tsukamoto, K. (2004) A quick method for species identification of the Japanese eel *Anguilla japonica* using real-time PCR, an onboard application for use during sampling surveys. *Marine Biotechnology* 6, 566–574.

Watanabe, S., Aoyama, J., Nishida, M. and Tsukamoto, K. (2005) A molecular genetic evaluation of the taxonomy of eels of the genus *Anguilla* (Pisces, Anguilliformes). *Bulletin of Marine Science* 76, 675–690.

Watanabe, S., Aoyama, J. and Tsukamoto, K. (2009) A new species of freshwater eel *Anguilla luzonensis* (Teleostei, Anguillidae) from Luzon Island of the Philippines. *Fisheries Science* 75, 387–392.

Wenner, C.A. and Musick, J.A. (1974) Fecundity and sex organ observations of American eel, *Anguilla rostrata*, migrating from Chesapeake Bay, Virginia. *Journal of the Fisheries Research Board of Canada* 31, 1387–1391.

Winter, H.V., Jansen, H.M. and Breukelaar, A.W. (2007) Silver eel mortality during downstream migration in the River Meuse, from a population perspective. *ICES Journal of Marine Science* 64, 1444–1449.

Yamada, Y., Zhang, H., Okamura, A., Tanaka, S., Hone, N., Mi-kawa, N., Utoh, T. and Oka, H.P. (2001) Morphological and histological changes in the swim bladder during maturation of the Japanese eel. *Journal of Fish Biology* 58, 804–814.

Yamamoto, K. and Yamauchi, K. (1974) Sexual maturation of Japanese eel and production of eel larvae in the aquarium. *Nature* 251, 220–222.

Yamauchi, K., Nakamura, M., Takahashi, H. and Takano, K. (1976) Cultivation of larvae of Japanese eel. *Nature* 263, 412.

Yokouchi, K., Aoyama, J., Miller, M.J., McCarthy, T.K. and Tsukamoto, K. (2009a) Depth distribution and biological characteristics of the European eel *Anguilla anguilla* in Lough Ennell, Ireland. *Journal of Fish Biology* 74, 857–871.

Yokouchi, K., Aoyama, J. and Tsukamoto, K. (2009b) Why hasn't the Japanese eel declined to the levels of the European and American eels? A possible explanation from the movement and age structure of growth-phase eels. In: Haro, A., Smith, K., Rulifson, R., Moffitt, C., Klauda, R., Dadswell, M., Cunjak, R., Cooper, J., Beal, K. and Avery, T. (eds) Challenges for diadromous fishes in a dynamic global environment. *American Fisheries Society Symposium* 69 Bethesda, MD, pp. 911–913.

Yoshinaga, T.M.J., Miller, M.J., Yokouchi, K., Otake, T., Kimura, S., Aoyama, J., Watanabe, S., Shinoda, A., Oya, M., Miyazaki, S., Zenimoto, K., Sudo, R., Takahashi, T., Ahn, H., Manabe, R., Hagihara, S., Morioka, E.L., Itakura, H., Machida, M., Ban, K., Shiozaki, M., Ai, B. and Tsukamoto, K. (2011) Genetic identification and morphology of naturally spawned eggs of the Japanese eel *Anguilla japonica* collected in the western North Pacific. *Fisheries Science* 77, 983–992.

CHAPTER 7

Growth, maturation, sex-ratio, length and age of European eel (*Anguilla anguilla*) in the Netherlands

Tessa van der Hammen[1], Karen E. van de Wolfshaar[1], Martin de Graaf[2] and Joep de Leeuw[1]

[1]Wageningen Marine Research, IJmuiden, the Netherlands; [2]Narrandera Fisheries Centre, Narrandera, NSW, Australia

Introduction

The EU Regulation for the Recovery of the Eel Stock (EC 1100/2007) has triggered the development of demographic assessment models based on biological information on life history characteristics, such as sex-ratio and growth rates observed in eel stocks. In this chapter we summarize key biological characteristics of the eel stock in the Netherlands based on annual biological market sampling of commercial catches. This information is used as input in current eel stock assessment models. Between 2009 and 2017, 12,480 biological samples from commercial eel catches were collected from different regions in the Netherlands, proportionate to the total commercial catch per region. The sex, length, weight and maturity status were assessed in the laboratory. In addition, from 348 individual eels, otoliths were analysed to assess cumulative inter-annual growth increments following the protocols set by the ICES workshop in Age Reading of European and American Eel 2009 (WKAREA). Length at 50% maturity was 40.4 cm for males and 73.8 cm for females. Sex ratios per length followed a logistic curve from 14% females at body lengths of 20 cm to >95% females at body lengths from 50 cm onwards. Individual back-calculated growth rates were highly variable, females showing higher growth rates than males. The results from the analyses presented here further broaden the use of the biological information for assessments of the European eel stock.

The eel (*Anguilla anguilla*) population in Europe has decreased starkly since the 1980s and has shown no sign of recovery (Dekker, 2003; ICES, 2017). ICES recommended that a recovery plan be developed for the European eel stock and that exploitation and other human activities affecting the fishery or the stock be reduced as close to zero as possible. In response to this advice the EU Regulation for the Recovery of the Eel Stock (EC 1100/2007) was adopted in 2007. In practice, this means that all member states have to present an eel stock assessment to the EU with a 3 year interval. As a consequence, models have been developed to estimate eel biomass and mortality rates (Åström and Dekker, 2007; Bierman et al., 2012; van de Wolfshaar et al., 2014; Westerberg and Wickström, 2015). Most of these models require biological information, such as the relationship between length and weight, the sex-ratio, maturity ogives and growth rates.

The objective of this chapter is to present life history characteristics for eels in Dutch waters, which are used in the Dutch eel stock assessment models to estimate stock indicators required for the reporting to the EU (van de Wolfshaar et al., 2018). In this way, the biological information becomes accessible for others for comparison.

Methods

Samples

All biological data were obtained from market sampling. Market samples were taken directly from fishermen from sorted (2009–2010) and unsorted (2011–2017) commercial retained catches from different areas in the Netherlands (van Keeken et al., 2011). The number of market samples per region was taken proportionate to the catch size of the region. This sampling scheme is representative for the retained catches (Bierman et al., 2012). The catches are not necessarily representative for the national eel population. However, assuming that the catches roughly follow the abundance of eel in Dutch waters, these are currently the best available data to represent the national eel population. From the measurement of 12,480 eels sex, length, weight and maturity were determined. The otoliths from 348 individual eels were analysed to assess inter-annual growth increments. Otolith analysis was done by Wageningen Marine Research (WMR) until 2013 and by the Swedish University of Agricultural Sciences (SLU) since 2014. The minimum landing size is currently 28 cm. However, smaller sizes occurred in the retained catches and sometimes a sample was taken from the total catches, resulting in sampled eel of sizes below 28 cm.

Length–weight

To estimate the length-weight relationship, a linear model (LM) was fitted to the data. The weight (W) was fitted as function of length (L): $log(W) = \beta 0 + \beta 1 \times log(L)$, with weight in grams and length in millimetres.

Maturation at length

To estimate maturity at length (maturity ogive), a binomial generalized linear model (GLM) with a logit link was fitted to the data for males and females separately. The proportion mature (p) at length (L) was modelled: $logit(p) = \beta 0 + \beta 1 \times L$.

Sex-ratio at length

To estimate the sex ratio at length, a binomial GLM with a logit link was fitted to the data. The sex could be assessed only for lengths longer than 20 cm, the determination of the sex was difficult and uncertain for smaller eel. The proportion of females (p) at length (L) was modelled: $logit(p) = \beta 0 + \beta 1 \times L$.

Growth

Individual growth curves were determined using the protocols set by the ICES workshop in Age Reading of European and American Eel 2009 (WKAREA, ICES, 2009). In short, the growth curves were constructed using the relative distances between annual growth rings and scaling these to the total length of the eel (van Keeken et al., 2011), by using the following formula: $L_{incr} = incr \times (L_{total} - 7.5)/\sum(incr)$, with L_{incr} being the growth in a year, *incr* being the distance between growth rings and L_{total} being the total length of the eel. It was assumed that glass eels enter the freshwater system at a length of 7.5 cm (Bierman et al., 2012). Two different sex-specific growth curves were calculated. The first is the cumulative average increment at age. This means that the average increase in length in an age class is added to the average increase in the previous age class. The second approach is the average length at age.

Results

Length–weight relationship

Length–weight relationships differed between males and females. Females were lighter at smaller sizes and heavier at larger sizes compared to males, however, these differences were very small (♀ $\beta 0 = -14.62$, $\beta 0 = 3.23$, $N = 7431$, $p < 0.001$; ♂ $\beta 0 = -12.79$, $\beta 1 = 2.93$, $N = 1587$, $p < 0.001$; ♀ and ♂ $\beta 0 = -14.29$, $\beta 1 = 3.18$, $N = 12461$, $p < 0.001$, Figure 7.1).

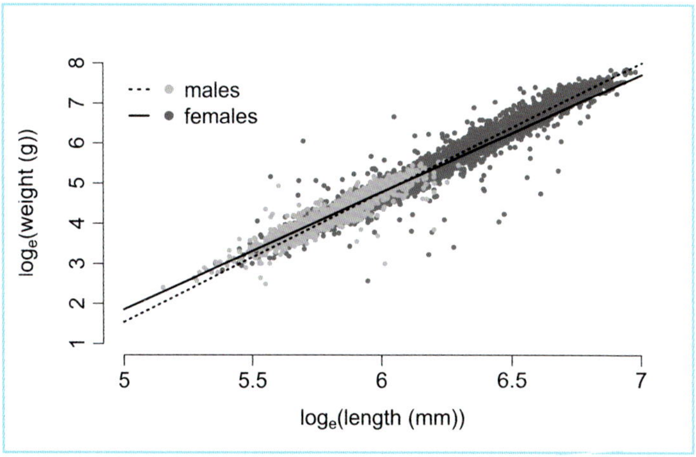

Figure 7.1 Length–weight relationship for males (light grey dots and dotted line) and females (dark grey dots and black line).

Maturation at length

As expected, females matured at larger sizes than males (♀ $\beta 0$ = –8.37, $\beta 0$ = 0.011, *L50%* = 73.8 cm, *N* = 7393, $p < 0.001$; ♂ $\beta 0$ = –13.23, $\beta 1$ = *0.033, L50%* = 40.4 cm, *N* = 1586, $p < 0.001$, Figure 7.2). The length at maturation of the males corresponds well with the mean length at metamorphosis based on 38 locations studied by Vollestad (1992, ♀ 62.3 cm, ♂ 40.6 cm), whereas the length at maturation of the females in our study appear in the higher end of the studies in Vollestad (1992).

Sex-ratio as function of length

With increasing length, the proportion of females increased from 14% females at a size of 20 cm to > 95% females from 50 cm onwards ($\beta 0$ = –5.05, $\beta 1$ = *0.016, L50%*= 31.2 cm, *N* = 9002, $p < 0.001$, Figure 7.3). The maximum length observed in males was 54 cm. At 31.2 cm the sex ratio was 50%–50%.

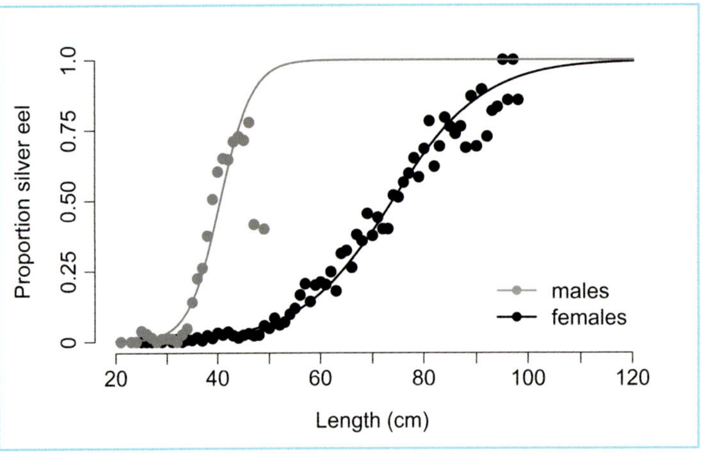

Figure 7.2 Observed (dots, proportion per cm-class) and predicted (lines) proportion of mature eel (silver eel) as a function of length. Notes: Black dots and line: females, grey dots and line: males. Observed proportions of length classes representing <5 individuals are not shown.

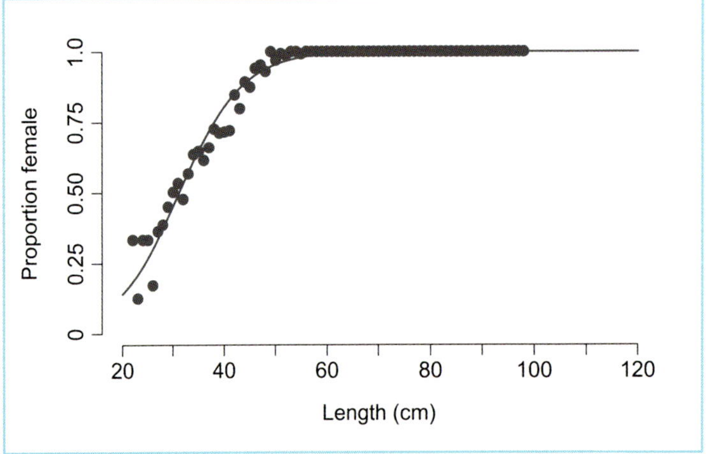

Figure 7.3 Observed (dots, proportion per cm-class) and predicted (line) proportion of females as a function of length. Notes: Observed proportions of length classes representing <5 individuals are not shown.

Growth

Growth rates are sex-specific with females growing faster than males (Figure 7.4a and 7.4b, Table 7.1). For females, the cumulative growth curve was similar as the one based on the average length at age, up to age 9. Thereafter the length-at-age curve is lower but catches up at age 18. For males, the two growth curves differed from age 5 onwards, with the length-at-age curve being substantially lower than the cumulative growth curve at older ages. Because eel migrate to the sea when they become silver eel, the data suggests that the eel transformation to silver eel is based on length and that other conditions determine at what age this length is reached (that is, growth rates depend on external factors such as food conditions, temperature and so on). Growth of individual eels was highly variable. For example, in the first year after transformation from glass eel to yellow eel, the fastest growing female was almost 29 cm, whereas the slowest growing female was 9.4 cm (Figure 7.4b). For males, the range was also very variable, between 10 cm and 27 cm in the first year after transformation (Figure 7.4a).

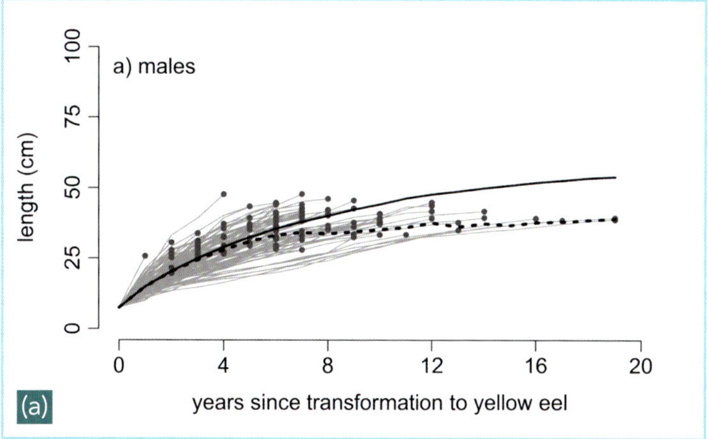

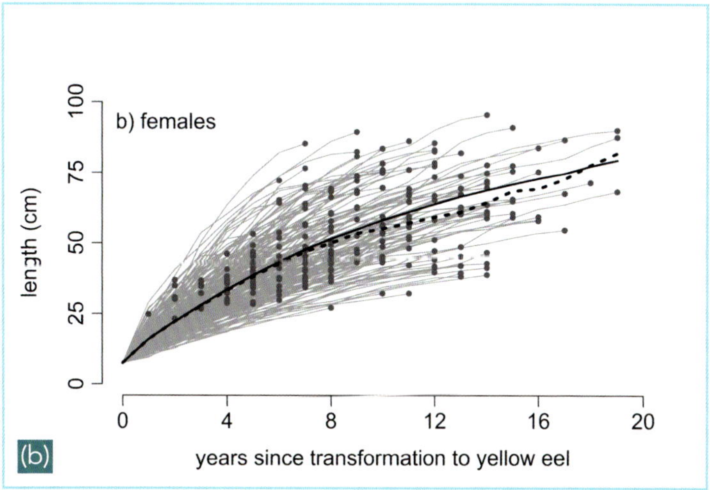

Figure 7.4 Back-calculated growth rates of individual eels (grey lines): (a) males (N = 113); (b) females (N = 234). Notes: The grey dots are the lengths and ages when the eels were caught. Black solid line: the cumulative growth curve. Black dotted line: the average length-at-age curve.

Table 7.1 Estimates for the back calculated sizes (cm) using two different methods (1) cumulative increment and (2) the average length at age.

Age	Cumulative increment		Average length at age	
	Males	Females	Males	Females
0	7.5	7.5	7.5	7.5
1	14.9	15.9	14.9	15.9
2	20.4	22.2	20.3	22.2
3	25.0	28.2	24.5	27.9
4	29.0	33.7	28.0	33.3
5	32.5	38.6	30.9	38.1
6	35.6	43.4	33.2	42.8
7	38.0	47.8	34.2	46.9
8	40.1	51.5	33.8	50.1
9	42.3	54.9	34.4	53.4
10	44.2	58.1	35.2	55.1
11	46.3	61.0	35.9	57.1
12	47.7	63.8	37.6	59.3
13	48.8	66.5	36.2	61.0
14	49.9	68.8	37.4	64.4
15	50.9	70.9	36.7	68.5
16	51.9	72.9	37.7	69.2
17	52.6	74.7	38.0	72.7
18	53.5	77.3	38.6	77.7
19	53.9	79.4	39.1	81.9

Discussion

The data presented here is based on market sampling with samples being taken from the fishermen directly. The market sampling is set up in a way that it is representative for the catches. However, this does not necessarily mean that it is representative for the national population. Bias could occur because the catches were sampled during the fishing season from May to August. Because mature eels (silver eels) migrate in autumn, peak migration months are underrepresented in the sampling. In contrast, over-representation of silver eels could occur because they may have higher catchabilities in passive gear (fyke nets) due to their migratory behaviour. Another reason for bias could be that catches are mostly from fyke nets, which are fully selective for eels from approximately 30 to 35 cm and longer. Thus, slow growing individuals will appear in the landings at older ages than fast growing individuals. In addition, small eels are underrepresented in the data because of the minimum landing size of 28 cm.

There are many European studies on biological parameters (Bevacqua et al., 2006; Boulenger et al., 2015; De Leo and Gatto, 1995; Daverat et al., 2012; Holmgren et al., 1997; Melià et al., 2006; Poole and Reynolds, 1996; Russell Poole and Reynolds, 1996; Silm et al., 2017; Simon, 2007, 2015; Vollestad, 1992). The results from the analyses presented here can be compared with already published work on biological parameters on which analyses on differences in the biological parameters could be carried out.

The biological information presented here fits within the trend of national governments to comply with the European legislation for national and international eel management. Our findings might help those countries where biological information is scarce, add to our growing knowledge of the biology of eel and contribute to a further understanding of the processes governing eel dynamics.

Acknowledgements

The research was financed by the Ministry of Agriculture, Nature and Food Quality as part of the WOT (Wettelijke Onderzoekstaken) programme.

Bibliography

Åström, M. and Dekker, W. (2007) When will the eel recover? A full life-cycle model. *ICES Journal of Marine Science* 64, 1491–1498.

Bevacqua, D., Melià, P., Crivelli, A.J., De Leo, G.A. and Gatto, M. (2006) Timing and rate of sexual maturation of European eel in brackish and freshwater environments. *Journal of Fish Biology* 69, 200–208.

Bierman, S.M., Tien, N., van de Wolfshaar, K.E., Winter, H.V. and de Graaf, M. (2012) Evaluation of the Dutch Eel Management Plan 2009–2011. ICES Document Report number C067/12.

Boulenger, C., Acou, A., Trancart, T., Crivelli, A.J. and Feunteun, E. (2015) Length-weight relationships of the silver European eel, *Anguilla anguilla* (Linnaeus, 1758), across its geographic range. *Journal of Applied Ichthyology*, 31, 427–430.

Daverat, F., Beaulaton, L., Poole, R., Lambert, P., Wickström, H., Andersson, J., Aprahamian, M., et al. (2012) One century of eel growth: changes and implications. *Ecology of Freshwater Fish* 21, 325–336.

De Leo, G.A. and Gatto, M. (1995) A size and age-structured model of the European eel (*Anguilla anguilla* L.). *Canadian Journal of Fisheries and Aquatic Sciences* 52, 1351–1367.

Dekker, W. (2003) Did lack of spawners cause the collapse of the European eel, *Anguilla anguilla*? *Fisheries Management and Ecology* 10, 365–376.

Holmgren, K., Wickström, H. and Clevestam, P. (1997) Sex-related growth of European eel, *Anguilla anguilla*, with focus on median silver eel age. *Canadian Journal of Fisheries and Aquatic Sciences* 54, 2775–2781.

ICES (2009) Workshop on Age Reading of European and American Eel (WKAREA), 20–24 April 2009, Bordeaux, France. ICES CM 2009\ACOM: 48.

ICES (2017) ICES Advice on fishing opportunities, catch, and effort. Ecoregions in the Northeast Atlantic. ele.2737.nea.

Melià, P., Bevacqua, D., Crivelli, A.J., de Leo, G.A., Panfili, J. and Gatto, M. (2006) Age and growth of *Anguilla anguilla* in the Camargue lagoons. *Journal of Fish Biology* 68, 876–890.

Poole, W.R. and Reynolds, J.D. (1996) Age and growth of yellow eel, *Anguilla anguilla* (L.), determined by two different methods. *Ecology of Freshwater Fish* 5, 86–95.

Russell Poole, W. and Reynolds, J.D. (1996) Growth rate and age at migration of *Anguilla anguilla*. *Journal of Fish Biology*, 48, 633–642.

Silm, M., Bernotas, P., Haldna, M., Järvalt, A. and Nõges, T. (2017) Age and growth of European eel, *Anguilla Anguilla* (Linnaeus, 1758), in Estonian lakes. *Journal of Applied Ichthyology* 33, 236–241.

Simon, J. (2007) Age, growth, and condition of European eel (*Anguilla anguilla*) from six lakes in the River Havel system (Germany). *ICES Journal of Marine Science* 64, 1414–1422.

Simon, J. (2015) Age and growth of European eels (*Anguilla anguilla*) in the Elbe River system in Germany. *Fisheries Research* 164, 278–285.

Van de Wolfshaar, K.E., Griffioen, A.B., Winter, H.V., Tien, N.S.H., Gerla, D., van Keeken, O. and van der Hammen, T. (2018) Evaluation of the Dutch Eel Management Plan 2018: Status of the eel population in 2005–2016. ICES Document CVO report: 18.009.

Van de Wolfshaar, K.E., Tien, N., Winter, H.V., de Graaf, M. and Bierman, S.M. (2014) A spatial assessment model for European eel (*Anguilla anguilla*) in a delta, the Netherlands. *Knowledge and Management of Aquatic Ecosystems* art. no. 130064, doi: 10.1051/kmae/2013083.

Van Keeken, O.A., Bierman, S.M., Wiegerinck, H., Goudswaard, K. and Kuijs, E. (2011) Proefproject marktbemonstering aal Voortgang 2010. IMARES Report C053/11.

Vollestad, L.A. (1992) Geographic variation in age and length at metamorphosis of maturing European eel: environmental effects and phenotypic plasticity. *Journal of Animal Ecology* 61, 41–48.

Westerberg, H. and Wickström, H. (2015) Stock assessment of eels in the Baltic: reconciling survey estimates to achieve quantitative analysis. *ICES Journal of Marine Science* 73, 75–83.

CHAPTER 8

Does a bigger glass eel mean better recruitment?

Cédric Briand[1], Carlos Fernández-Delgado[2], Lluís Zamora Hernández[3], Fernando Jiménez[4], Derek Evans[5] and Estibaliz Díaz[6]

[1]EPTB Vilaine, La Roche-Bernard, France; [2]Grupo de Investigación 'Aphanius' Departamento de Zoología, Universidad de Córdoba, Córdoba, Spain; [3]Universitat de Girona, Institut d'Ecologia Aquàtica, Girona, Spain; [4]Centro de Experimentación Pesquera, Escuela Náutico-Pesquera, Gijón, Spain; [5]Fisheries and Aquatic Ecosystems Branch, Agriculture Food and Environmental Science Division, Agri-Food and Biosciences Institute, Belfast, UK; [6]AZTI-Tecnalia Txatxarramendi ugartea z/g, Sukarrieta, Spain

Introduction

The length and condition of European eel (*Anguilla anguilla*, Anguillidae) glass eel spanning data for 20 years from 1997 to 2018 were examined from nine European locations, which included Northern Ireland, the Bay of Biscay, an estuary close to the Gibraltar Straight (the Guadalquivir) and the Mediterranean. Morphometric data were analysed by mixed models including both long-term and cyclic intra-seasonal trends. The long-term length trend extracted from these models helped to explain the recorded short-term variation in recruitment. The minimum in 2010 and largest recruitment values in 2014 were well explained by glass eel length, which in turn suggested that oceanic processes had a strong influence. However, over the long term, glass eel length has diminished despite returning several times to historical values, and so failed to explain the decline in recruitment. The long-term trend in body condition was less clear, but the intra-seasonal trend showed a marked difference between Atlantic and Mediterranean locations. This suggests a single wave of glass eel per season in Northern Ireland and the Bay of Biscay with a more complex recruitment process, including a secondary wave of migration, in the springtime, in the Guadalquivir and the Mediterranean.

Glass eel recruitment has decreased progressively since the 1980s (Dekker, 2004). Since 1998, the International Council for the Exploration of the Sea (ICES) has repeatedly stated that *A. anguilla* stock is 'outside of safe biological limits and lead the IUCN to classify the eel as critically endangered (Jacoby and Gollock, 2016). According to the last assessment (ICES, 2018), the status of *A. anguilla* remains critical: the inter-seasonal glass eel recruitment to European waters in 2018 was at 2.1% of the 1960–1979 level in the 'North Sea' series, and 10.1% in the 'Elsewhere Europe' series.

The causes for this decline in recruitment are not clearly identified or understood. Different anthropogenic factors impacting eel populations in continental waters have been proposed as probable causes of decline and include overfishing, habitat loss, dam and weir construction, hydropower plants, pumping stations and surface water abstractions. In addition, many studies have shown that oceanic and climatic factors are also associated with recruitment variability. Some authors have postulated that the ocean climate parameters (ocean circulation patterns, temperature or food availability) are directly related to the recruitment decline (Bonhommeau et al., 2009; Friedland et al., 2007; Knights, 2003). Westerberg et al. (2018) found that larval abundance in the Sargasso Sea seems to have decreased less than glass eel indices recorded for Europe. According to them, there might have been an increased leptocephalus mortality during their oceanic phase or a large geographic shift in glass eel arrival.

Mature silver eel from American (*A. rostrata*) and *A. anguilla* stocks reproduce in the Sargasso Sea in an area located 70–67°W in a narrow band of latitude at about 25°N, which corresponds to the temperature fronts in the South Tropical Convergence Zone (Miller, 2009). The eggs hatch into leptocephali larvae and are transported by oceanic currents from the Sargasso Sea to the American and European coasts. Hence, oceanographic conditions determine eel larvae transport and feeding (Miller, 2009), thus influencing their successful oceanographic migration and recruitment. Consequently, in those years with unsuitable oceanographic feeding conditions, larval mortality would increase. However, growth is dependent on food supply so even if the eel larvae survive, they would likely produce smaller (both in weight and length) glass eel with associated lower energy reserves. As such, our first hypothesis is that there may be a link between glass eel length, condition and subsequent recruitment at the glass eel stage. To test this hypothesis, we compared glass eel length and condition from nine locations in the Atlantic and the Mediterranean against the recruitment index from the respective areas.

Quantification of the effects of oceanographic factors on glass eel mortality and subsequent recruitment is hampered by leptocephali being located far out to sea. There is no consensus on the duration of larval migration, proposed periods include 1 year (Wang and Tzeng, 2000), 2 years (Kettle and Haines, 2006) and 3 years (Tesch, 2007). There is also uncertainty regarding the migration route leptocephali, and later glass eels, take to reach European and north African estuaries. Recent transport modelling studies have shown that *A. anguilla* larvae are transported through the Gulf Stream system and then follow the Azores Current or the North Atlantic Current (NAC) (Bonhommeau et al., 2009; Kettle and Haines 2006; Pacariz et al., 2014). However, another eastward route has also been proposed (Miller et al., 2015; Munk et al., 2010), whereby larvae on the eastern side of the spawning area might be drawn into eastwards-flowing frontal jets and transported to the northeast. Examination of eastwards-flowing frontal currents in the areas of leptocephalus distribution indicates that these larvae might be advected from the spawning area directly eastwards (Munk et al., 2010). According to Miller et al. (2015), it is not clear whether these eastward-flowing currents are present in conjunction with the thermal fronts serving as cues to eel spawning, or that they are sufficiently sustained to permit the drift of leptocephali to the Azores, from where a combination of oriented migration and currents would help leptocephali reach the coasts. As outlined above, the length and condition of both larvae and glass eel would mainly be the result of their surrounding oceanographic conditions influencing feeding and development during larval phase and subsequent oceanic migration. Furthermore, it has been shown that larvae using shorter routes produce smaller glass eels (Laflamme et al., 2012). Thus, larvae from a given spawning class using different migration routes could produce glass eels with different length and condition. Therefore, our second hypothesis is that there should be different trends in glass eel length or condition across locations depending on the migration route of leptocephali in conjunction with different recruitment waves of glass eels in the Mediterranean (originating from different cohorts or from different routes). To test this hypothesis, we collated a large dataset comprising data on glass eel length and weight from nine different European locations. We built models that integrated intra-seasonal and long-term trends, while comparing the response from the different locations against the recruitment trend. As weight and length are correlated, we also examined the length and associated condition factor of glass eels.

By examining the long-term dataset collected on glass eel metrics we hope to gain a better understanding of the ocean-related mechanisms influencing European glass eel recruitment. Such knowledge would help to fill some of the knowledge gaps that exist regarding leptocephali migration routes and in turn help to support the management of the wider stock by improving the assessment and adoption of management measures.

Material and methods

Data collection

Glass eel morphometrics have been collected from nine different locations in Europe ranging from Northern Ireland (NI from now on) to the Mediterranean (Figure 8.1, Table 8.1). The study includes three different coastline points in the Atlantic: (Strangford, Carlingford and Lagan) in NI that have been merged for statistical analysis; three estuaries in the Bay of Biscay (the Vilaine, Oria, and Nalón); one estuary close to the Gibraltar Strait (Guadalquivir) and one in the Mediterranean (Ter). Although the Guadalquivir River flows in to the Atlantic, in the following analyses and discussion it will be classified as 'Mediterranean' as it is located in the Mediterranean biogeographic area. Larvae arriving to the Mediterranean should follow the same route as those arriving to the Guadalquivir.

The sampling period varies across the nine locations, ranging from a series including 21 years in la Vilaine to 4 years in the Ter (Table 8.1). Similarly, in some of the locations the sampling was carried out all year round (Vilaine, Guadalquivir and Ter); while in others, the sampling effort was focused on the main recruitment period within a specific year (NI, Oria and Nalón).

Glass eel were measured to the nearest millimetre across the range of preservation methods: alive (Vilaine, Oria and Nalón), frozen (Ter and Guadalquivir), after preservation in 70% ethanol (NI) and formalin (Guadalquivir) (Table 8.1). In all estuaries (except for the three NI sites), glass eels were assessed for pigmentation stage and only lower stages (≤VIA3) (Elie et al., 1982) have been selected for length analysis to discount for stage-related differences. In the case of the Vilaine, some samples (2000–2001) have not been analysed for pigmentation stages. For weight analysis, only stages (<VIA2) have been used. Previous work has indicated an effect of feeding resumption on growth, starting at stage VIA3 (Lecomte-Finiger, 1983). However, a small proportion of glass eel may change their isotopic signature as early as VIA1, which would indicate an earlier feeding behaviour in some individuals (Bardonnet and Riera, 2005). Given this, selecting glass eel up to

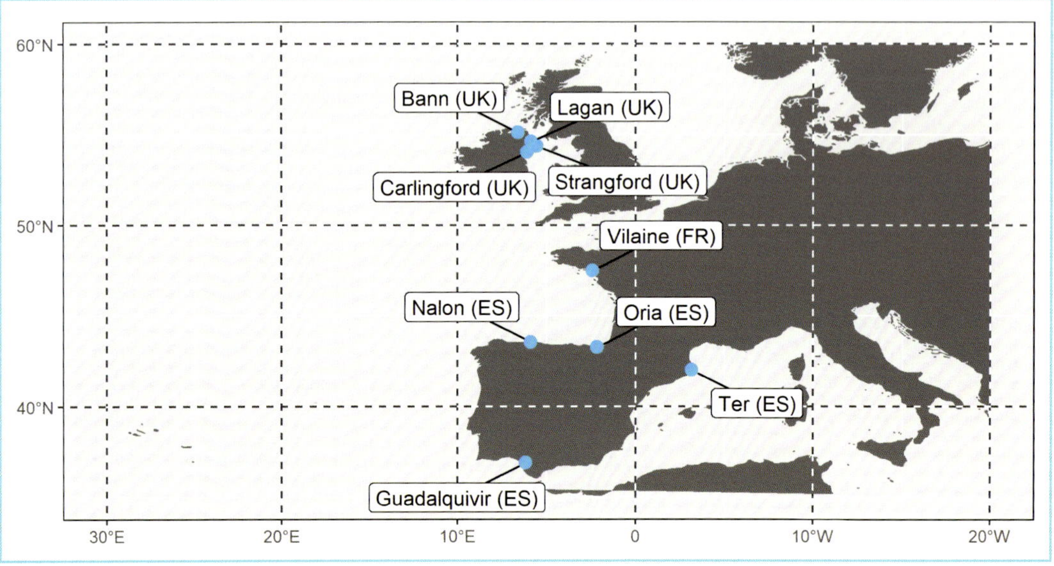

Figure 8.1 Location of glass eel collection points.

stage VIA2 for length and VIA1 for weight removes the risk of observing any growth resumption. In all locations, glass eels have been blotted on filter paper before weight measure.

Statistical analysis

Hypothesis 1

To analyse if recruitment is related to glass eel length and condition we first estimated inter-seasonal glass eel length per location and year. Glass eel length varies among years and locations, and throughout the season (Dekker, 1998; Désaunay and Guérault, 1997). In order to remove the bias linked with the intra-seasonal variation within one site, it is necessary to extract this information from a model integrating intra-seasonal trends. As a first step in the analysis of length and condition among locations, we used unpaired *t*-test to compare means, and *F* test to compare the variances of samples from normal populations. Then, to determine intra-seasonal trends in glass eel length and condition from each of the locations we used generalized additive models (GAM). We then checked if adding length or condition to a model improved the explanation of recruitment variability. To provide values consistent with previous analyses (Dekker, 1998), we performed a simple correlation between the obtained length and yearly recruitment for different periods of time.

Hypothesis 2

To test our second hypothesis, we compared length and weight of glass eel from the nine different sites. As above we compared means via a *t*-test, then used the same GAM models to compare intra-seasonal trends in length and condition across the different locations.

All analyses were performed using R (R Core Team, 2017) and associated packages ggplot2 (Wickham, 2009), mgcv (Wood, 2011), visreg (Breheny and Burchett, 2017), stacomiR

Table 8.1 Description of the glass eel sampling in each of the locations.

	Northern Ireland	Vilaine	Oria	Nalón	Guadalquivir	Ter
Sampling period and season	2003–2012 and 2015–2016, January–August	1997–1998 and 2000–2018, all year round	2006–2008 and 2018, October–February 2010–2012, November–January	2012–2016 and 2017–2018, November–April	1996–1999 and 2012–2015, all year round	2013–2016, all year round
Sampling method	1 m diameter drag nets deployed along the banks	Trapping ladder on the Arzal Dam, tidal limit (12 km to the sea) and fishery immediately below	Experimental fishery (sieve trawling)	2012–2016: Experimental fishery (sieve trawling) 2017–2018: from fishery catches	Experimental fishery (passive haul in an anchored boat)	Experimental fishery (fyke net)
Sampling location	Three coastline locations (Stanford, Carlingford, Lagan)	Trapping ladder and tidal limit (12 km to the sea)	Four points from the river mouth to the tidal limit (12 km)	River mouth	River mouth	River mouth
Glass eel status for measurement	70% ethanol	Alive	Alive	Alive	1997–1999: formalin 2003–2005: frozen	Frozen
Pigment stage	No	Yes	Yes	Yes	Yes	Yes
Number of glass eel for length modelling	2397	7583	5160	1291	29655	3962
Number of glass eel for condition modelling	0	2550	4791	1081	16206	3552
More detail about sampling		Briand et al., 2003	Aranburu et al., 2014		Sobrino et al., 2005	

(Briand et al., 2018) and xtable (Dahl, 2016). GAM models were checked for normality and homoscedasticity of residuals according to Zuur et al. (2009).

Inter-seasonal and intra-seasonal trends in glass eel length

We modelled glass eel length, *L*, variation using GAM, which fit a non-linear response of the dependent variable, in our case individual length of glass eels, against predictor variables (Wood, 2011). The inter-seasonal length will be that obtained by the model showing the best adjustment (the best model).

We started with a simple model that enabled splitting the variations into an inter-seasonal component (1), and an intra-seasonal component (2).

1 The inter-seasonal component was modelled using a *time* variable which was set to one at the beginning of our sampling (1-1-1997) and represents the number of days from that origin. We used a GAM smoother to model the variation of length over time. This choice enabled the introduction of length variation within the same year.
2 The intra-seasonal component was modelled against Julian days. The Julian date, *d*, was modelled with a GAM by a cyclic cubic spline, *cc*, allowing the intra-seasonal value to have a cyclic pattern, that is, returning to the value of $d = 0$ at $d = 365$.

Both variations were included in the model and tested for significance (Model 1).

$$\text{Model 1: } L \sim cc(d) + cr(time) + \varepsilon$$

where cc = cyclic cubic spline, d = Julian day, cr = cubic regression spline, $time$ = inter-seasonal component, ε = residuals.

In the second model, we introduced variation for one site. This variable actually provides a different response in the Guadalquivir according to the preservation method with a possible shrinking of formalin preserved samples. It also provides a different response between estuarine samples and those collected at the trapping ladder in the Vilaine as we know those are more advanced in stages (Briand, 2009).

$$\text{Model 2: } L \sim cc(d) + cr(time) + loc + \varepsilon$$

In the third model, we addressed the question of the possibility of a difference in intra-seasonal trends between locations. Different *intra_seasonal* trends were used for the NI estuaries, Vilaine, Oria, Nalón, Guadalquivir and Ter.

$$\text{Model 3: } L \sim cc(d)_{intra\_seasonal} + cr(time) + loc + \varepsilon$$

In the fourth model, we introduced a possibility of a different *intra-seasonal* trend from year to year (two-dimensional response using tensor product interaction).

$$\text{Model 4: } L \sim cc(d)_{estu} + cr(time) + ti(cc(d), cr(time))_{estu} + loc + \varepsilon$$

These four models provide the best likelihood of identifying any trend in the intra-seasonal component of the model. Désaunay and Guérault (1997) have provided a model using a simpler harmonic curve to describe the variation in glass eel length in the Vilaine from 1991 to 1993.

$$L = \gamma cos\left[\frac{2\pi}{365}(d - \phi)\right] + L_0$$

with $\phi = 319$, $L_0 = 6.981cm$, $\gamma = 0.111$ (Désaunay and Guérault, 1997)

We used a slight variation from this model (Model 5) as the cos can be expressed as a *cos* + *sin* with a phase shift. We also added the *time* trend and *loc* giving:

$$\text{Model 5: } L \sim \alpha sin\left(d\frac{2\pi}{365}\right)_{estu} + \beta cos\left(d\frac{2\pi}{365}\right)_{estu} + s(time) + loc + \varepsilon$$

To compare with historical results, we extracted range, γ, day of maximum length, ϕ, and intercepts, L_0, from Model 5.

To account for the structured nature of the data in samples, we used a GAMm (mixed model). The mixed model takes account of the structure per sample: glass eel coming from the same sample are grouped together in the random part of the model. The model selected (Model 6) is similar in structure to Model 3.

$$\text{Model 6: } L \sim cc(d)_{estu_{ij}} + cr(time)_{ij} + loc_{ij} + a_j + \varepsilon_{ij} \qquad a_j \sim N\left(0, \sigma^2\right)$$

$$i = glasseel,\ j = sample$$

We used the best model and predictions in the Vilaine to extract inter-seasonal average values.

Inter-seasonal and intra-seasonal trends in glass eel condition

As weight and length are correlated, we chose to model the condition index using a condition coefficient adapted specifically to eel body structure (Froese, 2006).

$$Condition = 100\frac{W}{aS^b}$$

where W = weight (g), S = length (cm).

The coefficients a (0.0012) and b (2.85) were extracted from the whole dataset by calibrating the model:

$$\log(W) = a' + b\ log(L)\ a = \exp(a')$$

To compare similarity between modelling protocols we then tested in turn GAM and GAMm models. GAMm models always produced a lower AIC (Akaike criteria) than their GAM counterparts. The variance of 'Mediterranean' glass eel condition was larger than that of glass eels from the Bay of Biscay. Hence separate models have been fitted for the Mediterranean (Guadalquivir and Ter) $fulton_{Med}$ and the Bay of Biscay $fulton_{Bisc}$ (Nalón, Oria and the Vilaine). As with the length models previously, we used intra-seasonal, loc and time trends as explanatory variables.

Model 7: $condition \sim cc(d)_{estu_{ij}} + a_{ij} \qquad a_j \sim N\left(0, \sigma^2\right)$

$$i = glasseel,\ j = sample$$

Model 8: $condition \sim cc(d)_{estu_{ij}} + loc_{ij} + a_{ij} \qquad a_j \sim N\left(0, \sigma^2\right)$

$$i = glasseel,\ j = sample$$

Model 9: $condition \sim cc(d)_{estu_{ij}} + cr(time)_{ij} + loc_{ij} + a_{ij} \qquad a_j \sim N\left(0, \sigma^2\right)$

$$i = glasseel,\ j = sample$$

Given that the glass eel weight is quite sensitive to formalin preservation we excluded some of the glass eels from the Guadalquivir. As a result, we did not have to include a *loc* variable in the Mediterranean and as such only compared these against Model 7 and Model 9. In the Atlantic, *loc* corresponds to a difference response between estuarine and trap samples in the Vilaine.

Relation of recruitment and length

We extracted inter-seasonal ($\widehat{length}$) value arbitrarily on 1 January from the best model from the inter-seasonal and intra-seasonal trends in glass eel length section as a proxy for inter-seasonal length. As a recruitment proxy we chose the Elsewhere Europe (EE) recruitment index developed by the Working Group on Eels (WGEEL) of ICES (ICES, 2017) as it is based on 25 recruitment series located in the same area as this study (the UK, the Biscay Gulf and the Mediterranean Sea). We used two different approaches to analyse the relationship between recruitment and length. First, we estimated a Pearson coefficient to determine the correlation between length and EE recruitment series over both a short (2001–2017) and a longer period (1998–2017). Second, we checked if adding the obtained length to a model improved the explanation of EE recruitment series variability. As the fit for long-term trends in length was common to all estuaries, the Vilaine was selected to predict this trend. Thus, we used the Vilaine trend in length as a covariate to explain the trend in recruitment using a low degree smoother (k = 3) and the length of glass eel:

Model 10: $recEE \sim s(year, k = 3) + (\widehat{length})$

This model contains the simpler model:

Model 11: $recEE \sim s(year, k = 3)$

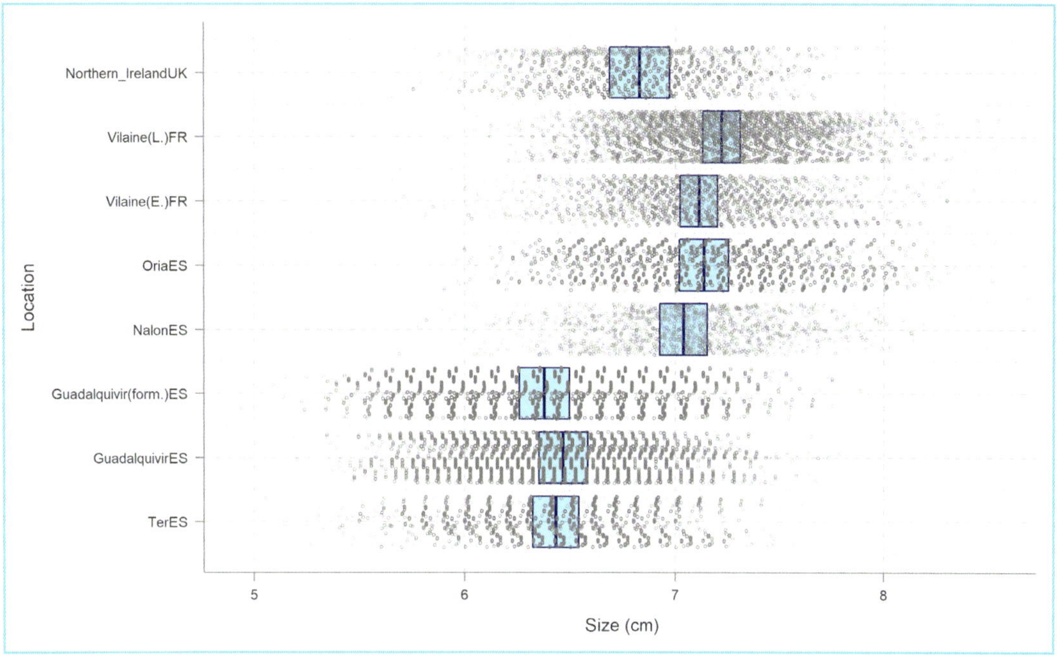

Figure 8.2 Glass eel length comparison between locations using GAM Model 6. Notes: The samples from the Vilaine are divided into glass eel collected at the trapping ladder or from the estuary while those from the Guadalquivir are divided into formalin preserved and fresh analysed glass eels. See Table 8.1 for sample number and time period detail per location.

We tested the two imbricated models for significance using ANOVA to check if adding glass eel length improved the model prediction for recruitment. This was repeated for all time periods assessed.

Results

Glass eel length differed significantly between locations (Figure 8.2). Mean glass eel length was larger in the Bay of Biscay 7.07 cm (SD = 0.46) than in the Mediterranean 6.39 cm (SD = 0.38) (Guadalquivir and Ter) (t-test, $p < 0.0001$) or at the Northern location (NI) 6.73cm (SD = 0.35) (t-test, $p < 0.0001$). Length variance was not significantly different between Mediterranean samples and others ($p > 0.1$) but the condition variance was different ($p < 0.001$).

Spatio-temporal trends in glass eel length

The structural components of the length models are summarized in Table 8.2. The best model fit according to the AIC was the GAMm model (Model 6) inclusive of the difference across locations for Guadalquivir and the Vilaine, an intra-seasonal trend per site, and a long-time trend common to all locations (Table 8.3). A similar conclusion would have been reached if we had

Table 8.2 Summary of the structural components of the models of glass eel length.

Model	Inter.	cr (time)	cc (d)	loc	cc (d)^{estu}	ti (cc(d) cr(time))^{estu}	sin^{estu}
6*	6.42	√		√	√		
4	6.37	√		√	√	√	
3	6.41	√		√	√		
5	6.46	√		√			√
2	6.42	√	√	√			
1	6.61	√	√				

Notes: √ indicates a component included in the model, * indicates a mixed model with a structure per sample in the random part. The best model (Model 6) is a mixed model with a random structure per sample). Inter: intercept, *cr*: cubic regression spline, *time*: inter-seasonal component, loc: location (estuary or ladder for the Vilaine), or preservation method (formalin preserved or not) for the Guadalquivir, *cc*: cyclic cubic spline, *d*: Julian day, ti: tensor interaction, that is, two-dimensional response on top of main effects, sin: harmonic response. The subscript$_{estu}$ implies a model with a different response per estuary.

Table 8.3 Results of the model of intra-seasonal trends of glass eel length (Models 1–6) sorted by AIC.

Model	df	LogLik	AICc	Delta
6*		−19152.91	38342	0
4	113	−19410.53	39048	706
3	49	−20032.14	40162	1820
5	29	−20437.55	40935	2593
2	25	−22266.22	44584	6242
1	18	−25598.43	51235	12893

Notes: *Model 6 is a mixed model with a random structure per sample. LogLik = log- likelihood, AICc = Akaike criteria, Delta = difference in AIC between the best model and the current model.

tested only mixed models with samples in their random part instead of GAM Models 1 to 3. A sequential testing of models with ANOVA for Models 1 to 4 showed a significant increase in the model fit (Table 8.3).

The following are the results for the period 1998–2018.

- There is both an intra-seasonal trend (Figure 8.3) and a long-time trend (Figure 8.4) in glass eel length.
- The intra-seasonal trend in length is different between locations (Figure 8.3). In the south of the Bay of Biscay, glass eel length decreases in the springtime, with this shifting to a later period in the north of the Bay of Biscay. In the Vilaine, where the trapping ladder is at a dam under strong tidal influence, and the local abundance of glass eel enables the collection of glass eel year-round, the minimum length is reached in July (Figure 8.3). In the Guadalquivir, the noted decrease in length and condition stops in April before beginning to increase again, while the decreasing trend seems to be most pronounced in the glass eels from the Ter.

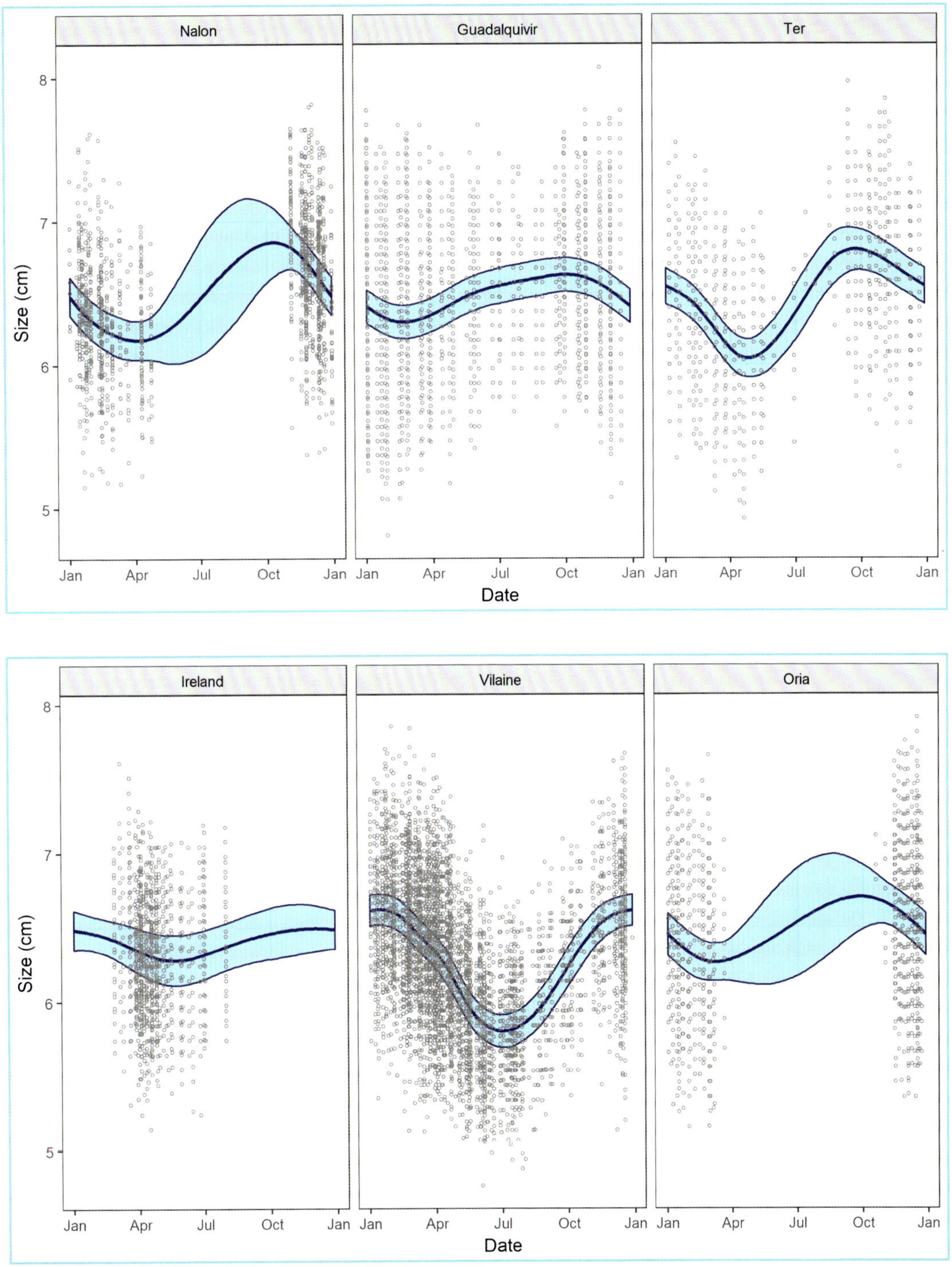

Figure 8.3a-b GAM Model 6 prediction curve of intra-seasonal length by location. Notes: Points indicate individual glass eel length. See Table 8.1 for sample number and time period detail per location.

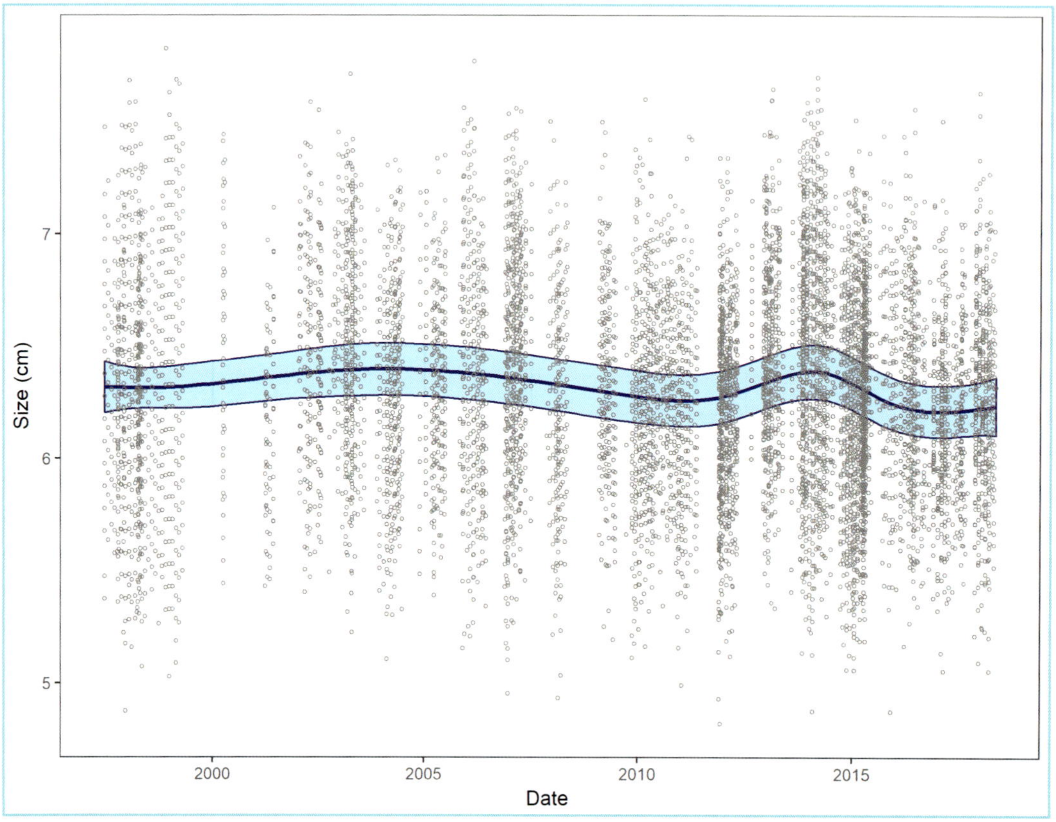

Figure 8.4 GAM Model 6 prediction of the long-term trend in glass eel length from the sampled estuaries. Notes: Points indicate individual glass eel length. See Table 8.1 for sample number and period detail per location.

- Glass eel sampled from the NI estuaries only exhibited an intra-seasonal trend which was less pronounced than that from the other sites. However, it should be noted that the sampling season was short, and corresponded to different sites.
- When using Model 6 there is an indication of a long-term trend in glass eel length over time we found the lowest values in seasons 2015–2016 (6.81 cm), 2016–2017 (6.80 cm) and 2010–2011 (6.84 cm), and the highest values in 2003–2004 and 2017–2018 (6.98 cm).
- Glass eel collected in the Guadalquivir, from 1996 to 1999 treated with formalin 6.45 (0.38) cm were shorter than more recent samples 6.38 (0.36) cm (*t*-test, $p < 0.0001$). The estimated shrinkage extracted from raw data is 7 mm. When using the coefficients from the model, which also take account of the effect of a long-term trend, a reduction of 9 mm is estimated. This corresponds to a 1.4 % lower length.
- Glass eel collected from the trapping ladder in the Vilaine have a longer length 6.93 (0.47) cm and lower condition, 0.88 (0.11), than their estuarine counterparts, 7.17 (0.43) cm and 0.95 (0.12), respectively.
- The observed better fit in Model 4 is likely due to an interaction between year and season. However, when tested as a mixed model, there was no convergence, most likely as a

Table 8.4 Comparison of the range (γ), intercepts (s_0) and day of maximum length (ϕ), obtained from Model 5 in different locations with those of Désaunay and Guérault (1997) (Vilaine 1991–1993 [DG]) and in the present study.

Location	γ	s_0 (cm)	ϕ
Vilaine 1991–1993[DG]	0.111	6.981	319
Northern Ireland UK *	0.270	6.934	361
FR	0.430	6.942	311
FR	0.430	7.046	311
Oria ES	0.290	7.130	332
Nalón ES	0.360	7.163	356
ES*	0.190	6.463	68
Guadalquivir ES	0.190	6.454	68
Ter ES	0.400	6.459	10

Note: *formalin (Guadalquivir) or ethanol (NI) preserved glass eel.

consequence that continuous data for both intra-seasonal and inter-seasonal trend were only available for the Vilaine. In the other estuaries, there are too many holes to allow for a continuous model over both dimensions *d* and *time*.

The average length s_0 (6.92 cm) collected in the Vilaine estuary (1996–2018) using our GAM harmonic model (Model 5) is lower than the value of 6.98 cm found in 1991–1993 by Désaunay and Guérault (1997) (Table 8.4). Mean length per season extracted from Model 5 varied from a minimum of 6.78 cm in 2010–2011 to a high of 7.03 cm in 2013–2014. The date corresponding to the maximum length was fitted, $d = 319$ (14 November) in 1991–1993 and $d = 311$ (15 November) for 1996–2018. However, the 1991–1993 variation in glass eel length was low (0.111 cm) when compared to the variation modelled for 1996–2018 in our dataset (0.430 cm). Thus, in Désaunay and Guérault (1997), the modelled shift from min to max length within the season was 0.2 cm while it was nearly 1 cm in our model. The length modelled for the Mediterranean are the lowest, the largest values are those for the Bay of Biscay and slightly shorter for NI estuaries. The long-term trend is very similar to that obtained with GAM models with a clear increase from 2011 to 2014.

Spatio-temporal trends in glass eel condition

The variance of glass eel condition was larger in the 'Mediterranean' than in the Bay of Biscay (F test, $p < 0.001$). The same model structure (Model 9) was selected in the Mediterranean and in the Bay of Biscay, which includes both the long-term trend and the intra-seasonal trend. In the Bay of Biscay area, the gain in AIC obtained by including the *time* trend was very low and the variation was very small, so the most parsimonious model is probably Model 8, which includes a lower condition in the Vilaine for glass eel ascending the ladder. For the Mediterranean series, the *time* was clearly significant, but its inclusion probably helped upon fitting the multiple wave pattern of the Guadalquivir. This multiple recruitment wave pattern observed in the Guadalquivir, and to a certain extent in the Ter, was more complex than that observed in the Bay of Biscay. The

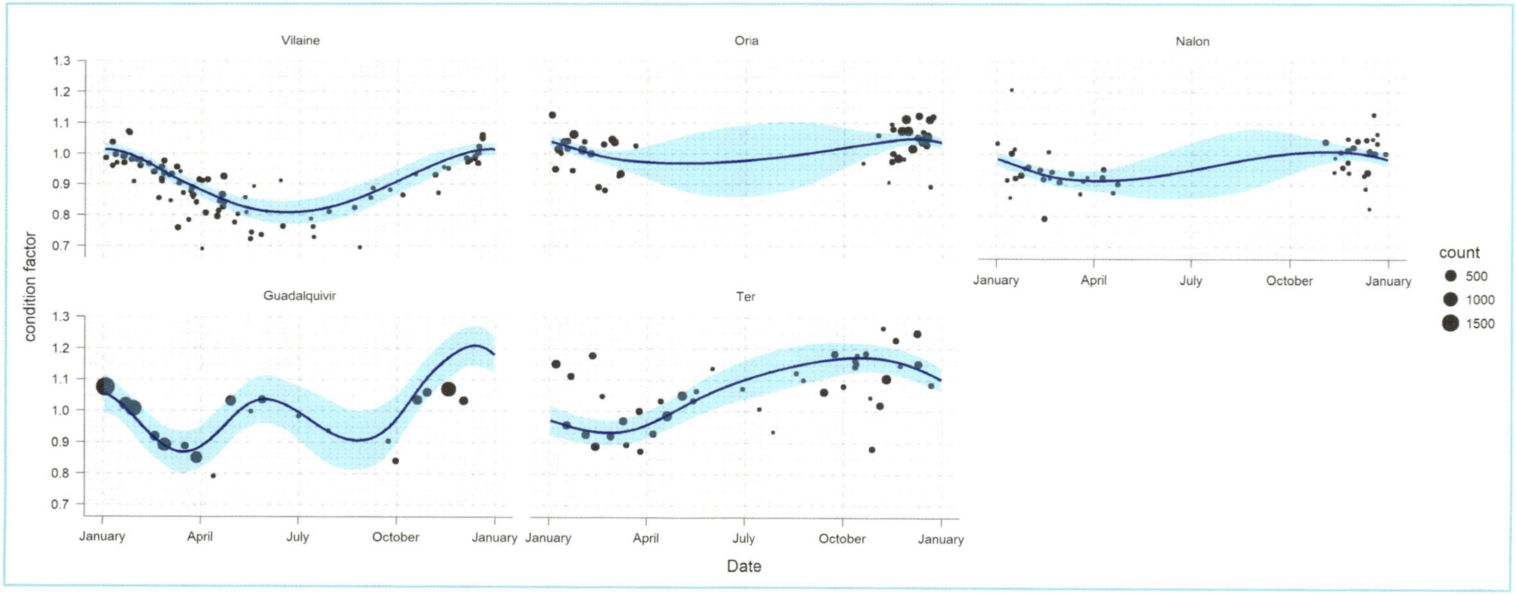

Figure 8.5 Trend in condition factor across different locations from the Atlantic and the Mediterranean. Notes: Points indicate mean values of the sample and point size indicates sample size. The curves correspond to the predicted response of Model 9 for Fulton coefficient. See Table 8.1 for sample number and period detail per location.

Table 8.5 Statistics for models on condition.

Area	Model	LogLik	AICc	Delta
Bay of Biscay	9	7133.68	−14245	0
	8	7130.25	−14242	3
	7	7112.6	−14213	32
Mediterranean	9	7999.24	−15984	0
	7	7992.14	−15974	10

Notes: As the variance was different between the Biscay area and the Ter and Guadalquivir, separate models have been tested. The best model for the Mediterreanean include a long-time trend in glass eel condition, the model in the Biscay does not. LogLik = log- likelihood, AICc =Akaike criteria, delta= difference in AIC between the best model and the current model. The Bay of Biscay includes la Vilaine, Oria and Nalón. The Mediterranean includes Guadalquivir and Ter.

same decline in length and condition was observed at the beginning of the season, but unlike the Bay of Biscay area, there is a clear return to increasing glass eel length and condition from April to June (Figure 8.5).

Correlation between recruitment and glass eel length

The trend in glass eel length was extracted from Model 6 (Figure 8.6a). These inter-seasonal estimates were then used as an additional variable to model the inter-seasonal EE recruitment (model plotted as a grey line Figure 8.6b). The model produced: $recrEE \sim year + length$ (Model 10 plotted as `S` joined by a dotted line Figure 8.6b) and was significantly better than the model without length (Model 11, ANOVA $p < 0.1$), which meant that some of the variation in glass eel recruitment could be accounted for by the trend in glass eel length. This was true in the short term (2001–2018) but not the long term (1996–2018), that is, when including data from 1997, this relationship was no longer significant (ANOVA $p > 0.1$). The correlation between the recruitment series and glass eel length was 0.44 ($p = 0.08$) for the period 2001–2018 and only 0.16 ($p = 0.48$) for 1996–2018.

Discussion

Correlation between recruitment and glass eel length

Our first hypothesis regarding the link between glass eel length, condition and recruitment is substantiated given that our study shows that there is a short-term relationship between glass eel length and recruitment, suggesting that oceanic processes have a strong influence on recruitment. The long-term trend in length extracted from the model can be used as a covariate to predict the short-term variations in glass eel recruitment. However, this correlation breaks down over the longer term as it is most likely superseded by the long-term decline in eel recruitment, thus implicating the influence of additional factors that may explain the stock current critical status.

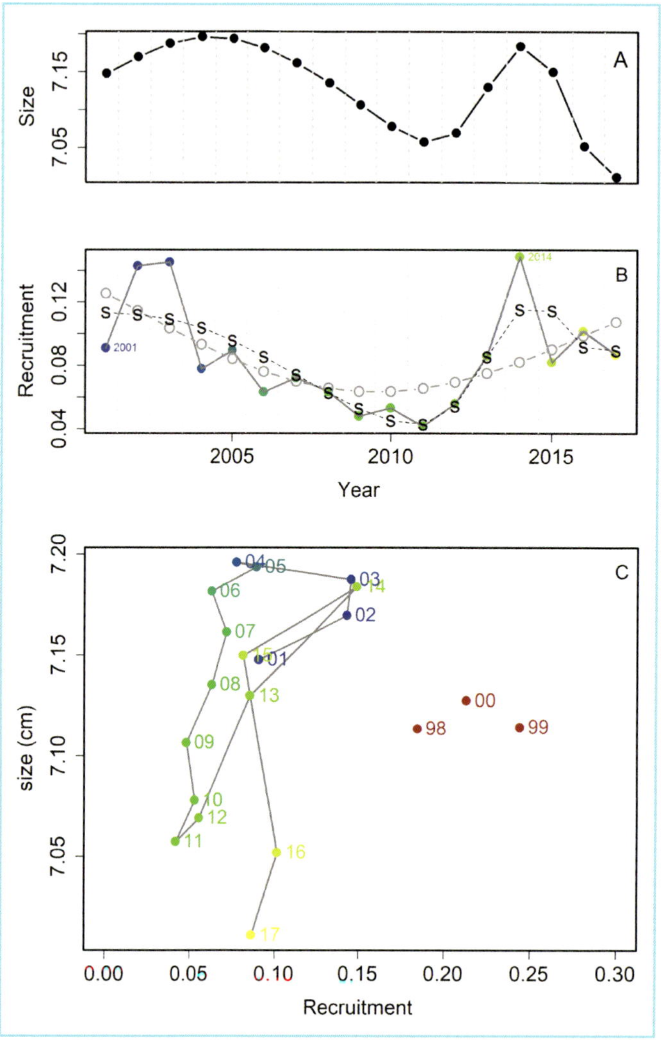

Figure 8.6a–c (a) Long-term trend in glass eel length extracted from the best model (Model 6), colour according to year. (b) Trend in the recruitment series 'Elsewhere Europe' as produced by the WGEEL (ICES, 2017), in grey simple GAM fitted on the trend without accounting for length (Model 11), in black -s- model including length (Model 10). (c) Correlation between recruitment and length, colour according to year.

It is interesting to note that glass eel minimal length from seasons 1990–1991 and 2009–2010 also correspond to local minima in recruitment levels (ICES, 2017; Figure 8.6a). In 1980, before the continuous drop in recruitment began, the onset of recruitment failure might correspond to lower survival rates of larvae crossing the Atlantic. Similarly, the large increase in recruitment from 2011 to 2014 also coincided with increasing length. Dekker (1998) found that glass eel lengths were more or less stable during the 1960s and 1970s, yet showed a prolonged decrease from 1983 to a minimum in 1991 before increasing to longer lengths coinciding with higher abundances at Den Oever (Netherlands). He showed a clear correlation (r = 0.7, p = 0.001) between glass eel length and glass eel abundance. In the current study, the correlation between the ICES WGEEL EE recruitment indices and glass eel inter-seasonal trend in length extracted from the model was not significant: r = 0.44 (p = 0.08) for 2001–2018 and r = 0.16 (p = 0.48). The low correlation is probably caused by higher values of recruitment in years 1998 to 2000

(Figure 8.6c, in brown) whereas the length is about average. It is possible that Dekker's analyses encompassed a period when larger changes occurred in the feeding conditions in the ocean that might have had a clearer effect on length and recruitment. Désaunay and Guérault (1997) also noted a decrease in length from 75 mm to 68 mm over the years 1980 to 1990. Similarly, in Le Grau de la Fourcade, in the Mediterranean, there was also a decline in glass eel length when comparing 1974–1975 measurements to 2000–2001 (Lefebvre et al., 2002). The average lengths we observed over 2001–2018 is similar to those from Désaunay and Guérault (1997) in years 1991–1993, so it is likely that the glass eel length has also decreased when compared to its former levels in the beginning of the 1980s.

Désaunay and Guérault (1997), Dekker (1998) and Lefebvre et al. (2002) ascribed the described changes in glass eel length to oceanic conditions. According to them, the food resources and energy reserves available to growing leptocephali probably determine the amount of surviving glass eels. The link we have found between length (a proxy for growth rate) and recruitment in the short term would support this hypothesis. The length of glass eel observed at European coasts is the result of growth periods occurring first in the Sargasso Sea and later during their trans-oceanic migration. In the Sargasso Sea atmospherically driven current variations induce variability in the transport timing and migration success from the Sargasso Sea. The correlation between glass eel transport and recruitment trend, which existed before 1980, is no longer evident and has been replaced by a continuous decrease unrelated to known fluctuations in the Sargasso (Baltazar-Soares et al., 2014). In addition, recruitment processes describing the success of transport from the Sargasso Sea seem poorly related to the current length trend, indicating that the main variation in glass eel length emanates from a later phase during their transatlantic migration.

A long-term trend in glass eel length has been observed in *A. rostrata* glass eels. So early events occurring during drift outside from the Sargasso Sea, and along the American coasts might influence growth. This suggests that ocean driven change in productivity might already be at play in the western Atlantic Ocean (Sullivan et al., 2006).

Westerberg et al. (2018) have shown that the levels of abundance of larvae in the Sargasso Sea is about 20–30% of the pre-1980 levels so this drop is lower than that in recruitment in the EE area which is about 8–10% of pre-1980s levels. This points to events occurring after the spawning during the leptocephalus migration outside from the Sargasso Sea. However, the use of a Lagrangian model does not point to a modification in passive drift pattern or speed that could explain the regime shift in recruitment over the last three decades (Pacariz et al., 2014). Including elements related to food availability and growth rate of leptocephali could possibly help to understand if these factors are an additional cause for recruitment variation.

Our finding that there is no relation between length and recruitment in the long term might provide a simple explanation. The recruitment drop has been a continuous process linked to the stock decline in continental waters (Dekker, 2004) and the relation between length and recruitment in the long-term is superseded by the continuous decline in eel recruitment.

Spatio-temporal trend in glass eel length and condition

Inter-seasonal trends in glass eel length are stable across the different locations. Building a mixed intra-seasonal and long-term model enabled the removal of intra-seasonal variation effects and extract an inter-seasonal trend in glass eel length. Whichever model was used, the trend showed a clear increase in length after 2010 (in all locations where data was available), coinciding with an increase in recruitment after a long period of decline. Given the dataset available, (some years were missing at some locations), the most parsimonious conclusion is that there is a common trend in inter-seasonal length for the different locations.

Conversely, intra-seasonal trends in length differ in time (over the years) and space (location). Our results indicate that the best GAM model (Model 4) was a model that included an interaction between year and time illustrating that trends vary from year to year, with some years having wider range in length than others. However, the fit for Model 4 gave rise to spurious values once extrapolated, as the data needed to fit a tensor interaction for the model were not available. On the one hand, only the Vilaine, the Guadalquivir and the Ter have all year-round sampling. The lack of samples from April in Spain (Nalón and Oria) makes it difficult to ascertain the date when the lowest length is reached. Even for those having all year-round sampling, glass eels are hard to collect from July to September for years with local very low recruitment, and for that reason the minimum value is probably not observed in some estuaries. On the other hand, except from the Vilaine (including continuous sampling from 1998 to 2018), the rest of the glass eel data series are short or discontinuous which makes it difficult to fit the more complex models.

The intra-seasonal trends in condition do not vary much over time in the Bay of Biscay and the time period available in the Mediterranean is too short to provide consistent results. This probably means that events related to the post metamorphosis period are not as variable as those driving their growth during the oceanic migration.

In spite of the inclusion of this variation in the range from year to year, the shape of the seasonal trend curve in glass eel length and condition in the Bay of Biscay is very stable. Glass eel with larger length and condition are always found at the beginning of a recruitment season. The maximum length is reached earlier at southern locations where recruitment starts earlier. Further north, the intra-seasonal trend for length modelled for NI seems to reach a peak before the Vilaine. However, given the information available, we can only say that the trend is similar to that in the Bay of Biscay. More widespread sampling would be necessary to obtain a more complete curve to cover the whole of the British Isles, but continuous glass eel samples are harder to collect from remote sites at this latitude.

Other authors have also found a similar intra-seasonal trend in length and provided suggestions of factors, which might explain these differences in length across a season.

- Differences in swimming capability of glass eel. Following metamorphosis, glass eel use a magnetic orientation linked with tidal cycles to migrate at sea (Cresci et al., 2017; Durif et al., 2013, 2017). The largest glass eel have better swimming capabilities, so it is expected that they would be the first estuarine arrivals. However, having observed a 6 month spread between the arrival of the longest and the shortest glass eel, we believe this is too long a period to be explained by differences in swimming performance.
- Other authors that have found the same trend, suggest that changes in planktonic availability linked to variable oceanographic conditions along the season might produce the observed changes in monthly patterns of recruitment (Dekker, 1998; Wang and Tzeng, 1998). Désaunay and Guérault (1997) believed that glass eel arriving in July would have experienced extended poor feeding conditions before their metamorphosis, compared to those

arriving in January. These poor feeding conditions would result in lower growth rate of pre-metamorphosing leptocephali and thus produce shorter glass eel. Therefore, oceanographic conditions would influence both the long-term inter-seasonal and the intra-seasonal trend in recruitment and glass eel length and condition.

- Miller et al. (2014) suggested that the glass eel arriving at the Biscay Bay belong to a mixture of different cohorts. However, the common intra-seasonal trends in length and condition in the estuaries of the present study from the Biscay region (Vilaine, Oria or Nalón) together with the above-mentioned studies showing the same trend, would point to a single cohort of glass eel that arrive at an earlier time along the Iberian Coast and migrate for a further two months to reach the Vilaine. Uncovering this common intra-seasonal trend in length would be the result of the previously explained mechanism influencing oceanographic condition related recruitment. It is important to recognize knowledge gaps remain into larval eel ecology making it difficult to probe the single cohort hypothesis in the Bay of Biscay. However, in the light of our results the mixed cohort hypothesis is more difficult to justify. First, different cohorts would arrive at different times and present a much more complex length structure than the clearly decreasing length curve that we have observed. Second, this would imply that the mixture of cohorts would be capable of producing the same regular curve year after year, which does not seem probable.

In the Mediterranean, the intra-seasonal trend in length and condition is different to that of the Bay of Biscay. In the Guadalquivir, the decline stops in April, whereas the decreasing trend seems to be more pronounced in the Ter. Furthermore, the intra-seasonal trend in length and condition in the Mediterranean suggests that glass eel recruitment comprises more than a single wave of arrival. C. Fernández-Delgado (Córdoba, 2018, personal communication) confirms that this is clearly the case in the Guadalquivir during the 2003–2005 period, with glass eels of longer length and better condition in April coinciding with increased recruitment levels. This, together with the wider variability in condition denotes that recruitment processes are probably more complex in the Mediterranean than in the Bay of Biscay. The second wave was not detected when analysing the historical dataset of the Guadalquivir (1997–1999: formalin preserved). Although the trend in length was also not modelled in the Ter, the observation of close intra-seasonal patterns on top of the modelled trend (Figure 8.6) shows that the same secondary pattern of several waves of recruitment might be present. The presence of an additional recruitment wave in April has already been recorded in the Mediterranean in the Grau de La Fourcade (Lefebvre et al., 2002). The observed two waves of recruitment do not seem to come from different cohorts, as the length-weight scatterplots of abundance do not show any bimodality (Figure 8.7).

Thus, the second wave of recruitment in April does not comprise very different (shorter) glass eels when compared to the beginning of the season.

For the years 1997–1999 and 2003–2005 any observed patterns do not support the existence of a separate migration route for the 'Mediterranean' series of glass eels. Glass eel coming from the Sargasso Sea from a putative route using an eastward-flowing Subtropical Counter Current (McCleave, 1993; Munk et al., 2010) would spend much less time in the ocean than the bulk of those using the Gulf Stream and would therefore be much shorter (Miller et al., 2014). The secondary wave of recruitment observed in the Guadalquivir corresponds to longer glass eel with a better condition, suggesting it is unlikely they come from this shorter route. However, they

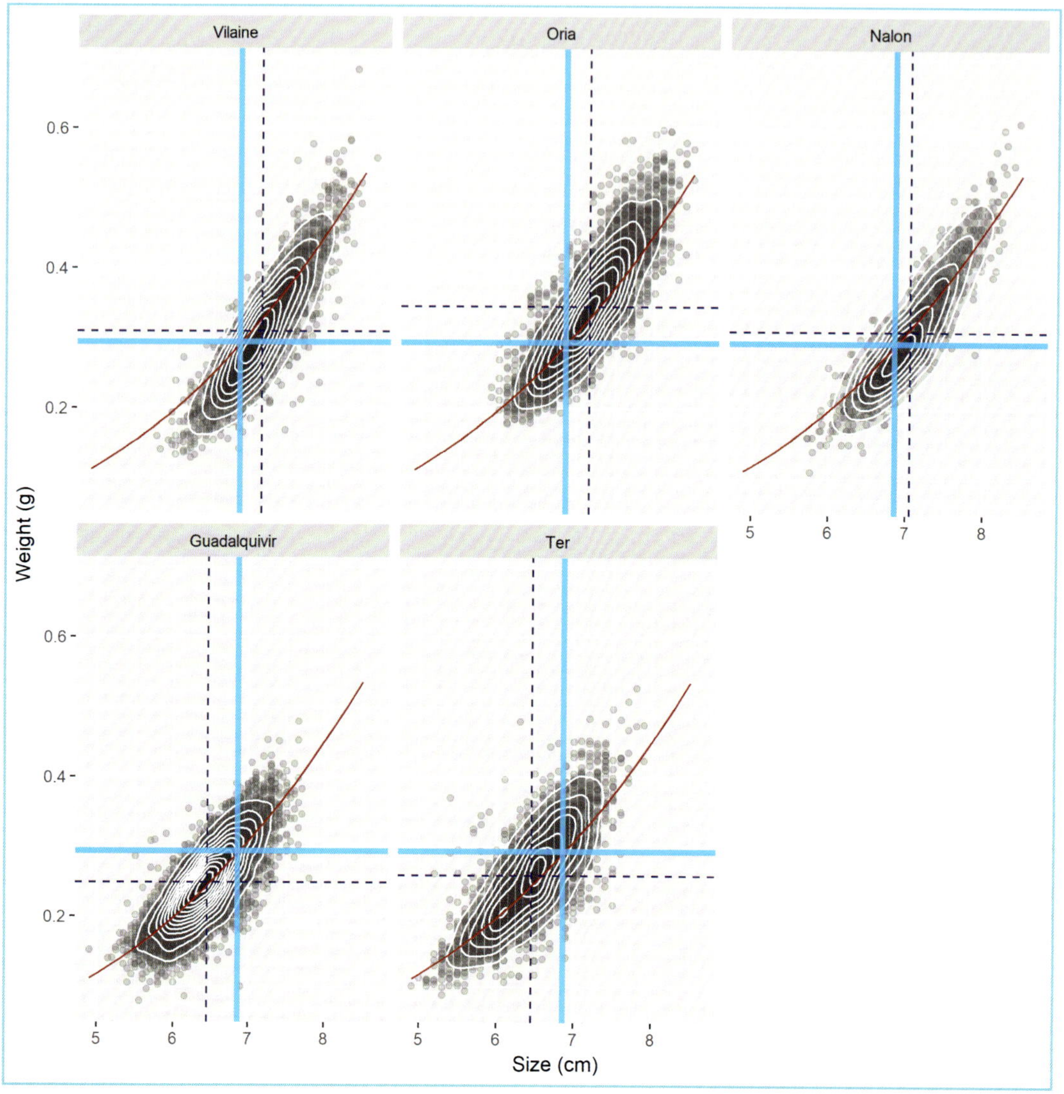

Figure 8.7 Relationship between length and weight for different sampling locations. Notes: The dotted blue line intersection corresponds to the average length and weights, the bold blue line intersection to the average of average location per site. The brown line corresponds to the log-log length-weight relationship modelled for all samples. See Table 8.1 for sample number and time period detail per location.

could be coming later from a more northerly location having been entrained in the Portugal current (McCleave, 1993) to arrive later near Gibraltar. As such, we are able to confirm our second hypothesis regarding that there are different trends in length or condition of glass eel between the Atlantic and the Mediterranean. These patterns suggest that glass eels arrive in a single recruitment wave to the Atlantic but in two recruitment waves in the Mediterranean. In neither the Atlantic nor the Mediterranean data was there a clear bimodality which would support the fact that glass eel belong to more than a single cohort or that some individuals are using a shorter route. Given the sparseness of our Mediterranean dataset (two locations; short time period) we would recommend additional sampling effort is devoted to the further monitoring of glass eel length and condition in this region.

The difference in length and condition between glass eel collected in the trap and those from the estuary of the Vilaine is explained by the time it takes for the glass eel to reach the trapping ladder (Briand et al., 2009). Within the pigmentation stage selected, glass eel arriving at the trapping ladder are older and may thus have arrived earlier. Our study also shows that the date when maximum length is attained in the Vilaine is comparable (8 days earlier) with that found previously for 1991–1993 by Désaunay and Guérault (1997).

The seasonal variations in glass eel length and condition described here have an important effect on the stock as it is well established that they are related to the fitness of glass eels and subsequently their capacity for migration (Bureau Du Colombier et al., 2007, 2011; Edeline et al., 2006).

Some authors have found an increase in length of leptocephali from south to north (Antunes and Tesch, 1997; Bast and Strehlow, 1990), which has been attributed to a possible oriented swimming behaviour of leptocephali. These results were later re-examined by McCleave et al. (1998) who identified a south to north cline in length with an additional cline from west to east, and only a small seasonal effect/influence. We have not found any obvious difference in length within the Bay of Biscay, probably because the latitudinal range across our study was too narrow. The lower length of glass eel in NI may be an effect of the preservation process (70% ethanol). Conversely, we can confirm that glass eels are much shorter in the Mediterranean compared to those in the Bay of Biscay.

Conclusion and way forward

Our results confirm that there is a relationship between length and recruitment. The long-term inter-seasonal trend of length extracted from our models explained short-term variation in recruitment and in turn can be used as a covariant to predict short-term variations in glass eel recruitment levels. Thus, glass eel length showed a link with the oceanic conditions they incurred during development and probably describe oceanic feeding conditions in the leptocephalus stage. The drop in the 1980s was accompanied by a reduction in glass eel length (Dekker, 1998) and is currently interpreted as a regime shift (Westerberg et al., 2018). A continuous decrease in glass eel length could therefore act as an early warning signal from an increased risk of stock collapse. However, over the longer term, glass eel length fails to explain the decline in recruitment, most likely because the decline is linked to continental rather than oceanic factors impacting the stock.

The modelling approach we have taken shows some advantages. Previously, the studies analysing European glass eel length up until now had focused on a single location (Dekker 1998; Désaunay and Guérault, 1997; Lefebvre et al., 2002). In this current study, we have compiled information collected from a range of locations and devised a method which incorporates glass eel length and condition with intra-seasonal variations at different locations within a single model. The relationship between season and length across all the locations in the present study has been fitted using a harmonic pattern, as was shown by Désaunay and Guérault (1997). However, the GAM models were more suited to describe the intra-seasonal trend than simple (Désaunay and Guérault, 1997) or mixture of sine curves analyses, as the intra-seasonal trends are not symmetrical (Figure 8.3) and GAM allow for complex or asymmetrical curves.

Extracting simple trends from this model would help to improve the knowledge about leptocephali migration routes. The coupling of Lagrangian ocean models that simulate the drift of leptocephali, with oceanic productivity dependent growth rates, (Melià et al., 2013), would produce models that could include the provision of the time spent by metamorphosing leptocephali and glass eel until they reach the coasts. The inclusion of the length and condition of glass eel once they reach estuaries at various points along the coasts all year round, would improve the calibration of the models which estimate the duration of the leptocephali migration, and perhaps test assumptions about the relationship between glass eel length, drift patterns and oceanic productivity.

Finally, we recommend that monitoring of glass eel length and weights should be set up within the Union framework for the collection, management and use of data (Regulation EU 2017/1004). Such additional monitoring would enable the adoption of new/improved management measures addressing impacts affecting the stock during its continental phase if a decrease in length and condition was detected. An increase of the number of sampling points in the Mediterranean is needed to help resolve questions about recruitment processes that have produced the findings from the Guadalquivir and Ter sites. Glass eel length and condition analysis should be done on fresh glass eels to avoid the problems we have found with preservation methods. It is also recommended that glass eel pigmentation stage is assessed in order to ensure that glass eel closest to their marine phase are included in the analyses.

Bibliography

Antunes, C. and Tesch, F.-W. (1997) Eel larvae (*Anguilla anguilla* L.) Caught by RV 'Heincke' at the European continent slope in autumn 1991. *Ecology of Freshwater Fish* 6, 50–52.

Arai, T., Otake, T. and Tsukamoto, K. (2000) Timing of metamorphosis and larval segregation of the Atlantic eels, *Anguilla rostrata* and *A. anguilla*, as revealed by otolith microstructure and microchemistry. *Marine Biology* 137, 39–45.

Aranburu, A., Díaz, E. and Briand, C. (2015) Glass eel recruitment and exploitation in a South European estuary (Oria, Bay of Biscay). *ICES Journal of Marine Science: Journal du Conseil* 73, 111–121.

Baltazar-Soares, M., Biastoch, A., Harrod, C., Hanel, R., Marohn, L., Prigge, E., Evans, D., Bodles, K., Behrens, E., Böning, C.W. and Eizaguirre, C. (2014) Recruitment collapse and population structure of the European eel shaped by local ocean current dynamics. *Current Biology* 24, 104–108.

Bardonnet, A., and Riera, P. (2005) Feeding of glass eels (*Anguilla anguilla*) in the course of their estuarine migration: new insights from stable isotope analysis. *Estuarine, Coastal and Shelf Science* 63, 201–209.

Bast, H.D. and Strehlow, B. (1990) Length composition and abundance of eel larvae, *Anguilla anguilla* (Anguilliformes: Anguillidae), in the Iberian basin (northeastern Atlantic) during July-September 1984. *Helgoländer Meeresuntersuchungen* 44, 353–361.

Boëtius, J. and Harding, E. (1985) A re-examination of Johannes Schmidt's Atlantic eel investigations. *Dana* 4, 129–162.

Bonhommeau, S., Castonguay, M., Rivot, E., Sabatié, R. and Le Pape, O. (2010) The duration of migration of Atlantic *Anguilla* larvae. *Fish and Fisheries* 11, 289–306.

Bonhommeau, S., Le Pape, O., Gascuel, D., Blanke, B., Tréguier, A.-M., Grima, N., Vermard, Y., Castonguay, M. and Rivot, E. (2009) Estimates of the mortality and the duration of the trans-Atlantic migration of European eel *Anguilla anguilla* leptocephali using a particle tracking model. *Journal of Fish Biology* 74, 1891–1914.

Bornarel, V., Lambert, P., Briand, C., Antunes, C., Belpaire, C., Ciccotti, E., Diaz, E., Diserud, O., Doherty, D. and Domingos, I. (2017) Modelling the recruitment of European eel (*Anguilla anguilla*) throughout its European range. *ICES Journal of Marine Science* 75, 541–552.

Breheny, P. and Burchett, W. (2017) Visualization of regression models using visreg. *The R Journal* 9, 56–71.

Briand, C., Fatin, D., Fontenelle, G. and Feunteun, E. (2003) Estuarine and fluvial recruitment of the European glass eel, *Anguilla anguilla*, in an exploited Atlantic estuary. *Fisheries management and Ecology* 10, 377–384.

Briand, C. (2009) Dynamique de population et de migration des civelles en estuaire de vilaine. Ph. D. Thesis Agrocampus Ouest.

Briand, C., Legrand, M. and Besse, T. (2018) *StacomiR: Fish migration monitoring*. Available at: https://R-Forge.R-project.org/projects/stacomir/ (accessed 9 May 2019).

Bureau Du Colombier, S., Bolliet, V., Lambert, J.G.D. and Bardonnet, A. (2007) Energy and migratory behaviour in glass eels (*Anguilla anguilla*). *Physiology and Behavior* 92, 684–690.

Bureau du Colombier, S., Bolliet, V., Lambert, P. and Bardonnet, A. (2011) Metabolic loss of mass in glass eels at different salinities according to their propensity to migrate. *Estuarine, Coastal and Shelf Science* 93, 1–6.

Cieri, M.D. and McCleave, J.D. (2000) Discrepancies between otoliths of larvae and juveniles of the American eel: Is something fishy happening at metamorphosis?. *Journal of Fish Biology* 57, 1189–1198.

Cresci, A., Paris, C.B., Durif, C.M., Shema, S., Bjelland, R.M., Skiftesvik, A.B. and Browman, H.I. (2017) Glass eels (*Anguilla anguilla*) have a magnetic compass linked to the tidal cycle. *Science Advances* 3, e1602007.

Dahl, D. B. (2016) Xtable: Export tables to latex or html. Available at: https://CRAN.R-project.org/package=xtable (last accessed 9 May 2019).

Dekker, W. (1998) Long-term trend in the glass eels immigrating at Den Oever, the Netherlands. *Bulletin Français de Pêche et de Pisciculture* 349, 199–214.

Dekker, W. (2004) *Slipping through our hands, Population dynamics of the European eel.* PhD thesis, University of Amsterdam.

Désaunay, Y. and Guérault, D. (1997) Seasonal and long-term changes in biometrics of eel larvae: A possible relationship between recruitment variation and North Atlantic ecosystems productivity. *Journal of Fish Biology* 51, 317–339.

Durif, C.M.F., Browman, H.I., Phillips, J.B., Skiftesvik, A.B., Vøllestad, L.A. and Stockhausen, H.H. (2013) Magnetic compass orientation in the European eel. *PLoS ONE* 8, e59212.

Durif, C.M., Bonhommeau, S., Briand, C., Browman, H.I., Castonguay, M., Daverat, F., Dekker, W., Díaz, E., Hanel, R. and Miller, M.J. (2017) Whether European eel leptocephali use the Earth's magnetic field to guide their migration remains an open question. *Current Biology* 27, 998–1000.

Edeline, E., Lambert, P., Rigaud, C. and Elie, P. (2006) Effect of body condition and water temperature on *Anguilla anguilla* glass eel migratory behavior. *Journal of Experimental Marine Biology and Ecology* 331, 217–225.

Elie, P., Lecomte-Finiger, R., Cantrelle, I. and Charlon, N. (1982) Définition des limites des différents stades pigmentaires durant la phase civelle *d'Anguilla anguilla* L. *Vie et Milieu* 32, 149–157.

European Union (2008) Regulation on the establishment of a Union framework for the collection, management and use of data in the fisheries sector and support for scientific advice regarding the common fisheries policy and repealing Council Regulation (EC) No. 199/2008.

Friedland, K.D., Miller, M.J. and Knights, B. (2007) Oceanic changes in the Sargasso Sea and declines in recruitment of the European eel. *ICES Journal of Marine Science* 64, 519–530.

Froese, R. (2006) Cube law, condition factor and weight–length relationships: history, meta-analysis and recommendations. *Journal of applied ichthyology* 22, 241–253.

ICES (2010) *Report of the 2010 session of the Joint EIFAC/ICES Working Group on Eels (WGEEL).* Hamburg.

ICES (2018) Report of the Joint EIFAAC/ICES/GFCM Working Group on Eels. (WGEEL), 5–12 Gdańsk.

Jacoby, D. and Gollock, M. (2014) *Anguilla anguilla.* The IUCN Red List of Threatened Species 2014: e.T60344A45833138. Available at: http://dx.doi.org/10.2305/IUCN.UK.2014-1.RLTS.T60344A45833138.en (accessed 23 April 2019).

Kracht, R. (1982) On the geographic distribution and migration of I- and II-group eel larvae as studied during the 1979 Sargasso Sea Expedition. *Helgoländer Meeresunters* 35, 321–327.

Kettle, A. and Haines, K. (2006) How does the European eel (*Anguilla anguilla*) retain its population structure during its larval migration across the North Atlantic Ocean? *Canadian Journal of Fisheries and Aquatic Sciences* 63, 90–106.

Knights, B.A. (2003) Review of the possible impacts of long term oceanic and climate changes and fishing mortality on recruitment of anguillid eels of the northern hemisphere. *Science of the Total Enviroment* 310, 237–244.

Laflamme, S., Côté, C., Gagnaire, P.-A., Castonguay, M. and Bernatchez, L. (2012) RNA/DNA ratios in American glass eels (*Anguilla rostrata*): evidence for latitudinal variation in physiological status and constraints to oceanic migration? *Ecology and evolution* 2, 875–884.

Lecomte-Finiger, R. (1983) Contribution à la connaissance de l'écobiologie de l'anguille, *Anguilla anguilla,* L. 1758, des milieux lagunaires méditerranéens du golfe du Lion: Narbonnais et Roussillon. Université de Perpignan.

Lefebvre, F., Sergent, E., Acou, A., Lecomte-Finiger, R. and Crivelli, A.J. (2002) Recrutement des civelles (*Anguilla anguilla*) sur la côte méditerranéenne française : analyse comparée des caractéristiques biométriques et pigmentaires des saisons 1974–75 et 2000–01. *Bulletin Français de Pêche et de Pisciculture* 368, 85–96.

McCleave, J. (1993) Physical and behavioural control on the oceanic distribution and migration of leptocephali. *Journal of Fish Biology* 43, 243–273.

McCleave, J.D. (2008) Contrasts between spawning times of *Anguilla* species estimated from larval sampling at sea and from otolith analysis of recruiting glass eels. *Marine Biology*, 155–249.

McCleave, J.D., Brickley, P.J., O'Brien, K.M., Kistner, D.A., Wong, M.W., Gallagher, M. and Watson, S.M. (1998) Do leptocephali of the European eel swim to reach continental waters? Status of the question. *Journal of the Marine Biological Association of the United Kingdom* 78, 285–306.

Melià, P., Schiavina, M., Gatto, M., Bonaventura, L., Masina, S. and Casagrandi, R. (2013) Integrating field data into individual-based models of the migration of European eel larvae. *Marine Ecology Progress Series* 487, 135–149.

Miller, M.J. (2009) Ecology of anguilliform leptocephali: Remarkable transparent fish larvae of the ocean surface layer. *Aqua-BioScience Monographs* 2, 1–94.

Miller, M.J., Bonhommeau, S., Munk, P., Castonguay, M., Hanel, R. and McCleave, J.D. (2015) A century of research on the larval distributions of the Atlantic eels: a re-examination of the data. *Biological Reviews* 90, 1035–1064.

Munk, P., Hansen, M.M., Maes, G.E., Nielsen, T.G., Castonguay, M., Riemann, L., Sparholt, H., Als, T.D., Aarestrup, K., Andersen, N.G. and Bachler M. (2010) Oceanic fronts in the Sargasso Sea control the early life and drift of Atlantic eels. *Royal Society of London. Proceedings. Biological Sciences* 277, 3593–3599.

Pacariz, S., Westerberg, H. and Björk, G. (2014) Climate change and passive transport of European eel larvae. *Ecology of Freshwater Fish* 23, 86–94.

R Core Team (2017) *R: A language and environment for statistical computing*. R Foundation for Statistical Computing, Vienna. Available at: https://www.R-project.org/.

Sobrino, I., Baldó, F., García-González, D., Cuesta, J.A., Silva-García, A., Fernández-Delgado, C., Arias, A. M., Rodríguez, A. and Drake, P. (2005) The effect of estuarine fisheries on juvenile fish observed within the Guadalquivir Estuary (SW Spain). *Fisheries Research* 76, 229–242.

Sullivan, M.C., Able, K.W., Hare, J.A. and Walsh, H.J. (2006) *Anguilla rostrata* glass eel ingress into two, U.S. East coast estuaries: patterns, processes and implication for adult abundance. *Journal of Fish Biology* 69, 1081–1101.

Tesch, F.W. (2007) The Eel. Springer, Berlin.

Westerberg, H., Miller, M.J., Wysujack, K., Marohn, L., Freese, M., Pohlmann, J.-D., Watanabe, S., Tsukamoto, K. and Hanel, R. (2018) Larval abundance across the European eel spawning area: an analysis of recent and historic data. *Fish and Fisheries* 00, 1–13.

Wang, C.H. and Tzeng, W.N. (2000) The timing of metamorphosis and growth rates of American and European eel leptocephali: a mechanism of larval segregative migration. *Fisheries Research* 46, 191–205.

Wickham, H. (2009) *ggplot2: Elegant graphics for data analysis*. Springer, New York.

Wood, S. N. (2011) Fast stable restricted maximum likelihood and marginal likelihood estimation of semiparametric generalized linear models. *Journal of the Royal Statistical Society (B)* 73, 3–36.

Zenimoto, K., Sasai, Y., Sasaki, H., Kimura, S. (2011) Estimation of larval duration in *Anguilla* spp., based on cohort analysis, otolith microstructure, and Lagrangian simulations. *Marine Ecology Progress Series* 438, 219–228.

Zuur, A.F., Ieno, E.N., Walker, N.J., Saveliev, A.A. and Smith, G.M. (2009) *Mixed Effects Models and Extensions in Ecology with R*. Springer, New York.

CHAPTER 9

Insight into Japanese eel divergence into sea-estuaries and rivers

Nobuto Fukuda[1], Kazuki Yokouchi[1], Toshihiro Yamamoto[1], Hiroaki Kurogi[1], Takashi Yada[2]

[1]National Research Institute of Fisheries Science, Japan Fisheries Research and Education Agency, Yokohama, Japan; [2]National Research Institute of Fisheries Science, Nikko, Japan.

Introduction

Anguillid eels are euryhaline fish that inhabit a wide range of salinities, from saline coastal waters and estuaries to freshwater rivers, during their development until maturation (McCleave and Edeline, 2009; Moriarty, 2003). Microchemistry analyses of strontium : calcium ratios in otoliths provided estimated eel migratory patterns; the studies revealed that some eels show a typical migration pattern of entering freshwater as juveniles, some remain in estuarine and coastal waters, and some move between freshwater and saline water during their growth phase (Daverat and Tomás, 2006; Tsukamoto et al., 1998; Tzeng et al., 2002; Yokouchi et al., 2012).

The stock of Japanese eels, as well as European (*A. anguilla*) and American eels (*A. rostrata*), has declined drastically since the 1970s (Dekker, 2003; Jacoby et al., 2015). Anthropogenic effects on eel habitats are a presumable cause; for instance, dams or weirs in rivers prevent the upstream migration of eels, resulting in them being unable to utilize all the available freshwater habitats. Insight into sea-estuary habitats is limited. The yellow phase sea-estuary residents of the Japanese eel appear in inner bays, such as Hamana Lake and Kojima Bay (Kaifu et al., 2010, 2013; Yokouchi et al., 2008), whereas they have not been caught offshore as seen in the statistics of 'Survey of Marine Fishery Production' (Ministry of Agriculture, Forestry and Fisheries of Japan, 1964–2017), implying that the habitat range of sea-estuary eels are restricted only in specific riverine-affected areas and do not extend to the offshore. The deterioration in the quality of sea-estuary habitats also cannot be overlooked. In the early 1960s, approximately 50 tonnes of eels were caught commercially in Tokyo Bay, Japan, but the catch became almost zero by the late 2000s (Mimura, 2005). Anthropogenic impacts, such as land reclamation or water pollution, has presumably affected the sea-estuary habitats of eels, but the specific conditions to diminish these habitats are still unclear.

The reproductive contribution of eels with saline-water history in addition to those with freshwater history cannot be neglected. Several analyses of the strontium : calcium otolith using mainly silver-phase eels collected along the Japanese coast and a Taiwanese river showed that eels with saline-water history were more prominent than eels with a freshwater history (Kotake et al., 2003, 2005; Tzeng et al., 2002; Yokouchi et al., 2012). Furthermore, 6 of 13 fully matured Japanese eels collected in the spawning area were those having a seawater history (Mochioka et al., 2010). Whether or not marine stocks decreased as well as riverine stocks is an important question. However, it is difficult to obtain historical demographic data to answer this question. Understanding the diverging mechanism into sea-estuaries and rivers may glean insight into population dynamics.

Eel movement in continental growth habitats can be realized as two distinct behaviours: migration and foraging (Edeline, 2007; McCleave and Edeline, 2009). Migration is a directed, long-distance movement between breeding and feeding areas initiated by internally cued behaviour. Foraging movements after the cessation of migration is not directed over long distance but is a habitat exploration driven by a search for resources (McCleave and Edeline, 2009). The early period after recruitment into their growth habitat is a transition from the migration phase to the foraging phase, and the behaviours of eels during the transition period is important to understand how individual eels enter freshwater or remain in the sea or an estuary. The behavioural traits with regard to habitat selection at each developmental stage (glass eels, elvers and yellow eels) is still scarce in Japanese eels. In this study, we introduced several behavioural experiments using Japanese eels from glass eels to young yellow eels, and herein, we proposed a hypothetical mechanism of eels diverging into different salinity habitats. Finally, we considered the eel stock management implicated by our results.

Eel entrance into freshwater or remaining in seawater

Salinity-choice tendency and locomotor activity at the glass eel stage

Several studies have reported on the salinity preferences of European eels at the glass eel stage. Tosi et al. (1988) reported that glass eels usually tend to prefer freshwater to saltwater, whereas Creutzberg (1961) reported only a slight preference for freshwater in one trial and no preference in another trial. Edeline et al. (2006) showed a shift in salinity preference of glass eels from freshwater to no preference during one week of rearing. These varying reports reflect the lack of comprehensive observations of salinity preference at the glass eel stage, and indicate possible relationships between their salinity preference and other factors such as the internal status of glass eels.

Edeline et al. (2006) proposed that the energy status of glass eels was linked to their salinity-choice behaviour because the condition factor (K) of individuals choosing freshwater were significantly higher than that of individuals choosing saltwater. However, it is not clear whether this tendency is common among Anguillid eels, because a link between K and salinity choice was not detected in a similar experiment using American eels (Boivin et al., 2015). These two studies used different methods to evaluate K values: Boivin et al. (2015) evaluated K values based on the morphology measurements of anaesthetized glass eel samples; whereas Edeline et al. (2006) evaluated K values using frozen glass eel samples.

On the basis of these previous studies using European and American eels, we examined the freezing effects on the K values of Japanese eels, and we investigated whether the K values differed

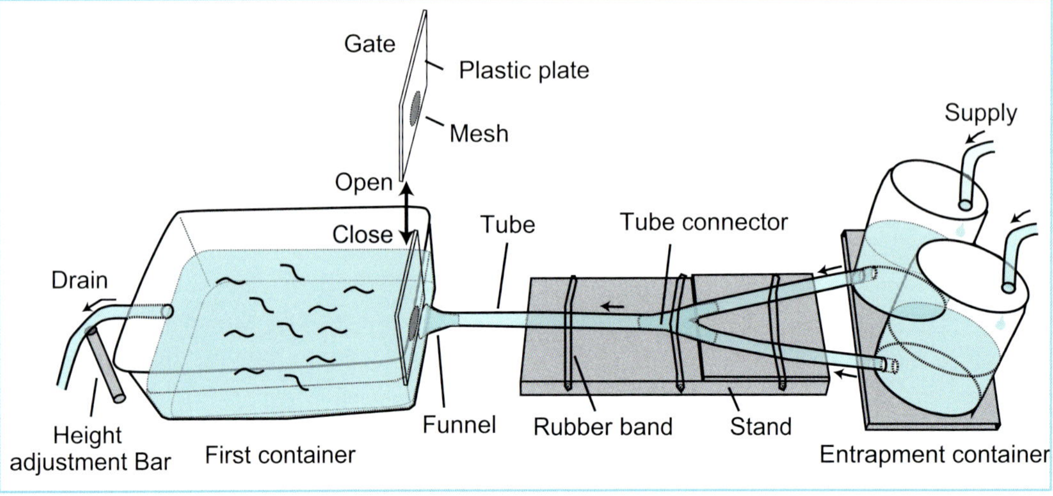

Figure 9.1 Y-maze apparatus used to test the preference of glass eels, elvers, and yellow eels (Fukuda et al., 2019a).

between glalss eels choosing seawater (SW) and freshwater (FW) (Fukuda et al., 2019b). We randomly transferred glass eels to seawater (SW-SW control), brackish water (BW) (SW-BW group) and freshwater (SW-FW group) from a rearing seawater tank and measured their morphology when anaesthetized. The K values were not significantly different among groups transferred to different salinities, whereas after freezing and thawing these samples, the K values were affected by the glass eel stage, time after thawing and salinity of surrounding water respectively. Subsequently, we compared the K values using anaesthetized samples from SW- and FW-choice groups from behavioural tests using a Y-maze behavioural apparatus (Figure 9.1). The result showed no significant differences between SW- and FW-choice groups, which was consistent with previous work done in American eels by Boivin et al. (2015). Edeline et al. (2006) had conducted experiments through the correct process ensuring that freezing was presumed to not likely affect the evaluation of K values in their study. However, the salinity effect of surrounding water, which might be contained when freezing the SW- and FW-choice groups, was possibly overlooked in their report.

From a methodological viewpoint, we also considered the comparison between the SW- and FW-choice groups after the behavioural tests (Fukuda et al., 2019b). Water-choice experiments are popular methods used to investigate fish preferences (Ford et al., 2018; Takahashi et al., 2010), and these experiments assume that the probability of each choice of water is affected by external cues (for example, stimulus type, salinity and temperature) or internal factors (for example, developmental stage, body condition and endocrine state). This type of experiment for glass eels also fits a probabilistic model, because eels tend to move up towards the flow of the attractive water type, but they could also move towards the other type because of their positive rheotaxis, even when they enter the unattractive water type by chance. Ideally, water-choice behavioural patterns of experimental fish should be approached on the basis of the operation of experimental parameters, but it is more difficult to treat internal factors (that is, condition factors) compared with external cues. Alternatively, internal factor effects are possibly examined indirectly by comparing the internal state of the individuals that chose each type after the completion of the choice trial, such as in the above-mentioned glass eel experiments (Boivin et al., 2015; Edeline et al., 2006; Fukuda et al., 2019b). These experiments are based on the assumption that FW-choice individuals have a behavioural propensity to choose FW, and SW-choice individuals have the opposite propensity. Therefore, it should be ensured at least that choices made by individuals do not merely result from probabilistic processes but also from individual internal factors.

If individual internal factors affect salinity-choice behaviour, each individual will tend to choose the same salinity when behavioural tests are repeated. In Edeline et al. (2005), when conducting repeated behavioural tests using the same glass eels (European eels), the frequencies of FW-choice, SW-choice and no choice in the second test were 45±14%, 22±16% and 33±10%, respectively, for glass eels showing FW-choice in the first test. These results were quite similar to the frequencies (FW-choice: 46±15%; SW-choice: 21±15%; no choice: 33±10%) observed in eels that exhibited SW-choice in the first test. We also carried out similar repeated behavioural tests using Japanese eels, which produced similar results, in that the frequency of individuals selecting FW and SW in the second test was not significantly different from selected choices in the first test (Fukuda et al., 2019b). These findings demonstrated that the salinity choice of each individual was not repeated and that choice was likely based mainly on the probabilistic process. Accordingly, the probabilistic salinity choice by glass eels also supports that differences in individual K values were hardly detected between resultant salinity-choice groups.

Regarding the locomotor activity of glass eels, its link to energy status has been reported in several studies of European eels, in which the impact of energy state on the migratory tactics in glass eels have been proposed. Active glass eels (European eels) under laboratory conditions exhibited higher dry weight, which is considered to represent the energy content of glass eels, than inactive ones (Bureau Du Colombier et al., 2007). The active glass eels also tend to show negative rheotaxis (Bureau Du Colombier et al., 2009) and lose less weight than inactive glass eels (Bureau Du Colombier et al., 2011). High locomotor activity may help glass eels enter freshwater sites of rivers during their landward and upstream migration, and low locomotor activity may cause settlement at sea or in an estuary.

Whether glass eels and elvers chose to enter freshwater or not during the migration phase was proposed as a conditional strategy (that is, a threshold reaction norm responding to internal energetic and ontogenetic status) based on data from European eels (Edeline, 2007; McCleave and Edeline, 2009). The subsequent experimental data examining the relationship between energetic status and Anguillid juvenile behaviour showed that the locomotor activity patterns corroborated the hypothesis; however, their salinity-choice behaviour did not thoroughly support it.

Ontogenetic shift of salinity-choice tendency

Behavioural change during the transition period from glass eels to young yellow eels is related to distribution across the sea, estuary or river. Eels that have fully reached the yellow eel stage but have not yet passed through the brackish zone do not develop any further impetus to enter freshwater (Tesch, 2003). Whether eels in the transition period change their salinity preference according to their developmental stage is unknown; we aimed to address this.

We investigated the water preferences of Japanese eels during their first year of growth in saline water after their recruitment via Y-maze behavioural trials (Fukuda et al., 2019a). In the water-choice trials between natural seawater and river water, significantly more eels at the glass eel and elver stage chose river water than seawater, whereas small yellow eels (shorter than 20 cm) showed an opposite tendency. The migratory behaviour of glass eels is associated with not only salinity (Tosi et al., 1988, 1990) but also riverine odour (Creutzberg, 1961; Sorensen, 1986; Tosi et al., 1990). Therefore, we also carried out trials to examine their odour and salinity preferences, respectively (Fukuda et al., 2019a). The results showed that all developmental stages from glass to young yellow eels tended to choose water with riverine odour, while their salinity preference shifted from having no preference or a preference for freshwater to saline water as they developed towards the yellow eel stage. These trials for each developmental stage during the first year after recruitment were conducted under seasonal ambient temperatures, and thus, the temperature also possibly affected water choices. At the least our experiments demonstrated a shift in salinity preference during development in saline waters from the glass eel stage in winter to the yellow eel stage in autumn and suggested that it would be a restricting factor for eel entry into freshwater.

Range and capacity of each salinity habitat for eels

Information on the distribution range of species is indispensable when considering the stock biomass or managing populations. Although Anguillid eels distribute from across seawater to freshwater, it is still unclear how far their seawater habitat extends offshore, what affects the range of their saline habitat or the number of inhabiting eels in each habitat.

Habitat range in seawater derived from eel behaviour

Japanese eels at any developmental stage tend to prefer riverine odour as described above (section 'Ontogenetic shift of salinity-choice tendency'), which implies the possibility that they use the riverine odour to distinguish their growth habitats. Studies of European and American eels provided insight on odour preference characteristics. Creutzberg (1961) reported that odour attraction in European eels is reduced by storage under aerated conditions and charcoal filtration. Sorensen (1986) showed that decaying leaf detritus and microorganisms responsible for decomposition are major sources of attractive odours for American eels. Amino acids, such as glutamate and threonine, the organic compound geosmin and other organic odorants are attractive to European eels (Sola et al., 1993; Sola and Tongiorgi, 1996; Tosi and Sola, 1993). However, attractive odours for Japanese eels have yet to be examined.

We examined the effect of filtration through glass fibre filter (GF; *0.7* μm pore size, Whatman), water circulation through activated carbon for 3 hours and biodegradation by aeration for 100 days under 20°C on Japanese glass eel attraction of riverine odour via the Y-maze test. GF filtration did not decrease odour attraction, indicating that attractants are included in the dissolved matter (Figure 9.2). Water circulation through activated carbon and biodegradation clearly decreased riverine odour attraction (Figure 9.2). These results suggested that the attractants are present in volatile and/or labile dissolved organic matters (DOMs) in river water.

Japanese eel juveniles use volatile and/or labile DOMs to seek their growth habitats. The odorous matters may have originated from the decomposition of leaf detritus by microorganisms in rivers as observed in studies with American eels (Sorensen, 1986). Assuming that the riverine

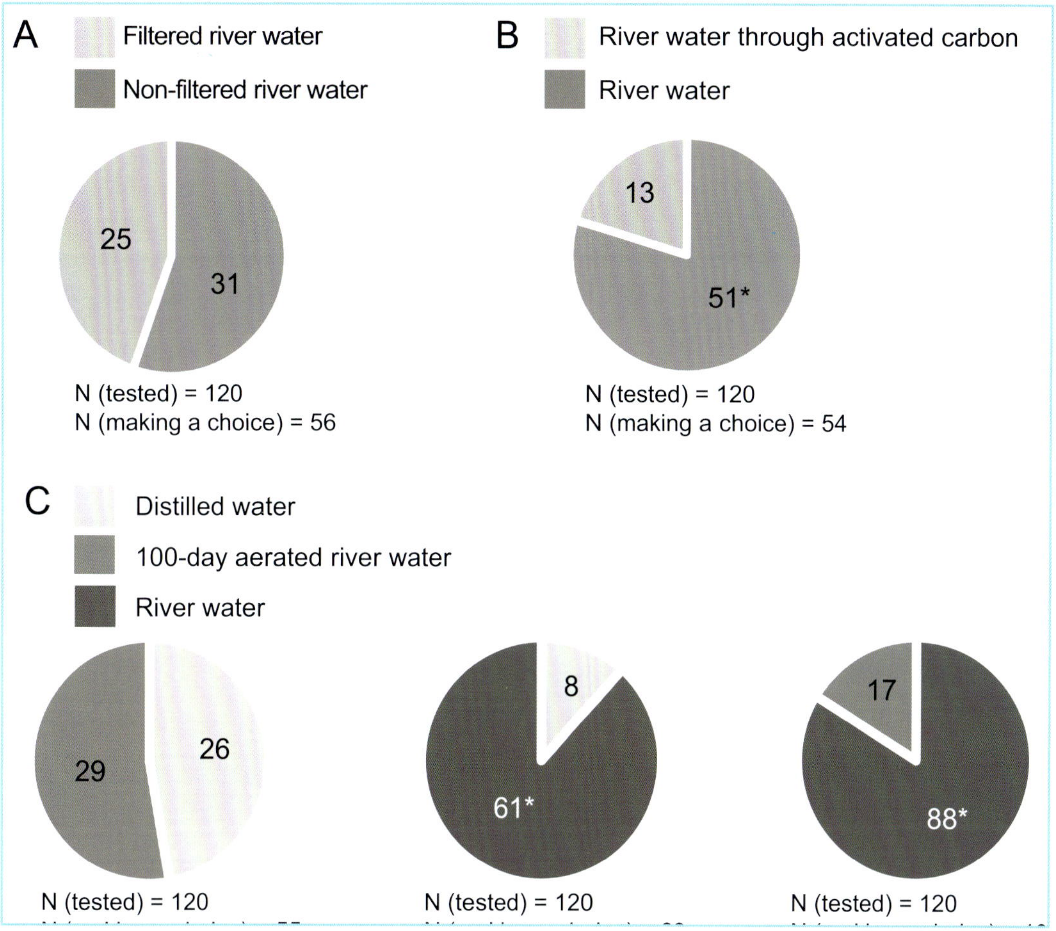

Figure 9.2 Effects of filtration by (A) glass fibre filters, (B) circulation through activated carbons and (C) biodegradation by aeration on the riverine odour attracting glass eels. River water was collected from the Maeda River in Yokosuka city, Japan. Notes: Water salinity was adjusted to salinity 34 by adding artificial marine salt (Rohto marine, Iwaki Co., Ltd. Tokyo, Japan) into each water and used for all behavioural trials. Water for the experiment (C) were filtered by glass fibre filters and the filtrates were kept frozen and used for the behavioural trials because it was assumed preliminarily that the freezing maintained the attractants of river water. *$p < 0.05$ in a chi-square test comparing to an expected value with a ratio of 1:1.

DOMs flowing into the sea gradually diffuse and are decomposed by organisms, there would be lower concentrations of DOMs further out from estuaries. The concentration of volatile and/or labile DOMs in saline areas could not only be orientation cues for glass eel migration, but it may also be related to the habitat range of eels in the sea-estuary. In fact, locations for sea-estuary eels in Japan are Hamana Lake, Kojima Bay and Mikawa Bay, which are inner bays with inflowing rivers (Kaifu et al., 2010, 2013; Kotake et al., 2006; Yokouchi et al., 2008). Therefore, the dynamics of volatile and/or labile DOMs from rivers may be critical in understanding the saline habitat range of Anguillid eels.

Capacity effect on settling eels in each salinity habitat

It is of importance to determine what factors affect the number or ratio of eels in each salinity habitat. Food and shelter availability are likely factors in addition to salinity concentration. Here we examined the effect of salinity and the area of each habitat on the number of inhabiting Japanese eels.

We used ten outdoor tanks, and three compartments in the tanks were partitioned by two concrete walls (Figure 9.3). Adjacent compartments were connected with a slope (4 cm in width, 10 cm in length and 1/5th gradient) on each wall. Because of the slopes, the supplied water always flowed downward and drained in the downstream compartment. There were five experimental settings, each conducted in duplicate (Table 9.1). In tanks with a salinity gradient, the upstream, midstream and downstream compartments were composed of freshwater (FW, mean ± SD salinity 0.5±1.6 psu), brackish water (BW, mean ± SD salinity 18.7±5.1 psu) and seawater (SW, mean ± SD salinity 27.4±3.2 psu), respectively. Underground water (salinity 0), natural seawater (salinity 34) drawn from a reef adjacent to the Yokosuka Laboratory of the National Research Institute of Fisheries Science in Japan and a mixture of the two waters were supplied into each compartment. The number of shelters and the amount of food (frozen *Chironomus* sp. larvae) per unit area (45 cm × 95 cm) were unified. For one unit area, three

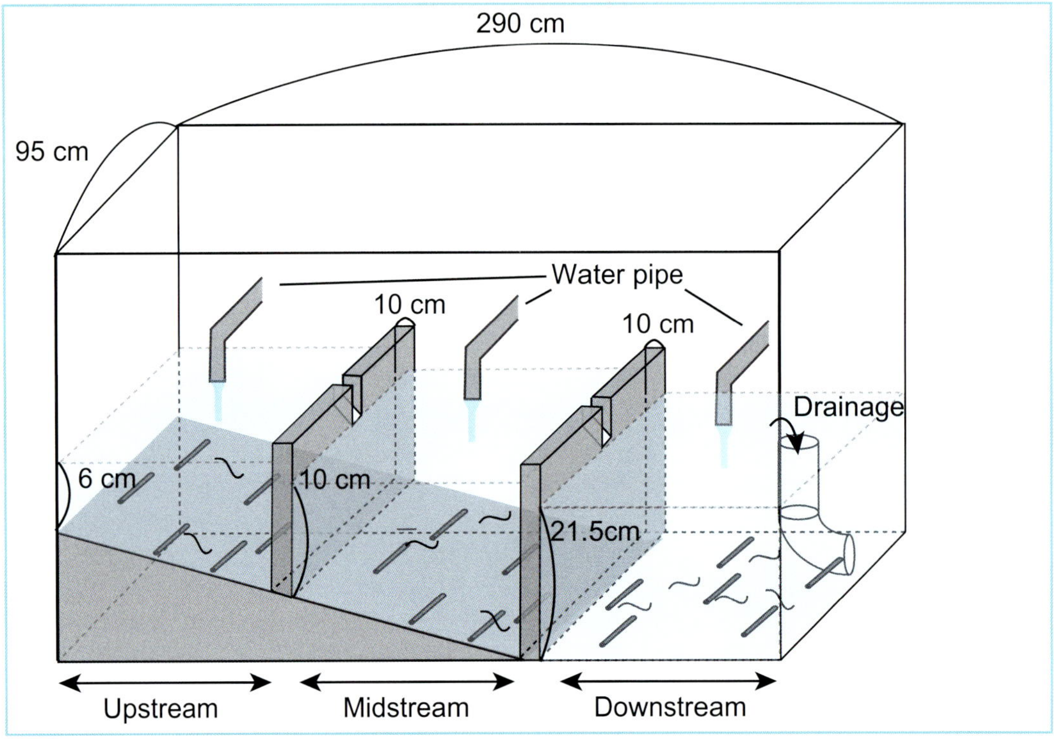

Figure 9.3 An experimental outdoor tank with three compartments partitioned by two concrete walls. Neighbouring compartments are connected with a slope (4 cm in width, 10 cm in length). An example of the tank with the same bottom area for each compartment is shown.

Table 9.1 Summary of experimental setting for outdoor tanks. Each compartment has 1–4 times unit area (45 cm × 90 cm) with seawater (SW), brackish water (BW) or freshwater (FW).

Kind of tank	Unit area of compartments for each salinity in a tank	Number of Japanese eels (at the beginning of experiment)	Number of tanks
A	(SW, SW, SW) = (2, 2, 2)	40	2
B	(SW, BW, FW) = (2, 2, 2)	40	2
C	(SW, BW, FW) = (4, 1, 1)	40	2
D	(SW, BW, FW) = (1, 4, 1)	40	2
E	(SW, BW, FW) = (1, 1, 4)	40	2

shelters (10 cm in length and 13 mm in diameter) were located, and 1.5 g of food was provided three times a week.

Glass eels were purchased from commercial fisheries in Tonegawa river, and raised in laboratory acrylic tanks with flowing seawater for 2 months until the experiment. At the start of the experiment, 40 elvers (pigmentation stage was mainly VI_B, including VI_{A4}, per classification outlined in Fukuda et al. [2013]) were released into each downstream compartment in May 2016. The number of inhabiting eels in each salinity compartment in each tank was counted every month from May to September 2016.

To examine the effect of a salinity gradient on eel distribution, the ratios of eels in upstream, midstream and downstream compartments in tank A and tank B (Table 9.1) were fitted to multinomial log-linear models using 'nnet' package in the R statistical software with presence or absence of salinity gradient and month as possible effect candidates. Furthermore, to investigate the effect of area of each salinity on eel distribution, the ratios of eels of SW, BW and FW compartment in tank B–E were fitted to the model with area of each salinity and month as possible effect candidates. The constructed models were evaluated by likelihood ratio tests.

The presence or absence of a salinity gradient improved slightly, but significantly, a model for the ratio of eels for each compartment (Figure 9.4, Table 9.2), suggesting that salinity gradient slightly affects the distribution of eel. In tanks with different bottom areas for SW, BW and FW compartments, the bottom area remarkably affected the ratio of eels for each compartment (Table 9.2). The number of eels settling in SW, BW and FW compartments was proportional to the bottom area (Figure 9.4). Furthermore, using DNA markers for identifying individuals (Yamamoto et al., 2019), individual movement between October and December was examined in two tanks with different salinity compartments (same bottom area for each). The proportion of individuals that stayed in the same compartment during the two months was 63±12% (mean ± SD), suggesting that individuals mainly tended to settle in the same compartment.

This observation implied that distribution during the transition period from elver to yellow eel is similar to ideally free distribution. The theory of ideally free distribution details that individuals will distribute in proportion to the amount of available resources (Fretwell and Lucas, 1969). Our experiment used the bottom area as an index of capacity with food and shelter proportional to the area, and we demonstrated that the number of individuals is strongly affected by the bottom area. Eels distributed evenly within an experimental tank across different salinities,

Table 9.2 Multinomial log-linear models for eel distribution in outdoor tanks under (a) the presence or absence of salinity gradient and (b) salinity gradient with different bottom area of SW, BW and FW compartment. Results of likelihood ratio tests between models are shown.

Model	Residual d.f.	Residual deviance	Model comparison	Likelihood ratio statistics (d.f.)	*p*
(a) Eel distribution under the presence or absence of salinity gradient					
1. Null	30	1192.9			
2. Presence or absence of salinity gradient	28	1184.4	M1 vs M2	8.49 (2)	0.014
3. Presence or absence of salinity gradient + Month	22	1174.8	M2 vs M3	9.57 (6)	0.14
(b) Eel distribution under salinity gradient					
1. Null	62	2374.2			
2. Bottom area of SW, BW and FW compartment	58	2043.7	M1 vs M2	330.5 (4)	$< 10^{-9}$
3. Bottom area of SW, BW and FW compartment + Month	52	2019.3	M2 vs M3	24.4 (6)	4.48×10^{-4}

and therefore, more eels appeared in the compartment with the larger bottom area. The field data of European eels reported that glass eels accumulate at the tidal limit and then disperse and distribute evenly across the saline estuary to freshwater sites at the young yellow stage (Edeline et al., 2007). These dispersions, occurring mainly during the elver stage, may help to increase individual fitness more evenly and make the most efficient use of potential habitats.

Time-window hypothesis

On the basis of our experiments, we developed a new hypothetical diverging mechanism of eels into sea, estuary and river waters during the transition period from glass eels to yellow eels. This transition period is the period from migration by glass eels and elvers to the foraging movement phase by yellow eels (McCleave and Edeline, 2009). In the migration phase, the energetic status of individuals is proposed as an internal cue affecting juvenile behaviour (Edeline, 2007; McCleave and Edeline, 2009). However, these relationships have not been fully researched in Japanese eels, especially about their salinity-choice behaviour (section 'Salinity-choice tendency and locomotor activity at the glass eel stage'). However, the ontogenetic preference shift from freshwater to seawater was observed in Japanese eels reared in seawater during the transition period (section 'Ontogenetic shift of salinity-choice tendency'). Their preference for the riverine odour was maintained during the growth phase (section 'Ontogenetic shift of salinity-choice tendency'), and the labile and/or volatile DOM included in river water were odour attractants (section 'Habitat range in seawater derived from eel behaviour').

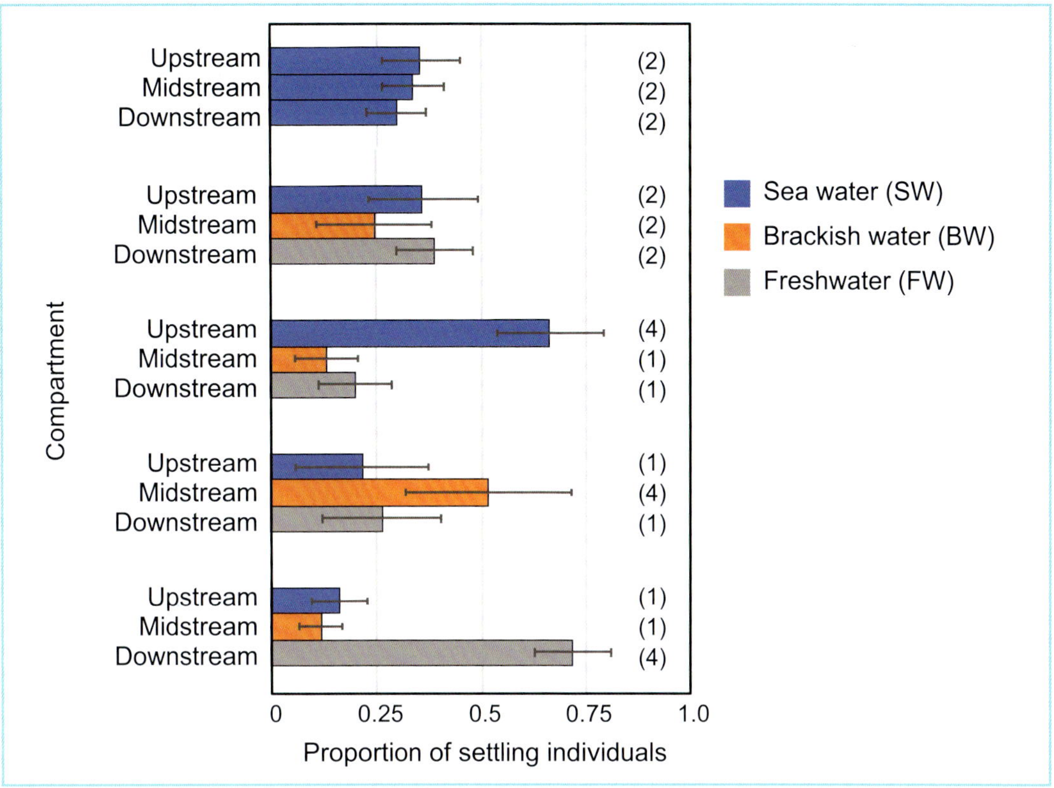

Figure 9.4 Mean ± SD proportions of eels in upstream, midstream and downstream compartments in outdoor tanks during June to September. The numbers in parentheses indicate bottom area (one unit = 45 cm × 95 cm) of each compartment with sea water (blue bar), brackish water (orange bar) or freshwater (grey bar).

The use of outdoor tanks provided more detailed information regarding distribution patterns during the transition period. The dispersion primarily during the elver stage would form the ideally free distribution according to the encountered environment (section 'Capacity effect on settling eels in each salinity habitat'). Our laboratory experiment showed that elvers did not have any preference for salinity (section 'Ontogenetic shift of salinity-choice tendency'), which would facilitate their dispersal across different salinity habitats. Settlement in each salinity habitat was observed in yellow eels in the autumn (section 'Capacity effect on settling eels in each salinity habitat'), indicating that developed eels settled down in the habitat used during the transition period from elver to yellow eel.

We propose this observed process in Japanese eels as the time-window hypothesis (Figure 9.5). This process assumes that glass eels choose their growth habitats using riverine odour and occasionally low salinity. Elvers disperse according to the capacity of each salinity habitat and settle down in a wide range of salinities as long as it has a rich riverine odour. If individuals grow in a particular salinity of water until the yellow eel stage, this causes them to remain there. The dispersion of elvers is consistent with the 'ideally free distribution', which would help eels maximize individual fitness equally. Subsequently, the 'time-window' to permit the movement

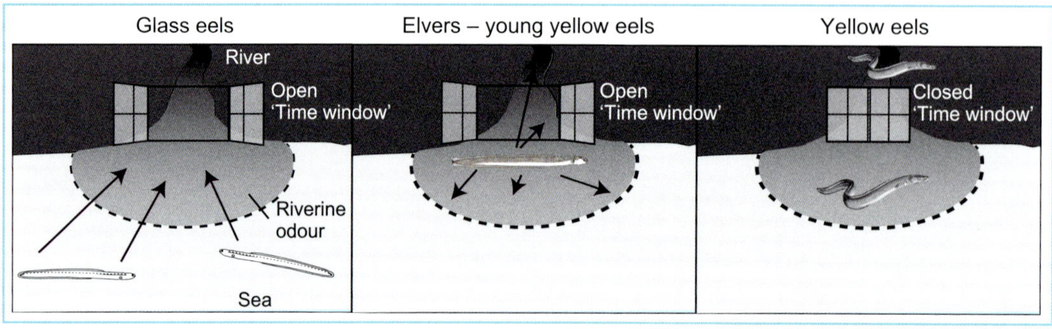

Figure 9.5 Conceptual diagram of time-window hypothesis. Arrows indicate the movement of individuals at glass eel and elver–young yellow eel stage. The dashed line represents the boundary of eel habitat and grey gradation shows the riverine odour. The conceptual time-window is open until young yellow eel stage and then closed afterward.

between saline water and freshwater will be closed biologically after juvenile eels spend a certain time in a particular salinity of water. During the subsequent period of foraging movement made by yellow eels, the freshwater entry from saline waters (upstream habitat shift) seemed to occur infrequently compared with movement from freshwater to saline waters (downstream habitat shift) (European eels [Daverat and Tomás, 2006], Japanese eels [Yokouchi et al., 2012]). This suggested that the freshwater entry may be restricted during the early period in continental habitats, which may help to better understand their migratory patterns and population dynamics.

Management implications

Elvers disperse in a wide range of salinities to make the most of their potential habitats; therefore, anthropogenic obstructions such as sluice gates or weirs near estuaries may have negative impacts on their ideal distribution. There are many sluice gates near estuaries in Japanese rivers to prevent salt damage to water intended for public, industrial and agricultural use. The fish pathways are usually equipped for strong swimmers like ayu (*Plecoglossus altivelis*) in these gates, but the intensive flow speed is often difficult to swim against for weak swimmers like elvers. Eels have a remarkable ability to climb up moist slopes; therefore, fishways specific to eels like eel ramps will be helpful to increase the connectivity between their upstream and downstream habitats (Briand et al., 2002; Knights and White, 1998).

In addition to keeping the connectivity from the sea to upstream river, increasing the potential capacity of each salinity habitat may positively affect the eel stocks during both the transition phase and the subsequent settlement phase. Riverine odour not only plays a role in external cues for glass eels seeking growth habitat but also possibly affects the range in sea-estuary habitats of eels. Labile and/or volatile DOMs in river water are sources of the odour (section 'Habitat range in seawater derived from eel behaviour'), and the presumable primary source is decaying leaf detritus due to microorganisms (Sorensen, 1986). Increasing riverside trees could be a practical way to enhance the capacity for eels across varied habitats. Increasing leaf drop in rivers could induce glass eels into rivers and extend the sea-estuary habitats. Furthermore, the fallen leaves would be shelters for eels, and the decaying leaf litter may also contribute to the productivity of an entire stream system (Wallace et al., 1999).

Gaining fundamental knowledge on eel behaviour and habitat utilization is pivotal in enhancing eel resources. Thus, there is a need to continuously investigate the means to refine eel management and utilize obtained knowledge of eel life history to form symbiotic relationships with eels and sustainably utilize them as a resource.

Acknowledgements

This study was financially supported by the Fisheries Agency of Japan. We thank the Chiba Prefectural Freshwater Fisheries Research Center and the Freshwater Fisheries Cooperative Associations of Chiba Prefecture for their assistance in the collection of glass eels. We thank all the other members in Yokosuka Laboratory, National Research Institute of Fisheries Science for supporting fish rearing.

Bibliography

Boivin, B., Castonguay, M., Audet, C., Pavey, S.A., Dionne, M. and Bernatchez, L. (2015) How does salinity influence habitat selection and growth in juvenile American eels *Anguilla rostrata*? *Journal of Fish Biology* 86, 765–784.

Briand, C., Fatin, D. and Legault, A. (2002) Role of eel odour on the efficiency of an eel, *Anguilla anguilla*, ladder and trap. *Environmental Biology of Fishes* 65, 473–477.

Bureau Du Colombier, S., Bolliet, V., Lambert, P. and Bardonnet, A. (2007) Energy and migratory behavior in glass eels (*Anguilla anguilla*). *Physiology & Behavior* 92, 684–690.

Bureau Du Colombier, S., Bolliet, V. and Bardonnet, A. (2009) Swimming activity and behaviour of European *Anguilla anguilla* glass eels in response to photoperiod and flow reversal and the role of energy status. *Journal of Fish Biology* 74, 2002–2013.

Bureau du Colombier, S., Bolliet, V., Lambert, P. and Bardonnet, A. (2011) Metabolic loss of mass in glass eels at different salinities according to their propensity to migrate. *Estuarine, Coastal and Shelf Science* 93, 1–6.

Creutzberg, F. (1961) On the orientation of migrating elvers (*Anguilla vulgaris* Turt.) in a tidal area. *Netherlands Journal of Sea Research* 1, 257–338.

Daverat, F. and Tomás, J. (2006) Tactics and demographic attributes in the European eel *Anguilla anguilla* in the Gironde watershed, SW France. *Marine Ecology Progress Series* 307, 247–257.

Dekker, W., Casselman, J., Cairns, D., Tsukamoto, K., Jellyman, D. and Lickers, H. (2003) Worldwide decline of eel ressources necessitates immediate action. Québec Declaration of Concern. *Fisheries* 28, 28–30.

Edeline, E. (2007) Adaptive phenotypic plasticity of eel diadromy. *Marine Ecology Progress Series* 341, 229–232.

Edeline, E., Dufour, S. and Elie, P. (2005) Role of glass eel salinity preference in the control of habitat selection and growth plasticity in *Anguilla anguilla*. *Marine Ecology Progress Series* 304, 191–199.

Edeline, E., Lambert, P., Rigaud, C. and Elie, P. (2006) Effects of body condition and water temperature on *Anguilla anguilla* glass eel migratory behavior. *Journal of Experimental Marine Biology and Ecology* 331, 217–225.

Edeline, E., Beaulaton, L., Le Barh, R. and Elie, P. (2007) Dispersal in metamorphosing juvenile eel *Anguilla anguilla*. *Marine Ecology Progress Series* 344, 213–218.

Ford, M.I., Elvidge, C.K., Baker, D., Pratt, T.C., Smokorowski, K.E., Sills, M., Patrick, P. and Cooke, S. J. (2018) Preferences of age-0 white sturgeon for different colours and strobe rates of LED lights may inform behavioural guidance strategies. *Environmental Biology of Fishes* 101, 667–674.

Fretwell, S.D. and Lucas, H.L. (1969) On territorial behavior and other factors influencing habitat distribution in birds. *Acta Biotheoretica* 19, 16–36.

Fukuda, N., Miller, M.J., Aoyama, J., Shinoda, A. and Tsukamoto, K. (2013) Evaluation of the pigmentation stages and body proportions from the glass eel to yellow eel in *Anguilla japonica*. *Fisheries Science* 79, 425–438.

Fukuda, N., Yokouchi, K., Yamamoto, T., Kurogi, H. and Yada, T. (2019a) Salinity and odor preferences of Japanese eel during the first year of post-recruitment growth in saline water. *Journal of Ethology* 37, 93–99.

Fukuda, N., Yokouchi, K., Yamamoto, T., Kurogi, H. and Yada, T. (2019b) Body condition evaluated following anesthesia implicates no relationship between body conditions and salinity-choice propensities in *Anguilla japonica* glass eels. *Zoological Science* DOI: 10.2108/zs180090.

Jacoby, D.M.P., Casselman, J.M., Crook, V., DeLucia, M.B., Ahn, H., Kaifu, K., Kurwie, T., Sasal, P., Silfvergrip, A.M.C., Smith, K.G., Uchida, K., Walker, A.M. and Gollock, M.J. (2015) Synergistic patterns of threat and the challenges facing global anguillid eel conservation. *Global Ecology Conservation* 4, 321–333.

Kaifu, K., Miller, M.J., Aoyama, J., Washitani, I. and Tsukamoto, K. (2013) Evidence of niche segregateion between freshwater eels and conger eels in Kojima Bay, Japan. *Fisheries Science* 79, 593–603.

Kaifu, K., Tamura, M., Aoyama, J. and Tsukamoto, K. (2010) Dispersal of yellow phase Japanese eels *Anguilla japonica* after recruitment in the Kojima Bay-Asahi River system, Japan. *Environmental Biology of Fishes* 88, 273–282.

Knights, B. and White, E.M. (1998) Enhancing immigration and recruitment of eels: the use of passes and associated trapping systems. *Fisheries Management and Ecology* 5, 459–471.

Kotake, A., Arai, T., Ozawa, T., Nojima, S., Miller, M.J. and Tsukamoto, K. (2003) Variation in migratory history of Japanese eels, *Anguilla japonica*, collected in coastal waters of the Amakusa Islands, Japan, inferred from otolith Sr/Ca ratios. *Marine Biology* 142, 849–854.

Kotake, A., Okamura, A., Yamada, Y., Utoh, T., Arai, T., Miller, M.J., Oka, H.P. and Tsukamoto, K. (2005) Seasonal variation in the migratory history of the Japanese eel *Anguilla japonica* in Mikawa Bay, Japan. *Marine Ecology Progress Series* 293, 213–221.

McCleave, J.D. and Edeline, E. (2009) Diadromy as a conditional strategy: patterns and drivers of eel movements in continental habitats. In: Haro, A.J., Avery, T., Beal, K., Cooper, J., Cunjak, R., Dadswell, M., Klauda, R., Moffitt, C., Rulifson, R. and Smith, K. (eds) *American Fisheries Society Symposium*, Vol. 69, Bethseda, MD, pp. 97–119.

Mimura, T. (2005) Eels: Past and current state of the fishery and stock. *Report on stock assessment and evaluation (fiscal year 2004/2005) (Shigen-hyoka-chosa-itaku-jigyo-houkokusho Heisei 16 nendo)*. Japan Fisheries Information Service Center, Tokyo, pp.153–154 (in Japanese).

Ministry of Agriculture, Forestry and Fisheries of Japan (1964–2017) Annual Statistics on Fishery and Aquaculture Production. Ministry of Agriculture, Forestry and Fisheries of Japan, Tokyo, Japan (in Japanese).

Mochioka, N., Wakiya, R., Kurogi, H., Chow, S., Morishita, K., Aoyama, J. and Tsukamoto, K. (2010) Environmental histories of adult eels collected from the spawning ground in 2008 and 2009. *Abstracts, the Japanese Society of Fisheries Science Autumn Meeting 2009*, pp. 17 (in Japanese).

Moriarty, C. (2003) The yellow eel. In: Aida, K., Tsukamoto, K. and Yamauchi, K. (eds) *Eel Biology*. Springer, Tokyo, pp. 89–105.

Sola, C. and Tongiorgi, P. (1996) The effects of salinity on the chemotaxis of glass eels, *Anguilla anguilla*, to organic earthy and green odorants. *Environmental Biology of Fishes* 47, 213–218.

Sola, C., Spampanato, A. and Tosi, L. (1993) Behavioural responses of glass eels (*Anguilla anguilla*) towards amino acids. *Journal of Fish Biology* 42, 683–691.

Sorensen, P. (1986) Origins of the freshwater attractant(s) of migrating elvers of the American eel, *Anguilla rostrata*. *Environmental Biology of Fishes* 17, 185–200.

Takahashi, K., Masuda, R. and Yamashita, Y. (2010) Ontogenetic changes in the spatial learning capability of jack mackerel *Trachurus japonicus*. *Journal of Fish Biology* 77, 2315–2325.

Tesch, F.W. (2003) Post-larval ecology and behaviour. In: Tesch, F.W. and Thorpe, J.E. (eds) *The Eel*, 3rd edn. Blackwell Science, Oxford, UK, pp. 181–183.

Tosi, L. and Sola, C. (1993) Role of geosmin, a typical inland water odour, in guiding glass eel *Anguilla anguilla* (L.) Migration. *Ethology* 95, 177–185.

Tosi, L., Sala, L., Sola, C., Spampanato, A. and Tongiorgi, P. (1988) Experimental analysis of the thermal and salinity preferences of glass-eels, *Anguilla anguilla* (L.), before and during the upstream migration. *Journal of Fish Biology* 33, 721–733.

Tosi, L., Spampanato, A., Sola, C. and Tongiorgi, P. (1990) Relation of water odour, salinity and temperature to ascent of glass-eels, *Anguilla anguilla* (L.): a laboratory study. *Journal of Fish Biology* 36, 327–340.

Tsukamoto, K., Nakai, I. and Tesch, W.V. (1998) Do all freshwater eels migrate? *Nature* 396, 635–636.

Tzeng, W., Shiao, J. and Iizuka, Y. (2002) Use of otolith Sr:Ca ratios to study the riverine migratory behaviors of Japanese eel *Anguilla japonica*. *Marine Ecology Progress Series* 245, 213–221.

Wallace, J.B., Eggert, S.L., Meyer, J.L. and Webster, J.R. (1999) Effects of resource limitation on a detrital-based ecosystem. *Ecological Monographs* 64, 409–442.

Yamamoto, S., Sekino, M. and Nagaiwa, R., (2019) A non-destructive method for individual identification in the Japanese eel *Anguilla japonica* based on DNA markers: a step-by-step protocol. Fisheries Agency, Japan, pp.1–60 (in Japanese).

Yokouchi, K., Aoyama, J., Oka, H.P. and Tsukamoto, K. (2008) Variation in the demographic characteristics of yellow-phase Japanese eels in different habitats of the Hamana Lake system, Japan. *Ecology of Freshwater Fish* 17, 639–652.

Yokouchi, K., Fukuda, N., Miller, M.J., Aoyama, J., Daverat, F. and Tsukamoto, K. (2012) Influences of early habitat use on the migratory plasticity and demography of Japanese eels in central Japan. *Estuarine, Coastal and Shelf Science* 107, 132–140.

CHAPTER 10

Behavioural patterns of migrating silver eel at a hydropower station on the River Meuse

A. Ben Griffioen, Olvin A. van Keeken and Hendrik V. Winter

Wageningen University, Wageningen, the Netherlands

Introduction

A telemetry experiment was carried out to study the behavioural patterns of downstream migrating silver eels confronted by a hydropower station (HPS)–weir–ship lock complex on the River Meuse. This study used four detection stations (NedapTrail-System®) covering the full river bed, two stations up and downstream of the HPS, two stations up and downstream of the weir and two stations in separated fish passes. One hundred and forty-eight silver eels were surgically implanted with Nedap-transponders in 2010. Of these, 123 eels migrated into the study area and showed a large variety of individual behavioural patterns. The majority, 68%, of the eels passed through the turbines either immediately or after hesitation or avoidance behaviour, 15% of the eels migrated over the weir and it was deduced that 14% of the eels migrated using ship locks in lateral canals either directly or after first returning to the HPS complex. A large number of silver eel (32%) passed the complex without apparent delay and 31–36% showed avoidance behaviour at greater distances (at least 50 m–1 km). Those eels did not follow the flow. An additional 28–32% passed the complex after hesitation at smaller distances or postponed migration. This study showed that eel migration does not always follow the flow when approaching a river barrier and altered behaviour may be exhibited. Some eels even moved upstream for considerable distances under high discharge conditions. Surprisingly, after initial arrival at the weir, most of these eels instead of passing the overflowing weir eventually passed through the turbines causing a high risk of lethal injuries. To be more effective in managing eel passage at barrier complexes a better understanding of the cues that trigger behavioural responses when they depart following the flow is needed.

European eel (*Anguilla anguilla*) has been in decline for several decades (Dekker, 2004). Many contributing factors have been suggested for this decline and most of these are anthropogenic. One important factor is migratory barriers, such as dams, hydropower stations (HPS), weirs or ship locks, which may cause high mortality rates, blockage or delay, and energy loss for silver eels during downstream migration to the sea (Acou et al., 2008; Calles et al., 2010; Jansen et al., 2007; Pedersen et al., 2012; Winter et al., 2007).

A better understanding of eel behavioural patterns and behaviour in the vicinity of river barriers is necessary in order to safeguard the efficient downstream passage for migrating silver eels and to reduce mortality rates at barriers. Timing of the downstream migration of silver eels in relation to environmental conditions may have a large influence on passage success of HPS–weir complexes. The peak migration activity occurs in autumn, when high precipitation levels and flood events trigger migration onset. Factors triggering onset might be higher or increasing discharge levels, rainfall, wind, atmospheric depressions and turbidity (Durif and Elie, 2008; Marohn et al., 2014; Tesch, 2003; Sandlund et al., 2017; Stein et al., 2016; Winter et al., 2006). River discharge level or a rapid increase of water discharge is suggested to be one of the most influential parameters that initiate silver eel migration (Behrmann-Godel and Eckmann, 2003; Durif et al., 2002; Jansen et al., 2007; Marohn et al., 2014; Sandlund et al., 2017; Vollestad et al., 1986). Other factors such as water temperature, luminosity, micro-seismic activity and lunar phase are also suggested to trigger migration (Durif and Elie, 2008; Durif et al., 2002; Marohn et al., 2014; Stein et al., 2016; Tesch, 2003).

Jansen et al. (2007) showed that of migrating eels many approached the trash rack of a HPS and at least 50% of those eels showed stationary, hesitation or recurrence behaviour. Other studies showed that some migrating eels swam upstream again approaching the HPS at close vicinity and made several consecutive attempts with varying timespans (Behrmann-Godel and Eckmann, 2003; Durif et al., 2002; Newbold et al., 2015). This behaviour is often called recurrence or hesitation behaviour (Jansen et al., 2007; Pedersen et al., 2012). To date, the number of detection stations in previous studies in the Netherlands per HPS–weir complex has been small, which has hampered the interpretation of precise movement patterns at these barriers in relation to environmental conditions (Jansen et al., 2007; Winter et al., 2007).

In this study, during which additional detection stations were added, we examined detailed eel behavioural patterns at a HPS–weir–fish pass–ship lock complex on the River Meuse. In order to direct eels away from hazardous routes to a safe passage route, and implement management measures to minimize mortality, obstruction or delay, more data on migratory behaviour near highly controlled water structures is necessary. Our focus is on quantifying specific behaviour shown by eels in the vicinity of an HPS –weir complex and how this relates to river discharge.

Material and methods

Study area

The River Meuse, originating in France and flowing through Belgium and the Netherlands, has a total length of 935 km, a catchment area of 36,000 km^3 and a mean discharge of 230 m^3/s, characterized by short peak flows following rainfall (Winter et al., 2006). Five weirs and two HPS–weir complexes are located in the Dutch section of the River Meuse. Fish passes and ship locks are located alongside each weir and each HPS–weir complex. The studied HPS–weir complex has four horizontal Kaplan-bulb turbines, which are switched on/off automatically at certain levels of river discharge (Bruijs, 2004; Jansen et al., 2007). The HPS–weir complex near the village of Linne (Figure 10.1), has two lateral canals with ship locks parallel to the River Meuse with their entrances situated just upstream of the complex (Figure 10.2).

Figure 10.1 The weir at the Linne hydropower station Linne (downstream).

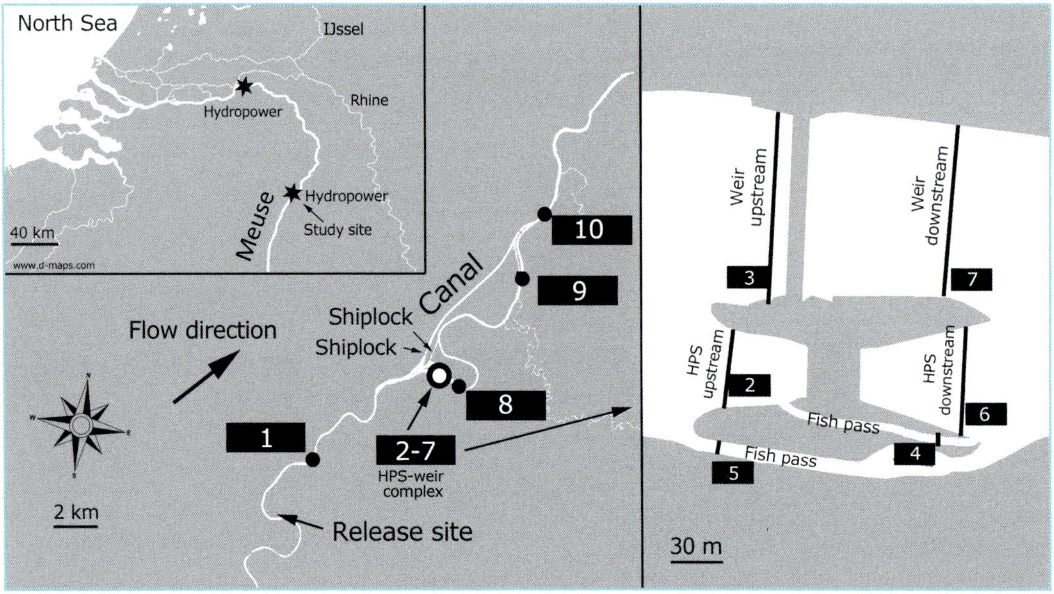

Figure 10.2 Left panel: Overview of the study area: a 15 km section of the River Meuse with ten detection stations and two lateral canals with ship locks. Right panel: detailed overview of the HPS–weir complex with six detection stations (detection stations numbers 2–7). Details of lakes and canals in the vicinity of the river are not shown in the figure.

Table 10.1 Management protocol of turbines (after Jansen et al., 2007).

Discharge	Protocol
0–30 m^3/s	Water is guided over the weir
>30–500 m^3/s	Water is guided through one or more turbines
>500–800 m^3/s	Water is equally divided over weir and turbines
>800 m^3/s	Water is guided over the weir

The basic HPS management protocol has four phases depending on the volume of river discharge (Table 10.1):

1. 0–30 m^3/s – all the water is guided over the weir
2. between 30 and 500 m^3/s – all the water is guided through one or more turbines depending on the discharge
3. above 500 m^3/s but lower than 800 m^3/s – the water is equally divided over the four turbines and the surplus water flows over the weir
4. above 800 m^3/s all the water flows over the weir.

Jansen et al. (2007) demonstrated that these management protocols are followed well in practice and that the actual flow through an HPS on any given day can be predicted accurately from the river discharge.

Telemetry experiments

The Nedap Trail™ telemetric system (NEDAP, Groenlo, the Netherlands) was used, for which an extensive infrastructure is available in the Dutch rivers (Breukelaar et al., 1998). Each detection station records the exact timing and unique ID of each transponder that passed the station within the detection range of approximately 50 m. Data on station detections were stored in a data-logger at each station and automatically retrieved daily via a telephone line connection. Each of the detection stations contains three parallel antenna cables on the river bottom, which send out interrogation signals every 4 s, activating each transponder that passes. When a silver eel remains present within the detection range, the transponder sends out a signal every 2 minutes. The battery life of these transponders is guaranteed for 2 years, if the total number of detections does not exceed 1000.

Compared to studies with a similar experimental setup carried out in 2002 and 2006 (Jansen et al., 2007; Winter et al., 2006, 2007) the number of detection stations was extended from two to six stations at the HPS–weir complex, allowing for discrimination between each potential route at the HPS–weir complex. For the lateral canals with ship locks passage was deduced from detection patterns in all other detection stations (Figure 10.2). Detection stations 2 (upstream) and 6 (downstream) covered passage through the HPS, stations 3 (upstream) and 7 (downstream) covered passage over the weir, and stations 4 and 5 covered passage through the fish passes (Figure 10.2). Detection stations 1, 8–10 covered the full river bed (Figure 10.2). If an eel was detected directly upstream the HPS at station 2, and the next detection was directly downstream at station 6, it was assigned migrated through the HPS. The same applies for migration over the weir (stations 3 and 7). In some cases, the last detection was upstream at station 2 or 3 at the complex

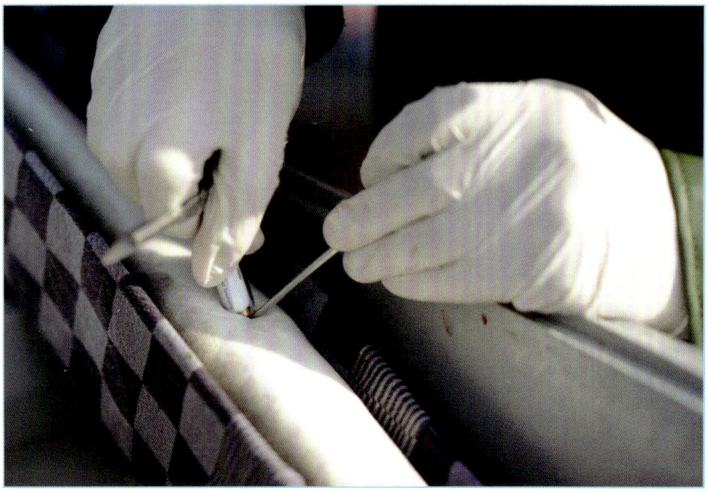

Figure 10.3 Implantation of NEDAP-transponder in a silver eel.

and the next detection was at stations 9 and 10, missing detection stations 6–8. In those cases, it was deduced that eels must have migrated through the lateral canals with ship locks.

For this study, 148 eels were caught by a local professional fisherman with fyke nets during September and October 2010 in the River Meuse at Ohé en Laak. Eels with completely silvery white ventral sides were used, rejecting individuals with yellow or partly yellow ventral sides. In total three batches were surgically implanted with Nedap-transponders (Figure 10.3) using the protocol described in Winter et al. (2006) and released in three batches into the River Meuse at Ohé en Laak, 13.5 km upstream of the HPS–weir complex on 14 September (n = 46), 23 September (n = 74) and 5 October (n = 28). The eels ranged 53.2–100 cm total length (mean 83 cm) and 660–879 g in weight (mean 1144 g). All individual eels were assumed to be females since males do not grow greater than 50 cm (Dekker, 2000).

Discharge at arrival of the HPS–weir complex

Discharge data at 10-minute intervals were available for the entire period at several locations along the Dutch sections of the river. Discharge was measured at Borgharen, 54 km upstream of the HPS–weir complex. In order to estimate the river discharge at which silver eels reached the detections station at the HPS–weir complex, a water velocity range of 0.5–1.5 m/s was assumed resulting in a range of river discharge delay of 10 hours at 0.5 m/s flows when using 0.5 m/s, and 30 hours when at 1.5 m/s flows. For eels that were detected at the HPS–weir complex, the discharge level when arriving was estimated using the time of detection and a discharge measurement delay range of 10–30 hours. This resulted in a 20 hour period of discharge data that encompassed the arrival time of an eel. The minimum (Q_{min}), maximum (Q_{max}) and average (Q_{avg}) discharge was calculated for this 20 hours period of discharge data. These values give an estimation of the discharge when an eel arrived at the HPS–weir complex. The variation of discharge levels within this 20h period was low: a peak in discharge higher than 800 m^3/s lasted for up to several days continuously while low discharge continued for up to several weeks, underlining that the variation in discharge levels within this 20 hour period is very low.

To determine which value would give the best fit for discharge estimates, a data check was performed. Detections of eels that were recorded at stations 2 and 6 (HPS), or stations 3 and 7 (weir), and which passed the complex without significant delay (immediately), were used to correspond with the management protocol (Table 10.1). In total 39 eels (n = 32 HPS and n = 7 weir) were used to check whether estimated discharge levels correspond to the water management protocol and successful passage. That is, only when discharge was between 30 and 800 m^3/s could eels pass the HPS and estimated discharge must be above 500 m^3/s in order to pass the weir.

The estimated Q_{min} of those 39 eels fell within the management protocol limits for 23 (59%) individuals. Therefore, 16 eels could never have passed the complex if Q_{min} was the discharge level at which they arrived. The management protocol does not allow for that. For Q_{max} this was correct for all 39 individuals (100%) and for Q_{avg} this was correct for 37 individuals (95%). Based on this data check, Q_{max} appears to be the best fit and was used for analyses.

Results

Overview of detections

At least 40 different patterns of subsequent order of detections at stations were observed, these could be summarized into ten categories (Table 10.2, cat. A–J), reflecting a large variation of behavioural patterns between individuals. In total 123 eels were detected at the HPS–weir complex, or further downstream, and were set at 100% as the batch starting migrating into the study area.

Non-migrating or missing eel

Of the 148 eels, 18 eels were never detected at any station (Table 10.2, cat. A) and must have remained in the upstream parts after release with an unknown fate. Seven eels did not pass the HPS–weir-ship lock complex after they were detected at station nr.1 (Table 10.2, cat. B1). Two eels swam upstream after reaching the HPS–weir complex, were again detected at station nr.1, and were not detected thereafter (Table 10.2, cat. B2). In total 14 eels (Table 10.2, cat. E) were never detected downstream at station nr.10 after passage of the HPS. It is possible that those eels died as a result of lethal injuries caused by the turbines while passing the HPS. It is unknown how many eels died due to lethal injuries after station number 10, further downstream.

Migrating eel

Of the 123 eels that were detected at the HPS–weir complex, 84 eels (68%) passed through the turbines of the HPS and 18 eels (15%) passed over the weir, while 17 eels (14%) migrated through the ship locks either directly during downstream movement (Table 10.2, cat. C) or after turning around at the HPS–weir complex and moving upstream to the canal entrances (Table 10.2, cat. D). The six eels that must have migrated through the lateral canals with ship locks directly, were never detected upstream of the HPS–weir complex, but were detected further downstream at stations 9 and/or 10 (Table 10.2, cat. C). Two eels swam upstream to station 1

Table 10.2 Overview of the results of the telemetry study of 148 eels.

Order of detection stations	Category	Total	< 10 min	< 1 day	1–7 days	> 7 days–months	Unknown	Passed station
Never detected	A	18						NP
1	B1	7						NP
1–10	C	4						SL
1–9–10	C	2						SL
Arrived at HPS (station nr.2)								
1–2	E	1					1	HPS
1–2–10	D	1					1	SL
1–2–3–10	D	1					1	SL
1–2–3–2	E	1		1				HPS
1–2–3–2–6–8–9–10	F	7		1	2	2	2	HPS
1–2–3–2–6 ↔ 7–8–9–10	G	2			2			HPS
1–2–3–7–8–9–10	F	1				1		WEIR
1–2–6–7	E, I	2	2					HPS
1–2–6–7–8	E, I	1					1	HPS
1–2–6–7–8–9–10	I	5	5					HPS
1–2–6–7–8–9	E, I	1	1					HPS
1–2–6–8–9–10	I	26	20		1		5	HPS
1–2–6	E	2	2					HPS
1–2–6–8	E	2	2					HPS
1–2–9 –10	D	1					1	SL
1–2 ↔ 3–1	B2, F	2					2	NP
1–2 ↔ 3–1–2–6–8–9–10	B3, F	1				1		HPS
1–2 ↔ 3–10	D, F	1					1	SL
1–2 ↔ 3–2–6–8–9–10	F	2			1	1		HPS
1–2 ↔ 3–4–8–9–10	F	1					1	FP
Total HPS first		61	32	2	6	5	16	
Arrived at weir (station 3)								
1–3–10	D	1					1	SL
1–3–2	E, F	1				1		HPS
1–3–2–3–7–8–9–10	F	1				1		WEIR
1–3–2–5–3 ↔ 2–1–3–2–3–7–8–9–10	B3, F	1					1	WEIR
1–3–2–6 –7–8–9–10	F	1	1					HPS
1–3–2–6–8–9–10	F	10	4	1	3	2		HPS
1–3–2–6 ↔ 7–8–9–10	G	1	1					HPS
1–3–2–6 ↔ 7	E, G	1		1				HPS
1–3–2–6 ↔ 7–8–9–10–3–10	G, H	1		1				HPS
1–3–7–8–9–10	J	12	7	1			4	WEIR
1–3–9–10	D	4					4	SL
1–3 ↔ 2–10	D, F	2					2	SL
1–3 ↔ 2–3–7–8–9–10	F	3				3		WEIR
1–3 ↔ 2–5–2 ↔ 3–5–2 ↔ 3–2–6–8– 9–10	F	1			1			HPS
1–3 ↔ 2–5–8–9–10	F	1					1	FP

Table 10.2 (Continued)

Order of detection stations	Category	Total	< 10 min	< 1 day	1–7 days	> 7 days–months	Unknown	Passed station
1–3 ↔ 2–6–8–9–10	F	13	1	1	4	7		HPS
1–3 ↔ 2–6	E, F	1	1					HPS
1–3 ↔ 2–6–8	E, F	1		1				HPS
Total weir first		56	15	6	8	14	13	
Total general		148	47	8	14	19	29	

Notes: For each eel, the order of stations (corresponding with Figure 10.2) and the time spent between first detection and the last detection upstream the HPS–weir complex was given, except for those that missed one or more detection stations. Each category (A–J) is pooled based on similar migration patterns. A = never detected. B1–3 = detected at station 1 either once (B1) or twice after upstream swimming (B2–3). C = lateral canal with ship locks direct. D = lateral canal with ship locks indirect after upstream swimming. E = not detected at (multiple) downstream stations. F = detections at both stations 2 and 3. G = detections at both stations 6 and 7. H = upstream lateral canal with ship lock passage. I = detected at HPS only. J = detected at weir only. NP = no passage. SL = ship locks. H = hydropower station. W = weir. FP = fish pass.

after multiple detections at the HPS–weir complex and continued migration after passing the complex (Table 10.2, cat. B3). Fifty-eight eels were detected at both detection stations 2 and 3 upstream of the complex (n = 53, Table 10.2, cat. F) or at stations 6 and 7 downstream of the complex (n = 5, Table 10.2, cat. G). One eel passed the HPS but swam upstream again through the lateral canals with ship locks, was detected at the complex, and swam downstream through the lateral canals with ship locks (Table 10.2, H). Forty-nine eels were either detected at only the HPS (n = 37, Table 10.2, cat. I) or only the weir (n = 12, Table 10.2, cat. J).

Eel that arrived at the HPS

In total 61 eels arrived at the entrance of the HPS. In total 52 eels (85%) out of 61 eels eventually passed the HPS of which the majority of this group (n = 32) passed the HPS within 10 minutes (Figure 10.4a). Of the eel that showed multiple detections as a result of recurrence behaviour (n = 29) the variation between individual patterns was large and resulted in multiple detections upstream of the complex at different times (Figure 10.4b). Eight eels that initially arrived at the HPS used other locations to pass the complex such as the weir (n = 1), fish pass (n = 1), lateral canals with ship locks (n = 4) or swam upstream and were never detected thereafter (n = 2). These behavioural patterns were summarized as 52% immediate passage (within 10 minutes) and 48% recurrence behaviour within the 'HPS group'.

Behavioural patterns after first arrival at the weir

In total 56 eels were first detected at the complex at the weir (Table 10.2). After reaching the weir seven of them passed the weir immediately thereafter (Figure 10.4a). The remaining 49 eels were detected at multiple stations or had multiple detections at a single station upstream of the HPS–weir complex (Figure 10.4c). Of the 49, 31 eels eventually passed the HPS. Of the remaining

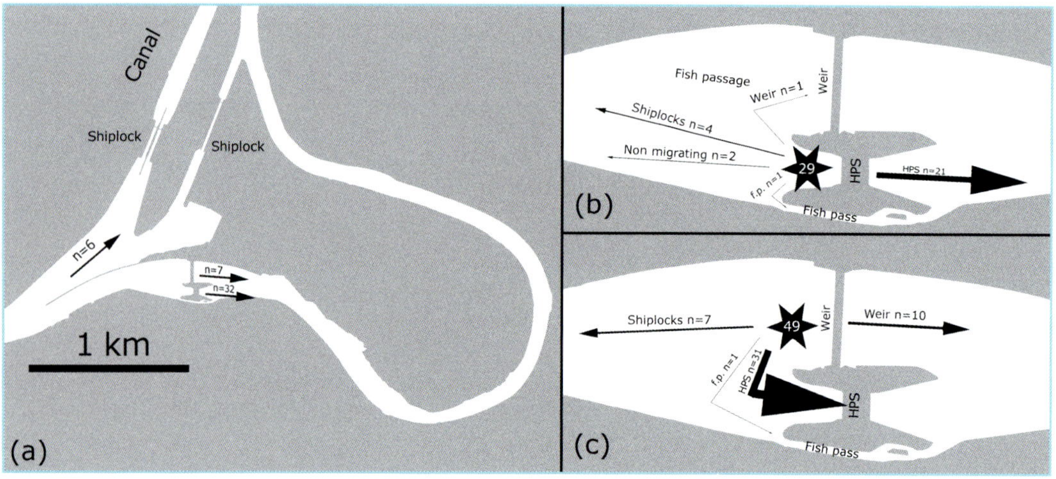

Figure 10.4 (a) Summarized behaviour of downstream migrating eel that showed immediate passage. (b) Eels that arrived at the entrance of the weir and showed recurrence behaviour thereafter. (c) Eels that arrived at the HPS and showed recurrence behaviour thereafter.

18 eels, 10 eventually passed the weir after some delay. One eel eventually passed the fish pass (station 5) and 7 eels turned around and passed the complex using the lateral canals with ship locks. These behavioural patterns were summarized as 12% immediate passage (<10 minutes) and 88% recurrence behaviour.

Discharge at arrival

In total 117 eels were detected at either the HPS or the weir (Table 10.2), 112 eels could be linked to a specific, estimated, discharge level (Q_{max}) since time of arrival was known (HPS n = 57 and weir n = 55). The other five missed a detection at stations 2 or 3, hence no discharge level could be determined since a time stamp was lacking. Estimated discharge (Q_{max}) levels at arrival varied between just below 60 m^3/s up to approximately 1800 m^3/s (Figure 10.5). The majority of the eels arrived with discharge levels of between 30 and 500 m^3/s (median = 176 m^3/s Q_{max}) for both arrival at the weir and for arrival at the HPS (Figure 10.5). The majority of the eels (n = 51) passed the HPS when all water was guided through the turbines (Table 10.3). The remaining six eels passed the weir, fish pass or lateral canals with ship locks, while water was guided through the turbines at the moment of arrival (Figure 10.6b–d).

Most eels arrived at the weir (n = 38) while all water was guided through the turbines (Table 10.3). If the discharge level was above 800 m^3/s, the detected eels arrived only at the weir (Figure 10.5 and Table 10.3). However, six eels that arrived at the weir at relatively high discharge levels turned around and did not pass the weir, but migrated through the lateral canals with ship locks (Figure 10.6h). Three eels arrived at the weir while there was water going through the turbines and over the weir (Figure 10.6e). Of the seven eels that passed the weir within 10 minutes, three arrived at discharge levels between 400 and 800 m^3/s and four passed it at levels higher than 800 m^3/s (Figure 10.6e). Thirty-five out of 55 eels (69%) arrived at the weir while all the river water was guided through the HPS turbines (Table 10.3).

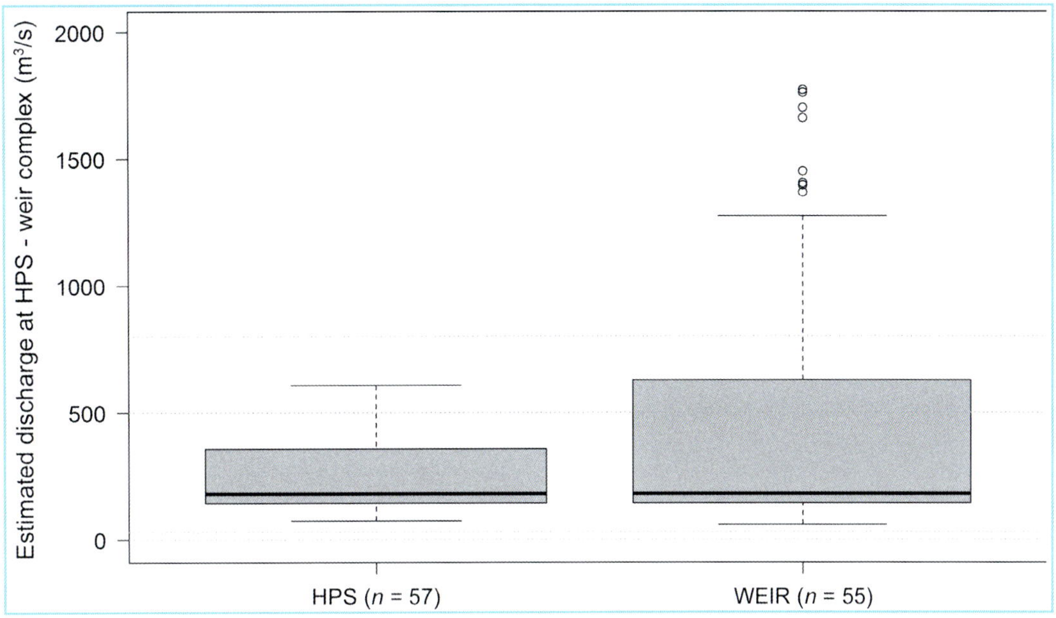

Figure 10.5 Estimated discharge levels at first arrival at the weir or the HPS (Q_{max}). Note: Dashed horizontal black lines indicate threshold value of management protocol according to Table 10.1.

Table 10.3 Estimated discharge (Q_{max}) of first arrival for 112 eels.

Discharge	Flow	Arrival at HPS	Arrival at weir	Total
0–30 m^3/s	weir	0	0	0
>30–500 m^3/s	HPS	51	38	89
>500–800 m^3/s	HPS and weir	6	6	12
>800 m^3/s	weir	0	11	11
Total		57	55	112

Note: Table shows number of eel of which discharge at arrival could be estimated.

Classification of movement patterns based on detection sequences

Behavioural patterns observed for eels that arrive at the HPS–weir complex can be classified into three different groups:

1. **Immediate passage.** The first group consist of the eels (32%, HPS n = 32 and weir n = 7) that passed the complex quickly after first arrival (<10 minutes).
2. **Avoidance.** A second group is defined as eels that showed avoidance behaviour at greater distances (50 m–1 km) The results showed that this may be applicable for 31% of the migrating eels (n = 38 weir Table 10.3). Those eels did not continue to follow the main flow

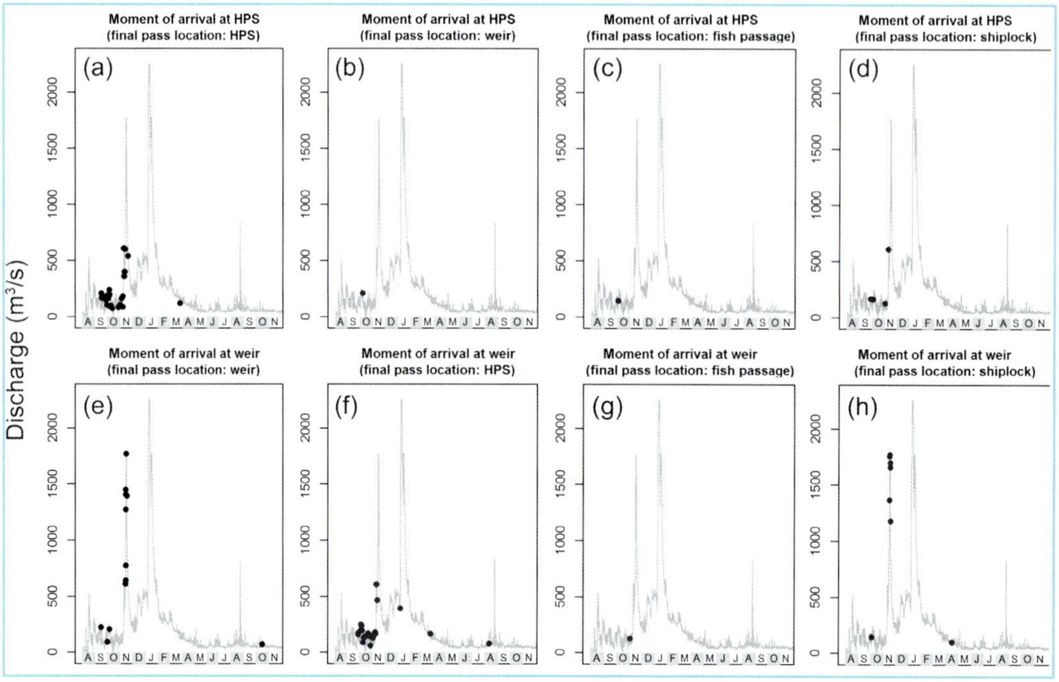

Figure 10.6 Moment of arrival of eels (black dots) at the HPS (a–d) or the weir (e–h) with corresponding discharge levels (grey lines, Q_{max}). Groups of different final passage locations are shown in separate panels. On the x-axis of each panel the months are shown starting at August 2010.

when approaching the complex, but arrived at the weir or ship locks while all the water was guided through the turbines of the HPS (>30–500 m^3/s). They turned around at least 50 m from the HPS. An additional six eels that passed through the lateral canals with ship locks directly should also be added to this group since they were never detected at the HPS–weir complex and chose a route with much lower flow than in the mainstream, potentially turning away from the HPS–weir complex at a greater distance of up to 1 km upstream (Figure 10.4a). In total 31–36% (n = 38–44) of the migrating eels showed route selections that lead to avoiding the HPS at greater distances (at least 50 m–1 km).

3 **Hesitation.** A third group of eels that initially followed the river flow, showed hesitation and/or searching behaviour after arrival near the entrance of the HPS or the weir. This is applicable to 28–32% (n = 34–39) of the migrating eels. This also includes a group of eels that arrived at the weir at relatively high discharge levels (>500 m^3/s), but turned around and swam upstream again to the ship locks or postponed further downstream migration.

Discussion

Although one-third of the migrating batch (32%, n = 32 HPS and n = 7 weir) passed the complex quickly after first arrival, the other two-thirds of the downstream migrating eel clearly responded to the barrier in the river with a large variety of movement patterns. In total, 68%

(n = 29 weir, n = 49 HPS and n = 6 lateral canals with ship locks) showed hesitation or avoidance behaviour. Moreover, for 38 eels the initial arrival point at the complex was not according to following the river flow and eels were first detected at the weir while the main river flow was guided through the turbines (Table 10.3), suggesting deviation from following the river flow at greater distances upstream from the complex. In addition, some eels swam several kilometres upstream again after first arrival at the complex. Others swam upstream during high discharge levels and eventually passed through the lateral canals with ship locks (Figure 10.6b–c) or passed these canals twice (Table 10.2, cat. H). This upstream migration behaviour was also observed in several other studies (Behrmann-Godel and Eckmann, 2003; Durif et al., 2002; Haro et al., 2000; Newbold et al., 2015; Piper et al., 2015).

The majority of the eels deviated from following the main flow and it is likely that those eels behaviourally responded to other trigger factors than river flow, which resulted in avoiding the HPS entrance at greater distances without having physical or visual contact to the trash rack or other structures of the HPS (Figure 10.6). A study by Piper et al. (2015) showed that eels swam along, or in the vicinity of, the river banks instead of using the main stream and also appeared to respond negatively to a floating debris boom without physical or visual contact (Piper et al., 2017). When eels showed similar positioning along the river banks this might explain the detections at the weir while the majority of the flow was at the opposite side of the river. However, eels in the study by Piper et al. (2015) still followed the stream up to several metres (<~10 m) before reaching the trash rack, while in our study eels appeared to have deviated from the river flow resulting in avoiding the HPS at greater distances. Piper et al. (2015) suggested that semi-passive drift probably accounts for the majority of downstream adult silver eel movement and that they reject abrupt changes in flow fields on the approach to structures. Another candidate trigger factor that can be noticed by eel upstream from the HPS–weir complex and may be used as a cue for altering their behaviour is sound. However, on the response of silver eel to sound not much is known, other than that they respond to low frequency infrasound (Sand et al., 2000).

There is no clear explanation as to what triggered eels to avoid, or why they hesitated when, approaching the complex. Sound produced by the turbines or overflowing weir and abrupt flow velocity gradients may cause alterations in 'normal' downstream swimming behaviour. These cues may be the reason that eel avoid the HPS at shorter or greater distances. Keefer et al. (2013) state that variability is a hallmark of animal behaviour and the degree of variability may fluctuate in response to environmental or biological gradients. They tagged and followed five species (lampreys, salmonids and shad) and showed that they were active during most of the diel cycle in low gradient and less hydraulically complex habitats. However, movements shifted predominantly to diurnal (salmonids and shads) or nocturnal (lamprey) at hydroelectric dam fishways where hydraulic complexity and predator density were high. Keefer et al. (2013) suggest that context-dependent behaviour is common during fish migrations, and that diel activity patterns vary with the degree of effort or predation risk required for movement. This is in line with the more nightly movement pattern for migrating eel near the HPS complex when compared to free-flowing river stretches in the Meuse, as observed by Winter et al. (2006).

Hesitation or avoidance behaviour will result in a migratory delay of up to several days and in some cases even up to several months (Table 10.2). Any kind of delay or blockage to reach the spawning grounds can cause loss in energy and subsequently in successful reproduction. In addition, there may be a higher risk of predation due to increased local densities. This may also result in a mismatch in timing when reaching the spawning grounds. River discharge and management protocols of barriers in the river play a crucial role in the success of downstream migration.

Our study showed that roughly one-third of the migrating eels showed behaviour that deviated from following the flow (for example, avoidance at greater distances) when approaching the HPS complex and another one-third showed hesitation behaviour. Moreover, some eels moved upstream after approaching the complex under high discharge conditions. Mortality rates may be reduced by providing alternative downstream passage opportunities that allows for recurrence behaviour and upstream swimming before the eels have become habituated to the situation or, are able to follow the flow despite the trigger factors that made them to deviate from following the flow and eventually end up in the HPS as seen in Figure 10.4c. Despite their initial arrival at the weir, most of these eels eventually did pass through the turbines causing a high risk of lethal injuries. The interaction of different trigger factors, for example, sound, underwater structures or topography, which cause them to deviate from following the river flow, what role habituation plays, and at what timescales the strong incentive to move downstream wins over factors causing recurrence or hesitation behaviour is still largely unknown. To be more effective in guidance of fish pass design a better understanding of cues that causes migration patterns independent to flow is needed. This may be reached by combining detailed 3D tracking of eels in relation to measuring environmental parameters such as flow, turbulence, sound, turbidity and temperature along with modelling approaches using different behaviour scenarios and responses to trigger factors.

Acknowledgement

The study was funded by the Dutch Ministry of Agriculture, Nature and Food Quality.

Bibliography

Acou, A., Laffaille, P., Legault, A. and Feunteun, E. (2008) Migration pattern of silver eel (Anguilla anguilla, L.) in an obstructed river system. *Ecology of Freshwater Fish* 17, 432–442.

Behrmann-Godel, J. and Eckmann, R. (2003) A preliminary telemetry study of the migration of silver European eel (*Anguilla anguilla* L.) in the River Mosel, Germany. *Ecology of Freshwater Fish* 12, 196–202.

Breukelaar, A.W., bij de Vaate, A. and Fockens, K.T.W. (1998) Inland migration study of sea trout (*Salmo trutta*) into the rivers Rhine and Meuse (the Netherlands), based on inductive coupling radio telemetry. *Hydrobiologia* 371, 29–33.

Bruijs, M.C.M. (2004) Effectiviteit visgeleidingssystemen bij de bestaande waterkrachtcentrales Linne en Alphen. KEMA-report 04–7019, Arnhem, the Netherlands.

Calles, O., Olsson, I.C., Comoglio, C., Kemp, P., Blunden, S.L., Schmitz, M. and Greenberg, L.A. (2010) Size-dependent mortality of migratory silver eels at a hydropower plant, and implications for escapement to the sea. *Freshwater Biology* 55, 2167–2180.

Dekker, W. (2000) Impact of yellow eel exploitation on spawner production in Lake IJsselmeer, the Netherlands. *Dana* 12, 17–32.

Dekker, W. (2004) Slipping through our hands: population dynamics of the European Eel. PhD thesis. University of Amsterdam.
Durif, C.M.F. and Elie, P. (2008) Predicting downstream migration of silver eels in a large river catchment based on commercial fishery data. *Fisheries Management and Ecology* 15, 127–137.
Durif, C., Elie, P., Gosset, C., Rives, J. and Travade, F. (2002) Behavioral study of downstream migrating eels by radio-telemetry at a small hydroelectric power plant. *American Fisheries Society Symposium* 33, 343–356.
Haro, A., Castro-Santos, T. and Boubée, J.A.T. (2000) Behavior and passage of silver-phase American eels, *Anguilla rostrata* (LeSueur), at a small hydroelectric facility. *Dana* 12, 33–42.
Jansen, H.M., Winter, H.V., Bruijs, M.C.M. and Polman, H.J.G. (2007) Just go with the flow? Route selection and mortality during downstream migration of silver eels in relation to river discharge. *ICES Journal of Marine Science* 64, 1437–1443.
Keefer, M., Caudill, C., Peery, C. and Moser, M. (2013) Context-dependent diel behavior of upstream-migrating anadromous fishes. *Environmental Biology of Fishes* 96, 691–700.
Marohn, L., Prigge, E. and Hanel, R. (2014) Escapement success of silver eels from a German river system is low compared to management-based estimates. *Freshwater Biology* 59, 64–72.
Newbold, L.R., Hockley, F.A. Williams, C.F., Cable, J., Reading, A.J., Auchterlonie, N. and Kemp, P.S. (2015) Relationship between European eel *Anguilla anguilla* infection with non-native parasites and swimming behaviour on encountering accelerating flow. *Journal of Fish Biology* 86, 1519–1533.
Pedersen, M.I., Jepsen, N., Aarestrup, K., Koed, A., Pedersen, S. and Økland, F. (2012) Loss of European silver eel passing a hydropower station. *Journal of Applied Ichthyology* 28, 189–193.
Piper, A.T., Manes, C., Siniscalchi, F., Marion, A., Wright, R.M. and Kemp, P.S. (2015) Response of seaward-migrating European eel (*Anguilla anguilla*) to manipulated flow fields. *Proceedings of the Royal Society of London B: Biological Sciences* 282, 282:20151098.
Piper, A.T., Svendsen, J.C., Wright, R.M. and Kemp, P.S. (2017) Movement patterns of seaward migrating European eel (*Anguilla anguilla*) at a complex of riverine barriers: implications for conservation. *Ecology of Freshwater Fish* 26, 87–98.
Sand, O., Enger, P.S., Karlsen, H.E., Knudsen, F. and Kvernstuen, T. (2000) Avoidance responses to infrasound in downstream migrating European silver eels, *Anguilla anguilla. Environmental Biology of Fishes* 57, 327–336.
Sandlund, O.T., Diserud, O.H., Poole, R., Bergesen, K., Dillane, M., Rogan, G., Durif, C., Thorstad, E.B. and Vollestad, L.A. (2017) Timing and pattern of annual silver eel migration in two European watersheds are determined by similar cues. *Ecology and Evolution* 7, 5956–5966.
Stein, F., Doering-Arjes, P., Fladung, E., Bramick, U., Bendall, B. and Schroder, B. (2016) Downstream migration of the European eel (*Anguilla anguilla*) in the Elbe River, Germany: movement patterns and the potential impact of environmental factors. *River Research and Applications* 32, 666–676.
Tesch, F.W. (2003) *The Eel.* Blackwell Science, Oxford.
Vollestad, L.A., Jonsson, B., Hvidsten, N.A., Naesje, T.F., Haraldstad, O. and Ruudhansen, J. (1986) Environmental-factors regulating the seaward migration of European silver eels (*Anguilla anguilla*). *Canadian Journal of Fisheries and Aquatic Sciences* 43, 1909–1916.
Winter, H.V. and Fredrich, F. (2003) Migratory behaviour of ide: a comparison between the lowland rivers Elbe, Germany, and Vecht, the Netherlands. *Journal of Fish Biology* 63, 871–880.
Winter, H.V., Jansen, H.M. and Bruijs, M.C.M. (2006) Assessing the impact of hydropower and fisheries on downstream migrating silver eel, *Anguilla anguilla*, by telemetry in the River Meuse. *Ecology of Freshwater Fish* 15, 221–228.
Winter, H.V., Jansen, H.M. and Breukelaar, A. (2007) Silver eel mortality during downstream migration in the River Meuse, from a population perspective. *ICES Journal of Marine Science* 64, 1444–1449.

CHAPTER 11

The chemical marking of eels – state of the art and application perspectives

Björn Kullmann and Ralf Thiel

University of Hamburg, Center of Natural History, Hamburg, Germany

Introduction

In 2003, eel researchers from around the world jointly published their findings of an obvious synchronous recruitment collapse of the three most economically important eel species, namely American eel (*Anguilla rostrata*), Japanese eel (*A. japonica*) and European eel (*A. anguilla*). Reasons for the drastic recruitment declines are poorly understood. Even though these three eel species have extended oceanic phases of migration during their catadromous life cycle, management measures are strictly limited to the continental phase. Since artificial breeding in captivity is not yet feasible, the reallocation of natural recruits, referred to as stocking, is the most commonly used option to supplement local eel stocks that chronically suffer from poor recruitment. There is a considerable gap of knowledge, however, concerning the benefit from these approaches as the removal of eels from the donor habitat may lead to a net loss in the reproductive output, and stocked recruits could show a decreased spawner quality. The International Council for the Exploration of the Sea (ICES) has therefore recommended the marking of eels prior to stocking in order to generate a comprehensive data basis for both a reliable stock assessment and the evaluation of potential benefit from stocking measures. As millions of eels are transferred from one water to another, including cross border relocations, chemical marking seems the only practical way to cope with these immense quantities of fish. Six different markers, namely alizarin complexone, alizarin red S (ARS), calcein, oxytetracycline, strontium and barium, have been used in eel science to produce detectable marks on otoliths of eight different species. ARS was found to be the best studied marker in terms of mass-marking practicability and marking efficiency while having an insignificant medium-term impact on growth and survival. Suggestions and recommendations for a standard ARS mass-marking protocol for stocked eels are presented and proposed to be integrated into eel management plans.

In general, the marking and tagging of fish is a widespread method in fisheries research with the first recorded studies in the second half of the 19th century (Jensen, 1962). Scientists mark fish both externally and internally to derive information about, for example, growth, migration and spatial distribution, age validation and population dynamics (Beamish and McFarlane, 1983; Thorsteinsson, 2002; Use of Fishes in Research Committee, 2014). The difference between marks and tags is that the former do not carry any individual information but allows only the discrimination between marked and unmarked individuals (Guy et al., 1996). By contrast, tags can individually store or record information, and hence have the potential to provide scientists with more detailed data (Thorsteinsson, 2002). Tagging, however, is also accompanied with difficulties and threats just as the rejection of tags, they might be overlooked by the observer, tags can have a certain size limit, they might injure the fish, influence the behaviour and might lead to increased mortalities (Dunn and Coker, 1951).

A large variety of marking and tagging methods are known in eel science (Nielsen, 1988) and have helped, for example, to elucidate parts of eels migration, their home range behaviour and have also revealed an extraordinary phenotypic plasticity (Jellyman et al., 1996; Oliveira, 1997; Righton et al., 2016; Stein et al., 2016; Tzeng, 1984). Furthermore, the development and evaluation of management measures can be based on mark and recapture experiments (Klein Breteler et al., 2007; Kullmann and Thiel, 2018), which gained in importance in recent decades as the recruitment of eel stocks worldwide have decreased dramatically (Chen et al., 2014; ICES, 2017; Québec Declaration of Concern, 2003) and the reasons are poorly understood (Castonguay and Durif, 2016).

Stocking of juvenile eels is a common approach to supplement local eel stocks that chronically suffer from poor recruitment (ICES, 2016b; Kaifu et al., 2018; Stacey et al., 2015). Since the artificial breeding of any eel species is not yet possible – at least not on a commercial scale (Masuda et al., 2012) – this management measure depends completely on wild catches, which are already scarce. Despite a considerable lack of knowledge concerning the proof of concept, eel stocking measures are being largely conducted to increase the silver eel escapement biomass (Couillard et al., 2014; ICES, 2016a; Lin et al., 2010) or to simply conserve an eel fishery (Righton and Walker, 2013). ICES has therefore recommended the marking of stocked recruits in order to allow estimations about their contribution to future recruitment (ICES, 2008, 2016b). The principle threats associated with stocking measures, which are potentially negating the management objectives, are: a potential life history effect (Couillard et al., 2014; Stacey et al., 2015); biosecurity and spreading of diseases (ICES, 2017; Kullmann et al., 2017a); corruption of the spatial population structure (Ward, 2006); unnaturally skewed sex ratios due to density dependant effects (Côté et al., 2015); and an increased mortality, often preceded by a farming process (Simon and Dörner, 2014).

Owing to the extraordinary longevity of anguillid eels (Chisnall and Kalish, 1993; Jellyman, 1995; Silm et al., 2017; Tesch and Thorpe, 2003) marks should be very durable to enable the monitoring of the entire life span. In this regard, chemical marks on otoliths are likely to be persistent over time as post-formational degeneration processes inside otoliths can be considered as negligible (Campana, 1999; Campana et al., 1993). This makes the chemical marking of otoliths a suitable method for the evaluation of large-scale stocking programmes (Pratt and Threader, 2011; Warren-Myers et al., 2018; Wickström and Sjöberg, 2014).

Nielsen (1988) has published a comprehensive review about marking and tagging methods applied to European eels, while the aim of this chapter is to provide an overview of usable chemical markers for large-scale eel monitoring programmes, and the application to a variety of different *Anguilla* species published since then. Numerous studies have been conducted on *A. anguilla*,

A. rostrata and *A. japonica*, whereby other species have received much less attention. Widening the focus, however, helps to reflect the diverse bandwidth of this special field of eel research and could be helpful to enhance research and by which the understanding of the massive decline of eel resources across all continents. As all species of *Anguilla* eels resemble each other in morphology and ecology, the described methods are likely to be applicable across the species range. Furthermore, specific recommendations for a harmonized mass-marking protocol are intended to be a contribution to potentially increase the efficiency of conservation measures towards the recovery of depleted eel stocks.

Chemical markers, application and detection

Six different substances, namely alizarin complexone, alizarin red S, calcein, oxytetracycline hypochloride, strontium chloride hexahydrate and barium chloride dihydrate, have been found in the literature as being used for the chemical marking of eel otoliths (see Table 11.1). Radioactive markers (for example, Europium-152) were used in the past (Andersson et al., 1991; Hansen and Fattah, 1986), however, this method has been discredited due to numerous reasons (for example, very harmful to eels and humans). Reviewed studies cover eight different eel species out of the 19 currently valid species within the genus *Anguilla* (Tseng, 2016). This is solely a selection which is intended to present as many different species per marker as possible and to provide information about the lowest necessary concentration, the application, mortality rates and marking success on otoliths. Species double entries occur only if more details (for example, mortality) or other aspects (other development stage) can be presented (see Table 11.1). The general marking process varies largely between chemical markers and can be applied via immersion bath in fresh- or saltwater, under an osmotic shock, or via individual injection either in the body cavity (intraperitoneal) or intramuscular. All substances have been applied to *A. anguilla* whereas for all other species, hitherto, applications of not more than one or two markers have been published.

Alizarin complexone

Alizarin complexone (ALC) is the most widely used chemical marker in eel science and has been applied successfully to seven different *Anguilla* species (see Table 11.1). The marking process was carried out through immersion bath throughout with concentrations used varying between 50 and 150 mg l^{-1}. A duration of between 16 and 24 hours was sufficient to achieve a marking success of approximately 100% on otoliths at mortality rates, if reported, of below 1%. Studies have been carried out at different temperatures (c. 10–26°C) and in fresh- or saltwater (up to 33 PSU) with no effect on marking success or mortality. ALC, therefore, was found to be a widespread and pronounced robust method to mark eels chemically. Since ALC is fluorochrome, marks can be detected by fluorescence microscopy on ground and polished otoliths (Holmgren, 1996). The microscope should be equipped with an appropriate light filter with respect to the excitation wave length between 530–580 nm and a maximum emission wave length of 600–645 nm (van Gaalen et al., 2010).

Alizarin red S

Alizarin red S (ARS) has been applied to *A. anguilla* glass and farmed eels only (see Table 11.1). A concentration of 150 mg l^{-1} dissolved in freshwater and an exposure time of 3 hours for glass and 9 hours for farmed eels (mean body weight 6.5–8.9 g) has led to a marking success of 100% and mortality rates below 1% throughout. Temperatures varied slightly (13–17°C) and appeared to have no impact on reviewed parameters. ARS is also a fluorochrome and has the identical excitation and emission wavelengths as ALC (van Gaalen et al., 2010). For detection purposes, Kullmann et al. (2018b) demonstrated that the ARS mark is clearly visible on the cutting surface of transversally cut otoliths without further grinding and polishing effort which tremendously increases the detection efficiency. Moreover, the position of the mark is unambiguously assignable to a respective stocking form making the time-consuming grinding process dispensable (Figure 11.1).

Calcein

Calcein (CAL) was successfully used to mark the otoliths of *A. anguilla* glass and on-grown eels (see Table 11.1). Alcobendas et al. (1991) have applied CAL as a 3.5 minutes immersion bath (20 g l^{-1}) at a simultaneously induced osmotic shock at 50 g NaCl l^{-1}. No mortality was observed, however, marking success was only 75% on otoliths. Alternatively, CAL can be injected intraperitoneally at a concentration of 40 mg per kg body weight which has resulted in 100% marking success while no mortality was reported. The fluorochrome nature of calcein allows the detection via fluorescence microscopy at maximum wave lengths of 494 and 517 nm for excitation and emission, respectively (Sabnis, 2010).

Oxytetracycline

The antibiotic oxytetracycline (OTC) has frequently been used to mark a total of six different *Anguilla* species, not only as an immersion bath with freshwater or in a hyperosmotic solution, but also as intraperitoneal or intramuscular injection (see Table 11.1). *Anguilla japonica* glass eels were kept for 24 hours in a 300 mg OTC l^{-1} freshwater solution at approximately 10°C. The marking success was 100% on otoliths, however, mortality was not reported. *A. anguilla* glass eels have been exposed to OTC at high concentrations of 10 g l^{-1} at osmotic shock conditions (50 g NaCl l^{-1}) for a very short period of time (3.5 minutes). The marking success was reported at 100% while no mortality was observed. Oftentimes OTC was injected into the body cavity or the musculature at doses of between 75 and 100 mg kg^{-1} body weight of different species, which most often led to clear marked otoliths (see Table 11.1). However, this procedure occasionally failed to produce visible marks on otoliths of *A. anguilla* (Dekker, 1986) and *A. reinhardtii* (Pease et al., 2003), which might have been a consequence of the accidental injection of OTC into the gut (Pease et al., 2003). Among ALC, ARS, and CAL, OTC is also a fluorochrome, hence detection is possible using a fluorescence microscope which should be equipped with light filters accounting for maximum excitation wavelengths of 365–490 nm and emission wavelengths of 520–570 nm (van Gaalen et al., 2010).

Table 11.1 Overview and usage of chemical markers for different *Anguilla* species.

Marker	Concentration	Application	Duration	Temperature (°C)	Mortality (%)	Marking success on otoliths (%)	Species	Development stage (approx. size)	Reference
Alizarin complexone	50 mg l^{-1}	IB in FreshW	24 h	17	NA	100	*A. anguilla*	On-grown (197–282 mm)	Holmgren (1996)
	150 mg l^{-1}	IB in SaltW (30–33 PSU)	12 h	NA	NA	98	*A. rostrata*	Glass eels	Cieri and McCleave (2001)
	100 mg l^{-1}	IB in FreshW	24 h	24–26	NA	100	*A. marmorata*	Glass eels	Sugeha *et al.* (2001)
	100 mg l^{-1}	IB in FreshW	24 h	24.5–26.4	NA	100	*A. celebesensis*	Glass eels	Arai *et al.* (2000)
	80 mg l^{-1}	IB in SaltW[d]	NA[e]	NA	None	100	*A. australis*	Glass eels	Silberschneider (2005)
	80 mg l^{-1}	IB in SaltW[d]	NA[e]	NA	None	100	*A. reinhardtii*	Glass eels	Silberschneider (2005)
	100 mg l^{-1}	IB in SaltW (32.5 PSU)	16 h	10–15	0.8	100	*A. japonica*	Glass eels	Fukuda *et al.* (2009)
Alizarin red S	150 mg l^{-1}	IB in FreshW	3 h	17	NA	NA	*A. anguilla*	Glass eels (mean of 70 mm)	Simon and Dörner (2005)
	150 mg l^{-1}	IB in FreshW[a]	3 h	14	0.56	100	*A. anguilla*	Glass eels (mean of 64 mm)	Kullmann *et al.* (2018)
	150 mg l^{-1}	IB in FreshW[c]	9 h	13–17	< 1.0	100	*A. anguilla*	On-grown (mean BW of 6.5–8.0 g)	Kullmann *et al.* (2017)
Calcein	20 g l^{-1}	SaltW (50 g NaCl l^{-1})[b]	3.5 min		0	75[f]	*A. anguilla*	Glass eels	Alcobendas *et al.* (1991)
	40 mg kg BW^{-1}	Individual IP	NA	NA	0	100	*A. anguilla*	On-grown (<350 – >450 mm)	Mounaix (1991)
Oxytetracycline	300 mg l^{-1}	IB in FreshW	24 h	10	NA	100	*A. japonica*	Glass eels (mean of 58 mm)	Umezawa and Tsukamoto (1991)
	10 g l^{-1}	SaltW (50 g NaCl l^{-1})[b]	3.5 min	NA	0	100	*A. anguilla*	Glass eels	Alcobendas *et al.* (1991)
	10 g l^{-1}	SaltW (50 g NaCl l^{-1})[b]	3.5 min	17	0	NA	*A. anguilla*	Glass eels (mean of 70 mm)	Simon and Dörner (2005)

	100 mg kg BW^{-1}	Individual IP	NA	NA	0	77–80	*A. reinhardtii*	On-grown (420–971 mm)	Pease *et al.* (2003)
	75 mg kg BW^{-1}	Individual IM	NA	NA	NA	NA	*A. reinhardtii*	On-grown (180–620 mm)	Hoyle *et al.* (2005)
	75 mg kg BW^{-1}	Individual IP	NA	NA	NA	100	*A. dieffenbachii*	On-grown (360–828 mm)	Chisnall and Kalish (1993)
	75 mg kg BW^{-1}	Individual IP	NA	NA	NA	100	*A. australis*	On-grown (343–586 mm)	Chisnall and Kalish (1993)
	75 mg kg BW^{-1}	Individual IP	NA	NA	NA	NA	*A. rostrata*	< 300 mm	Oliveira (1996)
Strontium chloride hexahydrate	1 g l^{-1}	IB in FreshW	24 h	20–23	NA	100	*A. anguilla*	Glass eels	Wickström and Sjöberg (2014)
	1 g l^{-1}	IB in FreshW	24 h	12–14	0.58	100	*A. anguilla*	Glass eels	Kullmann *et al.* (2018)
Barium chloride dihydrate	0.75 g l^{-1}	IB in FreshW	24 h	20–23	NA	85	*A. anguilla*	On-grown (mean of 111 mm)	Wickström and Sjöberg (2014)

Notes: HC, hydrochloric e; BW, body weight; IB, immersion bath; FreshW, fresh water; SaltW, salt water; NA, not available; IP, intraperitoneal injection; IM, intramuscular injection. [a]after double marking procedure with alizarin red S and strontium; [b]5 kg eels were put in an nylon pouch and submerged; [c]buffered with 150 mg TRIS l^{-1}; [d]not specified; [e]'over night'; [f]marking success was 100 % on vertebrae.

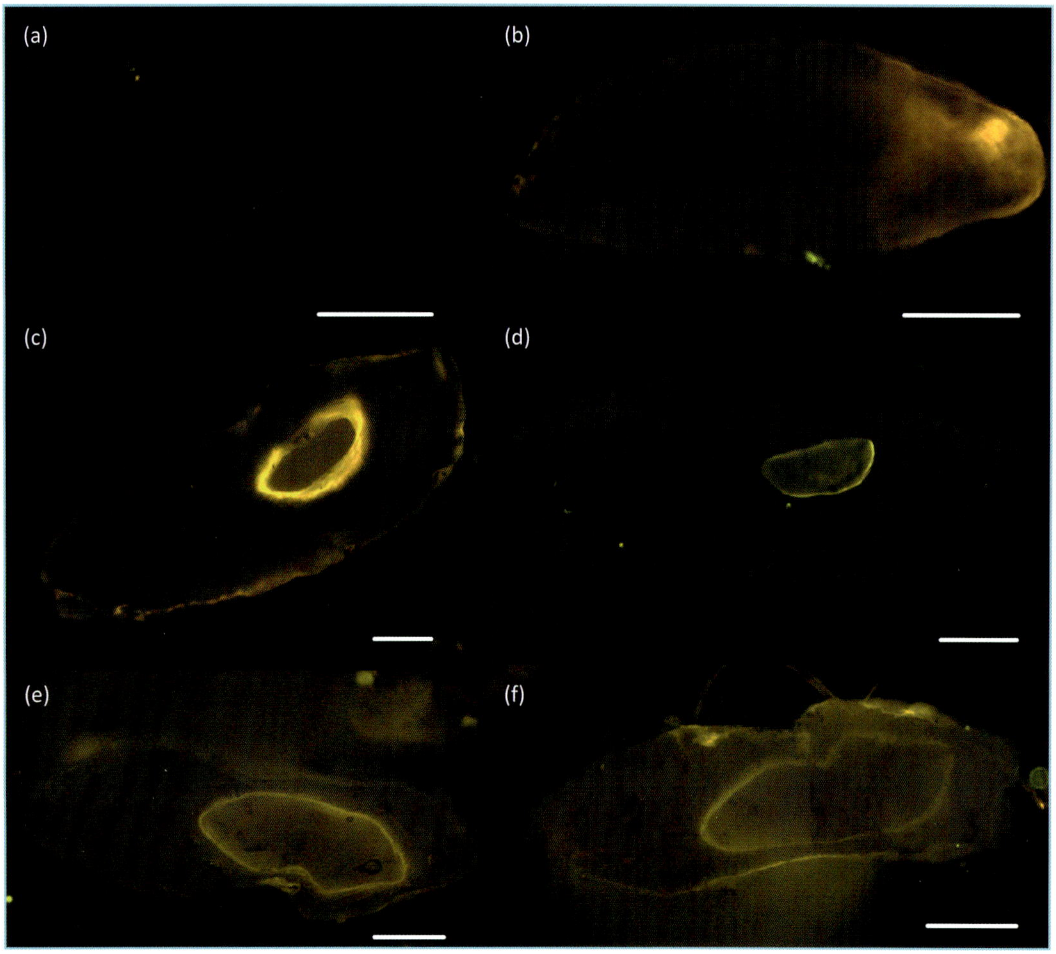

Figure 11.1 The cutsurface of unground *A. anguilla* otoliths showing no mark (a and b) and with clear alizarin red S marks (c–f). The position of the mark allows the dicrimination between marked glass (c and d) and farmed eels (e and f). All photographs have been taken under identical fluorescence microscope settings. The white bar equals 200 μm.

Strontium chloride

The marking of eels with strontium ($SrCl_2$) is presently only an established procedure for *A. anguilla*. Glass eels have been kept for 24 hours in a strontium bath (1 g l^{-1}) at temperatures between 12 and 23°C (see Table 11.1). The mortality was reported to be acceptably low (0.58%) and the immersion bath is considered a harmless procedure, thus strontium is already largely in use within the eel management plans of Sweden and Finland (Wickström and Sjöberg, 2014). With regard to detection, an on-grown phase of at least 14 days is recommendable to ensure sufficient otolith growth. The otoliths were removed and ground until the primordium was exposed. The preparations get carbon coated and can be checked for marks with an electron microscope by (1) an Electron Probe Microanalyser with a back-scattered electron detector

(Wickström and Sjöberg, 2014), (2) line scan using dispersive X-ray spectroscopy (Wickström and Sjöberg, 2014) or an X-ray microanalysis detector (Kullmann et al., 2018a) and (3) likely by laser ablation inductively coupled mass spectrometry (Marohn et al., 2011).

Barium chloride

The use of barium ($BaCl_2$) was successfully tested in Sweden on small on-grown eels (Wickström and Sjöberg, 2014). The eels have been exposed to 0.75 g $BaCl_2$ l^{-1} by bath immersion for 24 hours at temperatures between 20 and 23°C. The mortality, if any, was not reported while marking success was 85% on otoliths. The detection procedure is identical to strontium, however, the X-ray microanalysis method was found to be unsuitable for barium detection (personal unpublished findings). The authors have demonstrated that barium marks are much fainter than strontium marks and also mentioned an increased harmfulness compared to strontium. Wickström and Sjöberg (2014) concluded a basic suitability of barium to mark eels, however, suggested further improvement of this method.

Mass-marking of glass eels

There are very few studies describing mass-marking procedures for *Anguilla* eels, and the ones there are refer entirely to *A. anguilla*. Alcobendas et al. (1991) used oxytetracycline (OTC) to mark 500 kg of glass eels in 1.5 days. Five people were necessary to put 5 kg of glass eels at a time in nylon pouches to submerge the package for 3.5 minutes in a hyperosmotic OTC solution (10 g OTC l^{-1}; 50 g NaCl l^{-1}; see Table 11.1). They prepared a 100 l solution with 1 kg OTC and 5 kg NaCl 45 minutes in advance and reported renewing this solution for every 50 kg of glass eels. To account for a possible loss of OTC before the renewal, the last four pouches with eels have been submerged for 4 to 5 minutes. This leads to a total demand of 10 kg of OTC and 50 kg of NaCl. The authors reported the absence of mortality and a marking success of 100% on otoliths. The disposal has not been discussed, while this would have been of great interest since 10 kg of OTC could be described as an industrial quantity of an antibiotic which, nowadays, necessitates a disposal strategy and probably the approval of the authorities.

Caraguel et al. (2015) purchased 300 kg of glass eels and randomly selected 90 kg for marking with alizarin red S (ARS). This procedure was carried out in 26 single 70 l aquaria each filled with 40 l of rearing water. An amount of 150 g ARS was added per litre which resulted in a decrease of the pH value down to 6.9, which increased again after oxygenation. At an oxygen saturation level of 180%, 3.46 kg of glass eels were placed in each aquarium for 3 hours. Thereafter, eels were placed in fresh water for 'a few hours' and then stocked elsewhere. No mortality was reported. After the marking process the solution has been recovered and disposed of by a registered company. While a great focus was set on water parameters, no information is available about the actual handling of the glass eels, which, besides chemical tolerance, is also crucial for fish well-being. Instead, they assessed the short-term mortality using a net enclosure system at the stocking site and found no differences in marked and unmarked glass eels in survival. In contrast, by using the identical marking procedure, a significant impact of the chemical marking procedure on survival was found (Josset et al., 2016). However, this influence has been described as inexplicable (Josset et al., 2016) because other authors (Simon et al., 2009) found no evidence for an increased mortality of ARS marked eels compared to unmarked eels as previously reported from Caraguel et al. (2015). A possible reason, however, could be hidden in unreported details of the handling during the marking procedure and not the ARS treatment itself. Handling events are highly stressful for fish (Ashley, 2007; Davis, 2006) and in particular in combination with marking or tagging procedures in general (Murray and Fuller, 2000).

Kullmann et al. (2018a) have published a mass-marking procedure for European glass eels using ARS and also in combination with strontium chloride ($SrCl_2$). In contrast to Caraguel et al. (2015), one single rectangular tank (300 × 145 cm) with a maximum volume of 3.9 m^3 was used. 60 kg of glass eels arrived in cooled transport boxes, and were placed directly into the already prepared and aerated marking solution (1.5 m^3; 150 mg ARS l^{-1} or 1 g $SrCl_2$ l^{-1}). After 3 hours (ARS) or 24 hours ($SrCl_2$) the solution has been drained over a knotless net cage without any further use of dip nets. The ARS solution has been discharged into the canal with permission of the local waste water authorities while the strontium chloride containing water was caught up and disposed by a disposal company. The mortality was determined by weighing the dead individuals, which was between 0.56 and 0.58% of the total weight. To assess the marking success and a possible later mortality, a random sample of 60 glass eels has been held back for 14 days. No mortality was reported during that period of time and each glass eel showed a clear mark on the sagitta otoliths. The use of one single tank was discussed as potentially risky, however, fully drainable tanks and knotless net cages to separate the eels from the solution minimizes handling stress. Furthermore, the facility design allowed the efficient marking of glass eels by one single operator.

To mark the stocking material during transport to the stocking location, as described by Dorow and Schaarschmidt (2015), might be particularly advantageous when the release points are far apart from each other. A total of 120 kg of glass eels were separated into 1 kg fish transport bags filled with 30 l ARS solution (150 mg l^{-1}). Within 3–4 hours of transport to the recipient water the glass eels were separated from the marking solution on site and released. No mortality was reported and the disposal of the marking solution was described as appropriate. On the one hand, this approach might be time saving and splitting the total mass in many different portions is also minimizing the risk of accidental losses. On the other hand, a large volume of water, 3600 l in total, was used, which certainly necessitates higher staffing levels. Additionally, the on site separation of glass eels from the ARS solution which was recovered for disposal purposes must have been logistically complicated.

Mass-marking of farmed eels

Hitherto, only one publication deals with mass-marking of on-grown individuals from eel farms (Kullmann et al., 2017b), which is, compared to glass eels, a more demanding task both logistically and technically due to larger quantities of eels and volumes of marking solution. A maximum of 940 kg of farmed eels (mean body weight of 6.6 g) have been marked by two employees in four rectangular tanks (300 × 145 cm) each filled with 3.5 m^3 of ARS solution (150 mg l^{-1}). The eels arrived in a fish transport truck and were placed immediately into the aerated marking solution. After 9 hours of immersion, the marking solution was discharged in the canalization after been drained over net cages which can be moved by a forklift. Afterwards, the marked eels were placed for c. 10 hours in freshwater to provide a resting phase before being transported with a fish transport truck to the stocking site. The mortality was far below 1% and dead individuals showed external damages due to unknown reasons. A 100% marking success was assessed on a random subsample taken 60 days after the marking process.

Recommendations for standard mass-marking

Since all published mass-marking studies on *Anguilla* eels refer to ARS, OTC and strontium chloride no other chemicals can be taken into account. Mass-marking of eels with OTC was shown to be practicable (Alcobendas et al., 1991), however, this substance is an antibiotic and therefore unrecommendable for standard mass-marking protocols within eel management plans, particularly because of the risk of creating multi-resistant germs (Zheng and Zhou, 1999). Warren-Myers et al. (2018) have recommended the use of strontium chloride for standard mass-marking although the detection is comparably laborious and technically demanding. This leaves ARS as a suitable marker which has the advantage of being detectable by standard fluorescence microscopy even on otoliths that have been cracked but not further ground (Kullmann et al., 2018b; see Figure 11.1).

The following recommendations are intended to develop a harmonized ARS mass-marking protocol that could be integrated in eel management plans or large-scale monitoring programmes.

1. The exposure time of 3 hours for glass eels and 9 hours for larger farmed eels should not be reduced without further experiments, and a concentration of 150 mg l^{-1} was repeatedly shown to result in clear marks on otoliths.
2. In order to minimize handling effort and stress, the usage of fully drainable tanks is recommended to reduce handling (for example, the use of dip nets) as much as possible. If nets have to be used, for example for the separation of the eels from the marking solution, knotless material should be preferred.
3. The determination of mortality is also crucial to allow a valid conclusion of a potential net benefit from stock measures. It is proposed to measure mortality after transport, after the marking process, and after transport to the stocking site.
4. The marking solution should be aerated rather than oxygenated to avoid a hypersaturated solution and a potentially resulting gas bubble disease (Angelidis, 2011; Espmark et al., 2010).
5. A randomly taken representative sample should be held back to assess the marking success. An on-grown phase of 14 days for glass eels and 60 days for on-grown individuals is recommended.

After recapture of marked individuals, the separation of stocking cohorts, which have been marked identically, is only possible by counting annuli on cracked and burnt otoliths (ICES, 2009, 2011). This might be improved by combining ARS with other markers such as strontium chloride and barium chloride (Kullmann et al., 2018a; Wickström and Sjöberg, 2014). An alternating marking cycle might be used to mark for example, three consecutive stocking

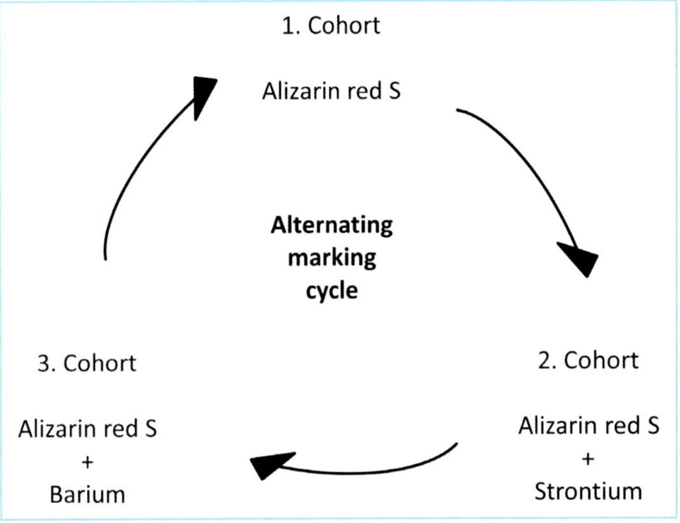

Figure 11.2 A proposed scheme to mark three successive stocking cohorts with alizarin red S (ARS), and ARS in combination with strontium chloride and barium chloride, respectively.

cohorts uniquely (Figure 11.2). To account for the technically demanding strontium chloride and barium chloride detection, all cohorts could be marked with ARS first to allow the quick separation of marked from unmarked eels, hence only a much smaller subsample would have been checked for an additional mark.

Conclusion and perspectives

Chemical marking is a rapid method to mark a large number of eels in a short time with reasonable requirements in time and manpower and, therefore, is the most promising option for the large-scale evaluation of stocking programmes. Warren-Myers et al. (2018) have identified minimum criteria that adequately reveal the limitations of a mass-marking technique: high ease and low cost of application, high mark retention and detectability, and no associated fish welfare issues whereby a high number of possibly marked fish could be added. For alizarin red S (ARS) several studies have demonstrated that ARS has hardly any negative impact on survival and growth (Caraguel et al., 2015; Josset et al., 2016; Kullmann et al., 2018a; Simon et al., 2009). Furthermore, OTC marked *A. rostrata* silver eels were observed in the St. Lawrence River (USA, Canada), which means that marked eels do mature and start the spawning migration (Verreault et al., 2010). No information, however, is available about potential sub-lethal long-term effects on on-grown and silvering individuals. It is still an open question whether ARS and other chemicals might accumulate in the liver, muscle tissue, the brain or elsewhere in eels other than calcified structures. In cod (*Gadus morhua*) larvae, for instance, chronic effects of an ALC treatment was found indicating that standard marking protocols could negate the conservation purpose (Meyer et al., 2012). Further research, therefore, on the biological effects of the chemical marking process is of vital importance due to the critical conservation status of many eel species. However, eel resources have declined and the reallocation of eels, referred to as stocking, is a widely accepted conservation measure which potentially increases pressure on eel stocks.

Mass-marking of stocked recruits therefore, should be temporarily mandatory to create a reliable data base to generate valid conclusions. In this regard, ARS is certainly the most comprehensively studied chemical marker in terms of mass-marking usability, medium term impact on growth, and survival making this substance the most promising option for the large-scale evaluation of eel stocking as conversation measure within the framework of management plans.

Bibliography

Alcobendas, M., Lecomte, F., Castanet, J., Meunier, F. J., Maire, P. and Holl, M. (1991) Technique de marquage en masse de civelles (*Anguilla anguilla* L.) par balnéation rapide dans le fluorochrome. Application au marquage à la tétracycline de 500 kg de civelles. *Bulletin Français de la Pêche et de la Pisciculture* 321, 43–54.

Andersson, J., Sandström, O. and Hansen, H.J.M. (1991) Elver (*Anguilla anguilla* L.) stockings in a Swedish thermal effluent-recaptures, growth and body condition. *Journal of Applied Ichthyology* 7, 78–89.

Angelidis, P. (2011) Improved system to normalize the oxygen distribution in the tank water in aquaculture. In: Salampasis, M. and Matopoulos, A. (eds) *Proceedings of the International Conference on Information and Communication Technologies for Sustainable Agri-production and Environment (HAICTA 2011).* CEUR-WS, http://ceur-ws.org/Vol-1152, pp. 473–485.

Arai, T., Limbong, D. and Tsukamoto, K. (2000) Validation of otolith daily increments in the tropical eel *Anguilla celebesensis. Canadian Journal of Zoology* 78, 1078–1084.

Ashley, P.J. (2007) Fish welfare: current issues in aquaculture. *Applied Animal Behaviour Science* 104, 199–235.

Beamish, R.J. and McFarlane, G.A. (1983) The forgotten requirement for age validation in fisheries biology. *Transactions of the American Fisheries Society* 112, 735–743.

Campana, S.E. (1999) Chemistry and composition of fish otoliths: pathways, mechanisms and applications. *Marine Ecology Progress Series* 188, 263–297.

Campana, S.E., Oxenford, H.A. and Smith, J.N. (1993) Radiochemical determination of longevity in flyingfish *Hirundichthys affinis* using Th-228/Ra-228. *Marine Ecology Progress Series* 100, 211–219.

Caraguel, J.-M., Charrier, F., Mazel, V. and Feunteun, E. (2015) Mass marking of stocked European glass eels (*Anguilla anguilla*) with alizarin red S. *Ecology of Freshwater Fish* 24, 435–442.

Castonguay, M. and Durif, C.M.F. (2016) Understanding the decline in anguillid eels. *ICES Journal of Marine Science* 73, 1–4.

Chen, J.-Z., Huang, S.-L. and Han, Y.-S. (2014) Impact of long-term habitat loss on the Japanese eel *Anguilla japonica. Estuarine, Coastal and Shelf Science* 151, 361–369.

Chisnall, B.L. and Kalish, J.M. (1993) Age validation and movement of freshwater eels (*Anguilla dieffenbachii* and *A. australis*) in a New Zealand pastoral stream. *New Zealand Journal of Marine and Freshwater Research* 27, 333–338.

Cieri, M.D. and McCleave, J.D. (2001) Validation of daily otolith increments in glass-phase American eels *Anguilla rostrata* (Lesueur) during estuarine residency. *Journal of Experimental Marine Biology and Ecology* 257, 219–227.

Côté, C.L., Pavey, S.A., Stacey, J.A., Pratt, T.C., Castonguay, M., Audet, C. and Bernatchez, L. (2015) Growth, female size, and sex ratio variability in american eel of different origins in both controlled conditions and the wild: implications for stocking programs. *Transactions of the American Fisheries Society* 144, 246–257.

Couillard, C.M., Verreault, G., Dumont, P., Stanley, D. and Threader, R.W. (2014) Assessment of fat reserves adequacy in the first migrant silver American eels of a large-scale stocking experiment. *North American Journal of Fisheries Management* 34, 802–813.

Davis, K.B. (2006) Management of physiological stress in finfish aquaculture. *North American Journal of Aquaculture* 68, 116–121.

Dekker, W. (1986) Age reading of eels using tetracycline labelled otoliths. *ICES C.M. 1086/M:16*, 14 pp.

Dorow, M. and Schaarschmidt, T. (2015) Besatz mit Glasaalen in Küstengewässern 2015. *Fischerei & Fischmarkt in Mecklenburg-Vorpommern* 1, 64–65 (in German).

Dunn, A. and Coker, C.M. (1951) Notes on marking live fish with biological stains. *Copeia* 1, 29–31.

Espmark, Å.M., Hjelde, K. and Baerverfjord, G. (2010) Development of gas bubble disease in juvenile Atlantic salmon exposed to water supersaturated with oxygen. *Aquaculture* 306, 198–204.

Fukuda, N., Kuroki, M., Shinoda, A., Yamada, Y., Okamura, A., Aoyama, J. and Tsukamoto, K. (2009) Influence of water temperature and feeding regime on otolith growth in *Anguilla japonica* glass eels and elvers: does otolith growth cease at low temperatures? *Journal of Fish Biology* 74, 1915–1933.

Guy, C.S., Blankenship, H.L. and Nielsen, L.A. (1996) Tagging and marking. In: Murphy, B.R. (ed.), *Fisheries Techniques*, 2nd edn. American Fisheries Society, Bethesda, MD, pp. 353–383.

Hansen, H.J.M. and Fattah, A.T.A. (1986) Long-term tagging of elvers, *Anguilla anguilla*, with radioactive europium. *Journal of Fish Biology* 29, 535–540.

Holmgren, K. (1996) Otolith growth scaling of the eel, *Anguilla anguilla* (L.), and back-calculation errors revealed from alizarin labelled otoliths. *Nordic Journal Freshwater Research* 72, 71–79.

Hoyle, S.D., Hutchison, M.J., Sellin, M.J., Peel, D., Mayer, D. and Sumpton, W.D. (1998) Biological data and model development for management of longfinned eel fisheries. *Project No. 98/128, 2005*, 125 pp.

ICES (2008) Report of the 2007 session of the Joint EIFAC/ICES Working Group on Eel Bordeaux, France, 3–7 September 2007. *ICES CM 2007/ACFM:23*, 156 pp.

ICES (2009) Workshop on Age Reading of European and American Eel (WKAREA) 20–24 April 2009 Bordeaux, France. *ICES CM 2011/ACOM:48*, 66 pp.

ICES (2011) Report of the Workshop on Age Reading of European and American Eel (WKAREA2) 22–24 March 2011 Bordeaux, France. *ICES CM 2011/ACOM:43*, 35 pp.

ICES (2016a) Report of the Joint EIFAAC/ICES/GFCM Working Group on Eel (WGEEL), 24 November–2 December 2015, Antalya, Turkey. *ICES CM 2015/ACOM:18*, 130 pp.

ICES (2016b) Report of the Workshop on Eel Stocking (WKSTOCKEEL): 20–24 June 2016 Toomebridge, Nothern Ireland, UK. *ICES CM 2016/SSGEPD:21*, 75 pp.

ICES (2017) Report of the Joint EIFAAC/ICES/GFCM Working Group on Eel (WGEEL), 3–10 October 2017, Kavala, Greece. *ICES CM 2017/ACOM:15*, 99 pp.

Jellyman, D.J. (1995) Longevity of longfinned eels *Anguilla dieffenbachii* in a New Zealand high country lake. *Ecology of Freshwater Fish* 4, 106–112.

Jellyman, D.J., Glova, G.J. and Todd, P.R. (1996) Movements of shortfinned eels, *Anguilla australis* in Lake Ellesmere, New Zealand: results from mark-recapture studies and sonic tracking. *New Zealand Journal of Marine and Freshwater Research* 30, 371–381.

Jensen, A.C. (1962) Marking and tagging fishes. *Fishery Leaflet* 534, 10 pp.

Josset, Q., Trancart, T., Mazel, V., Charrier, F., Frotté, L., Acou, A. and Feunteun, E. (2016) Pre-release processes influencing short-term mortality of glass eels in the French eel (*Anguilla anguilla*, Linnaeus 1758) stocking programme. *ICES Journal of Marine Science* 73, 150–157.

Kaifu, K., Itakura, H., Amano, Y., Shirai, K., Yokouchi, K., Wakiya, R., Murakami-Sugihara, N, Washitani, I. and Yada, T. (2018) Discrimination of wild and cultured Japanese eels based on otolith stable isotope ratios. *ICES Journal of Marine Science* 75, 719–726.

Klein Breteler, J.K., Vriese, T., Borcherding, J., Breukelaar, A., Jörgensen, L., Staas, S., de Laak, G. and Ingendahl, D. (2007) Assessment of population size and migration routes of silver eel in the River Rhine based on a 2-year combined mark-recapture and telemetry study. *ICES Journal of Marine Science* 64, 1450–1456.

Kullmann, B., Adamek, M., Steinhagen, D. and Thiel, R. (2017a) Anthropogenic spreading of *anguillid herpesvirus 1* by stocking of infected farmed European eels, *Anguilla anguilla* (L.), in the Schlei fjord in northern Germany. *Journal of fish diseases* 40, 1695–1706.

Kullmann, B., Hempel, M. and Thiel, R. (2018a) Chemical marking of European glass eels *Anguilla anguilla* with alizarin red S and in combination with strontium: in situ evaluation of short-term salinity effects on survival and efficient mass-marking. *Journal of Fish Biology* 92, 203–213.

Kullmann, B., Neukamm, R. and Thiel, R. (2017b) Mass-marking of farmed European eels (*Anguilla anguilla* (Linnaeus, 1758)) with alizarin red S. *Journal of Applied Ichthyology* 33, 914–917.

Kullmann, B., Pohlmann, J.-D., Freese, M., Keth, A., Wichmann, L., Neukamm, R. and Thiel, R. (2018b) Age-based stock assessment of the European eel (*Anguilla anguilla*) is heavily biased by stocking of unmarked farmed eels. *Fisheries Research* 208, 258–266.

Kullmann, B. and Thiel, R. (2018) Bigger is better in eel stocking measures? Comparison of growth performance, body condition, and benefit-cost ratio of simultaneously stocked glass and farmed eels in a brackish fjord. *Fisheries Research* 205, 132–140.

Lin, Y.-J., Chang, S.-L., Chang, M.-Y., Lin, S.-H., Chen, T.-I., Su, M.-S., Su, W.-C. and Tzeng, W.-N. (2010) Comparison of recapture rates and estimates of fishing and natural mortality rates of Japanese eel *Anguilla japonica* between different origins and marking methods in a mark-recapture experiment in the Kaoping River, Southern Taiwan. *Zoological Studies* 49, 616–624.

Marohn, L., Hilge, V., Zumholz, K., Klügel, A., Anders, H. and Hanel, R. (2011) Temperature dependency of element incorporation into European eel (*Anguilla anguilla*) otoliths. *Analytical and bioanalytical chemistry* 399, 2175–2184.

Masuda, Y., Imaizumi, H. and Oda, K. (2012) Artificial completion of the Japanese eel, *Anguilla japonica*, life cycle: challenge to mass production. The 39th Scientific Symposium of UJNR Aquaculture Panel: The Present and future of the Aquaculture Industry. *Bulletin* of *Fisheries Research Agency* 35, 111–117.

Meyer, S., Sørensen, S.R., Peck, M.A. and Støttrup, J.G. (2012) Sublethal effects of alizarin complexone marking on Baltic cod (*Gadus morhua*) eggs and larvae. *Aquaculture* 324–325, 158–164.

Mounaix, B. (1991) Validation de l'âge de l'anguille européenne, *Anguilla anguilla*, dans le bassin versant de la Vilaine (Bretagne): résultats préliminaires. *Otolithornétrie,* 109–117.

Murray, D.L. and Fuller, M.R. (2000) A critical review of the effects of marking on the biology of vertebrates. In: Boitani, L. and Fuller, T.K. (eds) *Methods and Cases in Conservation Science. Research Techniques in Animal Ecology. Controversies and Consequences.* Columbia University Press, New York, pp. 15–64.

Nielsen, J. (1988) Marking and tagging methods applied to eel, *Anguilla anguilla* (L.). *EIFAC Occasional Paper* 21, 1–24.

Oliveira, K. (1996) Field validation of annular growth rings in the American eel, *Anguilla rostrata*, using tetracycline-marked otoliths. *Fishery Bulletin* 94, 186–189.

Oliveira, K. (1997) Movements and growth rates of yellow-phase American eels in the Annaquatucket River, Rhode Island. *Transactions of the American Fisheries Society* 126, 638–646.

Pease, B.C., Reynolds, D.P. and Walsh, C.T. (2003) Validation of otolith age determination in Australian longfinned river eels *Anguilla reinhardtii. Marine and Freshwater Research* 54, 1–10.

Pratt, T.C. and Threader, R.W. (2011) Preliminary evaluation of a large-scale American eel conservation stocking experiment. *North American Journal of Fisheries Management* 31, 619–628.

CHAPTER 12

Estimation of downstream migrating Irish silver-phase eel *Anguilla anguilla* populations by removal sampling

T. Kieran McCarthy, Eamonn S. Lenihan and Colin Lawton

Zoology, School of Natural Sciences, Ryan Institute, National University of Ireland Galway, Ireland

Introduction

The decline of the European eel (*Anguilla anguilla*) and of other Anguillid eels has been a matter of concern for several decades (Dekker, 2003; Jacoby and Gollock, 2014; Poole et al., 2018). A variety of factors, including overfishing, habitat loss, barriers to river migration, hydropower passage mortalities, contaminants, introduced pathogens and climate change, have been implicated (Aarestrup, 2012; Bonhommeau et al., 2008; Calles et al., 2010; Geeraerts and Belpaire, 2010; Kettle et al., 2011; Pedersen et al., 2012; van den Thillart and Dufour, 2009). However, there may also be other undetected natural or anthropogenic causes of the sustained decline in the European eel stock and its reduced recruitment levels. Various measures have been adopted, at local, regional and international levels, in an attempt to restore the stock. These were generally initiated in response to an EU Regulation (EC No. 1100/2007), which required eel management plans (EMPs) to be adopted by member states to try to restore the panmictic stock. Evaluation of the success of local eel conservation measures, and of their contribution to the restoration of eel spawner escapement biomass to 40% of historical levels, requires effective monitoring of current silver eel population dynamics. Stock monitoring, especially at the silver-phase eel life cycle stage, is needed for the management of eel populations at local, river basin district and national levels and for overall management of the spawning stock. Sampling methodologies, analytical protocols and mathematical models (Aprahamian et al., 2007; Prigge et al., 2013; Walker et al., 2011) are being progressively refined. However, the overall objective of estimating the extent of successful migration to sea and to the spawning area still remains problematic (Jacoby et al., 2015).

Eel sampling, which varies with life-cycle stage and research objectives, includes qualitative, semi-quantitative and quantitative methodologies. They may involve active or passive sampling protocols and frequently are adapted from commercial fishing techniques (Koops, 1980; McGrath, 1971; Tesch and White, 2008). In Ireland, juvenile eels are trapped for assisted upstream passage at hydroelectricity dams. Yellow-phase eels were formerly captured by fishermen using either baited longlines or summer fyke nets (McCarthy, 2014; Poole, 1990). They are also sampled for research purposes using these methods and by electrical fishing (Callaghan and McCarthy, 1993; de Eyto et al., 2016; McCarthy et al., 2008a; Moriarty, 1996). Silver-phase eels were traditionally fished using winged river nets or at special eel fishing weirs that were equipped with arrays of coghill nets (McCarthy, 2014). Most Irish silver-phase eel research has been undertaken using these commercial fishing facilities (MacNamara and McCarthy, 2014; MacNamara et al., 2017; McCarthy et al., 1999, 2008b, 2014; Rosell et al., 2005) or at bespoke research traps (Poole et al., 2018).

Eels are often regarded as difficult to sample quantitatively and mostly populations of eels are monitored using fishery catch data or indices of relative abundance (Dekker, 2016). However, where removal sampling methods are possible, good estimates of yellow-phase eel density and biomass are frequently obtained. Electrical fishing protocols regularly involve repeat fishing of a defined river or stream section and estimation of eel population parameters by reference to the decline in catches (Acou et al., 2009; Aprahamian, 1986, 1988; Baldwin and Aprahamian, 2012; Callaghan and McCarthy, 1993; de Eyto et al., 2016; Lobon-Cervia et al., 1995). In analyses of fish populations, the analytical methods devised by Zippin (1956), Seber and Le Cren (1967) and Carle and Strub (1978) have been used for electrofishing surveys of yellow-phase eels (Acou et al., 2009; Callaghan and McCarthy, 1993; de Eyto et al., 2016) and other freshwater fishes (Guy and Brown, 2007). In some habitats, yellow-phase eels have also been investigated using mark-recapture techniques (Carss et al., 1999; Feunteun et al., 2000; Laffaille et al., 2005). Though mark-recapture experiments have frequently been used for monitoring migrating silver-phase eel populations (Amilhat et al., 2008; MacNamara and McCarthy, 2014; McCarthy et al., 2014; Rosell et al., 2005), removal methods have rarely been used. Removal methods, which are widely used for analyses of vertebrate and invertebrate animal populations, were initially developed by mammalogists in 1914 and there are numerous publications describing case studies and alternative analytical protocols (Seber, 1973; Southwood, 1978).

This chapter develops removal sampling protocols for monitoring silver-phase eel production and escapement in small rivers; provides quantitative information on downstream migrating silver-phase populations in two sub-catchments of the River Erne in Ireland; and compares the silver-phase eel production of these two tributary systems with that estimated annually in the entire river basin.

Materials and methods

Study area

The trans-boundary River Erne is located in the northwest of Ireland, and the river basin (4375 km^2) drains areas of both the Republic of Ireland and Northern Ireland. Information on the topography, soils, geology, land use, hydrochemistry and freshwater habitats of eels in the hydrosystem are available in many reports (for example, Matthews et al., 2001). The river is regulated for hydropower in the lower section, where mean annual discharge is 92 m^3·s^{-1} (range: 10–382 m^3·s^{-1}) and annual estimation of silver eel escapement to sea is undertaken using protocols outlined by McCarthy et al. (2014). A silver eel trap and transport conservation programme was adopted in 2009 as a mitigation measure for the adverse effects of hydropower on the spawner escapement biomass from the river system. Closure of commercial fisheries was specified in the Northwest International Regional River Basin Eel Management Plan. However, commercial fishermen were contracted by the Electricity Supply Board (ESB) to fish for silver eels at designated fishing locations from which eels are collected and taken in specially equipped trucks and trailers to a release location, close to the estuary in Ballyshannon, downstream of hydropower dams.

The present study was focused on the upper river basin and in particular in an area where the tributary River Annalee joins the Upper River Erne (Figure 12.1). Except for nights on which very low catches were expected, sequentially paired winged coghill nets were used on each tributary and a single net below the confluence. Daily catch weights were recorded for all nets fished. The bespoke nets, of dimensions suited to individual net locations, were set from a flat-bottomed boat by a two-man crew and the wings of the nets were attached to each bank. They were operated from dusk to dawn, with several lifts per night during migration peaks. The fishing season, under special authorization from ESB, extended from 27 October 2015 to 1 February 2016 at this site and at an experimental fishing weir (Roscor Bridge) in the lower part of the River Erne.

Preliminary fieldwork at the sites in the previous winter using Floy tags (N = 300) revealed that the capture efficiency of the bespoke nets at both sites was very high. Since screening catches for tagged eels was difficult, the observed recapture rates (mean = 83%), though high, were

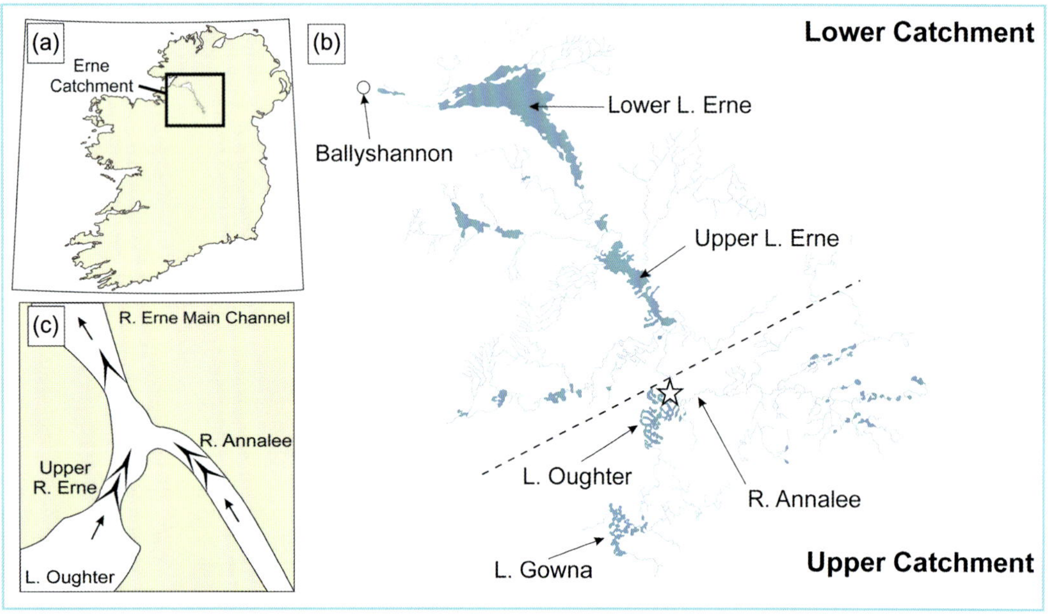

Figure 12.1 (a) Map of Ireland with the Erne catchment highlighted. (b) The upper and lower divisions of the Erne catchment, with various locations mentioned in the text indicated. The star indicates the location of the study site. (c) The study site, showing the location of paired nets in the River Annalee and Upper River Erne tributaries and a single net in the main River Erne channel.

recognized as being underestimates of the actual fishing efficiency. The 2015/16 research was therefore focused on investigating the migrating eel populations on these two tributaries using removal analyses.

Removal analysis

Fishing in the tributaries (n = 69 nights) consisted of either a single removal or a double removal using paired sequentially set nets. The pattern of catches in each tributary was compared using a Mann–Whitney *U*-test. On nights when both nets were fished, population size above each tributary was estimated using the weighted maximum-likelihood model of Carle and Strub (1978). On the Upper River Erne and River Annalee, when paired sequential nets were deployed, 84% and 86% of catch, on average, was caught in first net in the removal series, respectively, suggesting catchability was similar between the tributaries. Probability of capture was considered to be constant between nets in each tributary and thus cumulative capture efficiency for the river Erne was 97% and the River Annalee was 98%. This efficiency was applied to eels passing on double net nights. Only one net efficiency was applied on single net nights (84 and 86%). Daily catch from the first net was compared with population estimates from the Carle and Strub model using linear regression for both tributaries. On both the Upper River Erne, and River Annalee, there were a number of nights when only one net was used. If a sufficiently strong relationship existed between first net catches and population estimates, then this could be extrapolated to a Carle and Strub population estimate based on one net catch (Acou et al., 2009).

The single net located below the confluence in the River Erne main channel was used to validate the removal methodology. This net was deployed on 49 nights. Catches at this main channel net were very highly significantly correlated (Spearman's rs = 0.877, p < 0.001) with corresponding night catches with those in the first nets (two tributaries combined). Linear regression analysis ($y = 0.35 + 0.06x$, $R^2 = 0.676$, $p < 0.001$) was used to interpolate potential catch values for nights (n = 20) when the net was not deployed, but fishing occurred upstream in the tributaries. The capture efficiency of this net was estimated to have been 85% (mean value of Upper River Erne and River Annalee net efficiencies).

The Carle and Strub analyses were carried out using PISCES Removal Sampling software (v2.2.2.22). Sub-samples of eels were measured from the Upper River Erne (n = 194) and the River Annalee (n = 162). Their length frequencies were compared using a Mann–Whitney *U*-test. All statistics were carried out using R (R Core Team 2016, v3.3.1).

Results

The seasonal pattern of catch records is illustrated in Figure 12.2, together with information on the lunar phases and the variation in main channel discharge. As can be seen, though the effects of seasonal and environmental factors linked to lunar cycle are evident, increasing River Erne channel discharge at Belturbet was associated with higher catches during the main migration peaks.

The two tributary sites were fished by the experienced crew based on their interpretation of the variation in catches, which could be accurately predicted using information on peak

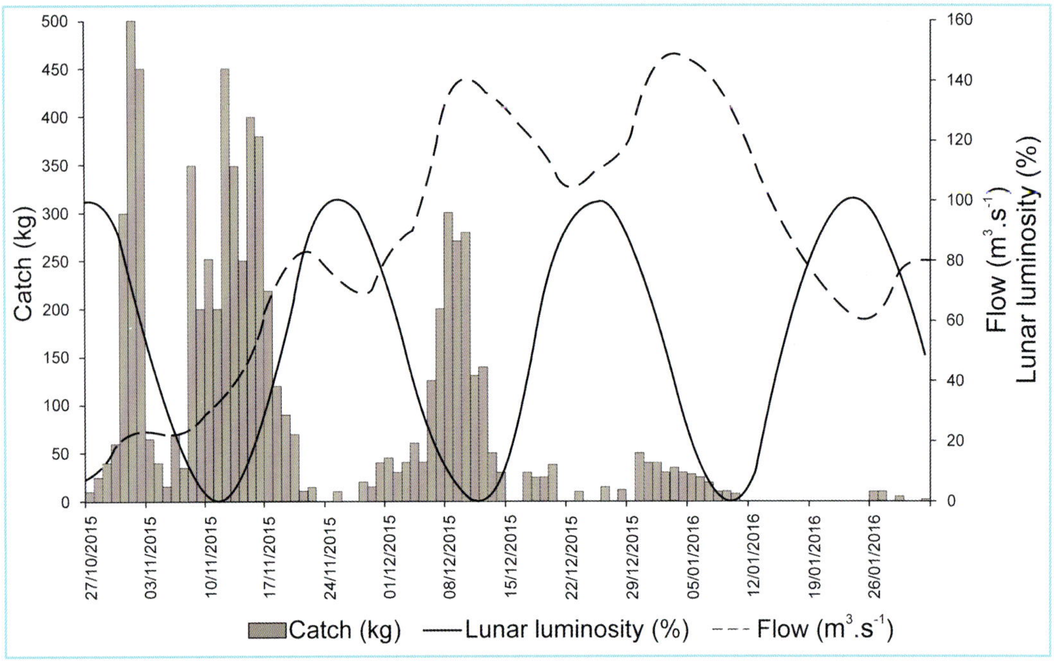

Figure 12.2 The pattern of eel catches in all nets, lunar luminosity, and discharge at the experimental fishing site during the fishing season.

migration periods and recent catches. Sequentially paired nets were used when significant catches were anticipated. On the River Annalee the full season catch was 4295 kg and median nightly catch was 20 kg, range 0–300 kg. Zero catch levels occurred on four nights. Two nets were used on 42 nights (61%) and 92% of the season's catch was caught in this time. In the Upper River Erne site, the full season catch was 2642 kg, with a median nightly catch of 14.5 kg and range 0–220 kg. Double nets were deployed on 59 nights (86%) in the Upper Erne site and 81% of the season's catch was recorded during this time. One night was observed with zero catch. The main River Erne channel was fished on 49 nights and on 5 of these no catch was obtained. A total of 361 kg was caught, with a median catch of 4 kg, and a range from 0–40 kg. The catch distribution pattern did not differ significantly between the two tributaries (Mann–Whitney *U*-test, $p > 0.05$) (Figure 12.3). Size frequencies (Figure 12.4) indicate that the two tributary systems produced both male and female eels and that there were significant differences in population structure (Mann–Whitney *U*, $p < 0.05$), with a higher proportion of males in the River Annalee.

Linear regression (Figure 12.5) revealed a highly significant relationship between first net catch and population estimate for both the Upper River Erne ($R^2 = 0.978$, $p < 0.001$) and the River Annalee ($R^2 = 0.956$, $p < 0.001$). For this reason, it was decided to use corrected single net catch data by extrapolating to a Carle and Strub population estimate based on linear regression results. From double net and corrected single net calculations, a total of 2815 kg of eels were estimated migrating to Upper River Erne nets, while 4587 kg were estimated migrating towards the River Annalee nets. Based on the Carle and Strub estimates, production and escapement was calculated

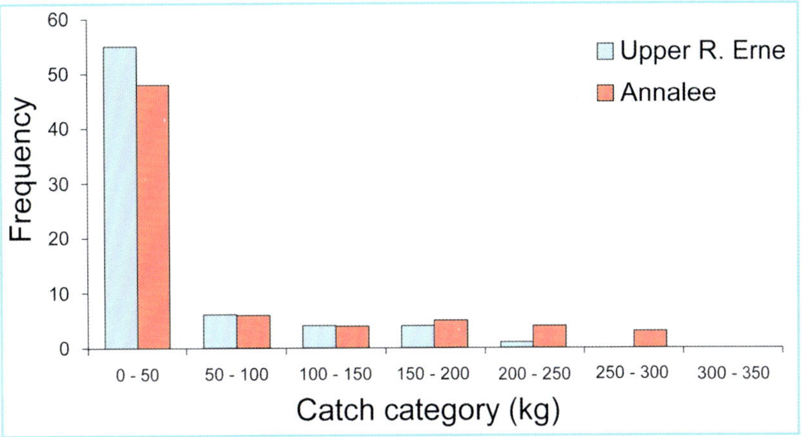

Figure 12.3 Silver eel catch frequencies for the Upper River Erne and the River Annalee.

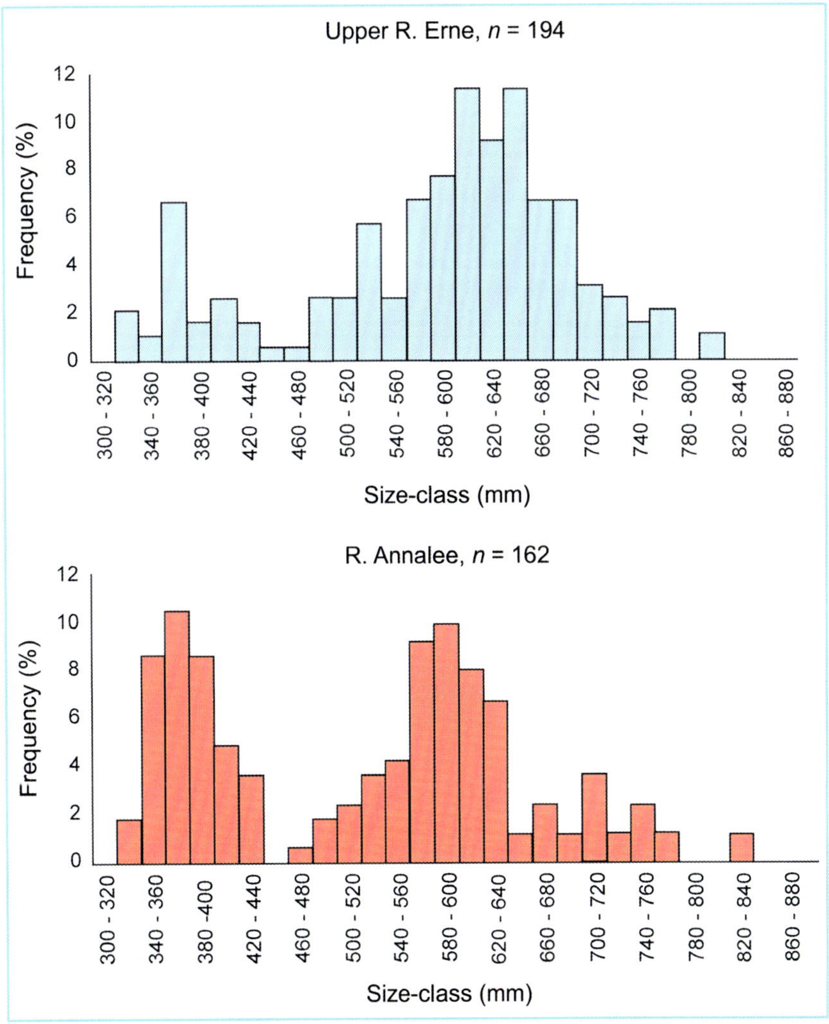

Figure 12.4 Size frequencies of silver eels caught in the Upper River Erne and River Annalee.

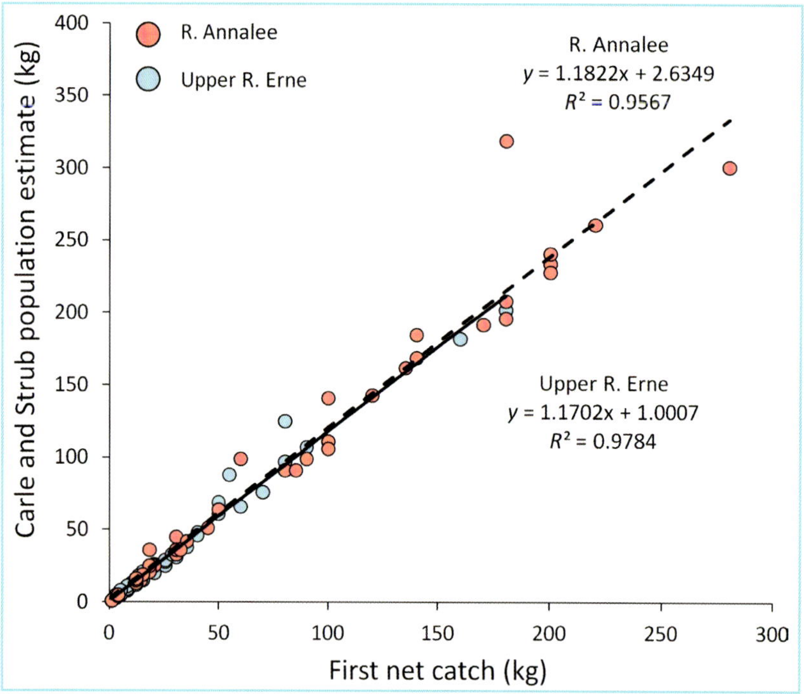

Figure 12.5 Linear regression between the biomass of silver eels caught in the first net and Carle and Strub population estimates for the Upper River Erne net array (n = 59, solid trend line) and the River Annalee net array (n = 42, dashed trend line).

for the entire upper catchment (5326 ha). Production was estimated to be 3.27 kg·ha^{-1}, while escapement was 3.23 kg·ha^{-1}. This estimate of production includes catches from conservation fishing sites located upstream of the Upper River Erne net array at Lough Gowna (4646 kg) and Lough Oughter (4835 kg) (Figure 12.1). The value for escapement includes trap and transport quantities and represents contribution to spawner biomass reaching the sea.

The single net located below the confluence of the tributaries caught 361 kg over the season (n = 49 nights, Figure 12.6). Interpolation based on linear regression analysis of the relationship

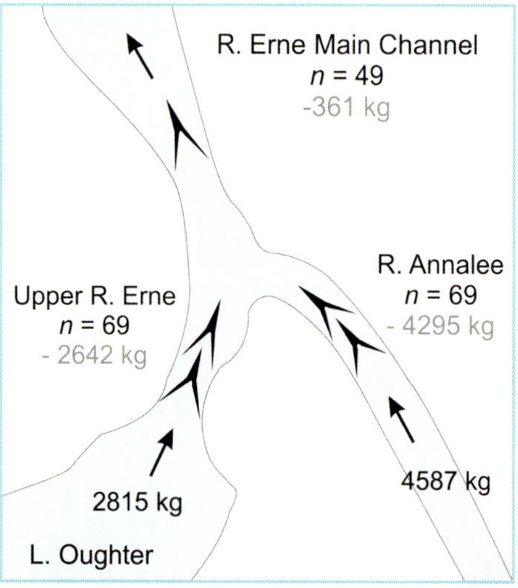

Figure 12.6 Summary of removal survey results. Grey figures are the observed catches. Figures in black are Carle and Strub population estimates. Number of fishing nights (n) indicated.

between main channel catches and those of the two upper tributaries indicated that the potential catch for the main channel during the 69-night fishing season was 408 kg. Based on this calculation and single net efficiency (85%), it was estimated that 480 kg was the biomass of eels that migrated from the two tributaries to the main channel. Likewise, it was estimated that the escapement downstream of the main channel net was 72 kg (15%). For the same period ($n = 69$), the difference between Carle and Strub estimates and the observed catches of eels in the tributaries, suggested that 465 kg of eels descended to the single main channel net below the confluence. To provide an alternative perspective, total nightly catches in each tributary and net efficiencies (Upper River Erne = 84%, River Annalee = 86%), were used to calculate escapement to the main channel. It was estimated that this was 267 kg and that a catch of 227 kg would be expected in the main channel net. However, the observed catch, 361 kg, was somewhat higher.

Discussion

Management of European eel stocks is generally now undertaken at a river basin district level. To monitor compliance with the EU Regulation, which aims for restoration of the stock, generally requires comparison of current spawner biomass escapement with levels that prevailed before human activities led to their reduction. Eel management plans seek to conserve eels by local management measures such as: reduction in fishing pressure; provision of improved eel passage facilities; reduction in adverse hydropower effects; improved habitat quality; and juvenile eel stocking. Robust annual estimates of silver-phase eel production and escapement are essential for evaluation of the effectiveness of eel management plans, at local, regional, national and EU levels. In the River Erne, confirmation that quantities captured in the trap and transport programme meet agreed annual targets also requires good knowledge of silver-phase eel population dynamics. In addition to providing information for eel management purposes, analysis of silver eel population dynamics provides good opportunities to research this important stage in the eel life-cycle.

Removal sampling has previously been used in analysis of eel populations in Ireland. Though widely used for the generation of semi-quantitative yellow-phase eel (catch per unit effort) data, fyke-netting has not been effectively used in removal sampling experiments. Such sampling for estimation of density and biomass is possible, as shown in a lake enclosure in Scotland (Carss et al., 1999). McCarthy et al. (2008a) used longline fishing as a removal sampling method in an Irish lake and for more extensive yellow-phase eel population studies (Grennan and McCarthy, 2013; McCarthy et al., 1999, 2008b). Removal fishing is used, in Ireland and elsewhere, in electrofishing eel surveys. This routinely involves double or triple pass fishings over a defined stretch of lotic or lake shore habitat (Callaghan and McCarthy 1993; de Eyto et al., 2016; McCarthy et al., 2008a). However, analyses of downstream migrating silver eel populations have generally relied on the use of catch records from established commercial fishing sites where mark-recapture experiments are used to provide information on the capture efficiency of the fishing operations. Irish eel fishing weirs, such as those on the River Shannon at Killaloe, the River Corrib at Galway, the River Erne at Roscor Bridge and on the River Bann, have been used to monitor the seasonal and between year populations of seaward migrating silver eel from important eel producing river basins (Bodles, 2016; MacNamara and McCarthy, 2014; McCarthy et al., 2014). Acoustic camera

(DIDSON) surveys have been used in support of this work in several Irish sites (Lenihan et al., 2019; MacNamara and McCarthy, 2014). However, at smaller tributary sites it is frequently not possible, or cost-effective, to undertake replicated tagging experiments and use of acoustic cameras may not be technically possible. In such situations, relatively inexpensive removal methods such as those used in the present study may be easily employed.

The fractal nature of the European eel stock, as recognized by Dekker (2000), contributes to the complexity of international eel management. The current conservation measures do not appear to be leading to restoration of the stock. Evaluation of the measures being applied to this widely distributed panmictic diadromous species is extremely difficult. Lack of resources often limits opportunities for assessment of spawner production and escapement. Consequently, there is widespread reliance on the use of mathematical models to estimate these sorts of eel population parameters in Europe (Aprahamian et al., 2007; Prigge et al., 2013). In Ireland, the EDA (v2.0) model has been used to provide estimates of spawner production in river ecosystems using extensive electrofishing survey results (de Eyto et al., 2016). However, due to the fact that many Irish river systems, including the Rivers Erne and Shannon, have extensive lake habitats available to their eel populations means that validation of model outputs is difficult. Though annual estimates of spawner production and escapement are possible for some of these large, hydropower regulated rivers, the research is expensive and is dependent on full season operation of fishing weirs (MacNamara and McCarthy, 2014; McCarthy et al., 2008b). The removal sampling protocols used in the present study could be cost effectively used for representative series of smaller Irish rivers, with and without lakes, not impacted by hydropower. Likewise, with EU support, they could be used in different parts of the European range of the species.

The results of the removal population surveys in the upper section of the River Erne catchment area (Figure 12.1) provided clear evidence of the high capture efficiency of the winged coghill nets used at the fishing locations in the experimental site. The present study showed that capture efficiencies were similar and high in both the Upper River Erne (84% single net, 97% double net) and River Annalee (86% single net, 98% double net) netting locations. Catch data for the River Erne main channel net below the confluence of the River Annalee and Upper River Erne, suggests that the capture efficiency of the single net used was similar to that of the upper nets in the two tributary locations. The annual pattern of silver eel migration was similar in all three locations and reflected seasonal factors, lunar periodicity and variation in discharge (Figure 12.2). Likewise, though sample sizes were relatively small, the size frequencies (Figure 12.4) indicate that both tributary systems produced both male and female eels and that there were significant differences in population structure. The high proportion of male eels (<430 mm), especially in the River Annalee (Figure 12.4), is interesting and contrasts with observations made on silver eels in the upper parts of the River Shannon system (MacNamara et al., 2017).

Using the combined results of the removal eel sampling in the two upper River Erne subcatchments, it was estimated that silver eel production was 3.27 kg·ha^{-1} and escapement was 3.23 kg·ha^{-1} for this upper part of the River Erne system. The escapement estimate was largely attributable to the safe transport of the catches to the estuary as part of the silver eel conservation. The intense fishing pressure at the experimental site illustrates how expert, hard-working, commercial fishing crews may have impacted previously on the eel populations prior to closure of commercial fishing following implementation of the eel management plan. The upper catchment production value observed during this survey is very similar to the value (2.98 kg·ha^{-1}) estimated for the entire River Erne for the same period. The production levels are estimated annually, following protocols adopted by McCarthy et al. (2014), using data from the trap and transport programme and from population estimates obtained by mark-recapture experiments at a fishing site in the Lower River Erne. The relatively uniform production levels in the river system suggest that elver stocking is allowing eels to exploit suitable habitat throughout the catchment. Except for two hydrodams in the very lowest part of the system, there appear to be few significant anthropogenic or natural obstacles to upstream migration of eels. The silver eel production level recorded in the River Erne is relatively high compared to those reported for some other Irish rivers. MacNamara and McCarthy (2014) described variations in production levels in the River Shannon (1.35–1.62 $kg \cdot ha^{-1}$), partly attributable to commercial fishery closure. MacNamara et al. (2017) estimated that in a small upper River Shannon tributary system, silver eel production was somewhat lower (0.79–1.84 $kg \cdot ha^{-1}$) and observed that silver eel size structure reflected eel population decline in local eel populations. Poole et al. (2018) reported that, despite variation in sex ratio, the silver eel production has remained relatively constant (1.1–1.4 $kg \cdot ha^{-1}$) over several decades. Much higher production levels (4–4.6 $kg \cdot ha^{-1}$) have been recorded in the River Bann–Lough Neagh system, which has been intensively stocked, with both Irish and imported glass eels (Rosell et al. 2005). Considerably higher production levels have also been reported from continental Europe habitats (Oír 13.2 $kg \cdot ha^{-1}$ [Charrier et al., 2012]; Frémur 17–42 $kg \cdot ha^{-1}$ [Acou et al., 2009]; Bages-Sigean lagoon 30 $kg \cdot ha^{-1}$ [Amilhat et al., 2008]).

The seasonal patterns of eel capture in the upper catchment area of the River Erne and some other Irish rivers (McNamara et al., 2017) are often characterized by clear lunar periodicity (Figure 12.2). However, other environmental factors such as increased discharge or rises in upstream lake water level are also important determinants of peak migration periods. This enables experienced fishermen to predict migration peaks with considerable accuracy. Migration patterns in lower river sections, especially where large lakes occur along the river continuum, are less clear cut. Also, in the case of some of the larger Irish rivers, river regulation for hydropower generation, flood control and waterway navigation purposes can obscure underlying lunar migration behaviour patterns of eels (McCarthy et al. 1999, 2008b). Opportunities for use of removal sampling are less easily accessed in such lower catchment areas, though one site may be available near Roscor Point in Lower Lough Erne.

The opportunities presented in Irish rivers, where hydropower mitigation involves silver-phase eels being trapped for safe release below hydropower dams, can allow for use of removal sampling protocols to be more widely employed in future. The sampling costs in such locations can be much lower than in rivers where the trap and transport conservation measure is not undertaken. In Ireland, the cumulative quantity of silver eels captured for hydropower mitigation since the national eel management plan was adopted in 2008 is now over 500 tonnes. Similar conservation measures are increasingly being implemented in other EU member states as well as in New Zealand and North America. However, even where trap and transport programmes are not involved, the overall cost of full season sampling can in some locations be partly offset by the sale of eels captured. Use of removal sampling on a more limited temporal scale can be effectively used to calibrate other population monitoring methods or to validate the accuracy of the more widely used mark-recapture methods. Reduction of the cost could also be achieved by limiting removal sampling to peak migration nights which may be predicted by reference to season, lunar phase and increases in river discharge.

The European eel occurs in over 90,000 km^2 of continental waters in Europe and North Africa. The annual yield was estimated to have been more than 20,000 tonnes and that the fisheries once provided substantial income to more than 25,000 people (Moriarty and Dekker, 1997). Its decline, following decades of diminishing recruitment, is therefore of great socioeconomic significance. The fisheries were the most extensive source of data for stock modelling. Therefore, reduction in fishing effort or complete closure of some fishing areas makes development of fishery independent eel population monitoring increasingly important. Employment of former commercial fishermen in research and conservation related tasks can ensure that important local knowledge continues to inform eel management decision making processes.

Acknowledgments

We wish to thank the following for help with this research: The ESB Fishery Conservation staff and their contract fishermen, especially Mr James Dalton; Mr Darek Nowak and Dr Anne Bateman, Ryan Institute, NUIG; Dr Ciara O'Leary, Fisheries Ireland.

Bibliography

Aarestrup, K. (2012) The current status of the European Eel (*Anguilla anguilla*). In: Gough, P., Philipsen, P., Scholema, P.P. and Wanningen, H. (eds) *From Sea to Source. International Guidance for Restoration of Fish Migration Highways.* Regional Water Authority Hunze en Aa's, Veendam, pp. 66–67.

Acou, A., Gaelle, G., Feunteun, E. and Laffaille, P. (2009) Differential production and condition indices of premigrant eels in two small Atlantic coastal catchments of France. *American Fisheries Society Symposium* 58, 157–174.

Amilhat, E., Farrugio, H., Lecomte-Finiger, R., Simon, G. and Sasal, P. (2008) Silver eel population size and escapement in a Mediterranean lagoon: Bages-Sigean, France. *Knowledge and Management of Aquatic Ecosystems* 5, 390–391.

Aprahamian, M.W. (1986) Eel (*Anguilla anguilla* L.) production in the River Severn, England. *Polskie Archiwum Hydrobiologii* 33, 373–389.

Aprahamian, M.W. (1988) Age structure of eel (*Anguilla anguilla* (L.)) populations in the Rivers Severn (England) and Dee (Wales). *Aquaculture and Fisheries Management* 19, 365–376.

Aprahamian, M.W., Walker, A.M., Williams, B., Bark, A. and Knights, B. (2007) On the application of models of European eel (*Anguilla anguilla*) production and escapement to the development of eel management plans: the River Severn. *ICES Journal of Marine Science* 64, 1472–1482.

Baldwin, L. and Aprahamian, M. (2012) An evaluation of electric fishing for assessment of resident eel in rivers. *Fisheries Research* 123, 4–8.

Bodles, K. (2016) The sustainability of the Lough Neagh eel fishery. PhD thesis, Queens University of Belfast.

Bonhommeau, S., Chassot, E., Planque, B., Rivot, E., Knap, A.H. and Le Pape, O. (2008) Impact of climate on eel populations of the Northern Hemisphere. *Marine Ecology Progress Series* 373, 71–80.

Callaghan, R.M. and McCarthy, T.K. (1993) Variations in population structures and growth rates of eels in the Dunkellin River system, western Ireland. *Irish Fisheries Investigations. Series A* 36, 59–67.

Calles, O., Olsson, I.C., Comoglio, C., Kemp, P.S., Blunden, L., Schmitz, M. and Greenberg, L.A. (2010) Size-dependent mortality of migratory silver eels at a hydropower plant, and implications for escapement to the sea. *Freshwater Biology* 55, 2167–2180.

Carle, F.L. and Strub, M.R. (1978) A new method for estimating population size from removal data. *Biometrics* 34, 621–830.

Carss, D.N., Elston, D.A., Nelso, K.C. and Kruuk, H. (1999) Spatial and temporal trends in unexploited yellow eel stocks in two shallow lakes and associated streams. *Journal of Fish Biology* 55, 636–654.

Charrier, F., Mazel, V., Caraguel, J.M., Abdallah, Y., Le Gurun, L.L., Legault, A. and Laffaille, P. (2012) Escapement of silver-phase European eels, *Anguilla anguilla*, determined from fishing activities in a Mediterranean lagoon (Oir, France). *ICES Journal of Marine Science* 69, 30–33.

De Eyto, E., Briand, C., Poole, R., O'Leary, C. and Kelly, F. (2016) Application of EDA (v2.0) to Ireland: Prediction of silver eel escapement. *Irish Fisheries Investigations* 27, 1–68.

Dekker, W. (2000) The fractal geometry of the European eel stock. *ICES Journal of Marine Science* 57, 109–121.

Dekker, W. (2003) Did lack of spawners cause the collapse of the European eel, *Anguilla anguilla*? *Fisheries Management and Ecology* 10, 365–376.

Dekker, W. (2016) Management of the eel is slipping through our hands! Distribute control and orchestrate national protection. *ICES Journal of Marine Science* 73, 2442–2452.

Feunteun, E., Acou, A., Laffaille, P. and Legault, A. (2000) European eel (*Anguilla anguilla*): prediction of spawner escapement from continental population parameters. *Canadian Journal of Fisheries and Aquatic Sciences* 57, 1627–1635.

Geeraerts, C. and Belpaire, C. (2010) The effects of contaminants in European eel: a review. *Ecotoxicology* 19, 239–266.

Grennan, J. and McCarthy, T.K. (2013) Observations on European Eel (*Anguilla anguilla*) populations in two small Irish lakes. *The Irish Naturalists' Journal* 33, 35–42

Guy, C.S. and Brown, M.L. (2007) *Analysis and Interpretation of Freshwater Fisheries Data*. American Fisheries Society, Bethseda, MD.

ICES (2016) Report of the Working Group on Eels (WGEEL), 15–22 September 2016, Cordoba, Spain. ICES CM 2016/ACOM:19.

Jacoby, D. and Gollock, M. (2014) Anguilla anguilla. The IUCN Red List of Threatened Species 2014: e.T60344A45833138. http://dx.doi.org/10.2305/IUCN.UK.2014-1.RLTS.T60344A45833138.en (accessed 26 April 2019).

Jacoby, D.M., Casselman, J.M., Crook, V., DeLucia, M.B., Ahn, H., Kaifu, K., Kurwie, T., Sasal, P., Silfvergrip, A.M., Smith, K.G. and Uchida, K. (2015) Synergistic patterns of threat and the challenges facing global anguillid eel conservation. *Global Ecology and Conservation* 4, 321–333.

Kettle, A.J., Asbjørn Vøllestad, L. and Wibig, J. (2011) Where once the eel and the elephant were together: decline of the European eel because of changing hydrology in southwest Europe and northwest Africa? *Fish and Fisheries* 12, 380–411.

Koops, H. (1980) Sampling eels. In: Backiel, T. and Welcome, R. (eds) *Guidelines for Sampling Fish in Inland Waters*. EIFAC Technical Paper No 33. Food and Agriculture Organization of the United Nations, Rome, pp. 159–166.

Laffaille, P., Acou, A. and Guillouët, J. (2005) The yellow European eel (*Anguilla anguilla* L.) may adopt a sedentary lifestyle in inland freshwaters. *Ecology of freshwater Fish* 14, 191–196.

Lenihan, E., McCarthy, T.K. and Lawton, C. (2019) Use of an acoustic camera to monitor seaward migrating silver-silver eels (*Anguilla anguilla*) in a regulated river. *Ecohydrology and Hydrobiology* 19(2), 289–295.

Lobon-Cervia, J., Utrilla, C.G. and Rincon, P.A. (1995) Variations in the population dynamics of the European eel *Anguilla anguilla* (L.) along the course of a Cantabrian river. *Ecology of Freshwater Fish* 4, 17–27.

Matthews, M.A., Evans, D.W., Rosell, R.S., Moriarty, C. and Marsh, I. (2001) The Erne eel enhancement programme. Northern Regional Fisheries Board. Ballyshannon, Co. Donegal.

McCarthy, T.K. (2014) Eels and people in Ireland: from mythology to international stock conservation. In: Tsukamotu, K. and Kuroki, M. (eds) *Eels and Humans*. Springer, Tokyo, pp. 13–40.

McCarthy, T.K., Cullen, P. and O'Connor, W. (1999) The biology and management of River Shannon eel populations. *Fisheries Bulletin* 17, 9–20.

McCarthy, T.K., Blaszkowski, M., Grennan, J., Egan, F., Cullen, P. and Nowak, D. (2008a) Population biology of European eel (*Anguilla anguilla*) in Lough Derg, an Irish mesotrophic lake ecosystem. In: Phillips, B., Megrey, B.A. and Zhou, Y. (eds) *Proceedings of the 5th World Fisheries Congress*, Yokohama, American Fisheries Society, Bethesda, MD, pp. 279–280.

McCarthy, T.K. Frankiwicz, P., Cullen, P Blaszkowski, O'Connor, W. and Doherty, D. (2008b) Longterm effects of hydropower installations and associated river regulation on River Shannon eel populations: mitigation and management. *Hydrobiologia* 609, 109–124.

McCarthy, T.K., Nowak, D., Grennan, J., Bateman, A., Conneely, B. and MacNamara, R. (2014) Spawner escapement of European eel (*Anguilla anguilla*) from the River Erne Ireland. *Ecology of Freshwater Fish* 23, 21–32.

McGrath, C.J. (1971) EIFAC consultation of eel fishing gear and techniques. EIFAC Technical Paper No. 14. FAO, Rome, Italy.

MacNamara R. and McCarthy, T.K. (2014) Silver eel (*Anguilla anguilla*) population dynamics and production in the River Shannon, Ireland. *Ecology of Freshwater Fish* 23, 181–192.

MacNamara, R. McCarthy, T.K. and Barry J. (2017) Estimation of production and longterm changes in the population of European eel *Anguilla anguilla* (L.) in an Irish lake. *Ecology of Freshwater Fish* 26, 602–608.

Moriarty, C. (1996) Variation in numbers of eel *Anguilla anguilla* caught by constant effort in an Irish lake, 1981–1994. *Ecology of Freshwater Fish* 5, 148–152.

Moriarty, C. and Dekker, W. (1997) Management of the European eel. *Fisheries Bulletin* 15, 1–110.

Pedersen, M.I., Jepsen, N., Aarestrup, K., Koed, A., Pedersen, S. and Økland, F. (2012) Loss of European silver eel passing a hydropower station. *Journal of Applied Ichthyology* 28, 189–193.

Poole, W.R. (1990) Summer fyke-nets as a method of eel capture in a salmonid fishery. *Aquaculture Research* 21, 259–262.

Poole, W.R., Diserud, O.H., Thorstad, E.B., Durif, C.M., Dolan, C., Sandlund, O.T., Bergesen, K., Rogan, G., Kelly, S. and Vøllestad, L.A. (2018) Long-term variation in numbers and biomass of silver eels being produced in two European river systems. *ICES Journal of Marine Science* 75(5), 1627–1637.

Prigge, E., Marohn, L., Oeberst, R. and Hanel, R. (2013) Model prediction vs. reality-testing the predictions of a European eel (*Anguilla anguilla*) stock dynamics model against the in situ observation of silver eel escapement in compliance with the European eel regulation. *ICES Journal of Marine Science* 70, 309–318.

Rosell, R., Evans, D. and Allen, M. (2005) The eel fishery in Lough Neagh, Northern Ireland – an example of sustainable management? *Fisheries Management and Ecology* 12, 377–385.

Seber, G.A.F. (1973) *The Estimation of Animal Abundance*. Hafner, New York.

Seber, G.A.F. and Le Cren, E.D. (1967) Estimating population parameters from catches large relative to the population. *Journal of Animal Ecology* 36, 631–643.

Southwood, T.R.E. (1978) *Ecological Methods with Particular Reference to the Study of Insect Populations*, 2nd edn. Chapman and Hall, London.

Tesch, F.W. and White, R.J. (2008) *The Eel.* John Wiley & Sons, Chichester.

Van den Thillart, G. and Dufour, S. (2009) How to estimate the reproductive success of European silver eels. In: van den Thillart, G., Dufour, S. and Rankin, J.C. (eds) *Spawning Migration of the European Eel.* Fish and Fisheries Series, 30. Springer, Dordrecht, pp. 3–9.

Walker, A.M., Andonegi, E., Apostolaki, P., Aprahamian, M., Beaulaton, L., Bevacqua, P., Briand, C., Cannas, A., De Eyto, E., Dekker, W., De Leo, G., Diaz, E., Doering-Arjes, P., Fladung, E., Jouanin, C., Lambert, P., Poole, R., Oeberst, R. and Schiavina, M. (2011) Pilot projects to estimate potential and actual escapement of silver eel. Final Report. The European Commission Directorate-General for Maritime Affairs and Fisheries. Service Contract No. SI2.539598.

Zippin, C. (1956) An evaluation of the removal method of estimating animal populations. *Biometrics* 12, 163–189.

CHAPTER 13

Impacts on European eels (*Anguilla anguilla*) at a Norwegian hydropower site: how operational changes can influence outcomes

Frode Kroglund

County Governor Office, Agder, Norway

Introduction

It is well documented that run-of-the-river hydropower (RORH) has a negative impact on all migratory fish species. Impacts can range from habitat loss to severe mortality. The damage can be reduced or minimized if appropriate measures are implemented. There are numerous guidelines suggesting appropriate measures (DWA, 2005; Environment Agency, 2011; Travade and Larinier, 2006). While mitigative measures are being implemented in many places, only a few of these have been validated and their effectiveness is unfortunately unproven (Roscoe and Hinch, 2010). A solution that does not work is not a solution.

Mitigation has both an investment cost and can create an energy loss. Many owners of hydropower plants will therefore be reluctant to implement relevant measures when the outcome is uncertain. It is therefore essential to collect evidence of both the impact of existing RORH sites and the efficacy of mitigative measures to demonstrate that fish mortality is often unnecessary and that it is possible to provide 'safe' migration at a relatively low cost.

While a RORH site will affect both upstream and downstream migration, the challenges fish experience will depend on migratory direction, fish species and size. Downstream migrants will generally follow the water flow, where the main flow under normal conditions will be towards the headrace or forebay and penstock. During flooding events, surplus water will spill over the dam or through dedicated spill gates, providing an alternative migration route. But under normal flow conditions, a RORH site can utilize 'all' water, thereby guiding fish towards the forebay. If there is no functional escape route in the forebay, the turbine route will present itself as the only available route for continued downstream migration. The turbine route poses several threats related to the physical properties of the forebay, the trash rack, the penstock pipes, the turbine blades and the tailrace. Migration over the dam can impact fish depending on fall height, water depth and so on.

There is normally a trash rack in front of the penstock. Fish will interact with this rack. Depending on rack properties, it can act as a physical barrier, a behavioural barrier, a migratory guide or have no clear effect on migration (Russon et al., 2010; Travade et al., 2006). Fish response to the rack depends on numerous factors including bar spacing (light opening), water depth, water velocity, fish species and size, and water temperature. While a rack with bar spacing smaller than the fish will act as a physical barrier, a rack with larger bar spacing can still act as a behavioural barrier, when or if the fish perceives the rack as a 'challenge'. Very large bar spacing relative to fish size will have little to no effect as a barrier. If fish cannot pass the trash rack, they can still be impinged on to the rack if water velocity exceeds swimming capacity. The trash rack can, as such, become a cause of fish mortality.

Fish that pass the trash rack can be killed by direct turbine strikes and/or by the various forces present in the penstock and tailrace. Fish surviving the turbine passage can still be injured resulting in reduced fitness and delayed mortality. Fish survival past a RORH site will depend on turbine type, water flow and fish size, and can range from being low to high. Long fish are more vulnerable to turbine strikes than short fish making eels more vulnerable than salmonid smolts. To reduce mortality, fish must be provided with an alternative route past the RORH site.

A case study involving fish kills and mitigation

A small scale 2 MW RORH site with an annual production of 8 GW was built on the River Storelva, Norway, in 2008. The power plant has a head of 14.5 m, turbine discharge is 16 m^3/s and is equipped with a four-bladed Kaplan turbine (1.65 m diameter, 0.6 m hub diameter and 330 rpm). Maximum theoretical water velocity in the forebay is 0.5 m/s. Surplus water is topped over the dam (Figures 13.1 and 13.2). The trash rack (α-type) stands at a 70°-angle to the flow and is submerged 60 cm under the water surface and raised 50 cm above the bottom. The trash rack (25 m^2) has a bar spacing of 50 mm with 10 mm bars. The first autumn the power plant was put into operation it killed eels, while smolts and kelts were killed the following spring. This was unsatisfactory for both the local community and for the power plant owner. We were therefore permitted to 'find' a solution, with the caveat that it should not cost money. The solution had therefore to be within the possibilities and limitations present in 2010.

Between 2010 and 2013 we documented the passage of smolts, kelts and eels past the RORH site. Salmonids were effectively bypassed when a surface-oriented bypass was provided next to the trash rack (Figures 13.1 and 13.2). When this escape route was fed 5–7% of the turbine discharge, >90% of the smolt and 100% of the kelt used this bypass (Haraldstad et al., 2018). While dead salmonids (300–500 smolt annually) were commonly observed during the first years (2008–10), these became equally scarce (<40 annually) after the bypass was put into operation, proving that it is possible to save fish. In addition, this bypass was used by a low number of spring migrating eels.

When river flow exceeds turbine discharge, excess water will spill over the dam (site 2 in Figures 13.1 and 13.2). Flow in spring is normally lower than the turbine discharge so downstream migration over the dam is not possible, and eels are led towards the headrace. The main silver eel migration occurs during the period September to November and around 30% of the days will have spill.

Figure 13.1 Run-of-river hydropower site in Storelva, Norway. Downstream migrants can pass using the fish ladder (1), go over the dam crest (2; DC), use a debris gate (3) or use the escape routes provided in the forebay (4ab; ER). 4a is for salmonids and 4b is for eels. Eels can also pass through the trash rack. The trash rack and turbine route (TR) is indicated.

Figure 13.2 The escape routes seen from inside (left) or outside (right) the forebay. Water level is indicated by the red line and escape routes by red squares. The size of the escape 'hole' was varied annually and within season to identify 'optimal' configuration. The trash rack or turbine route (TR) is encircled by green colour.

This implies that silver eels can migrate over the dam crest, where the availability of this route will vary within and between years. The remaining days, eels will be led towards the headrace.

We assumed that eels were primarily bottom oriented and/or would dive when encountering the trash rack (Russon et al., 2010). We therefore chose to use an undershot gate located adjacent to the trash rack as the escape route for eels (Figures 13.1 and 13.2). Eels using this bypass were either collected into a fyke net (2010) or collected in a holding pool (subsequent years). The holding pool acted also as an energy killer (6 m water pressure out of the undershot). The first 2 years we only documented the number of eels using the bypass as opposed to number of dead eels found by scuba diving below the RORH site. Starting in 2012, eels were PIT-tagged and released both upstream and downstream of the RORH site (see Table 13.1). The migratory route used by the eels was determined using PIT-antennas located 0.5 km upstream of, and 0.7 and 6 km downstream of, the RORH site and by manually checking all eels using the bypass and dead eels recovered downstream.

In 2012, 100 PIT-tagged eels were released upstream of the RORH site. In 2013, the number was increased to 170 eels. These eels could navigate past the RORH site by using one of three possible routes. The turbine route (TR), the escape route (ER) or by migrating over the dam crest (DC) (Figures 13.1 and 13.2). Apart from the fish ladder (site 1), site 2 was the main site for passage. Site 3 was not used to bypass eels.

The proportion of salmonids using the ER was related to water discharge in the bypass (*Qs*) relative to turbine flow (*Qt*) (Haraldstad et al., 2018). This also applies to eels (Travade et al., 2010). A spill ratio was calculated as *Qs*/*Qt*. The water flow through the ER was varied as we gained experience, but we could not vary turbine flow.

In 2012, 37 eels were PIT-tagged and released downstream of the RORH site. In 2013, 90 eels were released below the RORH site. Around 95% of these were detected on the antennas 0.7 and 6 km downstream of the RORH site within a week after release. Based on this, the tagged eels were most likely migrating silver eels.

The number of dead eels was calculated from the number of 'tails' recovered below the RORH site. Generally, we recovered twice as many tails as heads, also suggesting that heads can 'swim' further than tails. Based on PIT-tag eels migrating through TR, we calculated that we recovered around 60% of the dead eels.

Results

We did not detect any turbine killed PIT-tagged eels in 2012 (Table 13.1). This is likely because of a major flooding event (>100 m^3/s) on 25 September, when nearly all PIT-tagged eels migrated over the dam crest. Around 15% of the PIT-tagged eels used the ER, 85% the DC route assuming the undetected PIT-tagged eels also used the DC route. The ER migrants came prior to the flood, the DC migrants during the flood. The power plant had to shut down operation during the flood and turbine passage was therefore highly unlikely.

Although we did not observe the power plant killing any of the PIT-tagged eels in 2012, we periodically observed massive kills throughout the season (Figure 13.3). These occurred when the ER was fed less than 3% of the turbine discharge and there was no water flow over the DC, suggesting the ER flow was too small to attract fish and the TR became the favoured route.

Table 13.1 Number of PIT-tagged eels released upstream of the ROR and the number using one of three possible downstream routes. 'Unknown' in 2012 is most likely eels migrating over the dam crest that went undetected due to a major flood.

Year	*N*	Turbine route (TR)	Escape route (ER)	Dam crest (DC)	Unknown
2012	100	0	14	40	45
2013	170	57	41	72	0

Figure 13.3 Eels recovered by scuba diving below the RORH site on or after 25 September. This mortality came after the power plant started operation after a period of stand-still and the ER was not available.

In 2013, around 28% of the eels released above the RORH site out-migrated using the TR. Of these, 21% were detected 6 km downstream. The remaining 79% were assumed dead. Dead here implies that the eels were either found chopped in pieces and a PIT-tagged piece was recovered, or the eels were detected on the 0.7 km but not the 6 km downstream antenna. We frequently observed dead eels more than 1 km downstream of the RORH site, but do not believe that dead eels would progress further than 5 km along a slow meandering river and through a 1 km long lake where both birds and pike are likely predators (Havn et al., 2017).

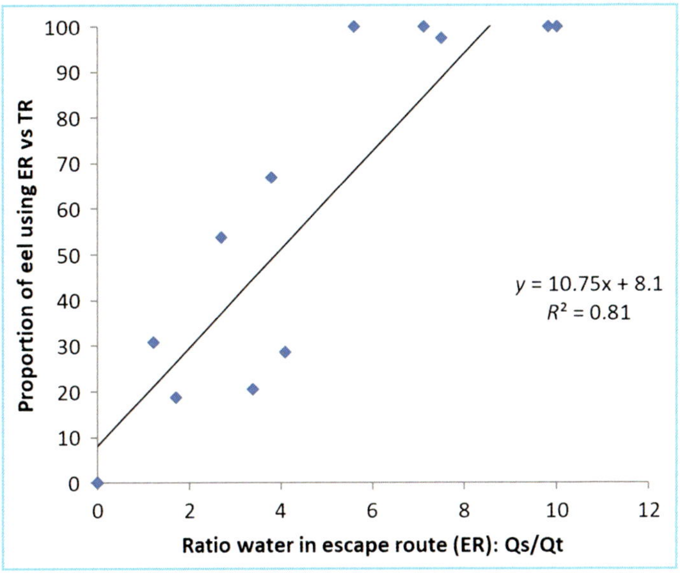

Figure 13.4 Relationship between proportion of eels using escape route (ER) relative to turbine route (TR) and associated *Qs/Qt*. Each dot represents a period with semi-stable flow conditions. Temperature decreases over the migration period, possibly explaining some of the variance.

Furthermore, >95% of the eels released below the RORH site were detected on both the 0.7 and 6 km antenna suggesting that 'live' eels passed the lake as opposed to assumed TR migrants. The differences in detection supports the assumption that eels undetected 6 km below the RORH site were dead.

Around 28% of the eel out-migrated through TR. The remaining used one of the two escape routes; ER or DC. Although the ER in the forebay was available at all times, only one-third of the eels used this route while two-thirds migrated over the dam crest. Although this could suggest that the ER did not function, the ER was the preferred route when water flow was below turbine flow. The importance of the ER will therefore depend on the total water flow, where the DC route appears to be most important when water is spilled and the ER when there is no spill.

There was a relationship between the proportion of eels using the ER to the TR based on *Qs*/*Qt* (Figure 13.4). When the ER was fed around 5% of the turbine discharge, most of the eels used this escape route. Based on this, it is possible to guide eels to an escape route, but at a cost of water, and hence power production.

Although the DC was the preferred route during flooding events, water was on average only spilled over the dam on around one-third of the days during the migration period. There is a large year-to-year variation in water flow, and hence number of days with spills. Likewise, the magnitude of the spill will vary within and between years. An alternative route is therefore needed to provide a safe migratory route when there is no spill. This ER must compete with the TR as the out-migration route.

Conclusion

Prior to opening the ER, eels could pass the RORH site by either using the TR or by moving over the DC. Around 80% of all eels using the TR died. As eels also used the DC route, this number does not represent true mortality. True mortality is lower and will depend on the ratio of eels using the TR versus the DC route.

Around two-thirds of the eels used the DC route during flooding events (>100 m^3/s), but we have no data suggesting how large the spill must be to compete with the TR. Increasing river flow from 16 to 32 m^3/s will give 10 cm water depth over the dam crest, while a 20 cm water depth will require a spill of around 50 m^3/s. Water flows exceeding 30 m^3/s are rare and will only occur a few days every year. Although an important route, the DC does not provide a reliable, safe route protecting silver eels every year. There was a need for an additional route that would benefit eels when the water flow is too low to make the DC route available.

Eels used an undershot sluice gate when this gate was fed 5% or more of the turbine flow. Similar solutions have been tested elsewhere (Egg et al., 2017). However, operating an undershot sluice gate was challenging as the solution required manual adjustments to adjust for changes in river water flow. The trials showed it was possible to guide eels this way, but it was not practical. Swedish experiments have shown good results using a low slanting trash rack with small bar spacing (Calles et al., 2012, 2013). This requires having a trash rack with an α-angle <35° and around 15–20 mm bar spacing. Here, the escape route can be placed closer to the water surface. This will normally be easy to operate, and avoids having to work with water under pressure. This solution will function both for salmonids and eels but should be built according to eel specifications (Larinier, 2001). Salmonids have fewer requirements regarding trash rack properties than eels (Dainys et al., 2018).

There is always a conflict between power production and protection of migrating fish. By understanding fish behaviour in general and site-specific variations, it is possible to retrofit solutions that can reduce mortality substantially. In the tests run in Storelva, the only cost was related to loss of water from power production (5% of turbine flow over some weeks). With the use of behavioural studies, the periods requiring water spill to protect eels can be reduced, making simple solutions more cost-efficient (Adam and Schwevers, 2006; Trancart et al., 2013). For new power plants, the solution should be an integrated part of the construction plan and not something to be solved later.

Bibliography

Adam, V.B. and Schwevers, U. (2006) Möglichkeit eines aalschonenden Betriebs von Wasserkraftanlagen mit dem Frühwarnsystem MIGROMAT®. Wasserwirtschaft 96(5), 16–21.

Calles, O., Karlsson, S., Hebrand, M. and Comoglio, C. (2012) Evaluating technical improvements for downstream migrating diadromous fish at a hydroelectric plant. *Ecological Engineering* 48, 30–37.

Calles, O., Karlsson, S., Vezza, P., Comoglio, C. and Tielman, J. (2013) Success of a low-sloping rack for improving downstream passage of silver eels at a hydroelectric plant. *Freshwater Biology* 58, 2168–2179.

Dainys, J., Stakėnas, S., Gorfine, H. and Ložys, L. (2018) Mortality of silver eels migrating through different types of hydropower turbines in Lithuania. *River Research and Applications* 34, 52–59.

DWA (2005) *Fish Protection Technologies and Downstream Fishways. Dimensioning, Design, Effectiveness Inspection.* German Association for Water, Wastewater and Waste (DWA), Hennef.

Egg, L., Mueller, M., Pander, J., Knott, J. and Geist, J. (2017) Improving European silver eel (*Anguilla anguilla*) downstream migration by undershot sluice gate management at a small-scale hydropower plant. *Ecological Engineering* 106, 349–357.

Environment Agency (2011) Screening at intakes and outfalls: measures to protect eel. *The Eel Manual.* Environment Agency, Bristol. Available at: https://assets.publishing.service.gov.uk/government/uploads/system/uploads/attachment_data/file/297342/geho0411btqd-e-e.pdf.

Haraldstad, T., H. Glund, E., Kroglund, F., Haugen, T.O. and Forseth, T. (2018) Common mechanisms for guidance efficiency of descending Atlantic salmon smolts in small and large hydroelectric power plants. *River Research and Applications* 34(9), 1179–85.

Havn, T.B., Økland, F., Teichert, M.A.K., Heermann, L., Borcherding, J., Sæther, S.A., Tambets, M., Diserud, O.H. and Thorstad, E.B. (2017) Movements of dead fish in rivers. *Animal Biotelemetry* 5, 1–9.

Larinier, M. (2001) Environmental issues, dams and fish migration. In: Marmulla, G. (ed.) *Dams, Fish and Fisheries. Opportunities, Challenges and Conflict Resolution.* FAO, Rome, pp. 45–90.

Roscoe, D.W. and Hinch, S.G. (2010) Effectiveness monitoring of fish passage facilities: historical trends, geographic patterns and future directions. *Fish and Fisheries* 11, 12–33.

Russon, I.J., Kemp, P.S. and Calles, O. (2010) Response of downstream migrating adult European eels (*Anguilla anguilla*) to bar racks under experimental conditions. *Ecology of Freshwater Fish* 19, 197–205.

Trancart, T., Acou, A., De Oliveira, E. and Feunteun, E. (2013) Forecasting animal migration using SARIMAX: an efficient means of reducing silver eel mortality caused by turbines. *Endangered Species Research* 21, 181–190.

Travade, F. and Larinier, M. (2006) French experience in downstream migration devices. In: *Free Passage for Aquatic Fauna in Rivers and Other Water Bodies.* International DWA-Symposium on water resources management. Deutsche Vereinigung für Wasserwirtschaft, Abwasser und Abfall, Hennef.

Travade, F., Gosset, C., Larinier, M., Subra, S., Durif, C., Rives, J. and Elie, P. (2006) Evaluation of surface and bottom bypasses to protect eel migrating downstream at small hydroelectric facilities in France. In symposium on Hydropower, Flood Control and Water Abstraction: Implications for Fish and Fisheries, 14–21 June 2006, Mondsee, Austria. FAO, Rome.

Travade, F., Larinier, M., Subra, S., Gomes, P. and De-Oliveira, E. (2010) Behaviour and passage of European silver eels (*Anguilla anguilla*) at a small hydropower plant during their downstream migration. *Knowledge and Management of Aquatic Ecosystems* 398, 1–19.

CHAPTER 14

New Zealand tuna (freshwater eels): Māori-driven management practices – past, present and future

Erica Williams[1], Anke Zernack[2], Jacques Boubée[3] and Erina Watene-Rawiri[4]

[1]National Institute of Water and Atmosphere, Wellington, New Zealand; [2]School of Agriculture and Environment, Massey University, Palmerston North, New Zealand; [3]Vaipuhi Freshwater Consulting Limited, Morrinsville, New Zealand; [4]National Institute of Water and Atmosphere, Hamilton, New Zealand

Introduction

Three tuna species occur in New Zealand; the longfin (*Anguilla dieffenbachii*), which is only found in New Zealand; the shortfin (*A. australis*), which also occurs in eastern Australia and the Pacific; and the Australian speckled longfin (*A. reinhardtii*), which has relatively recently been confirmed as present in New Zealand, and is also found in Australia and New Caledonia. The New Zealand longfin (Figure 14.1) is one of the largest and longest living eel species in the world (Tesch, 2003). Although longfins are still one of the most common freshwater fish in the country, there are concerns about the scarcity of large females, as well as elvers (Jellyman et al., 2000; PCE, 2013). The latest Threat Classification System assessment classified shortfins as being 'Not Threatened', while longfins are classified as 'At Risk–Declining' (Goodman et al., 2014).

The key pressures on tuna populations include habitat loss and degradation, interrupted connectivity for migrations between the sea and freshwater habitats, land and infrastructure management, and commercial, recreational and customary harvest. Other pressures may arise from competition with exotic fish species such as salmonids, catfish, and koi carp; exposure to algal toxins, diseases and parasites; as well as marine factors (for example, food availability) influencing the survival of larvae and juveniles. The full impact of these pressures on New Zealand tuna populations remains unknown.

Māori (the indigenous people of New Zealand) have an extensive knowledge of tuna that has been retained and actively practised over centuries by whānau (families), hapū (sub-tribes) and iwi (tribal groups). While tuna species are characterized in terms of biophysical science (family, genus, species), Māori have an extensive range of classifications for tuna related, for example, to appearance, coloration, season of the year, size, behaviour, locality and palatability. Tuna are a highly valued customary, recreational and commercial fishery for Māori. Many

Māori communities harvest tuna as part of their regular dietary intake, as well as for hui (meetings, gatherings) and tangi (funerals). In addition to concerns regarding the impacts of land use practices and the associated effects on water quality and instream fisheries habitat, Māori are concerned about the reduced availability of tuna stocks.

The Treaty of Waitangi is New Zealand's founding document and is an agreement, in Māori and English, that was made between the British Crown and Māori rangatira (chiefs, leaders) in 1840. After that, New Zealand became a colony of Britain and Māori became British subjects. However, Māori and Europeans had different understandings and expectations of the Treaty as the meaning of the English version was not the same as the meaning of the Māori translation. Following its signing, many of the rights guaranteed to Māori in the Treaty were ignored. To help rectify this, the Waitangi Tribunal was set up in 1975, to consider claims that the government had breached the Treaty. The Tribunal researches and makes legal decisions on cases where Māori land and other resources were taken illegally or unfairly in the past. Quite often this results in large settlements for iwi, including cash and land. One example of the Treaty in action is fisheries. Māori now have significant control of, and rights to, this important natural resource. Treaty settlements are playing a critical role in providing the legislative foundation for a range of new institutional arrangements for the improved governance and management of New Zealand fresh water ecosystems and the active implementation of rehabilitation strategies and actions to meet Māori aspirations.

In this chapter, we provide a few examples of the traditional and contemporary approaches that Māori are leading in tuna management and restoration. A glossary of Te Reo Māori (the Māori language) used is provided at the end of this chapter. New Zealand place names in Te Reo Māori are not included. More information about the Treaty of Waitangi and the Waitangi Tribunal is provided in the glossary.

Past fisheries management

There is no argument regarding the importance of tuna in the lives of Māori, where this taonga (highly prized) species permeates, for example, place names, whakataukī (proverbs), pakiwaitara (legends), waiata (songs) and artwork. Pre-European colonization, tuna were easy to catch and preserve, and provided a critical source of dietary fat for Māori in the absence of large land mammals. Further, they could be retained alive until needed in specially woven containers (McDowall, 2011), a practice that continues to be used by some iwi albeit with more modern materials (for example, Shortland, 2013). As well as providing a source of day-to-day sustenance, tuna are also important for trade, as koha (gifts), but also to provide for guests attending cultural occasions. For example, the Waahi Stream, which connects Lake Waahi to the Waikato River has been fished for migrant shortfin male eels (called puhi by the Waikato-Tainui people) for generations. Lake outlets were strategically chosen by the Waikato-Tainui people to build rauwiri/pā tuna (eel weirs) and capture puhi. During special occasions held at Waahi marae (meeting house), which is located on the banks of the Waahi Stream, it is still an expectation by those who attend that puhi will be served at meal times.

Māori have an extensive knowledge of the ecology of tuna and have used this knowledge to maintain their customary tuna fisheries for several centuries. Both migrating and feeder

Figure 14.1 New Zealand longfin tuna. Photographs: (a) Shannan Crow, NIWA, (b) Stu MacKay, NIWA.

(non-migrant) tuna are captured in the fishery. A wide range of methods were employed to catch tuna, with techniques varying by area, season and habitat. Examples of customary methods include hīnaki (eel pots), pā tuna, korapa (hand-netting), rama tuna (using torch light), matarau (spearing) and koumu (eel trenches) (Best, 1929; Marshall, 1987; McDowall, 2011). While variations of these methods may still be used in the present day, much customary fishing now uses modern equipment, such as fyke nets that are quick to deploy and retrieve. Māori would often undertake their fishing activities according to the maramataka (calendar that divides the Māori year into lunar months) (Potaka, 2008). In several areas tuna fishing activities would be focused on adult eels around the whakaheke (migration back to the sea) (Marshall, 1987; McDowall, 2011; MFish, 2009; Shortland, 2013). Occasionally, Māori transferred small eels to inland waterways or lakes with irregular connections to the sea, to create new populations for subsequent harvesting (McDowall, 2011).

The reliance on the environment for resources meant that Māori engaged with their waterways – in all weathers, in different seasons, and observed the environment in its many different states. This engagement, observation and reliance on food sources obtained from waterways provided the vector for transfer of intergenerational knowledge (Panelli and Tipa, 2007). Changing behaviours in response to changing conditions meant that Māori were well versed in the principles and practice of adaptive management. Māori regulated fresh waters and their fisheries through a series of adaptive and considered actions by tohunga (expert, specialist), kaitiaki (guardian), whānau and hapū (Te Wai Māori, 2008). These actions were developed in response to intergenerational observational learnings and practices to maintain sustainability of use for current and future generations. Complex relationships, tohu (environment indicators) and understandings of ecosystems were utilized to inform management options, including the use of rāhui (temporary restrictions) on the use of resources (Te Wai Māori, 2008). In the Waikato, for example, it is said that the availability of tuna in two main riverine lakes (Lakes Waahi and Hakanoa) had declined due to over-harvesting. In response to this, the local rangatira, Pōtatau Te Wherowhero, placed a rāhui (ritual prohibition) over the whole area for a period of time to prevent overfishing and enable the stocks to recover. When the rāhui was lifted there was a mass celebration and a haka (ceremonial dance) to lift the tapu (restriction) and make the lake noa (free from tapu) again, which is how Lake Hakanoa came to be named.

Present fisheries management

Tuna support valuable customary, commercial and recreational fisheries in New Zealand. The roles and responsibilities for the sustainable management of tuna populations and their associated habitats are currently spread across a wide range of agencies including Māori tribal entities, Department of Conservation (DOC), Ministry for Primary Industries (MPI), Fish and Game Council, regional councils, commercial fisher associations and conservation groups.

Commercial fisheries management

The New Zealand commercial tuna fishery began in earnest in the late 1960s and expanded rapidly until the early 1970s, peaking at slightly over 2000 tonnes in 1972. Jellyman (2013) describes the development of the commercial fishery in three phases: (1) exploitation phase (1965–80); (2) consolidation phase (1980–2000); and (3) rationalization phase (2000–12). The commercial fishery is currently managed under the Quota Management System (QMS) where individual transferrable quota for fish stocks are owned by private interests. The quantity of fish that can be taken by commercial fishers in one fishing year is called the Total Allowable Commercial Catch (TACC). On behalf of the government, MPI promotes policies and regulations to manage the commercial eel industry in a sustainable manner. MPI is responsible for administering the Fisheries Act 1996 and the QMS. It also monitors fishing activity and enforces fishing rules. In 1992, as part of the Treaty of Waitangi (Fisheries Claims) Settlement Act, Māori were given 20% of the overall commercial quota. Since 2004, much of the North Island tuna quota has been purchased by Māori-owned companies.

The tuna fishery was introduced into the QMS in stages between October 2000 to October 2004. The commercial tuna fishery is currently managed as 11 separate longfin and shortfin stocks (Figure 14.2) with the 2016–17 TACC set at 163 tonnes and 580 tonnes, respectively (Clements and Associates, 2016; MPI, 2016). New Zealand's commercial tuna fishery supplies both a domestic and export market and has an estimated export value of NZ$6.1 million (MFish, 2009).

Initially, South Island longfin and shortfin tuna stocks were combined, and the fishery was managed as 16 fish stocks nationwide. The review of South Island tuna stocks in 2016 led to the separation of the previously combined stocks into individual shortfin and longfin stocks (Figure 14.2), and the subsequent setting of precautionary catch limits and allowances for the separated stocks (Fisheries New Zealand, 2018). North Island tuna stocks were last reviewed in 2008, with significant reductions to the Total Allowable Catch (which includes recreational, customary and commercial catch limits) and TACCs of between 10–38% for shortfins and 35–78% for longfins. The government is currently developing a new package of management measures designed to at least maintain the current level of abundance. Fisheries New Zealand is currently seeking the views of Māori and stakeholders on this review of North Island freshwater eels (Fisheries New Zealand, 2018).

Three main harvest control measures are used by MPI to support the management of the New Zealand tuna fishery. The first is that there are upper and lower harvest size limits. This

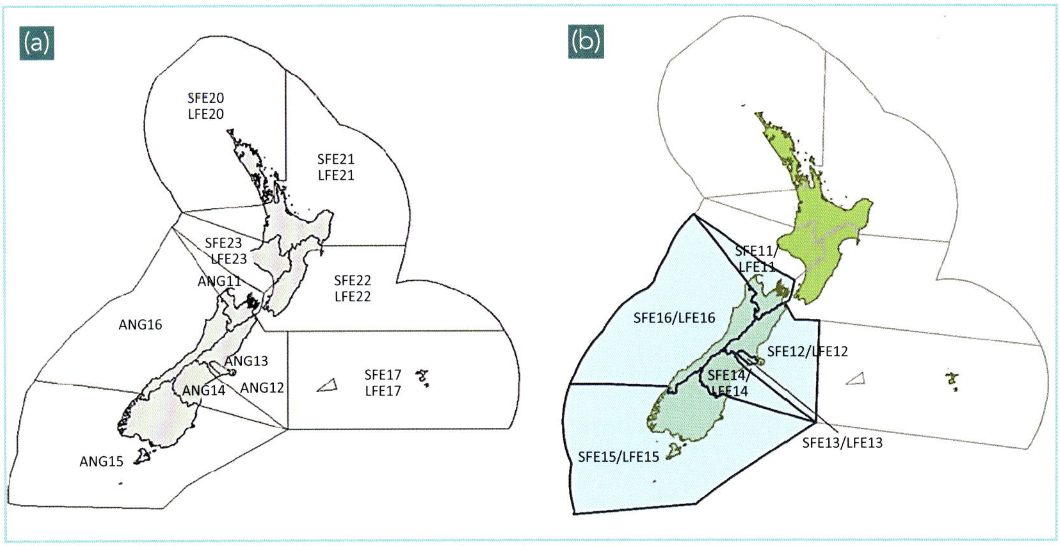

Figure 14.2 (a) Quota management areas for shortfin eel (SFE), longfin eel (LFE) and both species combined (ANG) prior to the review of the South Island tuna fishery in 2016. (b) New quota management areas for the South Island tuna fishery (SFE11–16 and LFE11–16). Credit: MPI (2016).

management measure has been in place for a number of years and essentially prevents the harvest and export of juvenile eels (notably glass eels and elvers) for aquaculture purposes, while large female tuna are also protected from harvest. The second is that tuna harvest is controlled by the QMS, which is adjustable across quota/fisheries management areas as the need arises. The third is the establishment of reserve areas that are free from commercial harvest, although DOC has the discretion to grant fishing concessions in reserve areas under their control (for example, national parks).

In response to growing concerns about the long-term sustainability of harvest levels, especially of longfins, in the rationalization phase of the commercial fishery MPI and commercial fishers have taken actions to reduce catches, including increases in the minimum size, an increase in reserve areas, removal of part-time fishers, a moratorium of fishing licences and reductions in TACCs. Several reviews and research programmes have highlighted the vulnerability of longfins, and the need to implement more conservative management practices to avoid substantial reductions in this fishery (Hoyle and Jellyman, 2002; Jellyman, 2009; Jellyman et al., 2000; McCleave and Jellyman, 2004). In 2013, the Parliamentary Commissioner for the Environment (PCE) undertook a review of the longfin fishery (PCE, 2013) in response to increasing public concerns about the sustainability of populations and the potential risk of extinction. An independent panel of experts was assembled to review the scientific information used to guide the management of New Zealand's tuna fisheries, particularly longfins. The panel recommended a comprehensive assessment that integrates conventional fisheries information with non-fisheries information, such as the impact of habitat degradation, pollution and other non-fisheries causes of mortality (for example, hydroelectric dams and flood pump stations) (Haro et al., 2015).

Māori fisheries management

Under Section 10 of the Treaty of Waitangi (Fisheries Claims) Settlement Act 1992 there is an obligation on the Crown to develop regulations that recognize and provide for customary food gathering by Māori and the special relationship between Māori and places of spiritual and cultural importance. One management mechanism that has enabled Māori to contribute to the management of tuna is establishing mātaitai reserves (areas where Māori manage all non-commercial fishing by creating bylaws). Several mātaitai have been established, with the first being created in the Mataura River. Other mātaitai reserves are now located in the Okarito Lagoon, Lake Wairewa, Temuka/Opihi River and Waikawa River. A constraint with this mechanism is that management is confined to specific species within a defined area. It is not a mechanism that enables management of the many factors that impact the fish/fishery. Non-commercial fishing in mātaitai is managed by mandated representatives through the creation of bylaws that can cover: species that can be taken; quantity of each species that can be taken; size limits relating to each species to be taken; the method by which each species can be taken; area or areas in which the species can be taken, and any other matters the tangata kaitiaki/tiaki (guardians of the fishery) considers necessary for the sustainable management of fisheries resources.

In both the North and South Islands, there are areas in which commercial fishing is prohibited in recognition of the special value of these areas for customary non-commercial purposes. In the South Island, Lake Waiwera and its tributaries have been set aside exclusively for Ngāi Tahu. Other areas, such as the lower Pelorus River, Taumutu (Te Waihora), Wainono Lagoon and its catchment, the Waihao catchment, the Rangitata Lagoon and the Ahuriri Arm of Lake Benmore, have been set aside as non-commercial areas. In the North Island, commercial fishing is prohibited from Lake Horowhenua, the Taharoa lakes, Whakakī Lagoon, Lake Poukawa, Arahura River, Lakes Kohangapiripiri and Kohangaterā and associated catchments (Jellyman and Sykes, 2009; McDowall, 2011; MFish, 2009). In addition, as a sustainability measure, the Mōhaka, Mōtu and parts of the Whanganui rivers have been closed to commercial fishing to aid spawning escapement (MFish, 2009). Some iwi have also placed a moratorium on the harvest of longfins, 'shelving' (not fishing) their quota, as they are concerned about declining stocks (Kawe, 2014; Te Kupenga o Maniapoto Limited, 2013). For example, the Maniapoto Māori Trust Board '...was concerned that the sustainability of this iconic species was under threat due to overfishing' (Te Kupenga o Maniapoto Limited, 2013, p. 2). In making their decision, they considered that 'tuna has a much higher cultural value to our people, far in excess of the earnings derived by the company from ACE (Annual Catch Entitlement) sales' (Te Kupenga o Maniapoto Limited, 2013, p. 2).

The rehabilitation of lost cultural, environmental and economic values and services is a key outcome of numerous Treaty settlements. Treaty settlements are driving a range of new governance and co-management models for the improved management of fresh waters, and the taonga species these ecosystems support. Declining tuna stocks have prompted Māori-driven initiatives to manage and restore critical eel habitats. In the following section, we provide two examples of how Māori are influencing change in tuna management and restoration in New Zealand.

Waikato River catchment

The Waikato-Tainui Raupatu Claims (Waikato River) Settlement Act was passed in 2010 with an overarching purpose to restore and protect the health and well-being of the Waikato River for future generations. This act established an iwi–Crown co-governance structure over the Waikato River, building on the Waikato-Tainui Raupatu Claims Settlement Act of 1995. Fisheries in the lower Waikato River catchment are now co-managed by Waikato-Tainui and the Crown, under the Waikato-Tainui (Waikato River) Fisheries Regulations 2011.

The Waikato River Authority (WRA) was established in 2010 as a co-management structure arising out of the Treaty settlements of a number of Waikato River iwi including Waikato-Tainui, Ngāti Tūwharetoa, Ngāti Raukawa and Te Arawa. The WRA has ten members, five each appointed by iwi and by the Crown, and is responsible for setting Te Ture Whaimana, the Vision and Strategy for the Waikato River. The overarching vision is for 'a future where a healthy Waikato River sustains abundant life and prosperous communities who, in turn, are all responsible for restoring and protecting the health and well-being of the Waikato River, and all it embraces, for generations to come' (Waikato River Authority, 2011, p. 4). Thirteen objectives have been designed to support the realization of this vision, including the following examples which are applicable to the improved management and restoration of Waikato River tuna populations:

1 the integrated, holistic and coordinated approach to management of the natural, physical, cultural and historic resources of the Waikato River
2 the protection and enhancement of significant sites, fisheries, flora and fauna.

To date 204 projects have received WRA funding of more than NZ$30.8 million (WRA, 2017). A large majority of the projects funded to date have focused on riparian, wetland and instream habitat restoration which will benefit tuna populations in the catchment. The WRA have also funded the Waikato-Tainui College for Research and Development to lead a major tuna restoration project in the lower Waikato River catchment, which includes collating Waikato-Tainui mātauranga (knowledge), the development of a tuna restoration plan, habitat restoration, tuna transfers, and improvements in fish passage.

As part of the Waikato River co-management agreements, each iwi that make up the region (Waikato-Tainui, Ngāti Tūwharetoa, Ngāti Raukawa, Te Arawa and Ngāti Maniapoto) is required to develop customary fisheries plans. For example, tuna fisheries in the lower Waikato River catchment are now co-managed by Waikato-Tainui and the Crown, under the Waikato-Tainui (Waikato River) Fisheries Regulations 2011. The regulations allow Waikato-Tainui to propose bylaws to MPI that may restrict or prohibit commercial, recreational and customary fishing. To date six bylaws pertaining to freshwater eels in the Waikato-Tainui Fisheries Area have been implemented (Waikato Raupatu River Trust, 2014).

- Bylaw 1: Allows for the incidental catch and release of tuna weighing more than the maximum size limit of 2 kg and female longfin eels that are in the migratory morphological phase so they can be released downstream of migration barriers.
- Bylaw 2: Temporary prohibition on the taking of fisheries resources from any waterways

within the Waikato-Tainui Fisheries Area due to the occurrence of a human death on that waterway.

- Bylaw 3: Allows for seasonal closure prohibiting the commercial harvest of any eels from specific streams and rivers of the Whangamarino wetland during the tuna heke (eel migration) from 1 March to 31 May.
- Bylaw 4: Minimum weight for commercially harvested eels. No commercial fisher may take or possess any shortfin eel weighing less than 300 g and any longfin eel weighing less than 400 g.
- Bylaw 5: Maximum weight for commercially harvested eels. No commercial fisher may take or possess any eel taken within the Waikato-Tainui Fisheries Area weighing more than 2 kg.
- Bylaw 6: Prohibition on taking of female migrant longfin eels. No commercial fisher may take or possess any female longfin eel taken within the Waikato-Tainui Fisheries Area that is in the migratory morphological phase.

Whanganui River

The Te Awa Tupua Whanganui Claims Settlement Act 2017 establishes a new legal framework for the Whanganui River, called Te Pa Auroa nā Te Awa Tupua. At the centre of the framework is the recognition of the Whanganui River as an integrated whole from the mountains to the sea, known as Te Awa Tupua. It is the first river in the world to be given the same legal status as a person. Te Awa Tupua will be recognized as a legal identity with legal standing, rights and an independent voice. Statutory recognition will also be given to a set of values for Te Awa Tupua, Tupua te Kawa (Te Awa Tupua values), which reflect the intrinsic attributes of Te Awa Tupua including the relationship of iwi and hapū with the river. The settlement provides for the establishment of the position of Te Pou Tupua (the human face of Te Awa Tupua), which will be comprised of one Crown appointee and an iwi appointee. Te Pou Tupua will be the 'human face' of Te Awa Tupua and advocate on behalf of the river's interests.

Mair (2017) commented that 'Tuna, in our indigenous world-view, are acknowledged as having status and personality due to their direct whakapapa [connection, genealogy] links with iwi and hapū.' Whanganui hapū are working together in the Whanganui River catchment to promote and protect the health and future for tuna by developing a strategy. The forthcoming tuna strategy for the Whanganui River seeks to support six mission statements (Potaka, 2017):

- healthy freshwater fishery that enables the hapū to exercise customary use
- restore important tuna habitat and water quality in the river
- revitalize traditional fishing practices
- develop tuna monitoring and management systems
- research and education
- establish governance and guardianship to ensure health and well-being of tuna.

Aquaculture

In New Zealand, several organizations sought to participate in land-based tuna aquaculture during the 1970–1980s, but they were not commercially successful (Jellyman and Coates, 1976; Sorrenson, 1981). While previous ventures/research has demonstrated the feasibility of tuna on-growing, one of the most significant bottlenecks to the development of an industry is the difficulty in obtaining an adequate supply of glass eels for on-growing. Currently, it is illegal to possess eels weighing less than 220 g, except under special permit, and commercial access to glass eels will require changes in fisheries regulations and the support and approval of Māori, fisheries managers, communities and other stakeholders (Quigley and Baines, 2014).

There are Māori organizations who are seeking to participate in tuna aquaculture at various scales and for different purposes (including restoration) ranging from a few small tanks for family use to large commercial-scale ventures for supplying international markets. The opportunity to develop a tuna aquaculture industry has been a priority for some parties for some time (Horizons Regional Council, 2016; MPI, 2015; Te Puni Kōkiri, 2009). Internationally, eel aquaculture relies on the collection of disease and parasite-free glass eels from the wild. Glass eel harvest (focusing on shortfins) potentially has implications for the sustainability of New Zealand tuna fisheries. In New Zealand, some Māori organizations have instigated research to assess the risks of harvesting glass eels and quantifying impacts on elvers (Williams et al., 2016; Zernack et al., 2017). This research is required to advise regulatory change and influence alternate fisheries management approaches, such as reducing the commercial catch of adult eels. However, the datasets required to responsibly inform this national discussion will need to be developed over multiple years.

Recognizing that the aspirations of hapū and iwi organizations are not always economically driven, the prospect of potentially securing ready access to farmed tuna is attractive and incentivizes some to investigate their options regarding small/local scale on-growing for customary benefit (for example, Muaupoko iwi [Collis, 2017]). At the other end of the scale, some iwi are opposed to shortfin glass eel-based aquaculture. For example, the Fisheries Accord between the government and Waikato-Tainui stipulates that the parties agree to establish mechanisms to provide for Waikato-Tainui to protect elvers and glass eels within the Waikato River from exploitation. The Waikato-Tainui Environmental Plan (Waikato-Tainui, 2013) re-enforces these protective measures for glass eels.

Future fisheries management

As part of the Fisheries Settlement, Te Wai Māori (http://waimaori.maori.nz/home.htm) was established to advance Māori interests in freshwater fisheries. The long-term outcomes Te Wai Māori is seeking to influence include:

- increasing iwi and hapū capacity and capability in freshwater fisheries and their ability to control their freshwater fisheries
- promoting and sharing indigenous fisheries expertise, knowledge and understanding
- increasing the quality and range of information to iwi and hapū on freshwater fisheries and their interests thereof
- enhancing the health and well-being of the indigenous fisheries and their environment.

The Wai Māori Strategic Plan states unequivocally that 'To lose our tuna is to lose our identity' and that Wai Māori is committed to working with iwi, hapū and whānau to enhance the status and aspirations of tuna for their future and find solutions to improve waterways to ensure changes are not at the expense of tuna (Te Wai Māori, 2017). To date Te Wai Māori has hosted two national conferences (held in 2014 and 2017) that have focused on the health and well-being of tuna. The conference provides a regular opportunity for iwi and hapū from all around New Zealand to come together and share their views on how Te Wai Māori can assist them to protect and enhance their interests in tuna that should take strategic priority.

Many Māori are unhappy with the way tuna are currently managed in New Zealand, because roles and responsibilities for the health and well-being of this species and the ecosystems supporting populations are spread across a wide range of agencies (including iwi, DoC, MPI, MfE, regional councils), with very little overarching strategy or leadership. As Ken Mair, chairman of Te Wai Māori, said in an address to the Māori Tuna Conference (Mair, 2017) 'We must find solutions to improve waterways to ensure changes are not at the expense of the status and the whakapapa of tuna. This is not a kaupapa [topic] we can afford to put off.'

Attendees of the 2017 Māori Tuna Symposium requested that existing legislative mechanisms that could be used to protect and enhance the status of tuna nationwide be examined, a task subsequently completed by Te Wai Māori in 2018 (Table 14.1). The assessment of options included the possible recognition of tuna as a legal person, following the example of the Te Awa Tupua Whanganui Claims Settlement Act 2017. This would involve granting tuna the right to exist and thrive, paired with the obligation to provide sustenance, economic development, and cultural well-being. However, tuna is currently classified as property under common law and cannot be considered as a legal person. Te Wai Māori consider that any agreement made would need to avoid establishing a precedent that undermined fishing rights as granted under the Fisheries settlement, and that in the process of drafting a bill that iwi would have limited control (Te Wai Māori, 2018). Te Wai Māori has proposed that a new National Eel Association be formed, including the implementation of a process to devolve eel management from MPI to the new association. A new Freshwater Fisheries Management Plan, that includes a more holistic assessment of habitat and other threats to tuna populations is also targeted to be delivered (Te Wai Māori, 2018).

Māori-driven research agendas

Outside of MPI-funded research into the state of the commercial tuna fishery, very little research is undertaken in New Zealand to increase our understanding of the health and well-being of tuna populations. For example, despite the numerous elver trap and transfer programmes occurring throughout New Zealand, the strong interest in the potential to use glass eels for aquaculture and/or on-growing for customary use and restoration initiatives, we know relatively little about the factors that influence juvenile eel survival, which can be a barrier (or used as a barrier by agencies) for the realization of Māori-driven outcomes. Māori resource managers need new knowledge and decision-support tools to inform forthcoming debates including: research that addresses uncertainties in access to tuna (for example, is glass eel harvest for aquaculture sustainable?); provision of knowledge to enhance the survival of juvenile tuna (for example, are the majority of the elvers transferred above hydro-dams surviving?); frameworks to assess the co-benefits and trade-offs between tuna conservation imperatives, consumption requirements, and commercial/ economic aspirations (for example, can we redistribute and translocate tuna to meet multiple

Table 14.1 Potential of legislative and policy tools to enhance the future status of New Zealand tuna as assessed by Te Wai Māori.

Legislative/policy tools	Description	Potential benefits for future tuna populations and management
Conservation Act 1987	The director general of DOC has the authority to draft Freshwater Fisheries Management Plans for approval by the minister. These plans are designed to implement policies and create detailed objectives 'for the management of freshwater fisheries within any area or areas'. These plans carry statutory weight, being binding on concession decisions and other Resource Management Act (RMA)-related decisions.	The statutory status of these plans can provide strong guidance and regulation to councils and other resource users whose actions may have an impact on tuna habitat viability while also contributing to existing fisheries management initiatives.
Fisheries Act 1996	This Act grants the governor-general, on the recommendation of the minister, the power to transfer any specified duty, power or function to an approved service delivery organization. This includes the ability to enact sustainability measures, set the TAC/TACC, and regulate the tuna fishery in general. To be eligible the service delivery organization must be incorporated under the Companies Act 1993, be representative of quota owners with an interest in the stock, and the organization has the capacity to carry out those functions, duties, or powers.	A balanced forum, combined with a decision-making protocol that promotes consensus, to provide balanced recommendations and advocacy regarding the national management of the fishery and develop management plans that reflect the needs of commercial and customary users.
Resource Management Act 1991	The RMA contains multiple provisions which could be utilized or improved to provide greater protection for tuna, e.g. Section 6, 'matters of national importance'. However, in practice, it appears that greater consideration is given to Section 7, 'other matters', which includes the protection of habitat for trout and salmon.	With a review of the RMA on the horizon, amending Section 7(h) that currently protects salmon and trout habitat to name specific indigenous species such as tuna will raise their profile. This will bring greater attention to the actions of authorities and how adequately they protect tuna habitat.

Source: Te Wai Māori (2018)

objectives?); and approaches that are able to draw upon multiple knowledge systems (including mātauranga Māori) to assess and communicate state and trends in tuna populations (for example, how do we measure and communicate the success of tuna restoration?).

A new research programme called 'Cultural Keystone Species (CKS): Co-management and Restoration' has been designed around the premise that indigenous communities must be actively involved in the management of CKS to balance competing values, needs and opportunities (Noble et al., 2016). This research is working alongside hapū who may be

Table 14.1 (Continued)

Legislative/policy tools	Description	Potential benefits for future tuna populations and management
Resource Management Act 1991	Local authorities have the ability to transfer one or more of its functions, powers, or duties under the RMA to another public authority. Mana Whakahono ā Rohe (Iwi Participation Arrangements) are a mechanism designed to assist councils in complying with requirements to have regard to kaitiakitanga (guardianship), Māori connection with their lands and taonga, and the Treaty of Waitangi.	It would theoretically be possible for regional councils to transfer management functions of significant waterways to iwi. This mechanism can provide iwi and hapū with the ability to have a greater voice in resource management decisions affecting their rohe (tribal area). These agreements can provide Maori with the necessary power to limit the impacts of land use practices and other resource use processes on tuna habitats.
Wildlife Act 1953	The Minister of Conservation has the authority to list a species as protected or partially protected. Full protection would prohibit harvesting while partial protection would allow the Minister to set conditions under which harvesting could take place.	Tuna would be given strong statutory protection, requiring councils to take greater care when making consenting and planning decisions. However, inclusion of tuna in the Wildlife Act would be unlikely to yield drastic change in tuna habitat management unless the strategy included lobbying for better implementation. Restricted harvesting would be at the discretion of the Minister with no requirement for consultation and would likely prohibit commercial take.
Freshwater Fisheries Regulations 1983	The Director General of DOC can gazette any waterway or body of water as a faunistic reserve. Faunistic reserves effectively function like Marine Protected Areas but for freshwater, prohibit the introduction of plants as well as the taking, killing or injuring of any fish.	This strategy could increase the number of migrant tuna making it to sea but fails to address habitat pressures. Further the indiscriminate barring of fishing in gazetted areas could negatively impact fishing rights.

monitoring glass eel recruitment, and hapū who are transferring elvers above a hydroelectric dam (Figure 14.3), as well as revisiting the Graynoth et al. (2008) study, which to date is the only attempt made to investigate juvenile eel mortality rates in New Zealand. The aim of the programme is to develop decision-support frameworks that will help to strengthen cross-cultural understandings of Māori values, aspirations and uses of CKS and develop tools that help communities prioritize co-management actions, assess co-benefits, and overcome barriers to CKS restoration.

Figure 14.3 Very little research has been undertaken in New Zealand to help explain what makes stream habitats suitable for elvers to thrive. The successful operation of the elver trap and transfer at the Wairua River power station over the past 6 years provides Ngā Kaitiaki o Ngā Wai Māori with an opportunity to investigate how well elvers and adult eels are surviving and growing in the different streams they have been transferred into. Credit: Stu MacKay, NIWA.

The tools considered here represent opportunities to influence change in tuna management at a nationwide scale and excludes (but would complement) iwi-specific mechanisms that are often catchment or regionally based.

Conclusion

Shortfin and longfin tuna are an important resource from both a human/cultural use and biodiversity perspective. Throughout New Zealand they continue to support valuable customary, commercial and recreational fisheries. In New Zealand, the roles and responsibilities for the management of tuna populations and their associated habitats are currently spread across a wide range of agencies including iwi, DOC, MPI, Fish and Game Council, regional councils, commercial fisher associations and conservation groups.

Because of their extensive migrations and relatively long life, tuna are a challenging fishery to manage and restore, notably because the relative importance and interaction between habitat, recruitment, harvest and other pressures have not been quantified. Furthermore, as there are no controls on the life stages of tuna migrating to and from their spawning grounds in the Pacific

Ocean, the maintenance and restoration of the tuna fishery has to rely on activities that enhance populations while in freshwater.

Through a variety of processes Māori have signalled to government that action needs to be taken to address the challenges tuna populations are facing. Many iwi and hapū across the country are implementing regional and catchment-scale initiatives to protect, enhance and restore local tuna populations. However, as tuna spawn well offshore and new recruits rely primarily on sea currents to return to New Zealand, recruitment and restoration needs to be managed on a national if not international scale. Te Wai Māori will be working towards the formation of a new National Eel Association to improve tuna management in New Zealand. A new Freshwater Fisheries Management Plan, that includes a more holistic assessment of threats to tuna populations will also be completed (Te Wai Māori, 2018).

Iwi and hapū have stated that there is real need for the identification and articulation of iwi and hapū values and perspectives of freshwater ecosystems to help build better understanding within resource management agencies. Effective integration of Māori interests and values into resource management requires full iwi and hapū participation and a commitment from agencies to collaborate and build strong, enduring relationships. This extends to organizations that wish to undertake research on freshwater ecosystems (Williams et al., 2017). The development of environmental reporting frameworks that draw on mātauranga Māori, acknowledges cultural values, and provides a role for iwi and hapū nationwide in implementation and interpretation would significantly improve the management of tuna in New Zealand.

Acknowledgements

This chapter was prepared with funding from Cultural Keystone Species: Co-management and Restoration (Ministry of Business, Innovation and Employment Contract Number: C01X1616).

Glossary of Te Reo Māori used in this chapter

haka	ceremonial dance
hapū	tribal grouping that consists of whānau who typically share descent from a common ancestor
hīnaki	eel pots (method of catching tuna)
hui	assembly, meeting, gathering
iwi	extended tribal grouping that consists of hapū or whānau who typically share descent from a common ancestor and associate with a distinct territory
kaitiaki	guardian
kaitiakitanga	guardianship
kaupapa	theme, philosophy, topic
koha	gift, present, offering, donation, contribution
korapa	hand netting (method of catching eels)
koumu	eel trenches (method of catching eels)
mana whakahono ā rohe	type of agreement between iwi and the Crown (http://www.mfe.govt.nz/rma/manawhakahono)
Māori	indigenous people of Aotearoa-New Zealand

marae	place typically in front of a wharenui (meeting house) where the members of whānau, hapū, or iwi meet and engage in pōwhiri (the ceremony of greeting and encounter), and includes associated buildings, such as the wharenui (meeting house) and wharekai (dining room), and surrounding land
maramataka	calendar that divides the Māori year into lunar months
mātaitai (of reserves)	areas where tangata kaitiaki/tiaki manage all non-commercial fishing by making bylaws
mātauranga Māori	holistic perspective encompassing all aspects of knowledge and seeks to understand the relationships between all component parts and their interconnections to gain an understanding of the whole system. It is based on its own principles, frameworks, classification systems, explanations and terminology. Mātauranga Māori is a dynamic and evolving knowledge system and has both qualitative and quantitative aspects
matarau	spearing (method of catching eels)
Muaupoko	name of an iwi
Ngā Kaitiaki o Ngā Wai Māori	collective of seven hapū with interests in the Wairua River catchment
Ngāti Maniapoto	name of an iwi
Ngāti Raukawa	name of an iwi
Ngāti Tahu	name of an iwi
Ngāti Tūwharetoa	name of an iwi
noa	free from tapu
pā tuna	eel weirs (method of catching tuna)
pakiwaitara	legends
puhi	migrant eel (Lake Waahi, Waikato region of New Zealand)
rāhui	prohibition
rama tuna	catching eels by torch light
rangatira	chief, leader
raupatu	confiscation, take without any right
rauwiri	eel weirs (method of catching tuna)
rohe	tribal area, district, region
tangata kaitiaki/tiaki	individuals involved in the guardianship of a resource/area
tangi	funeral
taonga	significant treasure, possessions, something prized
tapu	type of restriction, used as a way to control how people behaved towards each other and the environment
Te Arawa	name of an iwi
Te Reo Māori	Māori language
Te Pā Auroa Nā Te Awa Tupua	legal framework for the Whanganui River. The framework comprises the following elements: • Te Awa Tupua and its legal status • Tupua te Kawa (the Awa Tupua values) • Te Pou Tupua (the human face of Te Awa Tupua) • Te Karewao (the advisory group to Te Pou Tupua) • Te Heke Ngahuru ki Te Awa Tupua (the Te Awa Tupua strategy) • Te Kōpuka nā Te Awa Tupua (the Te Awa Tupua Strategy group) • Kia Matara Rawa (the vesting of Crown-owned parts of the bed of the Whanganui River in Te Awa Tupua); and • Te Korotete o Te Awa Tupua (the Awa Tupua Fund). (http://www.legislation.govt.nz/act/public/2017/0007/latest/whole.html)

Te Ture Whaimana o Te Awa o Waikato	Vision and Strategy for the Waikato River. (https://waikatoriver.org.nz/wp-content/uploads/2011/07/Vision-and-Strategy.pdf)
Treaty of Waitangi and the Waitangi Tribunal	The Treaty of Waitangi was a written agreement made in 1840 between the British Crown and more than 500 Māori chiefs.
tohu	sign, mark, symptom
tohunga	expert, specialist
tuna	freshwater eels
tuna heke	eel migration
waiata	songs
Waikato-Tainui	name of an iwi
whakaheke	eel migration back to sea
whakapapa	connection, lineage, genealogy between humans and ecosystems and all flora and fauna
whakataukī	proverb
whānau	family group that consists of individuals who typically share a common whakapapa and identify with a common living or recent ancestor

Bibliography

Best, E. (1929) Fishing Methods and devices of the Māori. *Dominion Museum Bulletin No 12.* National Museum, Wellington.

Clements and Associates Limited (2016) The Atlas of Area Codes and TACCs 2016–17. Clements and Associates Limited Report. Clements and Associates Limited, Nelson.

Collis, A. (2017) Keeping an eye on the eels in the spa. *New Zealand Herald* 16 August 2017.

Fisheries New Zealand (2018) Review of North Island eel sustainability measures for 2018/19. Consultation Document. *Fisheries New Zealand Discussion Paper No: 2018/04.* June 2018. Ministry for Primary Industries, Wellington.

Goodman, J.M., Dunn, N.R., Ravenscroft, P.J., Allibone, R.M., Boubée, J.A.T., David, B.O., Griffiths, M., Ling, N., Hitchmough, R.A. and Rolfe, J. (2014) Conservation status of New Zealand freshwater fish, 2013. *New Zealand Threat Classification Series 7.* Department of Conservation, Wellington.

Graynoth, E., Francis, R. and Jellyman, D.J. (2008) Factors influencing juvenile eel (*Anguilla* spp.) survival in lowland New Zealand streams. *New Zealand Journal of Marine and Freshwater Research* 42, 153–172.

Haro, A., Dekker, W. and Bentley, N. (2015) 2013 Independent review of the information available for monitoring trends and assessing the status of New Zealand freshwater eels. *New Zealand Fisheries Science Review 2015/2.* Ministry for Primary Industries, Wellington.

Horizons Regional Council. (2016) Accelerate 25. Manawatū-Whanganui Economic Action Plan. August 2016. Horizons Regional Council, Palmerston North, New Zealand.

Hoyle, S.D. and Jellyman, D.J. (2002) Longfin eels need reserves: modelling the effects of commercial harvest on stocks of New Zealand eels. *Marine and Freshwater Research* 53, 887–895.

Jellyman, D.J. (2009) Forty years on – the impact of commercial fishing on stocks of New Zealand freshwater eels (*Anguilla* spp.). Eels at the Edge. *American Fisheries Society Symposium* 58, 37–56.

Jellyman, D.J. (2013) The status of longfin eels in New Zealand – an overview of stocks and harvest. NIWA Client Report: CHC2012–006. January 2013 (revised). NIWA, Christchurch.

Jellyman, D.J. and Coates, G.D. (1976) The farming of freshwater eels in New Zealand. Paper presented at the Indo-Pacific Fisheries Council 17th Session, Symposium on the Development and Utilisation of Inland Fishery Resources. Colombo, Sri Lanka, 27–29 October 1976.

Jellyman, D. and Sykes, J. (2009) The eel (tuna) stocks of Lake Poukawa, Hawkes Bay. NIWA Client Report: CHC2009–172. October 2009. NIWA, Christchurch.

Jellyman, D.J., Graynoth, E., Francis, R.I.C.C., Chisnall, B.L. and Beentjes, M.P. (2000) A review of evidence for a decline in the abundance of longfinned eels (*Anguilla dieffenbachii*) in New Zealand. Final Research Report for Ministry of Fisheries Research Project EEL9802. February 2000. Ministry of Fisheries, Wellington.

Kawe, T.P. (2014) Feasibility study for assessment of customary harvest by Kaitiaki in the Te Tai Hauāuru and Tauranga Moana Regions. *New Zealand Fisheries Assessment Report 2014/61*. October 2014. Ministry of Primary Industries, Wellington, New Zealand.

Mair, K. (2017) Opinion: Tuna tied to our whakapapa. *Whanganui Chronicle* 21 July 2017. Available at: https://www.nzherald.co.nz/wanganui-chronicle/opinion/news/article.cfm?c_id=1503423&objectid=11893262 (accessed 19 July 2018).

Marshall, Y. (1987) Māori mass capture of freshwater eels: an ethnoarchaeological reconstruction of prehistoric subsistence and social behavior. *New Zealand Journal of Archaeology* 9, 55–79.

McCleave, J.D. and Jellyman, D.J. (2004) Male dominance in the New Zealand longfin eel population of a New Zealand river: probable causes and implications for management. *North American Journal of Fisheries Management* 24, 490–505.

McDowall, R.M. (2011) *Ikawai. Freshwater Fishes in Māori Culture and Economy*. Canterbury University Press, Christchurch.

MFish (2009) Fisheries Assessment Plenary. Stock Assessments and Stock Status. Ministry of Fisheries, Wellington, New Zealand.

MPI (2015) Tai Tokerau Northland Growth Study. Opportunities Report. February 2015. Ministry for Primary Industries, Wellington.

MPI (2016) Review of Management Controls for the South Island Longfin and Shortfin Eel Fisheries (LFE 11–16 and SFE 11–16) in 2016. MPI Discussion Paper No: 2016/15. Ministry for Primary Industries, Wellington.

Noble, M., Duncan, P., Perry, D., Prosper, K., Rose, D., Schnierer, S., Pittock, J., Woods, R., Williams, E. and Tipa, G. (2016) Culturally significant fisheries: social-ecological keystones in freshwater ecosystems. *Ecology and Society* 21, 22.

Panelli, R., Tipa, G. (2007) Placing well-being: a Māori case study of cultural and environmental specificity. *EcoHealth* 4, 445–460.

Parliamentary Commissioner for the Environment (PCE) (2013) On a pathway to extinction? An investigation into the status and management of longfin eel. April 2013. PCE, Wellington.

Potaka, B. (2008) Revitalisation of traditional fisheries management used by Whanganui Iwi. Customary Fisheries Research Fund CUS2007–04. Ministry of Fisheries, Wellington.

Potaka, B. (2017) Whanganui River Catchment Tuna Strategy. Invited speaker. Māori Tuna Conference, 17 July 2017. Whanganui War Memorial Centre, Whanganui. Available at: https://tuna.conference.maori.nz/assets/Uploads/Ben-Potaka.pdf (accessed 19 July 2018).

Quigley, R. and Baines J. (2014) How to improve your social licence to operate: a New Zealand industry perspective. MPI Information Paper No: 2014/05. March 2014. Ministry for Primary Industries, Wellington.

Shortland, T. (2013) Kete tangariki – pilot tuna enhancement project. Ngā Tirairaka o Ngāti Hine, Kawakawa.

Sorrenson, W.J. (1981) Future prospects for eel farming in New Zealand. Research Paper, 9/81. October 1981. Ministry of Agriculture and Fisheries, Wellington.

Te Kupenga o Maniapoto Limited (2013) Annual Report for the Year Ended 30 September 2013. Te Kupenga o Maniapoto Limited (Te Kupenga), Te Kuiti.

Te Puni Kōkiri (2009) Ahumoana Ahutāngata. Aquaculture Development in New Zealand: Scientific and Technical Information to Inform Māori. July 2009. Te Puni Kōkiri, Wellington.

Te Wai Māori (2008) Discussion on freshwater: a Wai Māori perspective. Te Wai Māori Trust, Wellington.

Te Wai Māori (2017) 3 Year Strategic Plan. Te Wai Māori Trust, Wellington, New Zealand.

Te Wai Māori (2018) Options to enhance the status of tuna: A primer. Questionnaire distributed at the Fisheries and Aquaculture Workshop. Māori Fisheries Conference: Tangaroa-ā-Mua – Future Māori Fisheries. 27–28 March 2018, Auckland.

Tesch, F.W. (2003) *The Eel*, 5th edn. Blackwell Science, Oxford.

Waikato Raupatu River Trust (2014) Waikato River Fisheries Bylaws for the Waikato Tainui Fisheries Area. Waikato Raupatu River Trust, Hamilton.

Waikato River Authority (WRA) (2011) Vision and Strategy for the Waikato River. Waikato River Authority, Hamilton.

Waikato River Authority (WRA) (2017) Annual Report 2017. Waikato River Authority, Hamilton.

Waikato-Tainui (2013) Tai Tumu Tai Pari Tai Ao: Waikato-Tainui environmental plan: Our plan our environment our future. Hamilton, Waikato-Tainui Te Kauhanganui Incorporated. August 2013.

Williams, E., Crow, S., Murchie, A., Tipa, G., Egan, E., Kitson, J., Clearwater, S. and Fenwick, M. (2017) Understanding taonga freshwater fish populations in Aotearoa-New Zealand. NIWA Client Report 2017326HN. September 2017. NIWA, Hamilton.

Williams, E., Crow, S., Graynoth, E., Baker, C., Boubée, J., Smith, J. and Zernack, A. (2016) Rangitaane Tuna Aquaculture: Overcoming the Hurdles – Field, Laboratory and Modelling Investigations: Year 3 Final Report. NIWA Client Report 2016087WN. September 2016. NIWA, Wellington.

Zernack, A., Williams, E., Heath, P., Jameson, J. and Horton, P. (2017) PROJECT TUNA – Overcoming the hurdles to establish an environmentally, economically and socially sustainable shortfin eel aquaculture industry in New Zealand. UK International Eel Science Symposium, 15 June 2017, London.

CHAPTER 15

Understanding eel behaviour to improve protection and passage at pumping stations

Jonathan D. Bolland[1] *and Rosalind M. Wright*[2]

[1]Hull International Fisheries Institute, University of Hull, Hull, UK; [2]Environment Agency, Feering, UK

Project summary

One of the major problems encountered at pumping stations is the entrainment of fish and eels during pump operation. The extent of the problem and the proportion of fish and eels that are damaged or killed by different types of pumps is largely unknown. Silver eels are especially vulnerable as they attempt to exit freshwater during their downstream spawning migration to the Sargasso Sea. The overall strategic objective for the REDEEM project (**Re**search and **D**evelopment of fish and **E**el **E**ntrainment **M**itigation) is to evaluate the impact of different types of pumping stations and investigate fish and eel behaviour to assess and develop solutions to minimize fish and eel entrainment at pumping stations without compromising pumping efficiency.

This project will use new technology with a combination of acoustic telemetry, sonar imaging and mark-recapture techniques to investigate eel migration, behaviour and entrainment at pumping stations. Understanding fish and eel behaviour in pumped catchments will lead to improved guidance on solutions for implementation of eel legislation and to minimize impact of pumping stations on eel and fish populations.

Introduction

Background

Stocks of the European eel (*Anguilla anguilla*) have declined markedly, with estimated recruitment reduced to between 4 and 12% of 1980s levels in many rivers. As a result, the European eel is designated as 'Critically Endangered' by the International Union for Conservation of Nature (IUCN) and the stock is considered outside safe biological limits.

Flows in many rivers in the UK, Europe and, indeed the world, are heavily modified to provide water for domestic supply, agriculture and industry as well as being controlled for flooding and hydropower generation. To achieve this, water is frequently pumped from or into rivers. Fish, including eels, can be drawn into pumps and water intakes and this may be a significant cause of anthropogenic mortality. Migrating silver eels are especially vulnerable because of their size and their need to migrate to the Sargasso Sea for spawning.

Legislation

In response to the population decline, the European Union adopted Council Regulation 1100/2007/EC in 2007, establishing measures for the recovery of stocks of European eel. The Regulation required all member states to produce eel management plans (EMPs) detailing actions to meet the target to permit with high probability the escape to sea of at least 40% of the silver eel biomass relative to the best estimate of escapement that would have existed if no anthropogenic influences had affected the stock.

The Eels (England and Wales) Regulations 2009 (Statutory Instrument 2009 No. 3344) requires that:

- no new infrastructure be constructed, or existing structures modified, without provision for eel
- appropriate physical exclusion or alternative measures be provided at all potentially harmful water abstraction or flow points (>20 m^3 per day), including pumping stations.

In addition, the Water Framework Directive (2000/ 60/EC) requires that water bodies meet good ecological status, of which unobstructed passage for fish is an important component. The presence of eel, where expected, contributes to the classification of fisheries when assessing ecological status.

Implementing protection measures at pumping station raised specific issues that required further research.

About the REDEEM project

The REDEEM project, a collaboration between the Environment Agency and the University of Hull, is underway to understand the impact of pumping stations on downstream migrating silver eels and river-resident coarse fish, and to identify adequate, cost-effective mitigation measures. The research focuses on understanding the spatial distribution of fish and eels in pumped catchments, the processes that lead to entrainment and the effectiveness of altered operating regimes, fish-friendly pumps and novel downstream bypass channels for maximizing safe eel escapement. The research will assess the effectiveness of existing technologies and develop innovative measures to ensure evidence based solutions for fish and eel protection are implemented at pumping stations.

The study has the following specific objectives.

1. A review of historic fish survey data upstream of pumping stations and an investigation of potential of using eDNA techniques to assess presence or absence of eels in pumped catchments to help prioritize eel protection measures.
2. An assessment of downstream migration of silver eels through the catchment in relation to environmental conditions (river flow, water temperatures, lunar phase and time of day) and pumping station operation.
3. An assessment of survival of eels passing through pumps of varying design and size, including pumps currently considered 'fish-friendly'.
4. A study of behaviour of coarse fish and silver eels at pumping stations with 'traditional' designs and those with 'best practice' or alternative eel protection measures, including any delays to migration and their ability to locate and utilize alternative migration routes.

The next section summarizes two of the initial peer-reviewed outputs from the ongoing research project, including the study background and outline, a list of key findings and the management implications.

Direct and indirect impacts of pumping station operation on silver eel migration – Bolland et al. (2019)

Background

Silver eel migration in unregulated rivers is typically induced by environmental cues including increased river discharge, a fall in water temperature and lunar phase. In pumped catchments river discharge is controlled by pumping regimes, rendering rivers with no flow when pumps are not operational (Figure 15.1), and this may delay or influence the timing of silver eels migrating. In addition, pumping stations are a physical obstruction and an intake where eels may be entrained during pump operation. This study investigated the direct and indirect impact of pumping station operation on downstream migrating of silver eels.

Figure 15.1 A river upstream of a pumping station without any flow because pumps were not operational. Photograph: Martin Redding.

Study outline

The pumping station in this study manages water level for drainage and flood defence of a 33,987 ha area and discharges into a tidal estuary; much of the catchment is below high tide level. There are two types of pumps in separate pumping station buildings at the site with a total combined capacity of 36.7 m^3/s (Figure 15.2). One pumping station has three mixed flow Allen T47 2 stroke 770 hp diesel powered pumps (installed in 1957). The intakes are screened (weedscreen) to prevent large debris, such as aquatic and riparian vegetation, from entering the pumps and causing damage. At the other pumping station there are four submersible axial flow Flygt 7100 200 kw (267 hp) electric powered pumps (installed in 1987); each has a 0.85 mm diameter, four-vane impellers, 850 rpm and a maximum capacity of 2.1 m^3/s. There was also a

Figure 15.2 Aerial view of the study pumping station. Photograph: Martin Redding.

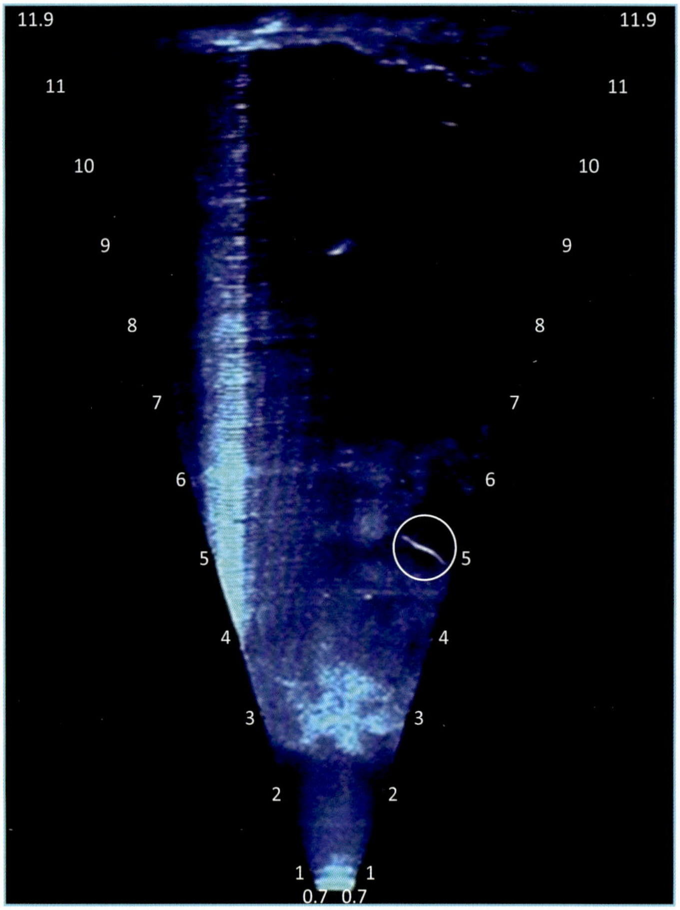

Figure 15.3 Example of multi-beam sonar (DIDSON and ARIS) imaging a pumping station weed screen and an eel approaching (inside white circle).

manually operated gravity sluice located adjacent to the mixed flow pumping station that was not screened.

A study was conducted during October to December 2015 in this pumped catchment. A total of 17 silver eels were captured, tagged with acoustic tags and released. Multi-beam sonars (ARIS) were used to image across the entrance to the mixed and axial flow pumping stations (Figure 15.3). Passive integrated transponder (PIT) tagged eels were released directly upstream of the studied pumping stations during operation and recaptured in a net downstream (Figure 15.4). Eels were weighed and measured and their behaviour and external condition visually reassessed.

Key findings

1 Silver eels quickly migrated downstream through unobstructed reaches, but the pumping station caused delays in migration.
2 A large proportion of eels were reluctant to pass through the weedscreen and retreated upstream.

Figure 15.4 Installation and operation of the eel recapture net. Photographs: Martin Redding.

3 Peaks in migration occurred during the nights preceding the new moon in October to December, with the majority (54.5%) approaching between 20:00 and midnight, peaking at 21:00.
4 All eels survived passage through a large (2.23 m diameter) mixed flow pump but 96.5% of eels had minor injuries, reduced physical condition and/or abnormal behaviour.
5 A large proportion of eels passing through a small (0.8 m diameter) axial flow pump died.
6 No eels passing through the electric pumping station were detected by receivers in the estuary, whereas 91% of eels released downstream of the pumping station were detected.

Management implications

1 The findings from the study will be used by the regulator (the Environment Agency) to assess prioritization of pumping stations for remedial works and make recommendations for operational changes to protect migrating silver eels.
2 Adapting pumping regimes to avoid pump operation during peaks in silver eel migration could be a cost-effective alternative for eel protection until pumping stations are refurbished with inclusion of eel protection measures. If combined with opening of gravity sluices it is likely to improve eel escapement.
3 Based on the recommendations of this research, operational changes are now being implemented by operators at the study pumping station.

Comparison of airlift and siphon deep entrance bypass systems – Baker et al. (2019)

Background

Multiple eel species worldwide are endangered and alleviating the impacts of barriers to downstream migration is essential to allow completion of the life cycle. Deep bypass systems with entrances located near the riverbed hold some promise for increased effectiveness compared to traditional downstream bypasses with entrances near the surface, as silver eels typically occupy the bottom of the water column.

Study outline

This investigation evaluated two deep entrance bypass designs; an airlift (the Conte Airlift) and a conventional gravity siphon of the same entrance dimensions. Tests were performed using migratory silver-phase American eels (*A. rostrata*), at night, in a simulated forebay environment. Passage performance was monitored over a 3 hour test period using both PIT tag and video recording equipment (Figure 15.5). Entrance velocity was fixed at 1.2 m/s in each of eight test runs with cohort size fixed in six runs at 14 and in two runs at 42.

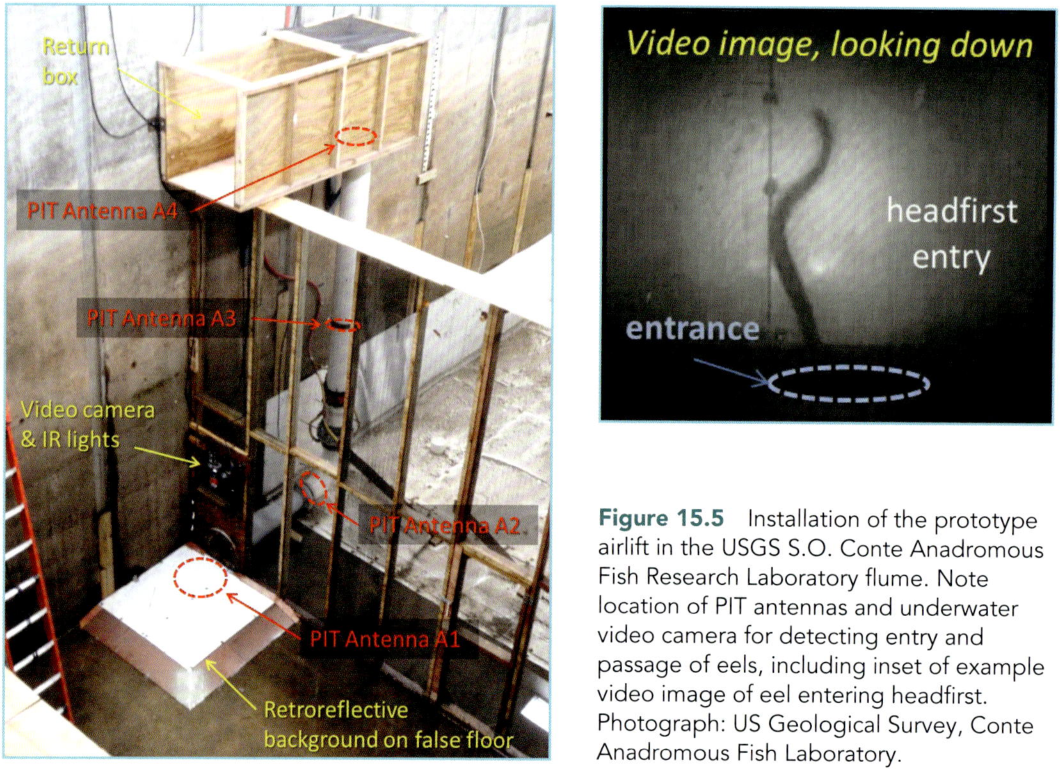

Figure 15.5 Installation of the prototype airlift in the USGS S.O. Conte Anadromous Fish Research Laboratory flume. Note location of PIT antennas and underwater video camera for detecting entry and passage of eels, including inset of example video image of eel entering headfirst. Photograph: US Geological Survey, Conte Anadromous Fish Laboratory.

Key findings

- Eels readily located, entered and passed both bypass designs.
- Eel length did not affect passage speed and fish density did not affect attraction and passage through the airlift or siphon.
- No mortality or signs of injury were observed on any of the test eels through a 48 hour post-test observation period.

Management implications

Both airlift and siphon downstream bypass systems show promise as effective technologies for protection of downstream migrating eels at water intakes that pose threats of impingement, entrainment, and turbine mortality.

Acknowledgements

Funding has been provided by the Environment Agency (FCRM and Fisheries), EU European Marine and Fisheries Fund (ENG2130), Internal Drainage Boards, Association of Drainage Authorities and the University of Hull. We would specifically like to thank the internal drainage boards for opening their doors to the research. We also acknowledge organizations and individuals in the peer-reviewed publications.

Bibliography

Baker, N., Haro, A., Watten, B., Noreika, J. and Bolland, J.D. (2019) Comparison of attraction, entrance and passage of downstream migrant American eels (*Anguilla rostrata*) through airlift and siphon deep entrance bypass systems. *Ecological Engineering* 126, 74–82.

Bolland, J.D., Murphy, L.A., Stanford, R.J., Angelopoulos, N.V., Baker, N.J., Wright, R.M., Reeds, J.D. and Cowx, I (2019) Direct and indirect impacts of pumping station operation on downstream migration of critically endangered European eel (*Anguilla anguilla*). *Fisheries Management and Ecology* 26, 76–85.

CHAPTER 16

Understanding eel behaviour to improve their protection and passage at river structures

Adam T. Piper[1,2], Rosalind M. Wright[3] and Paul S. Kemp[1]

[1]International Centre for Ecohydraulics Research, Faculty of Engineering and the Environment, University of Southampton, Southampton, UK; [2]Institute of Zoology, Zoological Society of London, London, UK; [3]Environment Agency, Feering, UK

Introduction

As a catadromous species, the European eel (*Anguilla anguilla*) typically migrates between marine and freshwater habitats to complete its lifecycle. Anthropogenic structures, such as weirs, dams, hydropower facilities and other water offtakes, can prevent eel accessing suitable habitats and may cause damage and mortality through impingement on screens or passage through pumps and turbines. This overview summarizes a programme of UK-based research conducted to quantify the movement and behaviour of European eel in relation to the conditions encountered at a range of engineered river structures. The research focused on the eel life stages most likely to encounter such structures: immigrating juveniles and emigrating adults. The topics covered include: delay and entrainment of silver eel during migration through a heavily regulated lowland river; eel behaviour and passage at an Archimedes screw hydropower turbine; silver eel response to stimuli (flow and infrasound) for protection and/or guidance, and use of attraction flow to enhance juvenile eel passage. Findings highlight ways to reduce barrier impacts through the manipulation of structures and abstraction regimes. Further, studies demonstrate how quantifying eel behavioural response to physical and hydrodynamic environments at biologically relevant scales provides potential to aid the development of attraction, guidance and passage technologies for the species.

Background

Stocks of European eel have declined markedly with estimated recruitment reduced to less than 5% of pre-1980s levels in many rivers (ICES, 2011a). As a result, the species was designated as 'critically endangered' and the stock considered outside safe biological limits (ICES, 2011b). As a facultative catadromous species, European eel typically requires access to both marine and freshwater habitats to complete its lifecycle. Barriers to migration such as weirs, dams and hydropower facilities which prevent eel accessing suitable habitats are thought to be a contributing factor in the population decline (Bruijs et al., 2009; Feunteun, 2002). Further, due to their elongated body shape and relatively poor swimming abilities, eel are particularly vulnerable to damage and mortality at screens, water pumps and turbines (Calles et al., 2010; Russon et al., 2010).

Legislation

In response to the population decline, the European Union adopted Council Regulation 1100/2007/EC in 2007, establishing measures for the recovery of stocks of European eel. The Regulation required all member states to produce eel management plans (EMPs) detailing actions to meet the target to permit with high probability the escapement to sea of at least 40% of the silver eel biomass relative to the best estimate of escapement that would have existed if no anthropogenic influences had affected the stock.

The Eel (England and Wales) Regulations 2009 (Statutory Instrument 2009 No. 3344) requires that:

- no new dams or other obstructions to be constructed, or existing structures modified, without provision for eel
- appropriate physical exclusion or alternative measures be provided at all potentially harmful water abstraction or flow points (>20 m^3 per day)

In addition, the Water Framework Directive (2000/60/EC) requires that water bodies meet good ecological status, of which unobstructed passage for fish is an important component. The presence of eel, where expected, contributes to the classification of fisheries when assessing ecological status.

Research overview

This chapter is intended as an overview of a larger body of peer-reviewed literature. We summarize a programme of research conducted between October 2009 to March 2015 to quantify the movement and behaviour of European eel in relation to the conditions encountered at riverine structures. The focus of the research was on the eel life stages most likely to encounter such structures, that is, immigrating juveniles and emigrating adults.

The main project partners and funders were the Environment Agency and the University of Southampton. Other collaborators included Cefas, Sembcorp, Essex and Suffolk Water, and partners within the Living North Sea project.

The programme involved separate studies at locations in Suffolk and Dorset. The principal topics covered were:

- the seaward migration of silver (adult) eel and their entrainment at water offtakes
- the influence of hydrodynamics on silver eel behaviour
- eel and coarse fish behaviour and passage at an Archimedes screw hydropower turbine
- using attraction flow to enhance juvenile eel passage at an intertidal barrier
- silver eel response to an infrasound device.

Each of the five sections gives the background to a particular study, an outline of the methods, a list of key findings, a summary of the management implications and a reference to the peer-reviewed publication resulting from the work.

The findings have implications for wider catchment management and highlight the potential to reduce barrier impacts through the manipulation of structures and abstraction regimes. Quantifying eel behavioural response to physical and hydrodynamic environments at biologically relevant scales provides potential to aid the development of attraction, guidance and passage technologies for the species.

Seaward migration of silver eel and entrainment at water offtakes – Piper et al. (2013)

Background

To meet EMP eel escapement targets, it is important to identify key locations of silver eel loss and delay during freshwater migration.

Silver eel migration typically occurs over short periods or 'runs' induced by environmental cues including increased river discharge, a fall in water temperature and dark lunar phase (Haro, 2003; Tesch, 2003). Barrier mitigation, for example, by opening spill gates, during these key periods can be effective in increasing escapement (Watene and Boubée, 2005).

The operation of sluices can influence route choice and the rate of eel migration. There is strong evidence that eel tend to make route selection choices based on those localities with the highest flow (Breteler et al., 2007; Jansen et al., 2007).

Knowledge of the physical and environmental conditions at structures that prevent or delay eel passage is crucial to identify and reduce restrictions to migration. In particular, there is limited understanding of the individual and cumulative effects of low-head structures that may be only intermittently passable.

Study outline

A study was conducted during 2009 and 2010 on the River Stour, Suffolk, a highly regulated lowland river. A total of 69 silver eel were captured, tagged with passive integrated transponder (PIT) and acoustic tags (Vemco) and released. Their movements were tracked through a 9.2 km reach that included:

- 12 cross-channel structures for water level management and navigation (weirs and locks)
- 2 water abstraction offtakes
- several points where the main channel bifurcates.

The results were used to quantify:

- successful migration from the study reach to the sea
- time taken to escape
- barrier delay

- migration velocity
- entrainment loss
- route choice.

Key findings

- Overall, 76% (in 2009) and 65% (in 2010) of tagged eel successfully migrated, taking between 5 and 113 days to reach the sea.
- The main cause of eel loss was entrainment at a water offtake (12% and 26% entrainment rate in 2009 and 2010, respectively). There was a positive correlation between loss and rapid increases in abstraction while eel were in the vicinity of the offtake.
- Migration speed was lower in obstructed reaches and eel were delayed at some structures for several days.
- The choice of route into the estuary depended on the height of spill over an intertidal weir combined with the opening regimes of a tidal gate at the termination of the alternative channel, and the abstraction rate at a nearby water offtake (Figure 16.1).

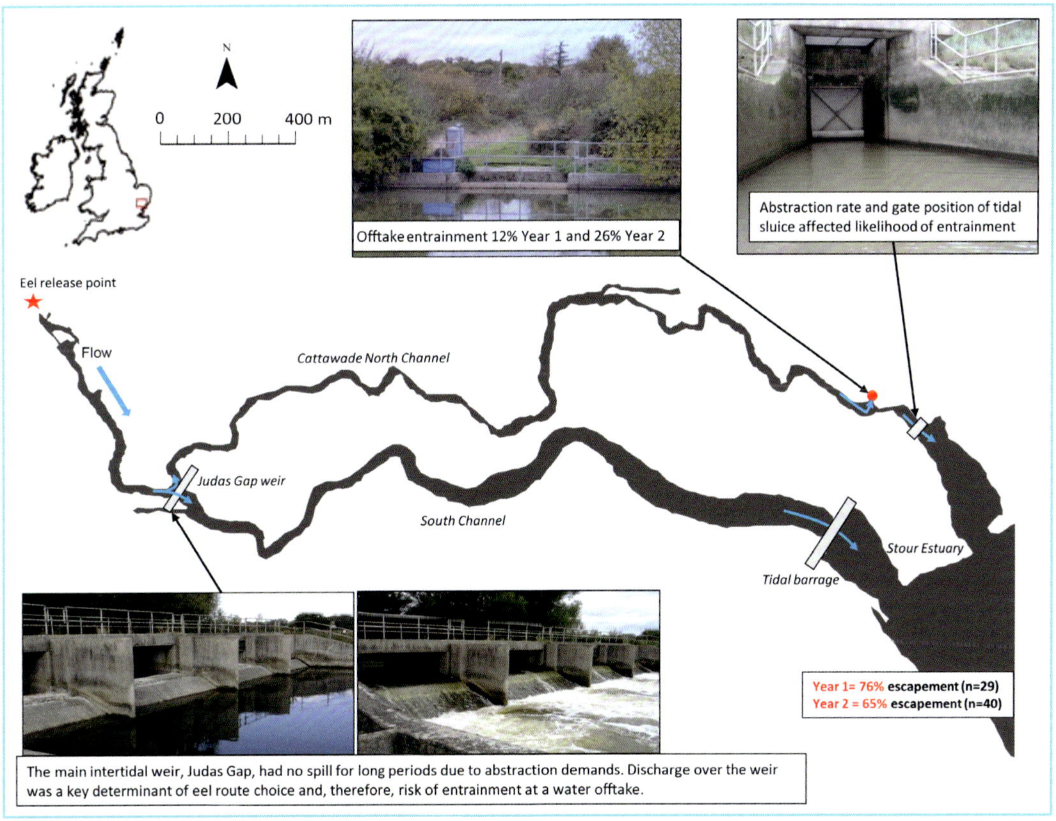

Figure 16.1 Study reach in the lower River Stour, Suffolk, indicating key locations of silver eel loss and delay.

Management implications

- The Environment Agency will use data from the study to assess whether eel escapement targets are being met for the River Stour.
- Catchment managers can use the outputs to inform how to modify the operation of structures to reduce the mortality and delay of adult eel during their seaward migration. Adapting management regimes can be a cost-effective alternative to installing fish passage facilities.
- Distinct peaks in eel migration are typically observed and strategic non-pumping during these short periods can be highly effective at improving escapement.
- The abstraction rate was an important determinant of entrainment loss. Cessation of abstraction during migration periods, combined with the opening of intertidal sluices, is likely to reduce eel loss.
- Where complete cessation of pumping is not viable, a slow start-up of pumps and provision of an alternate route of passage is likely to reduce entrainment loss at offtakes where eel entrance is volitional.
- Based on the recommendations of this research, adaptive management of water levels is now being implemented by the water company and screening of the deleterious offtake in accordance with Environment Agency best practice is planned.

The influence of hydrodynamics on silver eel behaviour – Piper et al. (2015, 2017)

Background

Fish passes are widely used to restore and maintain connectivity for fish at river structures such as dams, weirs and sluices. However, their effectiveness is variable and sometimes low, particularly for non-salmonid species such as the European eel (Boubée, 2014; Bruijs et al., 2009; Haro, 2014). To design effective fish passage solutions, fundamental knowledge of swimming capabilities is needed, combined with an understanding of behavioural response to environmental stimuli (both those that attract or repel fish) (Williams et al., 2012).

During downstream migration, fish tend to move with the flow so it is expected that they will rely more on behavioural responses than swimming capabilities (Williams et al., 2012). The high proportion of river flow diverted through water offtakes (such as at hydropower plants or other abstraction points) presents a strong directional cue. Seaward migrating adult eel tend to follow routes of bulk flow and are particularly vulnerable to being entrained at water offtakes (Breukelaar et al., 2009; Jansen et al., 2007). Mitigation efforts and research have focused on physically excluding eel from harmful areas (for example, by installing screens across offtakes), but eel may still suffer damage and mortality by being impinged on screens (Calles et al., 2010; Hadderingh and Jager, 2002).

Fish react to localized changes in flow-field characteristics including turbulence (Liao, 2007) and velocity acceleration (Enders et al., 2012). The rapid acceleration of flow at constrictions such as at offtake channels and downstream fish bypasses can elicit rejection behaviour among downstream migrating juvenile salmonids (Haro et al., 1998; Kemp et al., 2008); however, understanding is currently limited for other fish species. Eel are thought to be less sensitive to changes in velocity because previous flume studies have shown that they reject upstream after making physical contact with screens (Adam et al., 1999; Russon et al., 2010).

Greater understanding of how eel behave in the vicinity of structures such as water offtakes and respond to the hydrodynamics created around them is crucial for developing better methods to deter eel from entering harmful areas (that is, effective guidance), and for improving the design of bypass facilities.

Study outline

Acoustic positioning telemetry (HTI Hydroacoustic Technology) combined with high-resolution flow mapping and modelling were employed in a 2 year field study to investigate the route choice and responses of silver eel to accelerating flow at a water offtake.

A total of 65 eel were tagged and released in the forebay upstream of a water offtake at a redundant hydropower plant on the River Stour, Dorset. In Year 1, the distribution of eel over the five available passage routes was compared with that predicted by discharge alone. In Year 2, flow at the offtake was manipulated to create two treatments:

- high velocity constricted flow
- low velocity unconstricted flow.

Eel positions were logged approximately every second as they passed through the study site. The resultant movement paths were analysed in relation to underlying flow velocities to explore behavioural responses to the local hydrodynamics.

Key findings

- Eel showed clear rejection behaviour on encountering accelerating velocities due to flow constriction in the vicinity of the water offtake. All individuals abruptly changed direction and swam upstream before turning and ultimately passing through the offtake (Figure 16.2).
- When velocity acceleration was lower, most eel changed direction and exhibited slower 'searching' movements (milling) immediately upstream of the offtake (Figure 16.2).
- Route choice did not correspond to the proportion of discharge passing each route. Eel were reluctant to pass underneath a floating debris boom. Some individuals responded by swimming back upstream or tracing along the upstream edge of the boom before passing by an alternate route.
- Movement paths were non-randomly distributed across the forebay. Before encountering structures, eel aligned with the flow and tended to move within a band 2–4 m from the channel banks.

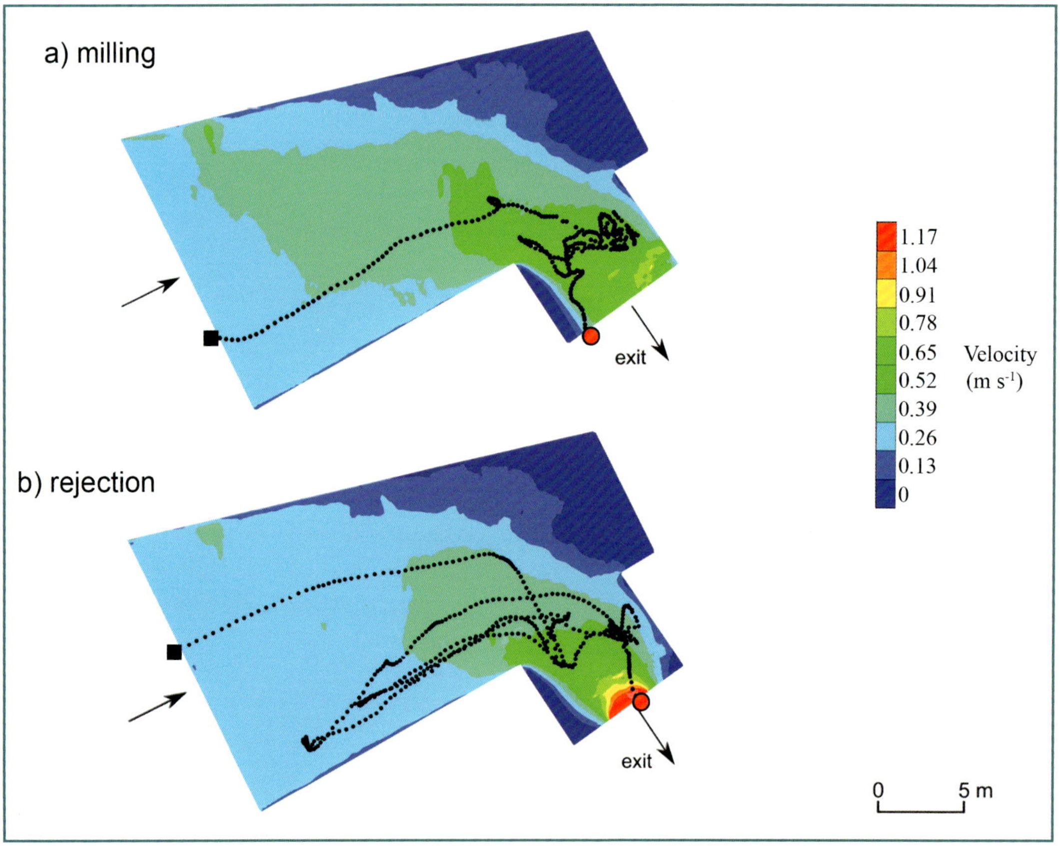

Figure 16.2 Example of trajectories of downstream migrating adult eel through the forebay of a redundant hydropower facility. Tracks show: (a) initial passive drift followed by milling/exploratory behaviour in the offtake channel under the unrestricted flow, low velocity treatment; and (b) initial passive drift followed by swimming back upstream at the offtake under the constricted flow, high velocity treatment. Black squares denote start locations of eel tracks; red circles denote point of passage. Flow direction is indicated by arrows.

Management implications

- The findings challenge historical preconceptions and reveal a strong behavioural component to downstream eel movements. This knowledge can be used to develop more effective guidance and passage solutions.
- Eel bypasses should be designed to avoid abrupt velocity acceleration at the entrance, as is currently advised for salmonids, with the aim of minimizing rejection.
- Conversely, avoidance behaviours present an opportunity to guide eel away from dangerous areas and towards safe passage routes. The observed avoidance behaviours indicate a clear potential for hydrodynamic-based guidance and/or surface guidance devices to enhance the effectiveness of traditional physical screens.

Eel and coarse fish behaviour and passage at an Archimedes screw hydropower turbine – Piper et al. (2018)

Background

Under EU renewable energy legislation to reduce dependence on fossil fuels, the UK is committed to increasing generation from renewable sources to 15% by 2020. Hydropower is expected to form a significant component of this. The most efficient types of turbines rely on substantial head drops and most opportunities for high-head hydropower in Europe have already been exploited (Paish, 2002). In response to government levies and feed-in tariffs, there has been a proliferation of small-scale, low-head hydropower installations (Nolden, 2015).

The impacts of high-head dams on fish have received much attention and include hindering or preventing migration, and modification of flow regimes (Kemp, 2015). Other potential impacts are damage and mortality caused by contact with the turbine blades or indirectly by cavitation (Kemp, 2015).

Although low-head hydropower facilities are generally perceived to be fish friendly, few studies have examined this, particularly for fish moving in the downstream direction (Abbasi and Abbasi, 2011; Robson et al., 2011). Information on potential impacts is urgently needed by regulatory bodies such as the Environment Agency to:

- underpin decisions within the consenting process
- effectively balance the need to increase renewable energy with legislative demands to maintain and improve connectivity for fish.

Study outline

A study was conducted between April 2013 and March 2015 to investigate fish passage and behaviour at an Archimedes screw hydroelectric power (HEP) turbine and associated structures at Flatford Mill on the River Stour, Suffolk. The Flatford complex consists of the hydropower turbine, six undershot penstock sluice gates, a Larinier fish pass and two trough-type elver passes.

Fish were captured, PIT tagged and released upstream (n = 442 coarse fish, n = 74 yellow eel) and downstream (n = 966 coarse fish, n = 216 yellow eel) of Flatford Mill during 2013 and 2014. Their movements past the structures were monitored using an array of PIT antennas installed on all the potential passage routes. In addition, 127 actively migrating silver eel were tagged with acoustic transmitters (Vemco) and released upstream of Flatford Mill during autumn 2013 and 2014 to assess passage through the complex and overall escapement rate to the estuary. Fine-scale fish behaviour in the vicinity of the structures was investigated using a high-resolution sonar camera (adaptive resolution imaging sonar, ARIS).

Key findings

- Overall, 239 of the tagged fish approached downstream of the structures.
- Attraction efficiency of the Larinier fish pass ranged from 47% for eel to 94% for chub. Passage efficiency ranged from 10% for roach to 81% for dace.
- Behavioural observations of coarse fish shoals near the downstream entrance of the Larinier pass showed they rapidly dispersed when the HEP started up but habituated and most returned to the area within 15 minutes.
- Silver eel successfully descended the HEP, but the majority of downstream migrating adult eel passed an alternate route through a navigation lock with overshot weir. River discharge and operation of the HEP were the principal determinants of passage route.
- Passage rate via the mill was slower with eel taking a median of 11.1 minutes (interquartile range = 7.4–16.4) to pass compared to a median of 4.3 minutes (interquartile range = 2.8–6.3) via the lock route.
- Footage of adult eel from the sonar camera showed clear milling and rejection behaviour immediately upstream of the HEP complex during operation (Figure 16.3).

Management implications

- There was no apparent mortality of those downstream-moving eel that migrated through the HEP complex before reaching the estuary. However, observed delay and milling/rejection in the vicinity of the complex suggests that such facilities have the potential to impose delay

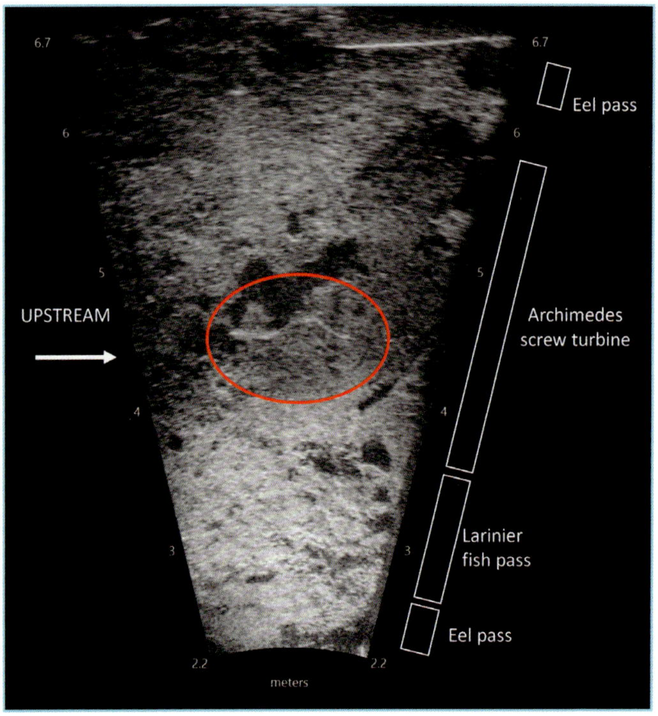

Figure 16.3 Sonar camera image showing adult eel (circled) swimming upstream after rejecting the structures at Flatford Mill. Arrow denotes direction of flow.

Figure 16.4 Woody debris and vegetation causing partial blockage in the lower section of the Larinier fish pass.

and energetic costs. This is particularly important in systems where there is no alternative migration route, that is, where schemes are 'run of river', and where there may be multiple HEPs in the same watercourse.

- In low flow systems such as the one studied, there may be limited opportunity for power generation while maintaining effective upstream fish passage routes during the summer. Observations of fish behaviour and the timing of upstream passage suggest that periods of non-operation would be preferable to the frequent stop–start generation caused by rapid flow fluctuation during low discharge.
- For small freestanding sectional Larinier passes such as the one studied, it is essential that regular checking and maintenance is carried out to prevent blocking by woody debris, particularly during peak migration periods (Figure 16.4).

Using attraction flow to enhance upstream eel passage at an intertidal barrier – Piper et al. (2012)

Background

European eel migrate upstream into freshwater systems as juveniles (glass eel and elvers). Glass eel move up tidal estuaries and rivers using coastal currents and tidal stream transport (Briand et al., 2005a; McCleave and Wippelhauser, 1987), while the larger pigmented glass eel and elvers swim actively upstream in response to flow and olfactory cues (Laffaille et al., 2007; Tesch, 2003).

Tidal barriers and other water control structures are widespread in European catchments for both flood alleviation and retention of water for abstraction, but they may severely restrict upstream eel migration (Feunteun et al., 1998; Knights and White, 1998). Upstream eel passes (or eel ladders) are widely installed to enhance immigration and have been shown to aid the recovery of freshwater eel stocks (Briand et al., 2005b; Laffaille et al., 2005). There is little consensus, however, as to the optimum pass positioning or attraction conditions that should be present to maximize efficiency.

For a fish pass to be effective, conditions should be such that fish are guided to the entrance (attraction) and then compelled and able to rapidly ascend (or descend) the pass (Castro-Santos et al., 2009). Given that flow and olfactory detection of salinity gradients are thought to form the principal navigation cues for upstream migrating juvenile eel, it has been suggested that delivery of a small quantity of plunging flow to the entrance of eel passes may enhance their effectiveness (Clay, 1995; Porcher, 2002). A field study was conducted to test the effect of providing plunging attraction flow on the efficacy of an upstream eel pass and to investigate movement patterns of juvenile eel at a tidal barrier.

Study outline

In a field study conducted over 11 weeks in 2010, four upstream eel traps were deployed across the main intertidal barrier on the River Stour, Suffolk. Attraction flow was delivered to each pass entrance either overhead as 'plunging flow' or underwater as 'streaming flow' (Figure 16.5). The numbers and sizes of eel that passed daily were analysed in relation to flow treatment and environmental factors (tide height, water temperature and weir spill).

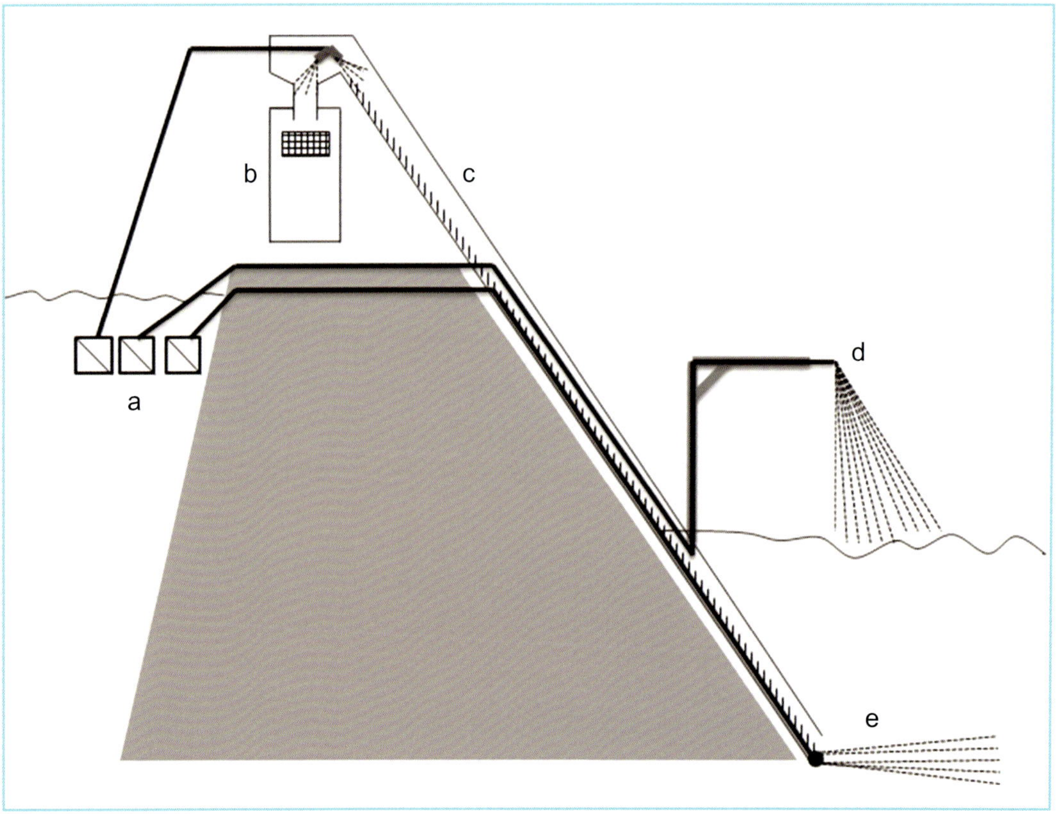

Figure 16.5 Schematic of the eel trap configuration used within the study. Submersible pumps (a) supply conveyance flow to the catch pot (b) and bristle-lined climbing trough (c), and flow to experimental treatments, plunging flow (d) and streaming flow (e).

Key findings

- Plunging flow was more than twice as effective at attracting juvenile eel to the pass than streaming flow.
- More eel, particularly the smallest size classes, ascended the passes located near the channel banks. Peak migration occurred during July (Figure 16.6).

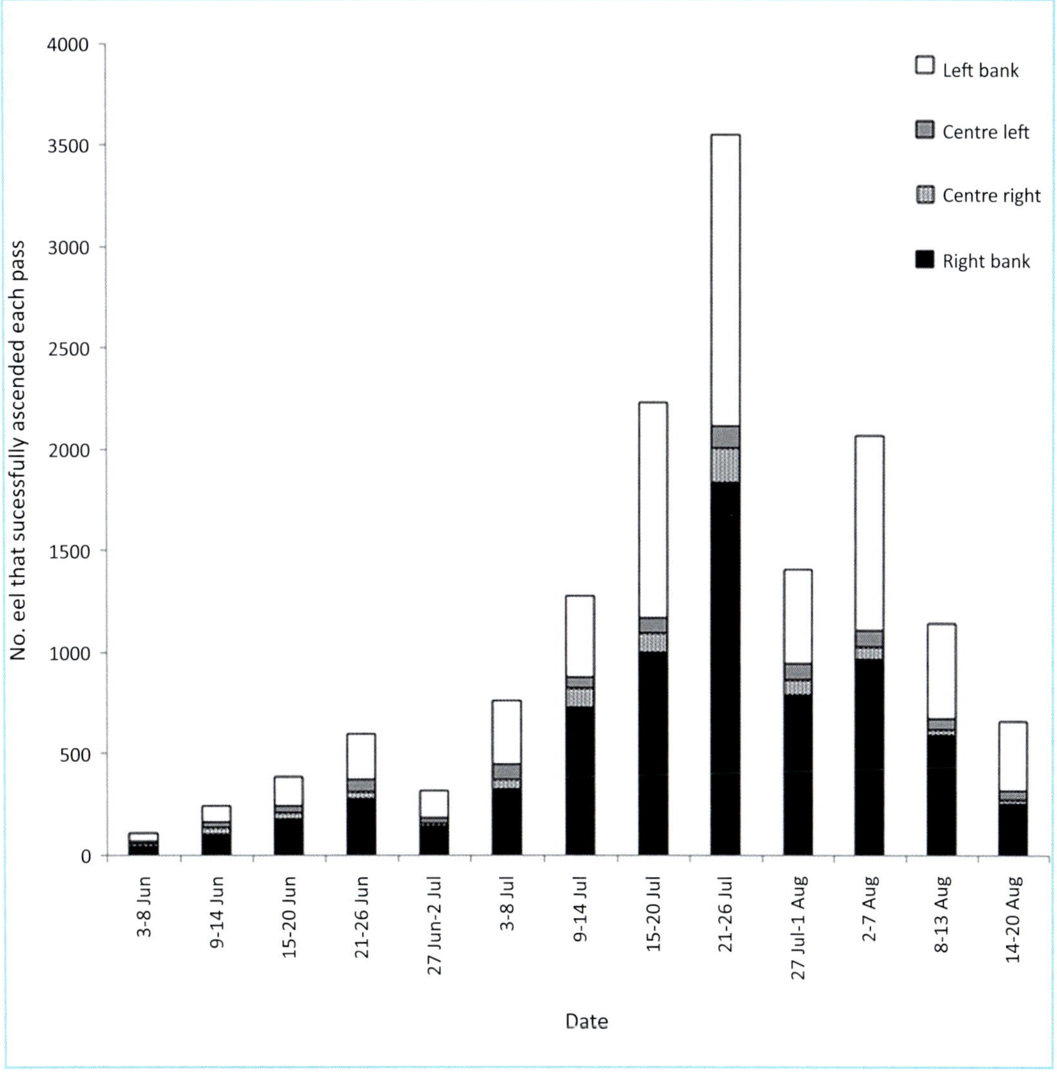

Figure 16.6 Total numbers of eel of five length classes captured in traps at four upstream eel passes (right bank, centre right, centre left and left bank) across Judas Gap weir, River Stour, June–August 2010.

Figure 16.7 Plunging attraction flow delivered to the base of an eel pass.

Figure 16.8 Recently constructed elver pass at Nieuwe Statenzijl, Netherlands, employing multiple plunging flow outlets at the downstream end of the pass to enhance attraction and passage. Photograph: Peter Paul Schollema.

Management implications

- The findings showed that simple manipulation of hydrodynamic conditions can significantly enhance passage efficiency at upstream eel passes.
- The provision of plunging attraction flow, which creates turbulent conditions near the pass entrance, should become a standard design feature for upstream eel passes (Figures 16.7 and 16.8). However, there is a trade-off. Turbulence and flow velocities within the pass itself should be within the swim capabilities of the smallest life stages and should not deter eel from ascending.
- To maximize recruitment, upstream eel passes should be preferentially located near the channel banks.

Silver eel response to an infrasound device – Piper et al. (2019)

Background

Physical screens are widely used to prevent entrainment of eel at water offtakes to hydropower or irrigation systems, and to divert them towards more favourable routes, for example, fish passes. However, installation and maintenance of screens is costly, and eel are particularly susceptible to impingement causing damage and mortality through abrasion and suffocation (Calles et al., 2010; Hadderingh and Jager, 2002).

Behavioural guidance devices employ stimuli such as sound (Ploskey and Johnson, 2001; Popper and Carlson, 1998), light (Brown, 2000; de Oliveira Mesquita et al., 2008), electric current (Pugh et al., 1970) or altered hydrodynamics (Amaral, 2003) to guide fish. There is considerable interest in developing such technologies because they may have potential to enhance, or even replace, physical screens and are generally less costly (Coutant, 2001). Previous studies indicate that eel are sensitive to low frequency sound in the infrasound range (<12 Hz) (Sand et al., 2001, 2000), hence behavioural guidance devices employing infrasound may have potential to guide eel at river structures. Using a controlled experiment under field conditions, this study tested silver eel response to an infrasound deterrent during downstream migration.

Study outline

The study was conducted in November 2013 on the River Stour, Dorset. Using acoustic positioning telemetry (HTI Hydroacoustic Technology), silver eel (N = 50) were tracked as they migrated through the forebay of a redundant hydropower facility. An infrasound deterrent (12 Hz) was positioned immediately upstream of the water offtake. The experiment ran over ten consecutive nights and the deterrent was switched ON and OFF on alternate nights. A batch of five eel was released into the upstream end of the forebay on each of the nights.

Eel trajectories through the forebay were constructed from the positioning data and analysed in relation to the mapped acoustic field (Figure 16.9). Flow through the site was monitored and controlled to minimize variation between nights.

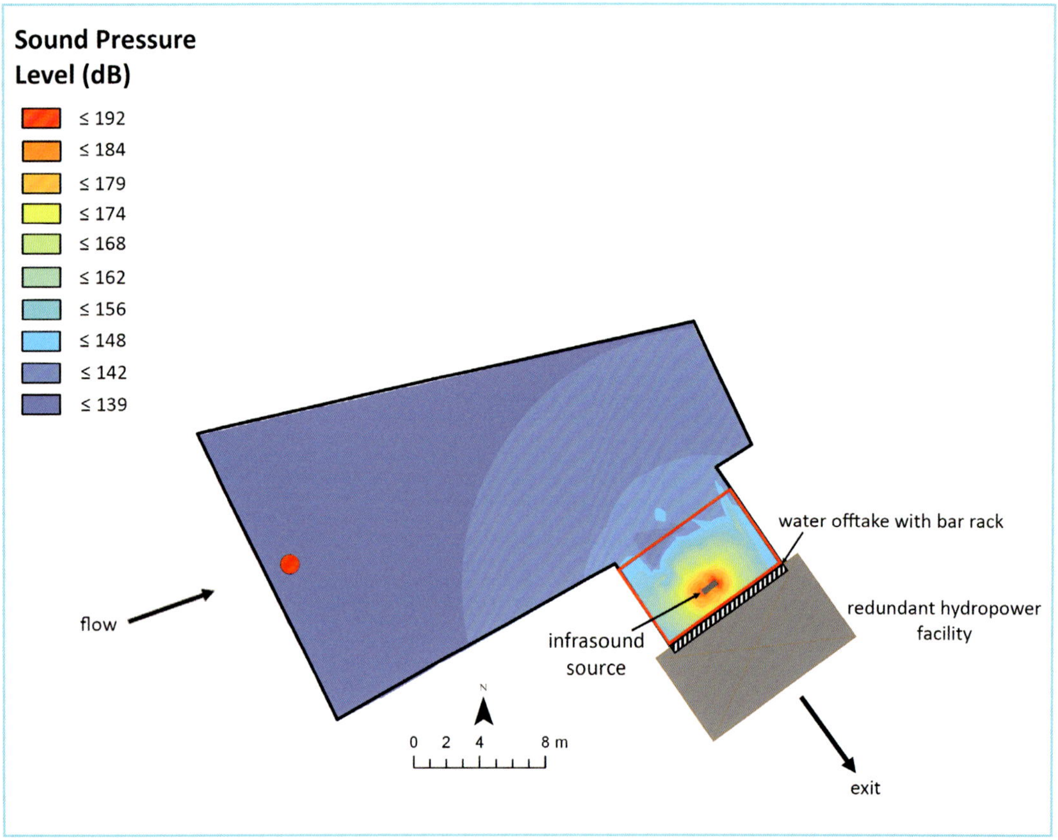

Figure 16.9 Sound pressure level (dB) in the forebay of a redundant hydropower facility when the infrasound deterrent was ON. Red circle denotes eel release location. Flow direction is indicated by arrows.

Eel behaviour was quantified to determine if the infrasound device:

- induced avoidance as they approached the offtake, indicated by modification of swim paths
- deterred them from passing the intake.

Key findings

- Of the eel that moved downstream into the forebay 78% passed the offtake. The infrasound deterrent did not affect how many individuals eventually passed.
- The deterrent modified eel behaviour. Eel paths were longer and more erratic when the device was switched ON (Figure 16.10).
- Eel showed avoidance of the infrasound and were less likely to remain in the vicinity of the offtake when it was operating, but the effect was not strong enough to deter them from passing through the offtake.

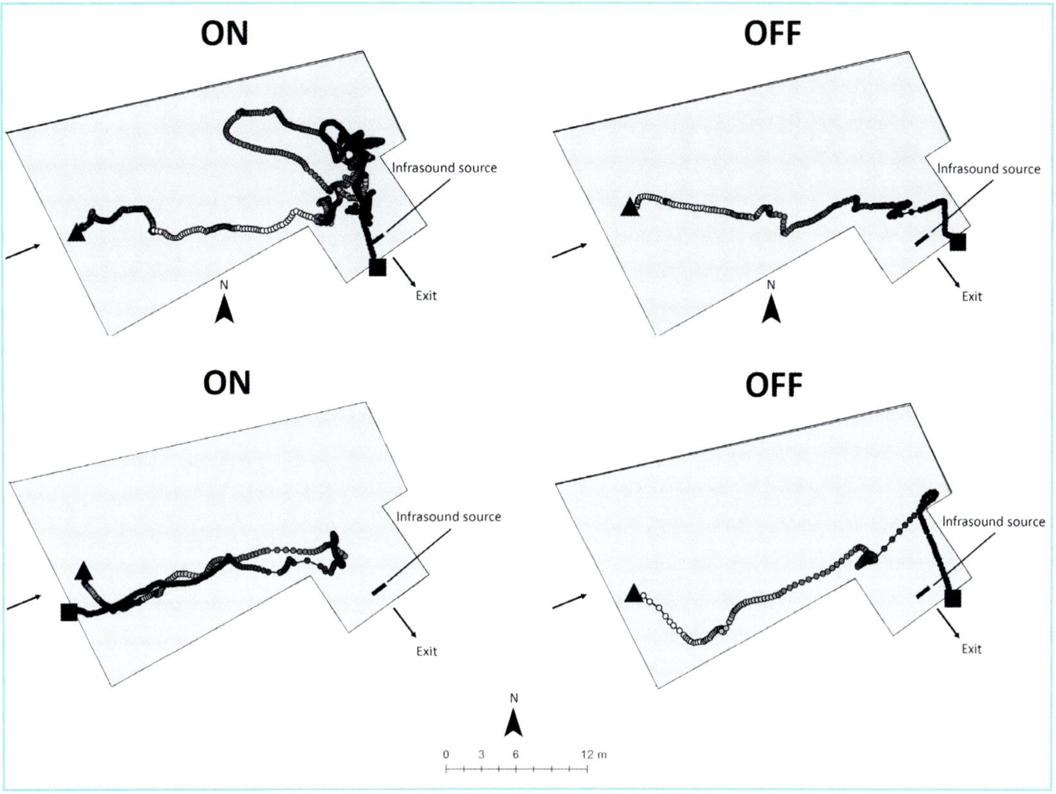

Figure 16.10 Trajectories of four eel moving through the forebay of a redundant hydropower facility and approaching an infrasound deterrent when it was ON and OFF. Black triangles denote the start of tracks and black squares the end. Flow direction is indicated by arrows.

Management implications

- Infrasound deterrent devices may have application at sites where a sufficiently intense acoustic field can be created around the source. This will depend on many factors including site geometry.
- To induce guidance, eel must both be able to detect and have sufficient time to respond to the deterrent. Infrasound may be most applicable where water velocities are relatively low and/or where lateral guidance is required.
- Behavioural guidance systems that provide multimodal cues (for example, sound and light) or that combine physical and behavioural guidance, that is, using deterrents to enhance the efficiency of physical screens such as bar racks, may be most effective. Such systems could more consistently outcompete both the inherent background noise around river infrastructure and the dominant hydrodynamic cues that mediate adult eel behaviour during migration.
- Use of fine resolution telemetry revealed behavioural responses to infrasound that would not have been detected by the traditional mark-recapture or coarse-scale telemetry methods often employed to assess the efficacy of behavioural guidance devices in the field.

Acknowledgements

This programme of research was jointly funded by the Environment Agency, the University of Southampton and the Interreg IVB Living North Sea project, which aims to improve access to migratory fish in the North Sea Region. Funding is also acknowledged from the EPSRC Doctoral Training Grant awarded to the University of Southampton (EP/P505119/1).

The authors thank all the participants in the various elements of this study and acknowledge organizations and individuals in peer-reviewed publications.

Bibliography

Abbasi, T. and Abbasi, S.A. (2011) Small hydro and the environmental implications of its extensive utilization. *Renewable and Sustainable Energy Reviews* 15(4), 2134–2143.

Adam, B., Schwevers, U. and Dumont, U. (1999) Planungshulfen fur den Bau funktionfahiger Fischaufstiegsanlagen. *Bibliothek Natur and Wissenschaft* 16(64), 1–63.

Amaral, S. (2003) *The use of angled bar racks and louvers for guiding fish at FERC-Licensed projects. FERC Fish Passage Workshop*. Alden Research Laboratory, Inc.

Boubée, J. (2014) Upstream and downstream passage of eels in New Zealand, 20 years on–lessons learned. In *144th Annual Meeting of the American Fisheries Society*.

Breteler, J.K., Vriese, T., Borcherding, J., Breukelaar, A., Jorgensen, L., Staas, S. and Ingendahl, D. (2007) Assessment of population size and migration routes of silver eel in the river Rhine based on a 2-year combined mark-recapture and telemetry study. *ICES Journal of Marine Science* 64(7), 1450–1456.

Breukelaar, A.W., Ingendahl, D., Vriese, F.T., de Laak, G., Staas, S. and Breteler, J.G.P.K. (2009) Route choices, migration speeds and daily migration activity of European silver eels Anguilla anguilla in the River Rhine, north-west Europe. *Journal of Fish Biology* 74(9), 2139–2157.

Briand, C., Fatin, D., Feunteun, E. and Fontenelle, G. (2005a) Estimating the stock of glass eels in an estuary by mark-recapture experiments using vital dyes. *Bulletin Francais de La Peche et de La Pisciculture* 378–79, 23–46.

Briand, C., Fatin, D., Fontenelle, G. and Feunteun, E. (2005b) Effect of re-opening of a migratory pathway for eel (Anguilla anguilla, L.) at a watershed scale. *Bulletin Francais de La Peche et de La Pisciculture* 378–79, 67–86.

Brown, R. (2000) The potential of strobe lighting as a cost-effective means for reducing impingement and entrainment. *Environmental Science & Policy* 3(Supplement 1), 405–416.

Bruijs, M., Durif, C. and Noakes, D.L.G. (2009) Silver eel migration and behaviour. In G. Thillart, S. Dufour and J.C. Rankin (eds) *Spawning Migration of the European Eel*. Springer, pp. 65–95.

Calles, O., Olsson, I.C., Comoglio, C., Kemp, P.S., Blunden, L., Schmitz, M. and Greenberg, L.A. (2010) Size-dependent mortality of migratory silver eels at a hydropower plant, and implications for escapement to the sea. *Freshwater Biology* 55(10), 2167–2180.

Castro-Santos, T., Cotel, A. and Webb, P. (2009) Fishway evaluations for better bioengineering - an integrative approach. In: Haro, A., Moffit, C. and Dadswell, M. (eds) *Challenges for Diadromous Fishes in a Dynamic Global Environment, American Fisheries Society Symposium 69*. Bethesda, MD, pp. 557–575.

Clay, C.H. (1995) *Design of Fishways and other Fish Facilities*, 2nd edn. Boca Raton, FL: Lewis Publishers.

Coutant, C.C. (2001) Behavioral technologies for fish guidance. Proceedings of the Symposium Behavioral Technologies for Fish Guidance held at Charlotte, North Carolina, USA, 30–31 August 1999 (Vol. 26). American Fisheries Society.

De Oliveira Mesquita, F., Godinho, H.P., de Azevedo, P.G. and Young, R.J. (2008) A preliminary study into the effectiveness of stroboscopic light as an aversive stimulus for fish. *Applied Animal Behaviour Science* 111(3), 402–407.

Enders, E.C., Gessel, M.H., Anderson, J.J. and Williams, J.G. (2012) Effects of decelerating and accelerating flows on juvenile salmonid behavior. *Transactions of the American Fisheries Society* 141(2), 357–364.

Feunteun, E. (2002) Management and restoration of European eel population (*Anguilla anguilla*): an impossible bargain. *Ecological Engineering* 18(5), 575–591.

Feunteun, E., Acou, A., Guillouet, J., Laffaille, P. and Legault, A. (1998) Spatial distribution of an eel population (*Anguilla anguilla* L) in a small coastal catchment of northern Brittany (France) – consequences of hydraulic works. *Bulletin Francais de La Peche et de La Pisciculture* 349, 129–139.

Hadderingh, R.H. and Jager, Z. (2002) Comparison of fish impingement by a thermal power station with fish populations in the Ems Estuary. *Journal of Fish Biology* 61(sA), 105–124.

Haro, A. (2003) Downstream migration of silver-phase anguillid eels. In: Aida, K. Tsukamoto, K. and Yamauchi, K. (eds) *Eel Biology*. Springer-Verlag, Tokyo, pp. 215–222.

Haro, A. (2014) Downstream passage and movements of silver-phase American eels at three hydroelectric projects on the Shetucket River, Connecticut. In *144th Annual Meeting of the American Fisheries Society*.

Haro, A., Odeh, M., Noreika, J. and Castro-Santos, T. (1998) Effect of water acceleration on downstream migratory behavior and passage of Atlantic Salmon Smolts and juvenile American shad at surface bypasses. *Transactions of the American Fisheries Society* 127(1), 118–127.

ICES (2011a) *Report of the 2011 Session of the Joint EIFAAC/ICES Working Group on Eels*. Lisbon, Portugal.

ICES (2011b) *Widely Distributed and Migratory Stocks European eel. ICES Advice* (Vol. Book 9)

Jansen, H.M., Winter, H.V, Bruijs, M.C.M. and Polman, H.J.G. (2007) Just go with the flow? Route selection and mortality during downstream migration of silver eels in relation to river discharge. *ICES Journal of Marine Science* 64(7), 1437–1443.

Kemp, P., Gessel, M. and Williams, J. (2008) Response of downstream migrant juvenile Pacific salmonids to accelerating flow and overhead cover. *Hydrobiologia* 609(1), 205–217.

Kemp, P.S. (2015) Impoundments, barriers and abstractions. In: Craig, J.F. (ed.) *Freshwater Fisheries Ecology*. John Wiley & Sons, Chichester, pp. 717–769.

Knights, B. and White, E.M. (1998) Enhancing immigration and recruitment of eels: the use of passes and associated trapping systems. *Fisheries Management and Ecology* 5(6), 459–471.

Laffaille, P., Acou, A., Guillouet, J. and Legault, A. (2005) Temporal changes in European eel, *Anguilla anguilla*, stocks in a small catchment after installation of fish passes. *Fisheries Management and Ecology* 12(2), 123–129.

Laffaille, P., Caraguel, J.M. and Legault, A. (2007) Temporal patterns in the upstream migration of European glass eels (*Anguilla anguilla*) at the Couesnon estuarine dam. *Estuarine Coastal and Shelf Science* 73(1 2), 81–90. https://doi.org/10.1016/j.ecss.2006.12.011

Liao, J.C. (2007) A review of fish swimming mechanics and behaviour in altered flows. *Philosophical Transactions of the Royal Society B-Biological Sciences* 362, 1973–1993.

McCleave, J.D. and Wippelhauser, G.S. (1987) Behavioral aspects of selective tidal stream transport in juvenile American eels. *Journal du Conseil* 40, 262–271.

Nolden, C. (2015) *Performance and Impact of the Feed-in Tariff Scheme: Review of Evidence*. Department of Energy and Climate Change.

Paish, O. (2002) Small hydro power: technology and current status. *Renewable and Sustainable Energy Reviews* 6(6), 537–556.
Piper, A.T., Wright, R.M. and Kemp, P.S. (2012) The influence of attraction flow on upstream passage of European eel (*Anguilla anguilla*) at intertidal barriers. *Ecological Engineering* 44, 329–336.
Piper, A.T., Wright, R.M., Walker, A.M. and Kemp, P.S. (2013) Escapement, route choice, barrier passage and entrainment of seaward migrating European eel, *Anguilla anguilla*, within a highly regulated lowland river. *Ecological Engineering* 57, 88–96.
Piper, A.T., Manes, C., Siniscalchi, F., Marion, A., Wright, R.M. and Kemp, P.S. (2015) Response of seaward-migrating European eel (Anguilla anguilla) to manipulated flow fields. *Proceedings of the Royal Society B: Biological Sciences* 282(1811), 1–9.
Piper, A.T., Svendsen, J., Wright, R.M. and Kemp, P.S. (2017) Movement patterns of seaward migrating European eel (*Anguilla anguilla*) at a complex of riverine barriers: implications for conservation. *Ecology of Freshwater Fish* 26(1), 87–98.
Piper, A.T., Rosewarne, P.J., Wright, R.M. and Kemp, P.S. (2018) The impact of an Archimedes screw hydropower turbine on fish migration in a lowland river. *Ecological Engineering*, 118, 31–42.
Piper, A.T., White, P.R., Wright, R.M., Leighton, T.M. and Kemp, P.S. (2019) Response of seaward-migrating European eel (*Anguilla anguilla*) to an infrasound deterrent. *Ecological Engineering* 127, 480–486.
Ploskey, G.R. and Johnson, P.N. (2001) Effectiveness of strobe lights and an infrasound device for eliciting avoidance by juvenile salmon. *Behavioral Technologies for Fish Guidance: American Fisheries Society Symposium.*
Popper, A.N. and Carlson, T.J. (1998) Application of sound and other stimuli to control fish behaviour. *Transactions – American Fisheries Society* 127(5), 10.
Porcher, J.P. (2002) Fishways for eels. *Bulletin Francais de La Peche et de La Pisciculture* 364, 147–155.
Pugh, J.R., Monan, G.E. and Smith, J.R. (1970) Effect of water velocity on the fish guiding efficiency of an electrical guiding system. *Fishery Bulletin* 68, 307–324.
Robson, A., Cowx, I.G. and Harvey, J.P. (2011) Impact of run-of-river hydro-schemes upon fish populations. Final Report. SNIFFER WFD114, Edinburgh.
Russon, I.J., Kemp, P.S. and Calles, O. (2010) Response of downstream migrating adult European eels (*Anguilla anguilla*) to bar racks under experimental conditions. *Ecology of Freshwater Fish* 19(2), 197–205.
Sand, O., Enger, P.S., Karlsen, H.E., Knudsen, F. and Kvernstuen, T. (2000) Avoidance responses of infrasound in downstream migrating European silver eels, *Anguilla anguilla. Environmental Biology of Fishes* 57(3), 327–336.
Sand, O., Enger, P.S., Karlsen, H.E. and Knudsen, F.R. (2001) Detection of infrasound in fish and behavioral responses to intense infrasound in juvenile salmonids and European silver eels: a minireview. *Behavioral Technologies for Fish Guidance* 26, 183–193.
Tesch, F.W. (2003) *The Eel*, 5th edn. Blackwell Science, Oxford.
Watene, E.M. and Boubée, J.A.T. (2005) Selective opening of hydroelectric dam spillway gates for downstream migrant eels in New Zealand. *Fisheries Management and Ecology* 12(1), 69–75.
Williams, J.G., Armstrong, G., Katopodis, C., Larinier, M. and Travade, F. (2012) Thinking like a fish: a key ingredient for development of effective fish passage facilities at river obstructions. *River Research and Applications* 28(4), 407–417.

PART IV

Eel health

CHAPTER 17

Major eel diseases in Europe: the past 30 years

Olga Haenen

Wageningen Bioveterinary Research, National Reference Laboratory for Fish, Shellfish and Crustacean Diseases, Lelystad, the Netherlands

Introduction

Since the decline of eel stocks in the early 1980s, several possible causes have been proposed including, commercial fisheries, pollution, migration barriers, predation by birds and/or diseases (Casselman and Cairns, 2003). This chapter addresses diseases, as a threat to wild and cultured eel.

Since 1986, the National Reference Laboratory for Fish Diseases of Wageningen Bioveterinary Research (WBVR) in the Netherlands has performed diagnosis of multi-disciplinary, notifiable and non-notifiable fish diseases, among which are eel diseases. The Netherlands has a long tradition of consuming fresh and smoked European eel (*Anguilla anguilla*). Historically, eels were recruited from wild fisheries only, then from the 1980s, as the Netherlands developed its eel farming practices, cultured eels featured more in the marketplace. The Netherlands is now Europe's leading producer of farmed eels.

Dutch eel farms have presented several eel disease problems to the Fish Disease Laboratory of WBVR for diagnosis. In the last 30 years the laboratory has diagnosed and discovered several eel diseases, followed up with research and publications and has, thereby, become a central European eel disease laboratory.

The multidisciplinary diagnosis of the laboratory includes necropsy, parasitology, bacteriology, virology, including also molecular techniques, and histopathology. Clients presenting eel disease cases vary, and include eel farmers and consulting veterinarians, eel fisheries, sport fishery boards, zoos, the government and (inter)national colleague institutes.

This chapter provides an overview of the major eel diseases found during the last 30 years in wild and farmed eels in the Netherlands, with reference to findings in some other European countries.

Development of eel diseases

What are the causes of eel diseases and how do they develop? We need to distinguish between wild eel and farmed eel, as they are subjected to different infection pressures. In wild eels, living in low densities, there is generally low infection pressure with regard to pathogens, like bacteria, viruses, and parasites. Wild eels may sustain a low number of parasites without becoming diseased, living in a kind of balance (Borgsteede et al., 1999). However, the introduction of a new parasite, with intermediate hosts in abundance, may bring about acute and severe infection, as we have seen with the swimbladder nematode *Anguillicoloides crassus* (Van Banning and Haenen, 1990). After a decade, this infection also changed into a milder infection, as is common with new parasites (PARA-SITE, 2018).

Eel farming in Europe has developed rapidly since the 1980s. Farms are mostly indoors, and intensive, although there is also less intensive pond culture in Italy and Spain. To date only wild eel is used as stocking animals (elvers or yellow eels). Recirculation of water in these farms means pathogens are pumped around, often in parallel tanks, and thereby may infect various tanks in the same system, starting from one tank with an infected eel.

As with other diseases, eel diseases may develop from the interaction of a fish with a low resistance (such as a mechanically injured eel right after grading) and a suboptimal environment, especially when water quality is also suboptimal (Figure 17.1) (Snieszko, 1974). The eel then is particularly vulnerable to opportunistic bacteria and eel viruses. In such cases, in addition to

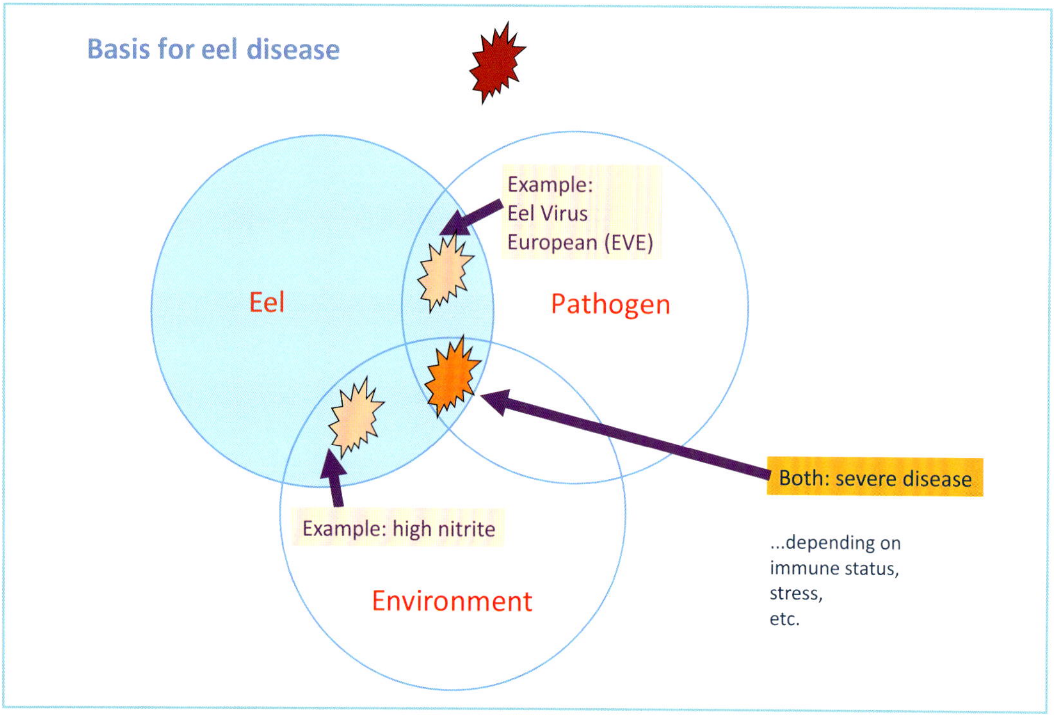

Figure 17.1 How eel diseases develop. After Snieszko (1974).

direct health effects from low oxygen, high ammonia, nitrite and nitrate levels, suboptimal pH and so on, opportunistic pathogens, for example, bacteria like *Aeromonas hydrophila*, which normally are ubiquitously present, might cause eel disease (Haenen, unpublished). Such bacteria are then secondary pathogens. In the case of primary pathogens, such as when an eel-virulent strain of *Vibrio vulnificus* is present, eels may become diseased directly. Incidences of disease are still dependent on the immune status of individual eel (Liewes and Haenen, 1993). When conditions are suboptimal, eels are mechanically injured and virulent pathogens are present, a more severe outbreak can be expected, again dependent on the immune status of the eel.

Eel parasites may cause anorexia. Intensively farmed eel may be subjected to increased parasite numbers, since eels are kept in high densities and infect each other through direct contact. In (warm water) eel farming, temperatures are permissive for the growth of various bacteria, among which are potential eel pathogens. Eel viral disease occur at particular water temperatures that are optimal for the virus. In eel farming, to cure the eels from a viral infection, the water temperature may be changed to a range suboptimal for the virus, but still suitable for the eels, this allows the eels to build up resistance against the virus. It is not uncommon for an eel to be infected with two, or even three, pathogens, for example, two viruses and a bacterium. In this chapter, some examples are presented.

Before 1980, there were only few papers on eel disease in Europe. This changed with the advent of eel farming, as more eels were then offered for diagnosis to fish disease laboratories, which in turn developed specific and sensitive diagnostic tests to detect viruses and bacteria. Moreover, eel farmers using stock other than European elvers (*Anguilla anguilla*) from the coasts of South West-Europe to populate their farms introduced other and new sources of pathogens. Eels were imported from East Asia, where Japanese eels (*A. japonica*) were utilized, introducing East Asian eel pathogens into Europe.

There are a few remarkable imported eel pathogens from East Asia, which include the monogenic trematode gill worms *Pseudodactylogyrus anguillae* (Yin and Sproston, 1948) and *P. bini* (Kikuchi, 1929), the swimbladder nematode *Anguillicoloides crassus* (Audenaert et al., 2003; Van Banning and Haenen, 1990) and anguillid herpesvirus AngHV1 (Davidse et al., 1999). These pathogens have caused moderate to severe disease in European eels (Buchmann et al., 1987; Haenen et al., 2002; Jensen, 2010; Van Beurden et al., 2012).

Summary

Since their introduction in the 1970s, *P. anguillae* (Yin and Sproston, 1948) and *P. bini* (Kikuchi, 1929) have gone on to cause problems in wild and, particularly, farmed eel (Buchmann et al., 1987; Jensen, 2010). In the 1980s, via imports of live eel, the swimbladder nematode *A. crassus* was introduced from East Asia into Europe, arriving first into the Netherlands and Belgium (Van Banning and Haenen, 1990) and has since caused disease problems all over Europe in wild and farmed eel. Since the 1980s, eel viruses have become more prominent across Europe. Some outbreaks of disease through Eel Virus European in farmed eel, and Eel Virus European X in farmed and wild eels has occurred (Castric and Chastel, 1980; Van Beurden et al., 2012). At the end of the 1980s, the bacterium *Pseudomonas anguilliseptica* was isolated from wild elvers from South Western France and caused disease problems in eel farms (Haenen and Davidse, 2001). The bacterium *Edwardsiella tarda* has been found in diseased wild and farmed eel since the 1980s

(Haenen, unpublished). The mid-1990s saw the start of several cases of bacterial infection by *Vibrio vulnificus* at Dutch eel farms, with high eel mortalities (Haenen et al., 2014b), and even a severe zoonotic case in an eel farmer (Dijkstra et al., 2009). Furthermore, in Denmark and Spain various *V. vulnificus* cases have been reported at eel farms, and also in humans (Dalsgaard et al., 1996, 1999; Fouz et al., 2006., Haenen et al., 2014a). In the mid-1990s, AngHV-1 was discovered in the Netherlands (Davidse et al., 1999), and soon after all over Europe in farmed and wild eel, with serious disease problems in farmed eel (Haenen et al., 2014a). It was also detected in wild elvers (Van Beurden et al., 2012). In the following sections details on these and other eel diseases are given. Standard recommendations regarding screening of wild eels for pathogens and diseases are given by Lewin et al. (2014) and by the Planning Group on the Monitoring of Eel Quality (WKPGMEQ) of ICES (2015).

Parasites

Parasites may be *protozoans* or *metazoans*. In nature, wild eels may carry some ecto- and endoparasites, without health problems when there is balance between host and parasite. However, since global trade in live eels started in the 1980s, the balance has been upset with the introduction of new eel parasites into naive European eel populations. The eels reacted both acutely and severely in some cases. Some examples of this response are described below. Kennedy (2007) stated that the following three introduced parasite species are important as pathogens of farmed European eel: *P. anguillae, P. bini* and *A. crassus.*

Few studies have been published in Europe on the parasitofauna of wild eels. Various protozoan parasites are known, many of which are were already known as common fish parasites. Up to 44 different species of metazoan eel parasites were found by Tesch (2003). Various parasites from different classes have been described in Denmark, the Netherlands, Germany, Spain, Poland, Iceland, Portugal, Hungary and other countries (Aguilar et al., 2005; Borgsteede et al. 1999; Hermida et al., 2008; Køie, 1988; Kristmundsson and Helgason, 2007; Maillo et al., 2005; Rolbiecki and Rokicki, 2006; Sures and Streit, 2001; Székely et al., 2009). Up to 63 species of metazoan eel parasites were found by Kennedy (2007) in European eels. In farmed eel, parasite numbers (intensity) may be higher, given the high densities in which fish are kept. Moreover, in the recirculation aquaculture systems of eel farms, the biofilter may act as a reservoir for zooplankton, of which several copepod species may act as intermediate hosts for the swimbladder nematode *A. crassus* (Haenen, 1995).

Protozoan parasites

Regarding protozoan parasites, most are not fish species-specific, and can be found as ectoparasite on the skin and gills of eels. They are diagnosed through fresh skin smears, and gill clippings at necropsy, which are directly examined by light microscopy.

Examples are ciliates, like *Trichodina* spp. (Figure 17.2), *Chilodonella* spp., *Ichthyophthirius multifiliis* (white spot) and flagellates, such as *Ichthyobodo* spp., which are regularly found on wild and farmed eel. In higher numbers, they cause the eel to exhibit a scratching behaviour and to cease feeding, which occurs especially in farmed eel (Haenen, unpublished).

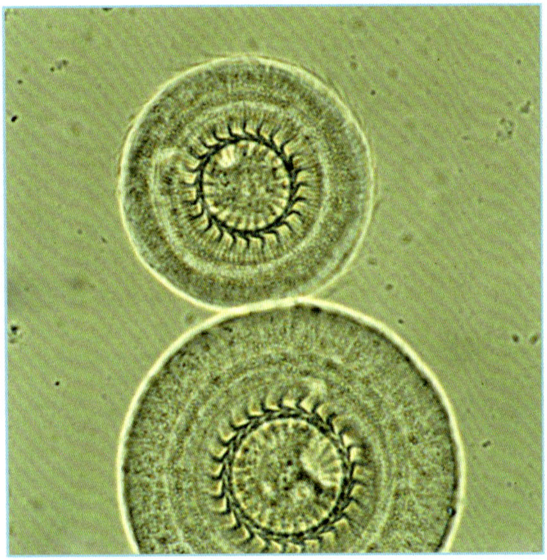

Figure 17.2 *Trichodina* spp., a parasitic ciliate from the skin of fish. Photograh: Haenen et al., WBVR.

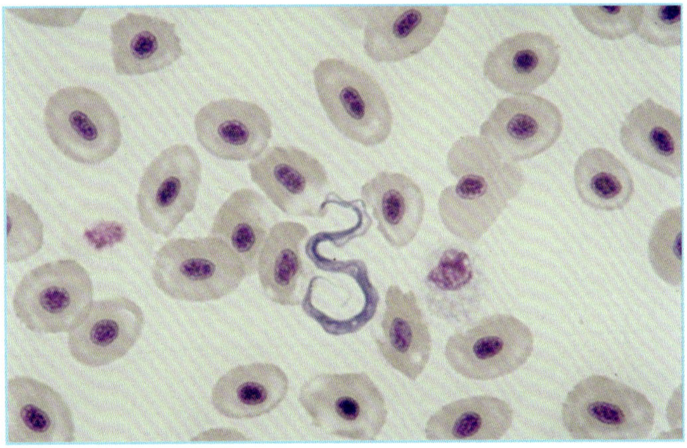

Figure 17.3 *Trypanosoma* in a Giemsa-stained eel blood smear. Photograph: Haenen et al., WBVR.

Wild eels may also suffer from a blood flagellate infection by *Trypanosoma granulosum*, which causes so-called 'sleeping disease' (Aguilar et al., 2005; Haenen et al., 2010; Zintl et al., 1997). Although this parasite was found in a high proportion of populations without clear signs of infection, in high parasite intensities only the eels at the bottom of the pond show this sleeping behaviour, and only elicit some activity when they are touched, after which they sink again to the bottom. Ultimately, they may die from this lethargy through anorexia. Diagnosis is carried out through Giemsa-stained blood smears (Figure 17.3). The trypanosome species infecting eel is not the same species that infects warm-blooded animals or humans.

A single and interesting case diagnosed at WBVR was a severe *Myxidium giardi* infection in wild eels in 2005, from Lake IJsselmeer, the Netherlands (Dekker and Engelsma, 2005). The eels showed lethargy and irregular swimming, because the lumen of their lateral line was totally filled up with this myxosporean parasite (Figures 17.4 and 17.5).

Figure 17.4 Wild eel with serious *Myxidium giardi* infection showing congested openings of side line channel (head). Photograph: M. Engelsma et al., WBVR.

Figure 17.5 Wild eel with serious *Myxidium giardi* infection showing congested openings of side line. Photography: M. Engelsma et al., WBVR.

Metazoan parasites

Regarding metazoan parasites, eels may carry mainly representatives of trematodes, nematodes, cestodes and acanthocephalans. In this chapter, we only deal with the most significant species, as diagnosed in the last 30 years in Western Europe, some trematodes and a nematode.

The external monogenic trematode gill worms *P. anguillae* (Yin and Sproston, 1948) and *P. bini* (Kikuchi, 1929) originate from East Asia, where they occur on Japanese eel. In 1977, they were introduced into Europe via imported eels from a farm in Russia (NOBANIS, 2018). Around 1985, they were detected in European eel in Denmark, *P. bini* (Buchmann et al., 1987; Køie, 1988, 1991), from that point the parasite rapidly spread all over Europe (Aguilar et al., 2005; Jensen, 2010; Saraiva, 1995; Škoriková et al., 1996; Sobecka and Pilecka-Rapacz, 2003). The trematode feeds on gill tissue and blood of eels (Buchmann et al., 1987). In eel culture, infected eels stop feeding, and through damage to the gills secondary pathogens may further infect the eel. These parasites are considered one of the most serious problems in eel farming. Chemical therapies are not always effective, and apart from toxicity problems may induce resistance formation (Buchmann and Bjerregaard, 1990). Regarding wild eel, it has been questioned if this parasite would have a negative effect on silvering of eels and spawning.

A third introduced metazoan parasite, which was imported into Europe in the start of the 1980s is the swimbladder nematode *A. crassus* (Figures 17.6 and 17.7). Originally *A. crassus* was described in Japanese eel by Kuwahara et al. (1974), in which it caused mild lesions. In Western Europe, since its introduction (Van Banning and Haenen, 1990), it has spread rapidly from the Netherlands and Belgium to the whole of Europe and beyond. So far it has been found that eels can be infected from the moment they begin feeding on invertebrate intermediate hosts, mostly copepods (Moravec, 2013). The rapid spread of the parasite was intensified partly by the presence of many copepod species, which act as intermediate hosts, and various fish species that act as paratenic hosts, such as smelt (*Osmerus eperlanus*) (Haenen et al., 1994) and ruffe (*Gymnocephalus cernuus*) (Haenen and van Banning, 1990). These fish species were proven to be paratenic hosts for *A. crassus* larvae by artificial infection experiments using infected swimbladders fed to eels (Haenen and van Banning, 1991). The lifecycle of the parasite is short, about 6 weeks, depending on the water temperature (De Charleroy et al., 1990; Haenen et al., 1989). The parasite caused an acute reaction in wild European eels, with many parasites and severe fibrosis of the swimbladder (Haenen et al., 1996; Molnár et al., 1993; Peters and Hartmann, 1986; Van Banning and Haenen, 1990). Gradually, from the mid-1990s, milder lesions and a lower number of parasites

Figure 17.6 *Anguillicoloides crassus* adult female parasite. Photograph: Haenen et al., WBVR.

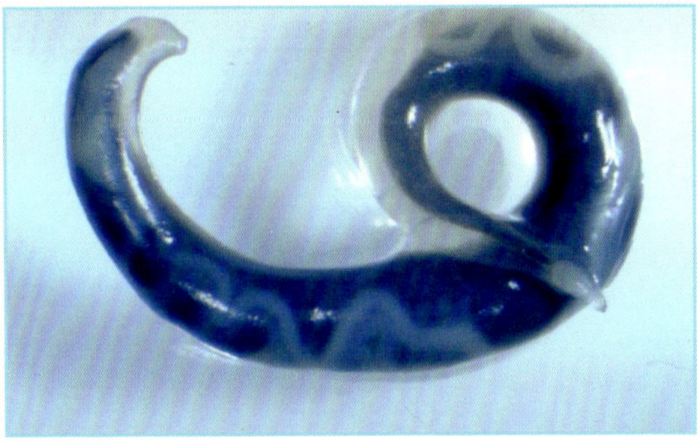

Figure 17.7 *Anguillicoloides crassus* adult male parasite. Photograph: Haenen et al., WBVR.

were seen in wild eels in Western Europe. The effect of the parasite on eels was shown to be related to the number of mature worms and migrating larvae (Székely et al., 2009). More than 120 papers have been published on this nematode, with a few reviews (Haenen, 1995; Moravec, 2006; Jakob et al., 2016).

There was concern as to whether the parasite would have a negative effect on eel silvering and spawning (Fazio et al., 2012). It was seen that the swimbladder function in silver eels decreased (Würtz et al., 1996), their swimming speed reduced (Palstra et al., 2007), their fitness and silvering was negatively affected (Fazio et al., 2012), and their stress tolerance was reduced (Kennedy, 2007; Kirk, 2003). The ability of these eels to reach the Sargasso Sea and spawn was brought in into question. This is still unknown. No reports exist on silver eels in the Sargasso Sea with this parasite. The decline of the European eel, both in Europe and on a global scale, is thought to be caused by several factors (Dekker, 2004, 2016), of which infection with *A. crassus* may be but one.

Bacteria

Eels may get secondary infections and disease from various opportunistic bacteria, especially when they have a low resistance, and if they are injured, for example, following netting. Bacteria like *Aeromonas hydrophila*, *A. sobria* and *A. veronii* occur, and may cause skin lesions or, in severe cases, full sepsis. However, there is only a short list of species of more or less specific eel pathogenic bacteria (ICES, 2015): *Pseudomonas anguilliseptica*, *Edwardsiella tarda*, certain strains of *Vibrio vulnificus*, *V. anguillarum* and atypical *Aeromonas salmonicida*.

Pseudomonas anguilliseptica

P. anguilliseptica occurs especially in brackish water and has been diagnosed since the 1980s (Stewart et al., 1983) in elvers in Northern Europe, and since then also in elvers originating from Southwest Europe (Haenen and Davidse, 2001). The disease has been evident since the 1970s in Japanese eel and known as 'sekiten-byo' (Wakabayashi and Egusa, 1972). Diseased elvers show a whitish slime layer and a congested posterior kidney. The bacterium has also been isolated, and is related to 'winter disease' in sea bream (*Sparus aurata*) (Doménech et al., 1997, 1999). The bacterium can be isolated almost pure from skin and internal organs (Haenen and Davidse, 2001). *Pseudomonas* is a cold-water bacterium, and by exception can be cured in eel farms by increasing the water temperature up to 27°C for 2 weeks, without using any antibiotics (Stewart et al., 1983).

Edwardsiella tarda

E. tarda is a pathogenic bacterium of eel, which may also cause disease in more than 18 other warm-water fish species (Hawke and Khoo, 2004), and in warm-blooded animals, such as swine (Owens et al., 1974). It occasionally causes severe disease in eels, and is known to be potentially zoonotic, that is, it may cause local inflammations in humans (Haenen et al., 2013). In eels, disease may occur with the following symptoms: bloody gut content, granuloma-like view of organs, congested spleen and deep skin ulcers (Figure 17.8) (ICES, 2015).

Figure 17.8 Wild eel with a deep ulcer by the bacterium *Edwardsiella tarda*. Photograph: Haenen et al., WBVR.

Vibrio vulnificus

V. vulnificus is present in seawater (Veenstra et al., 1993), and especially infects eels in brackish and marine water at a water temperature of 25°C or higher. In eel, *V. vulnificus* may cause severe disease, with haemorrhages in organs and skin, deep ulcerative skin lesions, abscesses, a congested spleen, ascites and high mortality (Amaro, et al., 1992; Esteve and Alcaide, 2009; Haenen et al., 2014b). The bacterium may also be zoonotic (Austin, 2010; Dijkstra et al., 2009). It has also often been isolated from eels in Spain (Esteve and Alcaide, 2009; Fouz et al., 2006). Haenen et al. (2014b) describe*s* 23 *V. vulnificus* cases from 1996 to 2008 at various eel farms in the Netherlands, including a zoonotic strain from an eel farmer (Dijkstra et al., 2009).

Vibrio (Listonella) anguillarum

V. (Listonella) anguillarum may cause 'red eel pest' in eel (Austin and Austin, 1999; Haenen et al., 2014a), and severe disease in various other fish species including salmonids and several marine fishes, especially in the Mediterranean (Larsen et al., 1994; Toranzo, 2004; Toranzo and Barja, 1990). It causes a haemorrhagic disease. Diseased eels may show haemorrhages in the skin and internal organs, popeye, ascites and high mortality. The causes of the fish disease *V. anguillarum* has been reviewed by Hickey and Lee (2017).

Aeromonas salmonicida

A. salmonicida atypical is originally known to cause carp erythrodermatitis in cyprinid fish (Bootsma et al., 1977). In the recent decades it appeared that strains of *A. salmonicida* atypical could also infect other cultured fish species like turbot (*Scophthalmus maximus*), and eels (Haenen et al., unpublished). Eels show deep skin lesions with a haemorrhagic border. The bacterium is best isolated from under the border of the lesion. The typical strain *Aeromonas salmonicida* subsp. *salmonicida* has also been isolated from eel (Pedersen et al., 1999), and it is questioned whether *A. salmonicida* is really a serious eel pathogen.

Contact-zoonotic bacteria

Potential contact-zoonotic bacteria, which have been isolated from eel and from the species described above, include certain genotypes of *V. vulnificus*, such as the sequence type (ST)112 and *Edwardsiella tarda*. In humans, who are severely immunocompromised, *V. vulnificus* ST112 may, after entering the blood stream via skin injuries or spine/punctures which get acutely infected, cause acute necrotic fasciitis, with risk of death within 48 hours if no acute multi-antibiotic therapy and surgery is practised (Dijkstra et al., 2009; Oliver, 2005). *E. tarda* may, especially in immunocompromised patients, cause necrotic skin lesions, gastroenteritis, and in severe cases, septicaemia (Gilman et al., 1971; Haenen et al., 2013).

Good hygiene practices are important in the avoidance of zoonosis and transfer of disease. Wash hands and skin with water and soap after contact with live fish, fish-holding water and fish materials that have had contact with fish-holding water. In cases of human disease, those infected should tell their medical doctor that they have had contact with fish to enhance the chances of an early and accurate diagnosis.

Viruses

There are three major viruses described in eels in Europe: Eel Virus European (EVE) a double stranded RNA virus; Eel Virus European X (EVEX, a rhabdovirus [RNA]); and anguillid herpesvirus 1 (AngHV1, formerly called Herpes virus anguillae, HVA), a herpes virus (DNA). The review by Van Beurden et al. (2012) presents details on these viruses. An outline is given below.

No therapy against viruses is available, and no eel viral vaccines exist. However, in the case of viral diseases in fish, water temperature may be changed to a suboptimal range to impede the replication of the virus. This may stop symptoms, and allow the eels to cope with the infection and recover, although the virus may still be present in the water system.

AngHV1

Anguillid herpesvirus 1 is an alloherpesvirus infecting eel species. Its old name is HVA, first isolated in Japanese and European eels farmed in recirculation systems at 30°C in Japan (Chen et al., 1982; Sano et al., 1990). Because the virus does not replicate in standard used fish cells, it was not until 1998 that it was first isolated in Europe, from farmed eels in the Netherlands by Davidse et al. (1999), using Eel Kidney-1 (EK-1) cells, kindly provided by M. Yoshimizu. Subsequently, PCR and qPCR tests were developed (Rijsewijk et al., 2005; Van Beurden et al., 2015). In the meantime, the virus has been detected all over Europe: in the UK, in 2009–10 in wild eels, mortalities at 17–19°C (Armitage et al., 2014); in Germany (Jakob et al., 2009); in Greece (Varvarigos et al., 2011); and Poland (Kempter et al., 2014; Tuan et al., 2016), among other countries. Research has provided insight into epidemiology, pathogenesis, pathogenicity, persistence, serology, molecular characteristics of the virus and infection, immunology (IL-10) and diagnostics (Haenen et al., 2002; Hangalapura et al., 2007; Van Beurden et al., 2010, 2011a, 2011b; Van Nieuwstadt et al., 2001). The virus is now endemic in wild eels in Europe, and causes disease (Figure 17.9) in wild eels when stress factors are in place, and eels are immunocompromised.

Table 17.1 The three major eel viruses in Europe: causing clinics in eel, especially when eels are stressed, at an optimum water temperature.

Virus name	Virus group	Clinics in eel	Water temp (optimum)	Reference
AngHV-1	Alloherpes virus (DNA)	Decreased growth rates, haemorrhages in skin and fins (Figure 17.9), pale liver, haemorrhages in liver, hepatic congestion, enteritis, enlarged spleen, pink fat, ascites, mortality may reach 50% at eel farms. Sometimes no disease signs.	26°C	Sano et al.,1990; Ueno et al., 1992; Davidse et al., 1999; Lee et al., 1999a; Haenen et al., 2002, 2010; Van Beurden, 2012
EVE (Eel Virus European)	Double stranded RNA (biRNA)	Abnormal shape of the trunk, congestion of the skin, fins and gills (Figure 17.10), enlarged kidney, necrosis or petechial haemorrhages in the liver and gill disease, mortality may reach 100%. Sometimes no disease signs.	20°C	Sano, 1976; Sano et al., 1981; Lee et al., 1999b; Okamoto et al., 1983 Van Beurden et al., 2012
EVEX (Eel Virus European X)	Anguillid rhabdovirus in genus Perhabdo-virus (bullet-shaped) RNA	Haemorrhagic disease: congestion of skin and fins, head downwards, haemorrhages, anaemia, variable mortality.	15°C	Sano, 1976; Van Ginneken et al., 2004; Haenen et al., 2009; Van Beurden et al., 2012; Galinier et al., 2012; Stone et al., 2013; Bellec et al., 2014

Figure 17.9 Eel with AngHV1, showing fine skin haemorrhages in the head region. Photograph: Haenen et al., WBVR.

EVE

Eel Virus European (EVE) was known in Japan in the 1970s (being transferred through the import of European elvers in 1968 probably) to cause branchionephritis in Japanese eels in culture ponds, with up to 50% mortality (Chen et al., 1985; Egusa, 1970; Hedrick et al., 1983;

Figure 17.10 Eel with an EVE infection, showing a congested skin and haemorrhages in skin and fins. Photograph: Haenen et al., WBVR.

Sano, 1976, Sano et al., 1981). The virus is antigenically related to IPNV (infectious pancreatic necrosis virus) strain Ab (Lee et al., 2001). In Europe, EVE has been found in diseased farmed eels since the 1990s (Bucke, 1981; Jørgensen et al., 1994). In the Netherlands, this virus has been detected subsequently in diseased eels (Figure 17.10), and identified as IPNV Ab or IPNV VR-299 (Haenen et al., own findings). Overall, EVE was detected in Japanese eel in Japan and Taiwan, in European eel in Japan, France, the UK, Germany, Denmark, the Netherlands and Greece, and in American eel (*A. rostrata*) in the USA (Van Beurden et al., 2012). Through research some insight on EVE in epidemiology, pathogenicity, pathogenesis and diagnostics has been gathered (Castric and Chastel, 1980; Van Beurden et al., 2012; Ueno et al., 1984).

EVEX

Eel Virus European X (EVEX) is a vesiculo-like rhabdovirus, a bullet shaped virus (Sano, 1976). The virus's optimum temperature is 15°C. It has been isolated from farmed and small and large wild European eels (Van Beurden et al., 2012; Van Ginneken et al., 2004, 2005). The virus was suspected to be a factor for the decline of eels, but since 2004 it has not been detected in silver eel (Haenen et al., own findings), so this is not conclusive. The virus is often detected in combination with an AngHV1 infection. Research has focused much on epidemiology, pathogenicity, molecular characterization, evolutionary dynamics and diagnostics (Amilhat et al., 2014; Bellec et al., 2014; Caruso et al., 2014; Galinier et al., 2012; Nishimura et al., 1981; Stone et al., 2013; Van Beurden et al., 2011b).

Regarding diagnostic samples, the WBVR Fish Diseases Laboratory has detected over 80 cases of AngHV1 (in elvers and bigger eels), over 50 cases of EVE (in farmed eel only), over 20 cases of EVEX (in wild and farmed eel) and about 10 cases of other viruses, since 1990 (Van Beurden et al., 2012).

Other eel viruses found so far in Europe include a reo-like virus, an orthomyxovirus (Van Beurden et al., 2012), a picornavirus (Fichtner et al., 2013) and a betanodavirus (Bandín et al., 2014). Furthermore, short-finned eel (*A. australis*) ranavirus (SERV) has been detected in imported eels in Italy (Subramaniam et al., 2016). Their significance is not yet clear (ICES, 2015). We need to continue to be cautious regarding the importation of eels from East Asia, as Viral Endothelial Cell Necrosis of Japanese eel (VECNE) (an adenovirus) has been detected in diseased Japanese eels in Japan (Ono et al., 2007) and polyomavirus (AmPyV) related to VECNE from diseased marbled eels (*A. marmorata*) was detected in Taiwan (Wen et al., 2015).

Double and triple infections

Through a joint monitoring project of the WBVR, with the Dutch Sport Fisheries Association OVB, and the German Institute LANUV (Landesamt für Natur, Umwelt und Verbraucherschutz), wild caught silver eel from River Rhine (Germany and the Netherlands) and Lake IJsselmeer (the Netherlands) were screened for health. Their condition, necropsy, blood smear, parasites, bacteria, viruses loads were investigated. *A. crassus* was detected in 50–90% of the eels, *Trypanosoma* in over 70%, AngHV1 herpesvirus anguillae in 13–44%. Most silver eels showed an abnormal haematology, although they were in good condition (Haenen et al., 2010). It was questioned, whether AngHV1 could also be partly responsible for the decline in wild eel, given the fact that silver eels migrate to the Sargasso Sea, and experience a lot of stress, passing through water temperatures that are sometimes ideal for AngHV1 (20–26°C). However, during an expedition to the Sargasso Sea, in 2014, Danish researchers did not detect any virus in the cell cultures from a low number of wild caught leptocephali of European eels (N.J. Olesen, 2018, personal communication), so this was not proven.

Various double and triple infections were reported in mostly farmed eel (Van Beurden et al., 2012), with EVE and AngHV1 in the Netherlands (Haenen et al., 2002) (Figure 17.11), and in Greece (Varvarigos et al., 2011), and double infections with EVE and EVEX in Germany (Ahne and Thomsen, 1985) and in Italy (van Ginneken et al., 2004). Further combinations of viruses with bacteria, like *E. tarda* or *P. anguilliseptica* have been diagnosed in the Netherlands in diseased farmed eel (Haenen et al., unpublished).

Interaction of wild and farmed eel

As all eel farming is based on using stock gathered from the wild, that is, elvers or yellow eels, this means there is an interaction between wild and farmed eels at the start of the farming phase. Additionally, latterly, to help to solve the problem of the decline of wild eel populations, farmed larger, often silver, eels are restocked from North Western European eel farms into the wild, to help restore the wild silver eel spawning populations, and allow them to migrate to the

Figure 17.11 Eel suffering a double viral infection with AngHV1 and EVE, showing petechial haemorrhaging in the skin and a congested skin. Photograph: Haenen et al., WBVR.

Table 17.2 Characteristics of wild and farmed eels in their habitat.

Wild eel	Farmed eel
Open water	Indoor recirculation (or ponds)
Start with glass eels	Start with glass eels
Seasonal temperatures	At constant temp of 25°C
Wild feed	Fish eggs/pelleted feed
Low densities	High densities
Natural water quality	Artificial water quality

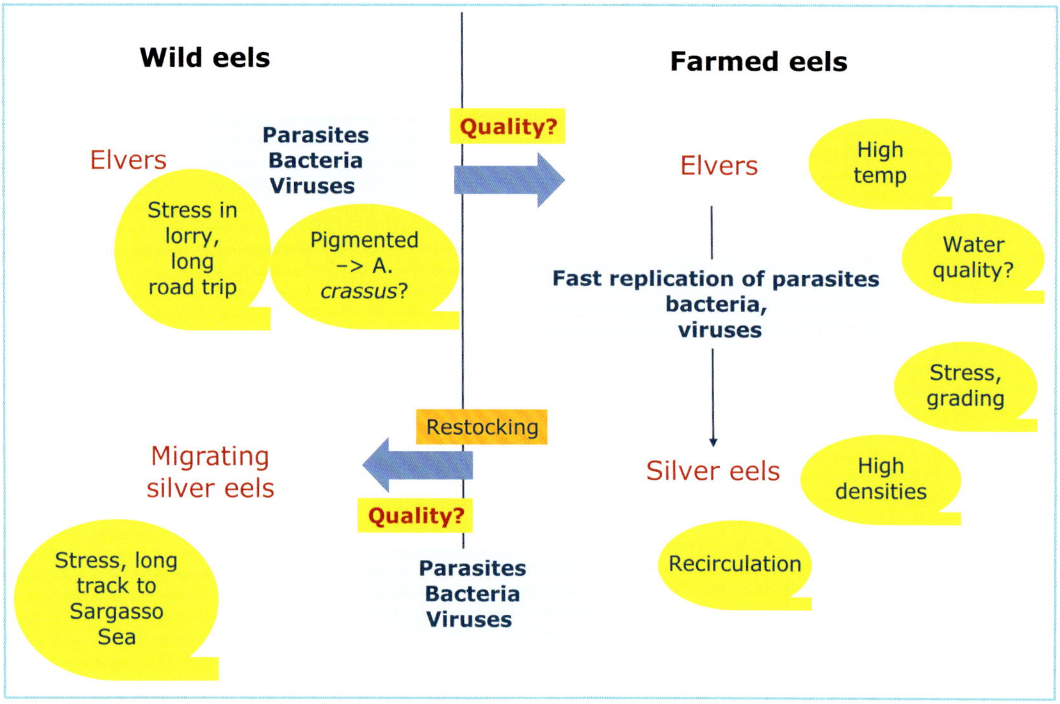

Figure 17.12 Interactions between wild and farmed eel and their pathogens. Credit: Haenen, WBVR.

Sargasso Sea to spawn. Table 17.2 gives an idea about the fundamental differences between wild and farmed eels in their habitat. Wild eels are unprotected, whereas farmed eels are living in a protected environment, although they both face risks of diseases affecting eels: the wild eels in their natural environment with natural pathogens, and the farmed eels through their high densities, the recirculation system and high temperatures that are permissive for growth of certain pathogens.

Therefore, transferring eels from the wild to eel farms, and vice versa, brings risks regarding pathogen and disease transfer. Figure 17.12 depicts some aspects involved in the risk of pathogen and disease transfer between wild and farmed eels and vice versa in a later phase.

Apart from the parasites, bacteria and viruses that are transferred in every movement of eels, we also need to be aware in case eels have been treated with antibiotics at an eel farm. Antimicrobial resistance might also be transferred into wild eels, and thereby into the environment. Risk assessment related to these transfers is recommended, using the ICES (2015) guidelines.

Conclusion

There are various eel pathogens that cause disease in wild and farmed eels in Europe, partly introduced by global imports from East Asia. Some pathogens may cause severe eel disease and mortality, depending on the virulence of the pathogen and the immunity of the eel, and sometimes the water temperature. Most diseases at eel farms start with management problems. Sometimes eels have double or triple infections (viruses and bacteria, and/or parasites). In Dutch and German wild eels, predominantly AngHV1, *Trypanosoma* and *A. crassus* were found. These pathogens seem to have no direct relationship with the decline in the eel population, but do show the potential to severely hamper eel health. Alongside the use of wild eels as stocking animals in eel farming, currently there is an even closer interaction of wild and farmed eel through restocking of larger (silver) eels into the wild. It is understood that eel diseases may affect eel reproduction. However, to what extent is not clear. It is recommended that to avoid introduction and spread of (new) pathogens transfers of eels should be limited to the region in which they are produced and that transfers do not move eels across international borders. Additionally, that the eel quality is checked before restocking (Haenen et al., 2012; ICES, 2015). We need to develop and disseminate our knowledge of eel diseases and to take precautions against eel disease risks in order to help restore the eel population.

Acknowledgements

I thank my team at the WBVR Fish Diseases Laboratory and all external co-workers for many years of good eel diagnosis and research: Aart Davidse, Françoise Keuzenkamp, Betty van Gelderen, Grytsje Wybenga, Sonja Dijkstra, Peter van Tulden, Ton van Nieuwstadt, Frans Rijsewijk, Rob Zwart, Basav Hangalapura, Frans Wagenaar, Lisette Ruuls, Henk Dorsman, Johan Meijer, Sonja Kerkhoff, Ineke Roozenburg, Michal Voorbergen, Steven van Beurden, Monique Oosterhof, Rianka Vloet, Mirjam Boonstra, and Marc Engelsma, and all involved students, Bart Jansen, Annette Boerlage, Annemieke Botter, Madelon Willemsen, Marco de Mik and Jurjen Tellingen. I thank the Ministry of Agriculture, Nature and Food Quality for their financial support.

Bibliography

Aguilar, A., Álvarez, M.F., Leiro J.M. and Sanmartín, M.L. (2005) Parasite populations of the European eel (*Anguilla anguilla* L.) in the Rivers Ulla and Tea (Galicia, northwest Spain). *Aquaculture* 249, 85–94.

Ahne, W. and Thomsen, I. (1985) The existence of three different viral agents in tumor bearing European eel (*Anguilla anguilla*). *Zentralblatt für Veterinärmedizin. Reihe B.* 32, 228–235.

Amaro, C., Biosca, E.G., Esteve, C., Fouz, B. and A. E. Toranzo (1992) Comparative study of phenotypic and virulence properties in *Vibrio vulnificus* biotypes 1 and 2 obtained from a European eel farm experiencing mortalities. *Diseases of Aquatic Organisms* 13, 29–35.

Amilhat, E., Fazio, G., Simon, G., Manetti, M., Pari, S., Delahaut, L., Furrugio, H., Lecomte-Finiger, R., Sasal, P. and Faliex, E. (2014) Silver European eels health in Mediterranean habitats. *Ecology of Freshwater Fish* 23(1), 49–64.

Armitage, J., Hewlett, N.R., Twigg, M., Lewin, N.C., Reading, A.J., Williams, C.F., Aprahamian, M., Way, K., Feist, S.W. and Peeler, E.J. (2014). Detection of Herpesvirus anguillae during two mortality investigations of wild European eel in England: implications for fishery management. *Fisheries Management and Ecology* 21(1), 1–12.

Audenaert, V., Huyse, T., Goemans, G., Belpaire, C. and Volckaert, F.A.M. (2003) Spatio-temporal dynamics of the nematode *Anguillicola crassus* in Flanders, Belgium. *Diseases of Aquatic Organisms* 56, 223–233.

Austin, B. (2010) Vibrios as causal agents of zoonoses. *Veterinary Microbiology* 140, 310–317.

Austin, B. and Austin, D.A. (1999) *Bacterial Fish Pathogens: Disease in Farmed and Wild Fish,* 3rd edn. Springer-Verlag, Berlin.

Bandín, I., Souto, S., Cutrín, J.M., López-Vázquez, C., Olveira, J.G., Esteve, C., Alcaide, E. and Dopazo, C.P. (2014) Presence of viruses in wild eels *Anguilla anguilla* L, from the Albufera Lake (Spain). *Journal of Fish Diseases* 37, 597–607.

Bellec, L., Cabon, J., Bergmann, S.M., Deboisseson, C., Engelsma, M., Haenen, O., Morin, T., Olesen, N., Schütze, H., Toffan, A., Way, K. and Bigarré, L. (2014) Evolutionary dynamics and genetic diversity from three genes of Anguillid Rhabdovirus. *Journal of General Virology* 95, 2390–2401.

Bootsma, R., Fijan, N. and Blommaert, J. (1977) Isolation and preliminary identification of the causative agent of carp erythrodermatitis. *Veterinary Archives* 47(6), 291–302.

Borgsteede, F.H.M., Haenen, O.L.M., De Bree, J. and Lisitsina, O.I. (1999) Parasitic infections of European eel (*Anguilla anguilla* L.) in the Netherlands. *Helminthologia* 36(4), 251–260.

Buchmann, K. and Bjerregaard, J. (1990) Comparative efficacies of commercially available benzimidazoles against *Pseudodactylogyrus* infestations in eels. *Diseases of Aquatic Organisms* 9, 117–120.

Buchmann, K., Mellergaard, S. and Køie, M (1987) *Pseudodactylogyrus* infections in eel: a review. *Diseases of Aquatic Organisms* 3, 51–57.

Bucke, D. (1981) A note on IPNV from eels in the UK. *Bulletin of the European Association of Fish Pathologists* 1, 12.

Caruso, C., Peletto, S., Gustinelli, A., Arsieni, P., Mordenti, O., Modesto, P., Acutis, P.L., Masoero, L., Fioravanti, M.L. and Prearo, M. (2014) Detection of a phylogenetically divergent eel virus European X (EVEX) isolate in European eels (*Anguilla anguilla*) farmed in experimental tanks in Italy. *Aquaculture* 434, 115–120.

Casselman, J.M. and Cairns, D.K. (2003) Worldwide decline of eel resources necessitates immediate action: Québec declaration of concern. *Fisheries* 28(12), 28–30.

Castric, J. and Chastel, C. (1980) Isolation and characterization attempts of three viruses from European eel, *Anguilla anguilla*: preliminary results. *Annales de l'Institut Pasteur Virology* 131, 435–448.

Chen, S.N., Ueno, Y. and Kou, G.H. (1982) A cell line derived from Japanese eel (*Anguilla japonica)* kidney. *Proceedings of the National Science Council, Republic of China,* Part B 6, 93–100.

Chen, S.N., Kou, G.H., Hedrick, R.P. and Fryer, J.L. (1985) The occurrence of viral infections of fish in Taiwan. In: Ellis, A.E. (ed.) *Fish and Shellfish Pathology.* Academic Press, London, pp. 313–319.

Dalsgaard, A., Frimodt-Moller, N., Bruun, B., Hoi, L. and Larsen, J.L. (1996) Clinical manifestation and molecular epidemiology of *Vibrio vulnificus* infections in Denmark. *European Journal of Clinical Microbiology & Infectious Diseases* 15, 227–232.

Dalsgaard, I., Høi, L., Siebeling, R.J. and Dalsgaard, A. (1999) Indole-positive *Vibrio vulnificus* isolated from disease outbreaks on a Danish eel farm. *Diseases of Aquatic Organisms* 35(3), 187–194.

Davidse, A., Haenen, O.L.M., Dijkstra, S.G., Van Nieuwstadt, A.P., Van der Vorst, T.J.K., Wagenaar, F. and Wellenberg, G.J. (1999) First isolation of herpesvirus of eel (Herpesvirus anguillae) in diseased European eel (*Anguilla anguilla* L.) in Europe. *Bulletin of the European Association of Fish Pathologists* 19(4), 137–141.

De Charleroy, D., Grisez, L., Thomas, K., Belpaire, C. and Ollevier, F. (1990) The life cycle of *Anguillicola crassus*. *Diseases of Aquatic Organisms* 8, 77–84.

Dekker, W. (2004) Slipping through our hands – population dynamics of the European eel. PhD thesis, University of Amsterdam, Amsterdam. Available at: http://www.diadfish.org/doc/these_2004/dekker_thesis_eel.pdf (last accessed 8 December 2018).

Dekker, W. (2016) Management of the eel is slipping through our hands! Distribute control and orchestrate national protection. *ICES Journal of Marine Science* 73, 10, 2442–2452.

Dekker, W. and Engelsma, M.Y. (2005) Zieke aal op het IJsselmeer een raadsel: geen gevaar voor de volksgezondheid. *Visserijnieuws* 25 June (in Dutch).

Dijkstra, A., Van Ingen, J., Lubbert, P.H.W., Haenen, O.L.M. and Möller, A.V.M. (2009) Fasciitis necroticans ten gevolge van een *Vibrio vulnificus* infectie in een palingkwekerij. *Nederlands Tijdschrift voor Geneeskunde* 153, B157 (in Dutch).

Doménech, A., Fernández-Garayzábal, J.F., García, J.A., Cutuli, M.T., Blanco, M., Gibello, A., Moreno, M.A. and Domínguez, L. (1999) Association of *Pseudomonas anguilliseptica* with 'winter disease' in sea bream, *Sparus aurata* L. *Journal of Fish Diseases* 22, 69–71.

Doménech, A., Fernández-Garayzábal, J.F., Lawson, P., García, J.A., Cutuli, M.T., Blanco, M., Gibello, A., Moreno, M.A., Collins, M.D. and Domínguez, L. (1997) Winter disease outbreak in sea-bream (*Sparus aurata*) associated with *Pseudomonas anguilliseptica* infection. *Aquaculture* 156, 317–326.

Egusa, S. (1970) Branchionephritis prevailed among eel populations in farm-pond in the warm water 1969–1970. *Fish Pathology* 5, 51–56 (in Japanese).

Esteve, C. and Alcaide, E. (2009) Influence of diseases on the wild eel stock: the case of Albufera Lake. *Aquaculture* 289, 143–149.

Fazio, G., Sasal, P., Mouahid, G., Lecomte-Finiger, R. and Moné, H. (2012) Swim bladder nematodes (*Anguillicoloides crassus*) disturb silvering in European eels (*Anguilla anguilla*). *Journal of Parasitology* 98, 695–705.

Fichtner, D., Philipps, A., Groth, M., Schmidt-Posthaus, H., Granzow, H., Dauber, M., Platzer, M., Bergmann, S.M., Schrudde, D., Sauerbrei, A. and Zell, R. (2013) Characterization of a novel picornavirus isolate from a diseased European eel (*Anguilla anguilla*). *Journal of Virology* 87, 10895–10899.

Fouz, B., Larsen, J.L. and Amaro, C. (2006) *Vibrio vulnificus* serovar A: an emerging pathogen in European anguilliculture. *Journal of Fish Diseases* 29(5), 285–91.

Galinier, R., Van Beurde, S., Amilhat, E., Castric, J., Schoehn, G., Verneau, O., Fazio, G., Allienne, J.F., Engelsma, M., Sasal, P. and Faliex, E. (2012) Complete genomic sequence and taxonomic position of eel virus European X (EVEX), a rhabdovirus of European eel. *Virus Research* 166, 1–12.

Gilman, R.H., Madasamy, M., Gan, E., Mariappan, M., Davis, C.E. and Kyser, K.A. (1971) *Edwardsiella tarda* in jungle diarrhoea and a possible association with *Entamoeba histolytica*. *Southeast Asian Journal of Tropical Medicine and Public Health* 2(2), 186–9.

Haenen, O.L.M. (1995) *Anguillicola crassus* (Nematoda, Dracunculoidea) infections of European eel (*Anguilla anguilla)* in the Netherlands: epidemiology, pathogenesis and pathobiology. Thesis, LU Wageningen, the Netherlands.

Haenen, O.L.M. and Van Banning, P. (1990) Detection of larvae of *Anguillicola crassus* (an eel swimbladder nematode) in freshwater fish species. *Aquaculture* 87, 103–109.

Haenen, O.L.M. and Van Banning, P. (1991) Experimental transmission of *Anguillicola crassus* (Nematoda, Dracunculoidea) larvae from infected prey fish to the eel *Anguilla anguilla. Aquaculture* 92, 115–119.

Haenen, O.L.M., Van Banning, P. and Dekker, W. (1994) Infection of eel *Anguilla anguilla* L. and smelt *Osmerus eperlanus* L. with *Anguillicola crassus* (Nematoda, Dracunculoidea) in the Netherlands from 1986–1992. *Aquaculture* 126, 219–229.

Haenen, O.L.M. and Davidse, A. (2001) First isolation and pathogenicity studies with *Pseudomonas anguilliseptica* from diseased European eel *Anguilla anguilla* (L.) in The Netherlands. *Aquaculture* 196, 27–36.

Haenen, O.L.M., Van Wijngaarden, T.A.M., Van der Heijden, M.H.T., Höglund, J., Cornelissen, J.B.J.W., Van Leengoed, L.A.M., Borgsteede, F.H.M. and van Muiswinkel, W.B. (1996) Effects of experimental infections with different doses of *Anguillicola crassus* (Nematoda, Dracunculoidea) on European *eel* (*Anguilla anguilla). Aquaculture* 141, 41–57.

Haenen, O.L.M., Grisez, L., De Charleroy, D., Belpaire C. and Ollevier, F. (1989) Experimentally induced infections of European eel *Anguilla anguilla* with *Anguillicola crassus* (Nematoda, Dracunculoidea) and subsequent migration of larvae. *Diseases of Aquatic Organisms* 7, 97–101.

Haenen, O.L.M., Dijkstra, S.G., Van Tulden, P.W., Davidse, A., Van Nieuwstadt, A.P., Wagenaar, F. and Wellenberg, G.J. (2002) Herpesvirus anguillae (HVA) isolations from disease outbreaks in cultured European eel, *Anguilla anguilla* in the Netherlands since 1996. *Bulletin of the European Association of Fish Pathologists* 22(4), 247–257.

Haenen, O.L.M., Van Ginneken, V.J., Engelsma, M.Y. and van den Thillart, G.E.E.J.M. (2009) Impact of eel viruses on recruitment of European eel. In: Van Den Thilart, D., Dufour, S. and Rankin, J.C. (eds) *Spawning Migration of the European Eel.* Springer, Berlin, pp. 387–400.

Haenen, O.L.M., Lehmann, J., Engelsma, M.Y., Stürenberg, F.-J., Roozenburg, I., Kerkhoff, S. and Klein Breteler, J. (2010) The health status of European silver eels, *Anguilla anguilla*, in the Dutch River Rhine watershed and Lake IJsselmeer. *Aquaculture* 309, 15–24.

Haenen, O.L.M., Mladineo, I., Konecny, R., Yoshimizu, M., Groman, D., Muñoz, P., Saraiva, A., Bergmann, S.M. and Van Beurden, S.J. (2012) Diseases of eels in an international perspective: Workshop on Eel Diseases at the 15th International Conference on Diseases of Fish and Shellfish, Split, Croatia, 2011. *Bulletin of the European Association of Fish Pathologists* 32, 109–115.

Haenen, O.L.M., Evans, J.J. and Berthe, F. (2013) Bacterial infections from aquatic species: Potential for and prevention of contact zoonoses. *Scientific and Technical Review of the Office International des Epizooties* 32(2), 497–507.

Haenen, O.L.M., Fouz, B., Amaro, C., Isern, M.M., Mikkelsen, H., Zrnčić, S., Travers, M.A., Renault, T., Wardle, R., Hellström, A. and Dalsgaard, I. (2014a) Vibriosis in aquaculture. *Bulletin of the European Association of Fish Pathologists* 34(4), 138–148.

Haenen, O.L.M., Van Zanten, E., Jansen, R., Roozenburg, I., Engelsma, M.Y., Dijkstra, A., Boers, S.A., Voorbergen-Laarman, M. and Möller, A.V.M. (2014b). *Vibrio vulnificus* outbreaks in Dutch eel farms since 1996, strain diversity and impact. *Diseases of Aquatic Organisms* 108, 201–209.

Hangalapura, B.N., Zwart, R., Engelsma, M.Y. and Haenen, O.L.M. (2007) Pathogenesis of Herpesvirus anguillae (HVA) in juvenile European eel *Anguilla anguilla* after infection by bath immersion. *Diseases of Aquatic Organisms* 78, 13–22.

Hawke, J.P. and Khoo, L.H. (2004) Infectious diseases. In: Tucker, C.S. and Hargreaves, J.A. (eds) *Biology and Culture of Channel Catfish*. Elsevier, Amsterdam, pp. 387–443.

Hedrick, R.P., Okamoto, N., Sano, T. and Fryer, J.L. (1983) Biochemical characterization of Eel Virus European. *Journal of General Virology* 64(Pt 6), 1421–1426.

Hermida, M., Saraiva, A., and Cruz, C. (2008) Metazoan parasite community of a European eel (*Anguilla anguilla)* population from an estuary in Portugal. *Bulletin of the European Association of Fish Pathologists* 28, 35–40.

Hickey, M. and Lee, J.-L. (2017) A comprehensive review of *Vibrio* (*Listonella) anguillarum*: ecology, pathology and prevention. *Reviews in Aquaculture* 10, 585–610.

ICES (2015) Report of the Workshop of a Planning Group on the Monitoring of Eel Quality under the subject 'Development of standardized and harmonized protocols for the estimation of eel quality' (WKPGMEQ) ICES. 20–22 January 2015, Brussels, Belgium.

Jakob, E., Neuhaus, H., Steinhagen, D., Luckhardt, B. and Hanel, R. (2009) Monitoring of Herpesvirus anguillae (HVA) infections in European eel, *Anguilla anguilla* (L.), in northern Germany. *Journal of Fish Diseases* 32, 557–561.

Jakob, E., Walter, T. and Hanel, R. (2016) A checklist of the protozoan and metazoan parasites of European eel (*Anguilla anguilla)*: checklist of *Anguilla anguilla* parasites. *Journal of Applied Ichthyology* https://doi.org/10.1111/j.1439-0426.2009.01345.x.

Jensen, K.R. (2010) NOBANIS – Invasive Alien Species Fact Sheet – *Pseudodactylogyrus anguillae* and *Pseudodactylogyrus bini*. Available at: https://www.nobanis.org/globalassets/speciesinfo/p/pseudodactylogyrus-anguillae/pseudodactylogyrus-anguillae.pdf (accessed 2 December 2018).

Jørgensen, P., Castric, J., Hill, B., Ljungberg, O. and De Kinkelin, P. (1994) The occurrence of virus infections in elvers and eels (*Anguilla anguilla)* in Europe with particular reference to VHSV and IHNV. *Aquaculture* 123, 11–19.

Kempter, J., Hofsoe, P., Panicz, R. and Bergmann, S. M. (2014) First detection of anguillid herpesvirus 1 (AngHV-1) in European eel (*Anguilla anguilla)* and American eel (*Anguilla rostrata)* in Poland. *Bulletin of the European Association of Fish Pathologists* 34, 87–94.

Kennedy, C.R. (2007) The pathogenic helminth parasites of eels. Review. *Journal of Fish Diseases* 30, 319–334.

Kikuchi, H. (1929) Two new species of Japanese trematodes belonging to Gyrodactylidae. *Annotationes Zoologicae Japonensis* 12, 175–186.

Kirk, R. (2003) The impact of *Anguillicola crassus* on European eels. *Fisheries Management and Ecology* 10, 385–394.

Køie, M. (1988) Parasites in European eel *Anguilla anguilla* (L.) from Danish freshwater, brackish and marine localities. *Ophelia* 29, 93–118.

Køie, M. (1991) Swimbladder nematodes (*Anguillicola* spp.) and gill monogeneans (*Pseudodactylogyrus* spp.) parasitic on the European eel (*Anguilla anguilla). Journal du Conseil International pour l'Exploration de la Mer* 47, 391–398.

Kristmundsson, A. and Helgason, S. (2007) Parasite communities of eels *Anguilla anguilla* in freshwater and marine habitats in Iceland in comparison with other parasite communities of eels in Europe. *Folia Parasitologica* 54, 141–153.

Kuwahara, A., Niimi, A. and Itakaki, H. (1974) Studies on a nematode parasitic in the air bladder of the eel. 1. Description of *Anguillicola crassa* n. sp. (Philometridea, Anguillicolidae). *Japanese Journal of Parasitology* 23(5), 275–279 [Japanese summary].

Larsen, J.L, Pedersen, K. and Dalsgaard, I. (1994) *Vibrio anguillarum* serovars associated with vibriosis in fish. *Journal of Fish Diseases* 17, 259–267.

Lee, N.S., Kobayashi, J. and Miyazaki, T. (1999a) Gill filament necrosis in farmed Japanese eels, *Anguilla japonica* (Temminck and Schlegel), infected with Herpesvirus anguillae. *Journal of Fish Diseases* 22, 457–463.

Lee, N.S., Nomura, Y. and Miyazaki, T. (1999b) Gill lamellar pillar cell necrosis, a new birnavirus disease in Japanese eels. *Diseases of Aquatic Organisms* 37, 13–21.

Lee, N.S., Miyata, M. and Miyazaki, T. (2001) Genome sequence of a VP2/NS junction region of pillar cell necrosis virus (PCNV) in cultured Japanese eel *Anguilla japonica*. *Diseases of Aquatic Organisms* 44, 179–182.

Lewin, N.C., Reading, A.J., Hockley, F.A., Cable, J., Turnbull, J.T., Davies, G.D., Feist, S.W., Evans, D., Belpaire, C., Walker, A.M., Way, K., Kemp, P.S., Aprahamian, M.W., Haenen, O.L.M., Hoole, D., Dufour, S., Hickley, P. and Williams, C.F. (2014) Silver service: a standard protocol for eel health examinations. Poster presentation, European Association of Fish Pathologists (EAFP), UK Meeting, Keele, 15–16 September 2014.

Liewes, E. and Haenen, O.L.M. (1993) The Veterinary approach to eels. In: Brown, L. (ed.) *Aquaculture for Veterinarians: Fish husbandry and Medicine*. Pergamon Press, Oxford, pp. 311–326.

Maillo, P.A., Vich, M.A., Salvadó, H., Marqués, A. and Gracia, M.P. (2005) Parasites of *Anguilla anguilla* (L.) from three coastal lagoons of the River Ebro delta (Western Mediterranean). *Acta Parasitologica* 50(2), 156–160.

Molnár, K., Baska, F., Csaba, G., Glávits, R. and Székely, C. (1993) Pathological and histopathological studies of the swimbladder of eels *Anguilla anguilla* infected by *Anguillicola crassus* (Nematoda: Dracunculoidea). *Diseases of Aquatic Organisms* 15, 41–50.

Moravec, F. (2006) *Dracunculoid and Anguillicoloid Nematodes Parasitic in Vertebrates*. Academia, Prague.

Moravec, F. (2013) *Parasitic Nematodes of Freshwater Fishes of Europe*. Academia, Prague.

Nishimura, T., Toba, M., Ban, F., Okamoto, N. and Sano, T. (1981) Eel rhabdovirus, EVA, EVEX and their infectivity to fishes. *Fish Pathology* 15, 173–184.

NOBANIS (2018) Available at: https://www.nobanis.org/marine-identification-key/parasites/pseudodactylogyrus-anguillae/ (last accessed 7 December 2018).

Okamoto, N., Sano, T., Hedrick, R.P. and Fryer, J.L. (1983) Antigenic relationships of selected strains of infectious pancreatic necrosis virus and European eel virus. *Journal of Fish Diseases* 6, 19–25.

Olesen, N.J. (2018) Facts about the Danish Eel Expedition 2014. Available at: https://www.fishsec.org/app/uploads/2014/05/Fact-sheet-Danish-Eel-Expedition-2014.pdf (last accessed 15 May 2019).

Oliver, J.D. (2005) Wound infections caused by *Vibrio vulnificus* and other marine bacteria. *Epidemiology and Infection* 133, 383–391.

Ono, S.-I., Wakabayashi, K. and Nagai, A. (2007) Isolation of the virus causing viral endothelial cell necrosis of eel from cultured Japanese eel *Anguilla japonica*. *Fish Pathology* 42, 191–200.

Owens, D.R., Nelson, S.L. and Addison, J.B. (1974) Isolation of *Edwardsiella tarda* from swine. *Applied Microbiology* 27, 703–705.

Palstra, A.P., Heppene, D.F.M., Van Ginneken, V.J.T., Szekely, C. and Van den Thillart, G. (2007) Swimming performance of silver eels is severely impaired by the swim-bladder parasite *Anguillicola crassus*. *Journal of Experimental Marine Biology and Ecology* 352, 244–256.

PARA-SITE (2018) http://parasite.org.au/para-site/introduction/introduction-essay.html (accessed 7 December 2018).

Pedersen, K., Garcia, J.A. and Larsen, J.L. (1999) *Aeromonas salmonicida* a potential pathogen in modern eel (*Anguilla anguilla*) farming? *Bulletin of the European Association of Fish Pathologists* 19(3), 127–129.

Peters, G. and Hartmann, F. (1986) *Anguillicola*, a parasitic nematode of the swimbladder spreading among eel populations in Europe. *Diseases of Aquatic Organisms* 1, 229–230.

Rijsewijk, F., Pritz-Verschuren, S., Kerkhoff, S., Botter, A., Willemsen, M., Van Nieuwstadt, A. and O. Haenen (2005) Development of a polymerase chain reaction for the detection of Anguillid herpesvirus DNA in eels based on the herpesvirus DNA polymerase gene. *Journal of Virological Methods* 124, 87–94.

Rolbiecki, L. and Rokicki, J. (2006) Parasite fauna of the eel, *Anguilla anguilla* (Linnaeus, 1758), from the Polish part of the Vistula Lagoon. *Wiadomości Parazytologiczne* 52(2), 115–119.

Sano, T. (1976) Viral diseases of cultured fishes in Japan. *Fish Pathology* 10, 221–226.

Sano, T., Okamoto, N. and Nishimura, T. (1981) A new viral epizootic of *Anguilla japonica* Temminck and Schlegel. *Journal of Fish Diseases* 4, 127–139.

Sano, M., Fukuda, H. and Sano, T. (1990) Isolation and characterization of a new herpesvirus from eel. In: Perkins, F.O. and Cheng, T.C. (eds) *Pathology in Marine Sciences.* Academic Press, San Diego, CA.

Saraiva, A. (1995) *Pseudodactylogyrus anguillae* (YIN and SPROSTON, 1948) GUSSEV 1965 and *P. bini* (KIKUCHI, 1929) GUSSEV 1965 (Monogenea: Monopisthocotylea) in Portugal. *Bulletin of the European Association of Fish Pathologists* 15, 81–83.

Škoriková, B., Scholz, T. and Moravec, F. (1996) Spreading of introduced monogeneans *Pseudodactylogyrus anguilla* and *P. bini* among eel populations in the Czech Republic. *Folia Parasitologica* 43, 155–156.

Snieszko, S.F. (1974) The effects of environmental stress on outbreaks of infectious diseases of fishes. *Journal of Fish Diseases* 6(2), 197–208.

Stewart, D.J., Woldemariam, K., Dear, G. and Mochaba, F.M. (1983) An outbreak of 'Sekiten-byo' among cultured European eels, *Anguilla anguilla* L., in Scotland. *Journal of Fish Diseases* 6, 75–76.

Sobecka, E. and Pilecka-Rapacz, M. (2003) *Pseudodactylogyrus anguillae* (Yin et Sproston, 1948) Gussev, 1965 and *P. bini* (Kikuchi, 1929) Gussev, 1965 (Monogenea: Pseudodactylogyridae) on gills of European eel, *Anguilla anguilla* (Linnaeus, 1758) ascending rivers of the Pomeranian coast, Poland. *Acta Ichthyologica et Piscatoria* 33(2), 137–143.

Stone, D., Kerr, R.C., Hughes, M., Radford, A.D. and Darby, A.C. (2013) Characterization of the genomes of four putative vesiculoviruses: tench rhabdovirus, grass carp rhabdovirus, perch rhabdovirus and eel rhabdovirus European X. *Archives of Virology*, DOI 10.1007/s00705–013–1711-x.

Subramaniam, K., Toffan, A., Cappellozza, E., Steckler, N. K., Olesen, N. J., Ariel, E., and Waltzek, T. B. (2016) Genomic Sequence of a Ranavirus Isolated from Short-Finned Eel (*Anguilla australis*). *Genome Announcements* 4(4), e00843–16.

Sures, B. and Streit, B. (2001) Eel parasite diversity and intermediate host abundance in the River Rhine, Germany. *Parasitology* 123, 185–191.

Székely, C., Palstra, A., Molnár, K. and Van den Thillart, G. (2009) Impact of the swim-bladder parasite on the health and performance of European eels. In: van de Thillart, G., Dufour, S. and Rankin, J.C. (eds) *Spawning Migration of the European Eel.* Springer, Amsterdam, pp. 201–226.

Tesch, F.-W. (2003) *The Eel*, 5th edn. Blackwell Science, Oxford.

Toranzo, A.E. (2004) Report about fish bacterial diseases. *Options Mediterraneennes*: Serie B. *Etudes et Recherches*, 49, 49–89.

Toranzo, A. E. and Barja, J.L.(1990) A review of the taxonomy and sero epizootiology of *Vibrio anguillarum,* with special reference to aquaculture in the northwest of Spain. *Diseases of Aquatic Organisms* 9, 73–82.

Tuan, N., Kempter, J. and Panicz, R. (2016) Presence of Herpesvirus anguillae (AngHV1) DNA in the native ichthyofauna of North-Western Poland, Electronic *Journal of Polish Agricultural Universities* 19(4). Available at: http://www.ejpau.media.pl/volume19/issue4/art-06.html

Ueno, Y., Chen, S.N., Kou, G.H., Hedrick, R.P. and Fryer, J.L. (1984) Characterization of virus isolated from Japanese eel (*Anguilla japonica)* with nephroblastoma. *Bulletin of the Institute of Zoology, Academia Sinica* 23, 47–55.

Ueno, Y., Kitao, T., Chen, S.-N., Aoki, T. and Kou, G.-H. (1992) Characterization of a herpes-like virus isolated from cultured Japanese eels in Taiwan. *Fish Pathology* 27, 7–17.

Van Banning, P. and Haenen, O.L.M. (1990) Effects of the swimbladder nematode *Anguillicola crassus* in wild and farmed eel *Anguilla anguilla*. In: Perkins, F.O. and Cheng, T.C. (eds) *Pathology in Marine Science*. Academic Press, New York, pp. 317–330.

Van Beurden, S.J., Bossers, A., Voorbergen-Laarman, M.H.A., Haenen, O.L.M., Peters, S., Abma-Henkens, M.H.C., Peeters, B.H.P., Rottier, P.J.M. and Engelsma, M.Y. (2010) Complete genome sequence and taxonomic position of anguillid herpesvirus 1. *Journal of General Virology* 91, 880–887.

Van Beurden, S.J., Leroy, B., Wattiez, R., Haenen, O.L.M., Boeren, S.A., Vervoort, J.J.M., Peeters, B.P.H., Rottier, P.J.M., Engelsma, M.Y. and Vanderplasschen, A.F. (2011a) Identification and localization of the structural proteins of anguillid herpesvirus 1. *Veterinary Research* 42, 105.

Van Beurden, S. J., Voorbergen-Laarman, M.A., Roozenburg, I., Boerlage, A.S., Haenen, O.L.M. and Engelsma, M.Y. (2011b) Development and validation of a two-step real-time RT-PCR for the detection of eel virus European X in European eel, *Anguilla anguilla*. Journal of Virological Methods 171, 352–259.

Van Beurden, S., Engelsma, M.Y., Roozenburg, I., Voorbergen-Laarman, M.A., Van Tulden, P.W., Kerkhoff, S., Van Nieuwstadt, A.P., Davidse, A. and Haenen, O.L.M. (2012) Viral diseases of wild and farmed European eel (*Anguilla anguilla)* with particular reference to the Netherlands. *Diseases of Aquatic Organisms* 101, 69–86.

Van Beurden, S.J., Voorbergen-Laarman, M.A., Roozenburg, I., Van Tellingen, J., Haenen, O.L.M. and Engelsma, M.Y. (2015) Development and validation of a real-time PCR assay for the detection of anguillid herpesvirus 1. *Journal of Fish Diseases* doi:10.1111/jfd.12330, 1–10.

Van Ginneken, V., Haenen, O., Coldenhoff, K., Willemze, R., Antonissen, E., Van Tulden, P., Dijkstra, S., Wagenaar, F. and Van den Thillart, G. (2004) Presence of virus infections in eel species from various geographic regions. Short communication, *Bulletin of the European Association of Fish Pathologists* 24(5), 268–272.

Van Ginneken, V., Ballieux, B., Willemze, R., Coldenhoff, K., Lentjes, E., Antonissen, E., Haenen, O. and Van den Thillart, G. (2005) Hematology patterns of migrating European eels and the role of EVEX virus. *Comparative Biochemistry and Physiology Part C* 140, 97–102.

Van Nieuwstadt, A.P., Dijkstra, S.G., Haenen, O.L. (2001) Persistence of herpesvirus of eel Herpesvirus anguillae in farmed European eel *Anguilla anguilla*. *Diseases of Aquatic Organisms* 45, 103–107.

Varvarigos, P., Vendramin, N., Cappellozza, E. and Bovo, G. (2011) First confirmation of Herpes Virus Anguillae (HVA) and Infectious Pancreatic Necrosis (IPN) virus infecting European eel *Anguilla anguilla* farmed in Greece. *Bulletin of the European Association of Fish Pathologists* 31, 101–111.

Veenstra, J., Rietra, P.J.G.M., Coster, J.M., Stoutenbeek, C.P., Ter Laak E.A, Haenen, O.L.M., de Gier H.H.W. and Dirks-Go, S. (1993) Human *Vibrio vulnificus* infections and environmental isolates in the Netherlands. *Aquaculture and Fisheries Management* 24, 119–122.

Wakabayashi, H. and Egusa, S. (1972) Characteristics of a *Pseudomonas* sp. from an epizootic of pond-cultured eels (*Anguilla japonica). Bulletin of the Japanese Society of Scientific Fisheries* 38(6), 577–587.

Wen, C.-M., Chen, M., Wang, C., Liu, P. and Nan, F.-H. (2015) Isolation of a novel polyomavirus, related to Japanese eel endothelial cell-infecting virus, from marbled eels, *Anguilla marmorata* (Quoy & Gaimard). *Journal of Fish Diseases* 39, 10.1111/jfd.12423.

Würtz, J., Taraschewski, H. and Pelster, B. (1996) Changes in gas composition in the swimbladder of the European eel (*Anguilla anguilla)* infected with *Anguillicola crassus* (Nematoda). *Parasitology* 112, 233–238.

Yin, W.-Y. and Sproston, N. G. (1948) Studies on the monogenetic trematodes of China. Parts 1–5. *Sinensia* 19, 57–85.

Zintl, A., Poole, W.R., Voorheis, H.P. and Holland, C.V. (1997) Naturally occurring *Trypanosoma granulosum* infections in the European eel, *Anguilla anguilla* L. from County Mayo, western Ireland. *Journal of Fish Diseases* 20, 333–341.

CHAPTER 18

Eel toxicology: a review of exposure to trace elements and results of research on mother–egg transfer, accumulation and effect

Annalisa Zaccaroni, Antonio Casalini, Albamaria Parmeggiani, Rebecca Andreini, Pietro Emmanuele and Oliviero Mordenti

Department of Veterinary Medical Sciences, University of Bologna, Bologna, Italy

Introduction

Owing to its commercial, scientific and naturalistic value, the decline in European eel (*Anguilla anguilla*) numbers generated significant international interest. Its presence in water bodies has dramatically reduced (Feunteun, 2002) making it a key species for biodiversity conservation. The continuous decline of the eel population, mainly in the last 30 years, lead the IUCN to classify the species as critically endangered (Jacoby and Gollock, 2014). The prolonged decline in recruitment throughout Europe, and in capture of adults in many European and Italian water bodies, lead to increased concern (Dekker, 2003a, Dekker, 2003b, Dekker, 2003c, Dekker, 2004, ICES, 2014) and many researchers began to investigate the environmental and physiological reasons for the reduction (van den Thillart and Dufour, 2009). Generally speaking, it is considered that the causes of this decline are many and combined they increase the risk of extinction. The potential reasons for the decline of the European eel encompass both natural and anthropogenic causes.

Among natural causes, the most probable hypothesis is the change in oceanic currents, which seems to be correlated to the reduction in recruitment (Knights, 2003, Friedland et al., 2007). Reproduction itself can be considered as an issue for conservation: being a semelparous species, adults die just after spawning, while larvae are highly predated during their way back to Europe (Dollerup and Graver, 1985). Among the other natural causes of decline is predation by fish-eating birds and parasites. In the last 20 years, parasite infestations have been more and more frequently reported, mainly by *Anguillicola crassus*, whose adults live in the swim bladder, thus impacting the swimming efficiency of silver eels, reducing their migratory and reproductive capacity (Székely et al., 2004, Székely et al., 2005).

Among anthropogenic impacts excessive commercial fishing plays a major role (Ciccotti, 2006). Catching of glass eel for export, to fulfil the growing demand from Eastern countries,

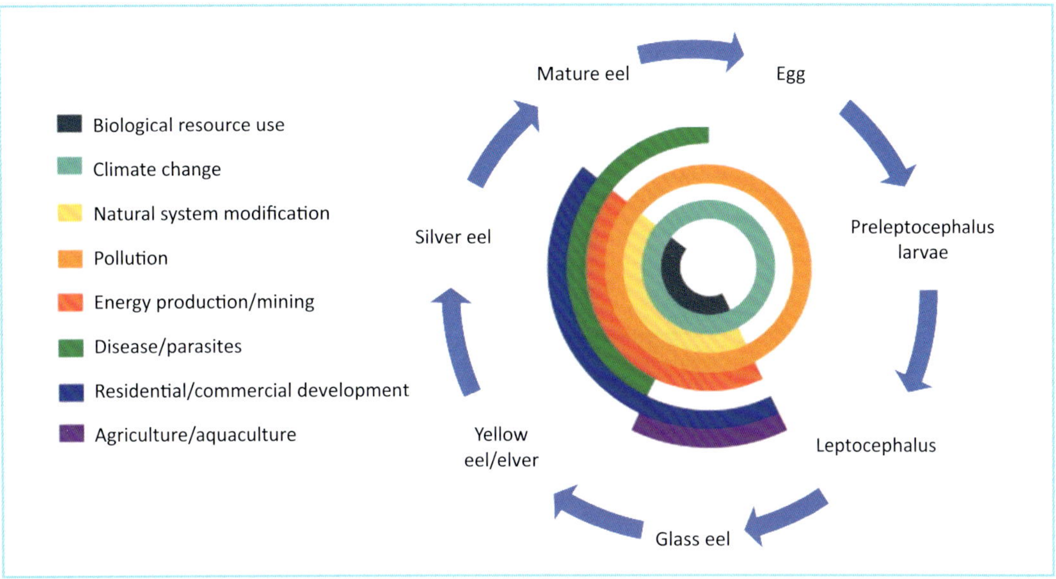

Figure 18.1 Relationship between natural cause of decline and anthropogenic impacts on life cycle of the European eel.

is also very important; the real number of eels collected for export is at the moment difficult to quantify, hence the real impact of this practice is at the moment not so understood. During the second half of the 20th century, reclamation of inland wet areas and depletion of coastal areas lead to the loss of important habitat, reducing the extent of estuarine and wet areas (ICES, 2002, Feunteun, 2002). Finally, the construction of artificial dams and the cementification of rivers made the upriver migration of eels almost impossible, reducing their chances of reaching growing areas (Figure 18.1).

More recently, it has become more and more evident that contaminants may have a dramatic impact on the eel population, not only by direct toxicity (acute and chronic) but also by degrading their habitat and reducing available and suitable areas for the species (Palstra et al., 2006). Most of the available literature on contaminant exposure and its effect in eels focuses on organic contaminants, that is, polychlorinated biphenyls (PCBs), dichloro-diphenyl-trichloroethane (DDT), Polybrominated diphenyl ethers (PBDEs) and dioxins (Belpaire et al., 2016). Less research has focused on the impact of heavy metals, and most studies represent a monitoring of environmental pollution using this species.

In this chapter, we provide a short review of the current knowledge on exposure to, and toxic effects of, trace elements and report on some preliminary studies performed at the Department of Veterinary Medical Sciences of the University of Bologna, Italy, concerning the modification in trace element content following a standardized induced reproduction protocol (Mordenti et al., 2012; Mordenti et al., 2018) and the eventual mother–egg transfer of metals.

Bioaccumulation and the effect of contaminants in eels, with a focus on trace elements

One essential contributor to the decline of the European eel is the poor quality of broodstock and their progeny due to the impact that pollution by cumulative compounds can have on reproductive success of the species.

Contaminants can induce impairment of immune, reproductive, nervous and endocrine systems, and the deriving effects have an impact on different levels of organization, that is, cells, organs, individuals, up to populations.

The effect induced varies depending on the function of the contaminant, on the combination with other chemicals (which can potentiate their effect), on the development stage and on environmental parameters, that is, temperature, salinity, oxygen and pH (Lawrence and Elliott, 2008, Geeraerts and Belpaire, 2010). One of the most important aspects is the degree of accumulation of pollutants in the body of fish, as this can modulate their (mainly) chronic effects. Some of the ecological and physiological traits of eels make this species particularly prone to bioaccumulation of pollutants. Eels live in or nearby sediments, which are usually richer in contaminants than the water column, and are apparently more resistant than other species to degraded waters and can thus live in highly polluted areas. The species is considered as a large, long-lived top predator, with a higher muscle lipid content compared to other fish species, thus experiencing a noticeable degree of biomagnification. Females reproduce only once in their life at an old age (from 5 to 20 years) and cannot reduce their body burden of contaminants with multiple reproduction (Goemans and Belpaire, 2007).

Eels spend most of their life (up to 5.9 years for males and 8.7 years for females) (Geeraerts and Belpaire, 2010, Vøllestad, 1992) in fresh waters, which are usually more contaminated than sea water, and can consequently accumulate great amounts of xenobiotics. There is an age-dependent increase in contaminant burden, and the peak is reached at migration. At this moment of their life the lipid content in muscle ranges between 27% and 29%, and this can allow the bioaccumulation of many lipophilic compounds, with a reduced clearance (Geeraerts and Belpaire, 2010, van Ginneken et al., 2009, Larsson et al., 1991, de Boer and Hagel, 1994, de Boer et al., 1994, Tulonen and Vuorinen, 1996, Knights, 1997, Daverat et al., 2006).

All these characteristics make eels good environmental pollution sentinels, and they are used mainly, as previously stated, for PCBs, pesticides and, to a lesser extent, metals (Geeraerts and Belpaire, 2010, de Boer and Hagel, 1994, de Boer et al., 1994, Goemans and Belpaire, 2007, Geeraerts et al., 2008). These studies show that in many European areas contaminants in eel tissues reached levels well above the thresholds defined for human consumption (Geeraerts and Belpaire, 2010, European Commission, 2008, European Commission, 2006, Buslé, 1991, Buslé, 1990, FAO, 2006). In 2007 a European database (EEQD) was created to collect information on the exposure of eels to pollution in 14 European countries (FAO, 2007, FAO, 2009). The analysis of the data collected provides evidence of persistently high levels of contamination, also a noticeable variability among and within studied areas due to local anthropogenic pollution (FAO, 2009).

The acute effects of pollutants, leading to mortality events, are connected to point source releases of toxicants at high concentration, that is, spills on a local scale or leakage/discharge from pesticide producing plants (Belpaire et al., 2016). These events are relatively rare, and can be easily identified and recognized, but are also hard to resolve, due to the acute toxicity and general rapidity in the onset of symptoms and fatalities.

Most common episodes are linked to chronic, low-level exposure to contaminants. Induced effects include infection of wounds, tissue damage, impairment of immune, endocrine, reproductive system, and so on; these effects have been correlated to pesticides, PCBs, polycyclic aromatic hydrocarbons, phenols and trace elements.

Among trace elements, most studied are cadmium (Cd), lead (Pb) and mercury (Hg), which often reach the maximum residue level as set by the EU (European Commission, 2008, European Commission, 2006). Few studies consider other elements, which are important contaminants for aquatic environments or essential elements, that is, arsenic (As), chromium (Cr), nickel (Ni), copper (Cu) and zinc (Zn).

Cadmium

Cadmium is a largely diffused heavy metal, because it is used in pesticides, fertilizers and issued by industrial emissions. It stays in the ground and in the water, and it can be accumulated along the food chain. Cadmium has a long half-life, up to 20 years, as it is chelated by metallothionein, that accumulates in the kidney and thereby reduces its elimination. It has been proven to induce branchial alterations, with swelling of primary and secondary lamellae, epithelial hypertrophy and melanism of branchial blood vessels. Epithelial exfoliation and fusion of lamellae have been observed, as well as hepatic lesions (Geeraerts and Belpaire, 2010, Gony, 1990). This element also proved to inhibit in vitro in a dose-dependent manner the activity of Na^+/K^+-ATPase and carbonic anhydrase in the intestine and gills of eels: this is suggestive of an impairment of both acid-base equilibrium and of osmoregulation in the species (Geeraerts and Belpaire, 2010, Lionetto et al., 1998). Additionally, when combined with mercury, it can compromise the intracellular transduction pathway involved in adrenergic control of glucose metabolism and impair endocrine control of hepatic function (Geeraerts and Belpaire, 2010, Fabbri et al., 2003).

The effect of cadmium was also studied in glass eels in terms of alteration of the mitochondrial respiratory chain (as mitochondrial gene expression), cellular response and oxidative stress. Cadmium seems to be also able to mimic the effects of hypoxia at the gene level, reducing the expression of those genes involved in the respiratory chain and in antioxidant action (Geeraerts and Belpaire, 2010, Pierron et al., 2007b).

Cadmium appears to modulate lipidic storage in yellow eels: after 1 month exposure to 5 μg/; Cd, an alteration in the activity and the level of expression of enzymes involved in lipolysis and lipogenesis, with a final dose-dependent reduction in lipid deposition in muscular tissue potentially compromising the reproductive success of animals (Geeraerts and Belpaire, 2010, Pierron et al., 2007a).

Another characteristic of cadmium is its capacity to accumulate in different tissues, for example, the kidney, liver, gills and intestine. Tissues tropism is not only dependent on the chemical characteristics of cadmium, but also on the physiological status of eels: following a 30 day exposure period to 15 μg/l Cd prior to maturation, a redistribution of the elements to kidney and gonads when eels were moved to clean water to reach maturation, was associated with a net reduction in gills and intestinal concentration (Geeraerts and Belpaire, 2010,

Pierron et al., 2008). The cumulative effect of endocrine disruptors is paradoxical in relation to reproduction: in females, low concentrations can have an oestrogenic effect, in fact Cd^{2+} stimulates ovarian progesterone synthesis, increasing the conversion of cholesterol in pregnenolone (Henson and Chedrese, 2004) stimulating also the steroidogenesis in pituitary, the vitellogenesis and premature ovarian growth. In the testis of *Salvelinus fontinalis*, after 24 hours exposure to 10 µg/l, tissue lesions could be seen; whereas after 7 days exposure at 15 µg/l Cd, the result was necrosis, bleeding and destruction of lobules (Sangalang and O'Halloran, 1973, Sangalang and O'Halloran, 1972).

Lead

Lead is the most diffused polluting metal in the world, due to its past and present use in many sectors: siderurgic and electronic industry, hunting and fishing, in metal alloys for food conservation, paintings and water pipes. Given its readiness to disintegrate, particles and ions can diffuse in the environment, especially in soils and water. Toxicity is caused by ionic forms; organic ones are not dangerous.

Inorganic lead was shown to modify blood composition in elvers exposed for 30 days at a concentration of 300 µg/l Pb, increasing the number of lymphocytes and the plasma lactate level, due to induced oxidative stress. Lymphocytes are key components of immune defences, transforming into antibody-producing cells; an important increase in their number can represent a marker of the effect of contaminants, lead included, on the immune system (Geeraerts and Belpaire, 2010, Santos and Hall, 1990). Being a known neurotoxicant, lead can also impair the behaviour of eels, as shown in glass eels: the concentration of the metal was higher in those glass eels that showed reduced activity, with respect to those more actively swimming. These body concentration differences were explained due to different environmental exposures of the eels in the collection areas and/or different accumulation levels when exposed to micro-pollutants during their estuarine migration (Geeraerts and Belpaire, 2010, Du Colombier et al., 2007).

Lead acting as an endocrine disruptor has a bimodal profile: at low concentrations, it stimulates steroids (cortisol, testosterone or 17β estradiol) synthesis, while it has an inhibitory effect at high concentrations. Contamination by lead causes an inhibition of granulosa cells of follicles, while in males lead induces damage similar to that of cadmium: an exposure of 15 mg/l Pb causes a reduction of spermatic activity, dissolution of germ cells, necrosis and vasodilatation (Srivastava, 1987).

Mercury

Mercury has a wide distribution in soils and waters all over the world. It is largely used in UV lamps (using mercury vapours), thermometers and also in batteries. Metallic mercury is toxic only when vapours are inhaled, while inorganic and organic compounds, that is, methylmercury or dimethylmercury, are more toxic and are both largely diffused in waters. When in water, inorganic mercury is anaerobically transformed by sediment microflora into methyl- and dimethylmercury, which is highly lipophilic and bioaccumulates and biomagnifies along food chains. Once absorbed by aquatic organisms, it diffuses throughout the whole body, and can easily pass the blood–brain barrier and the placenta, exerting its toxic effect on the brain and the foetus egg. Mercury concentrations in tissues of fish species decreases. Following the rank muscle > liver > intestine > spleen > brain > gonads, depending of the content of functional, cysteine-rich protein, with high affinity for mercury, of each compartment (Geeraerts and Belpaire, 2010, Golovanova, 2008).

Mercury shows a dose-dependent immunotoxic effect: low doses induce a reduction in haematopoiesis and enzymatic activity, and the toxic effects increase till the death of immune cells. Mercury exhibites a biphasic action on the immune system of fish, with low doses acting as stimulants and high doses inhibiting it. Increased mitotic activity and intracellular calcium levels have been observed following exposure to low concentrations of mercury, while an elevated concentration suppresses DNA synthesis, induces a rapid calcium influx in the cell and impairs proteins phosphorilation (Geeraerts and Belpaire, 2010, Carlson and Zelikoff, 2008). The injection of mercury in eels induced micronuclei expression at a concentration of 1.7 μg/body weight Hg, at an expression rate of 2.64–2.35 micronuclei/1000 cells (Geeraerts and Belpaire, 2010, Sanchez-Galan et al., 2001).

Exposure to mercury for 180 days caused inhibition of gonadotropin secreting cells in the pituitary, while it induced thyroid stimulating hormone (TSH) overproduction in *Clarias batrachus* (Kirubagaran and Joy, 1989). Owing to inhibitory activity on gonadotropin secreting cells, mercury causes a decrement of steroidogenesis, hormone synthesis and fertility, causing a reduction of gonadal activity, and a reduction of seminiferous tubules and degeneration of interstitial cells in testes.

Arsenic

Arsenic is a metalloid largely used in pesticides. It is also a contaminant of carbon and after its combustion it can diffuse into the environment; arsenic is also bound to gold and lead and is released in the environment after metal extraction. Main concentrations are reported in water and food. Generally, in marine food webs, arsenic is bioconcentrated but not biomagnified, and the accumulations are large in creatures lower in food chains (Lindsay and Sanders, 1990, Francesconi and Edmonds, 1993, Francesconi and Edmonds, 1998, Eisler, 1988, Lam et al., 2004, Saeki et al., 2000).

Arsenic has various toxicological effects: it is a cancerogenic substance and, like endocrine disruptors, interferes with vitellogenesis, binding to oestrogenic receptors. A study by Shukla and Pandey (1984) showed that exposure to a concentration of 14 mg/l for 30 days in male *Colisa fasciatus* caused degeneration of testicular lobules and DNA damage. In females of this species, at the same concentration, there was an increase of intrafollicular space of oocytes (Shukla and Pandey, 1984).

Maes et al. (2005) assessed whether the genetics of yellow eels under natural conditions could be linked to fitness and pollution pressure. A significant negative correlation was observed between the level of several trace elements (arsenic, cadmium, chromium, copper, lead, mercury, nickel, selenium and zinc) and eel condition suggesting that they had an impact on the health of sub-adult eels. In general, there was a reduced genetic variability in strongly polluted eels, as well as a negative correlation between bioaccumulation level and allozymatic multi-locus heterozygosity.

Mother–egg transmission of trace elements and correlated effects

Trace elements, when considered as environmental pollutants, are highly dangerous compounds; they biodegrade slowly and can spread across vast areas once released to the environment. When trace elements reach water bodies, they can be taken up by aquatic living organisms and accumulate along food chains (Brown and Austin, 2012). Being part of the Earth's crust composition, they are to be considered as ubiquitous elements and can be found in all organisms, aquatic species included. Among the most toxic elements are the mercuric compounds, which play a major role in reproductive toxicity (Brown and Austin, 2012, Depew et al., 2012, Kim et al., 2016, Lesniewska et al., 2009, Pannetier et al., 2016). Trace elements can also adversely affect eel populations through their toxicity to developing embryos by accumulating in eggs during their maturation. Interestingly, the only element studied from this point of view is mercury, while no data is available for other elements.

As mentioned in previous sections, mercury has a deleterious impact on fish reproduction (Penglase et al., 2014, Scheuhammer et al., 2015). The toxic effects of mercury include inhibition of growth and of gonadic development (Baillon et al., 2015), impairment of endocrine balance and sex inversion (Crump and Trudeau, 2009, Depew et al., 2012). It can also reduce vitality and motility of sperm (Dietrich et al., 2010, Hatef et al., 2011) and the percentage of fertilized eggs, thus reducing fish fertility. This negative impact worsens gamete quality, leading to a reduced number of live larvae and an increase in larval deformities (Dietrich et al., 2010, Ismail and Yusof, 2011, Dong et al., 2016). The toxic effect is more severe in highly contaminated areas, and the increasing level of mercury pollution in the habitats of eels raises concerns around the reproductive efficiency of the species and their real chances of recovery (Brown and Austin, 2012). Being bioaccumulative and biomagnificating, mercury shows the highest concentrations in larger eels, accumulating in muscle and ovaries (Figure 18.2) (Esteve et al., 2012, Nowosad et al., 2018). Interestingly, there seems to be no transfer of mercury from muscle and ovaries to eggs, with a possible protective, still unknown mechanism of protection of progeny; indeed, while muscular and ovarian mercury concentrations drastically increase after spawning, mercury egg content was eight times lower than in ovaries and 78 times lower than in muscle (Nowosad et al., 2018).

In as yet unpublished research performed by our group at the Department of Veterinary Medical Sciences of University of Bologna, the mother–egg transfer of four trace elements (arsenic, cadmium, lead and mercury) has been studied, also taking into account the protocol applied for induced artificial reproduction (Mordenti et al., 2012, Mordenti et al., 2018).

Three experimental groups of wild females have been set, undergoing different hormonal treatments:

1. **control** (CTRL): females not treated for reproduction
2. **treated control** (CPE): females induced with increasing doses of carp pituitary extract (CPE).
3. **implant** (IMP): females induced with CPE treatment and implant of 17-metiltestosterone (17-MT).

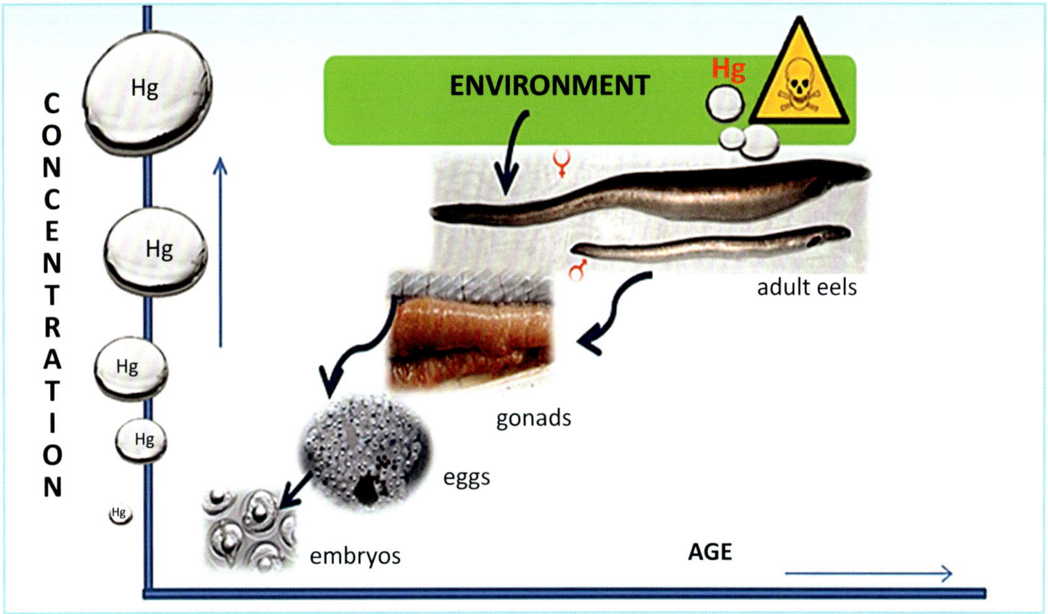

Figure 18.2 Vertical transmission of mercury in European eel.

Cutaneous mucus, blood and eggs (for CPE and IMP groups only, as CTRL did not reproduce) were analysed for arsenic, cadmium, lead and mercury (Table 18.1).

When mucus is considered, the lowest concentrations were observed for arsenic in the CPE group and the highest in the CTRL group. Arsenic mucous excretion seems to be related to reproduction, as already reported by (Uthus, 2003, Uthus, 1992), who proved the importance of arsenic in methionine's metabolism, particularly during critical phases of reproduction, that is, egg production. It is possible then that mucous excretion for the metalloid might have been reduced in females under hormonal treatment, to support egg production and methionine's

Table 18.1 Mean concentrations of arsenic, cadmium, lead and mercury in treated eels.

Treatment	Matrix	As (mg/kg)	Cd (mg/kg)	Hg (mg/kg)	Pb (mg/kg)
CTRL	Mucus	1.394±1.262	0.003±0.002	0.429±0.328	0.185±0.186
	Blood	0.215±0.046	0.027±0.001	1.022±0.215	0.302±0.043
CPE	Mucus	0.445±0.292	0.047±0.095	1.055±0.489	0.855±1.227
	Blood	0.632±0.056	0.029±0.001	1.126±0.171	0.370±0.061
	Eggs	0.317±0.051	0.009±0.001	1.398±0.100	0.327±0.121
IMP	Mucus	0.914±1.032	0.015±0.018	0.941± 0.789	0.488±0.479
	Blood	0.470±0.160	0.029±0.003	1.192±0.138	0.336±0.063
	Eggs	0.401±0.074	0.011±0.002	0.309±0.154	1.597±0.062

Note: mg/kg ww ± SD.

synthesis (fundamental for normal embryos development). On the contrary, CPE showed highest hematic levels of arsenic and CTRL showed the lowest, confirming this hypothesis of a certain retainment of arsenic to be used in metabolic processes (Uthus, 2003, Uthus, 1992).

Mercury, lead and cadmium presented highest concentrations in the CPE group, in both mucus and blood, suggesting an active passage from blood to mucus.

No clear correlation was observed between blood and egg trace element concentrations in the CPE and IMP groups, making it possible to exclude an active mother–egg transfer of these pollutants. The levels of arsenic, mercury and lead observed in eggs did not differ between the CPE and IMP groups. The mercury levels detected were high enough to impair fertilization by inducing micropyle occlusion, as reported for *Fundulus heteroclitus* (Khan and Weis, 1993). Nevertheless, the high fertilization rates in both groups make it possible to hypothesize that eels could be more resistant to the adverse effects of mercury on reproduction. Alternatively, wild eels grown in mercury polluted ponds might have developed resistance mechanisms to counter the toxic effects on fertilization (Khan and Weis, 1993).

Conclusion

Trace elements are ubiquitous contaminants of water bodies and can be bioaccumulated by aquatic species. Owing to their ecological and physiological characteristics eels are particularly prone to exposure and accumulation of these pollutants and are therefore adversely affected. Although pollution might be considered as one of the causes contributing to the persistent decline in eel populations, few studies are available reporting on the effects of trace elements, and almost no data exists on their endocrine disruption activity.

Data presented here are a first step in defining the reproductive and endocrine effects of trace elements in eels experiencing artificially induced reproduction. Also, they provide evidence of the need for further, more structured studies, which could help identify the eventual endocrine disruptive effect of trace elements in the species and the risks posed by trace elements to eel conservation.

Bibliography

Baillon, L., Oses, J., Pierron, F., Bureau Du Colombier, S., Caron, A., Normandeau, E., Lambert, P., Couture, P., Labadie, P., Budzinski, H., Dufour, S., Bernatchez, L. and Baudrimont, M. (2015) Gonadal transcriptome analysis of wild contaminated female European eels during artificial gonad maturation. *Chemosphere* 139, 303–309.

Belpaire, C., Pujolar, J.M., Geeraerts, C. and Maes, G.E. (2016) Contaminants in eels and their role in the collapse of the eel stocks. *J Biol. Ecol. Anguillid Eels* 225.

Brown, I.A. and Austin, D.W. (2012) Maternal transfer of mercury to the developing embryo/fetus: is there a safe level? *Toxicological and Environmental Chemistry* 94, 1610–1627.

Buslé, J. (1990) Effects of heavy metals on eels, Anguilla sp. *Aquatic Living Resources* 3, 131–141.

Buslé, J. (1991) The eel (Anguilla sp) and organic chemical pollutants. *Science of The Total Environment* 102, 1–19.

Carlson, E.A. and Zelikoff, J.T. (2008) The immune system of fish: a target organ of toxicity. In: Di Giulio, R.T. and Hinton, D.E. (eds) *The Toxicology of Fishes*. CRC Press, Boca Raton, FL.

Ciccotti, E. (2006) Problematiche di conservazione dei pesci su scala continentale: il caso dell'Anguilla Europea. Workshop Fauna Ittica delle acque interne, 22 September 2006, Pontevecchio di Magenta (MI).

Crump, K.L. and Trudeau, V.L. (2009) Mercury-induced reproductive impairment in fish. *Environmental Toxicology and Chemistry* 28, 895–907.

Daverat, F., Limburg, K.E., Thibault, I., Shiao, J.-C., Dodson, J.J., Caron, F., Tzeng, W.-N., Iizuka, Y. and Wickström, H. (2006) Phenotypic plasticity of habitat use by three temperate eel species, Anguilla anguilla, A. japonica and A. rostrata. *Marine Ecology Progress Series* 308, 231–241.

De Boer, J. and Hagel, P. (1994) Spatial differences and temporal trends of chlorobiphenyls in yellow eel (Anguilla anguilla) from inland waters of the Netherlands. *Science of The Total Environment* 141, 155–174.

De Boer, J., Van Der Valk, F., Kerkhoff, M.A.T., Hagel, P. and Brinkman, U.A.T. (1994) An 8-year study on the elimination of pcbs and other organochlorine compounds from eel (Anguilla anguilla) under natural conditions. *Environmental Science & Technology* 28, 2242–2248.

Dekker, W. (2003a) Did lack of spawners cause the collapse of the European eel, Anguilla anguilla? *Fisheries Management Ecology* 10, 365–376.

Dekker, W. (2003b) On the distribution of the European eel (Anguilla anguilla) and its fisheries. *Canadian Journal of Fisheries and Aquatic Sciences* 60, 787–799.

Dekker, W. (2003c) Status of the European eel stock and fisheries. In: Aida, K., Tsukamoto, K. and Yamauchi, K. (eds) *Eel Biology*. Springer Japan, Tokyo.

Dekker, W. (2004) What caused the decline of the Lake Ijsselmeer eel stock after 1960? *ICES Journal of Marine Science* 61, 394–404.

Depew, D.C., Basu, N., Burgess, N.M., Campbell, L.M., Devlin, E.W., Drevnick, P.E., Hammerschmidt, C.R., Murphy, C.A., Sandheinrich, M.B. and Wiener, J.G. (2012) Toxicity of dietary methylmercury to fish: derivation of ecologically meaningful threshold concentrations. *Environmental Toxicology and Chemistry* 31, 1536–1547.

Dietrich, G.J., Dietrich, M., Kowalski, R.K., Dobosz, S., Karol, H., Demianowicz, W. and Glogowski, J. (2010) Exposure of rainbow trout milt to mercury and cadmium alters sperm motility parameters and reproductive success. *Aquatic Toxicology* 97, 277–284.

Dollerup, J. and Graver, C.M. (1985) Repeated induction of testicular maturation and spermiation, alternating with periods of feeding and growth in silver eels, Anguilla anguilla (L.). *Dana* 4, 19–39.

Dong, W., Liu, J., Wei, L., Jingfeng, Y., Chernick, M. and Hinton, D.E. (2016) Developmental toxicity from exposure to various forms of mercury compounds in medaka fish (Oryzias latipes) embryos. *PeerJ* 4, e2282.

Du Colombier, S.B., Bareille, G., Bolliet, V., Lambert, P. and Bardonnet, A. (2007) Micro-pollutant content in Anguilla anguilla glass eels and relationship with migratory behaviour. *Science of The Total Environment* 57, 223–227.

Eisler, R. (1988) *Arsenic Hazards to Fish, Wildlife, and Invertebrates: a Synoptic Review*. US Fish and Wildlife Service, Patuxent Wildlife Research Center, Laurel, MD.

Esteve, C., Alcaide, E. and Ureña, R. (2012) The effect of metals on condition and pathologies of European eel (Anguilla anguilla): in situ and laboratory experiments. *Aquatic Toxicology* 109, 176–184.

European Commission (2006) Commission Regulation (EC) No 1881/2006 of 19 December 2006 setting maximum levels for certain contaminants in foodstuffs (Text with EEA relevance). *EC No 1881/2006.*

European Commission (2008) Commision Regulationn (EC) No 629/2008 of 2 July 2008 amending Regulation (EC) No 1881/2006 setting maximum levels for certain contaminants in foodstuffs. *(EC) No 629/2008*

Fabbri, E., Caselli, F., Piano, A., Sartor, G. and Capuzzo, A. (2003) Cd2+ and Hg2+ affect glucose release and cAMP-dependent transduction pathway in isolated eel hepatocytes. *Aquatic Toxicology* 62, 55–65.

FAO (2006) Report of the 2006 session of the Joint EIFAC/ICES Working Group on Eels. Rome, 23–27 January 2006.

FAO (2007) Report of the 2007 session of the Joint EIFAC/ICES Working Group on Eels. Bordeaux, France, 3–7 September 2007.

FAO (2009) Report of the 2008 session of the Joint EIFAC/ICES Working Group on Eels. Leuven, Belgium, 3–9 September 2008.

Feunteun, E. (2002) Management and restoration of European eel population (Anguilla anguilla): an impossible bargain. *Ecological Engineering* 18, 575–591.

Francesconi, K.A. and Edmonds, J.S. (1998) Arsenic species in marine samples. *Croatica Chemica Acta* 71, 343–359.

Francesconi, K.A., Edmonds, J.S.J.O. and Review, M.B.A.A. (1993) Arsenic in the sea. *Oceanography Marine Biology: An Annual Review* 31, 111–151.

Friedland, K.D., Miller, M.J. and Knights, B. (2007) Oceanic changes in the Sargasso Sea and declines in recruitment of the European eel. *ICES Journal of Marine Science* 64, 519–530.

Geeraerts, C. and Belpaire, C. (2010) The effects of contaminants in European eel: a review. *Ecotoxicology* 19, 239–266.

Goemans, G. and Belpaire, C. (2007) Eels: contaminant cocktails pinpointing environmental contamination. *ICES Journal of Marine Science* 64, 1423–1436.

Geeraerts, C., Goemans, G., Quataert, P., Belpaire, C. and Parmentier, K. (2008) Pollution fingerprints in eels as models for the chemical status of rivers. *ICES Journal of Marine Science* 65, 1483–1491.

Golovanova, I.L. (2008) Effects of heavy metals on the physiological and biochemical status of fishes and aquatic invertebrates. *Inland Water Biology* 1, 93–101.

Gony, S. (1990) Short note on the effects of cadmium on the gill of the glass eel (Anguilla anguilla). *Internationale Revue der gesamten Hydrobiologie und Hydrographie* 75, 835–836.

Hatef, A., Alavi, S.M.H., Butts, I.A.E., Policar, T. and Linhart, O. (2011) Mechanism of action of mercury on sperm morphology, adenosine triphosphate content, and motility in Perca fluviatilis (Percidae; Teleostei). *Environmental Toxicology and Chemistry* 30, 905–914.

Henson, M.C. and Chedrese, P.J. (2004) Endocrine disruption by cadmium, a common environmental toxicant with paradoxical effects on reproduction. *Experimental Biology and Medicine* 229, 383–392.

ICES (2002) Report of the ICES/EIFAC Working Group on Eels. Available at: www.ices.dk/sites/pub/CM%20Doccuments/2002/ACFM/ACFM0302.PDF (accessed 13 Mary 2019).

ICES (2014) Report of the Joint EIFAAC/ICES/GFCM Working Group on Eels, 3–7 November 2014, Rome, Italy. http://www.ices.dk/sites/pub/Publication%20Reports/ICES%20Annual%20Report/Annual%20Report%202014.pdf#search=Report%20of%20the%20Joint%20EIFAAC%2FICES%2FGFCM%20Working%20Group%20on%20Eels (accessed 13 May 2019).

Ismail, A. and Yusof, S. (2011) Effect of mercury and cadmium on early life stages of Java medaka (Oryzias javanicus): a potential tropical test fish. *Marine Pollution Bulletin* 63, 347–349.

Jacoby, D. and Gollock, M. (2014) Anguilla anguilla. The IUCN Red List of Threatened Species 2014: e.T60344A45833138. http://dx.doi.org/10.2305/IUCN.UK.2014-1.RLTS.T60344A45833138.en.

Khan, A.T. and Weis, J.S. (1993) Differential effects of organic and inorganic mercury on the micropyle of the eggs of Fundulus heteroclitus. *Environmental Biology of Fishes* 37, 323–327.

Kim, K.-H., Kabir, E. and Jahan, S.A. (2016) A review on the distribution of Hg in the environment and its human health impacts. *Journal of Hazardous Materials* 306, 376–385.

Kirubagaran, R. and Joy, K.P. (1989) Toxic effects of mercurials on thyroid function of the catfish, Clarias batrachus (L.). *Ecotoxicology and Environmental Safety* 17, 265–271.

Knights, B. (1997) Risk assessment and management of contamination of eels (Anguilla spp.) by persistent xenobiotic organochlorine compounds. *Chemistry and Ecology* 13, 171–212.

Knights, B. (2003) A review of the possible impacts of long-term oceanic and climate changes and fishing mortality on recruitment of anguillid eels of the Northern Hemisphere. *Science of the Total Environment* 310, 237–244.

Lam, J.C.W., Tanabe, S., Chan, S.K.F., Yuen, E.K.W., Lam, M.H.W. and Lam, P.K.S. (2004) Trace element residues in tissues of green turtles (Chelonia mydas) from South China Waters. *Marine Pollution Bulletin* 48, 174–182.

Larsson, P., Hamrin, S. and Okla, L. (1991) Factors determining the uptake of persistent pollutants in an eel population (Anguila anguilla L.). *Environmental Pollution* 69, 39–50.

Lawrence, A.J. and Elliott, M. (2008) Introduction and conceptual model. In: Lawrence, A.J. and Hemingway, K.L. (eds) *Effects of Pollution on Fish: Molecular Effects and Population Responses.* John Wiley & Sons, Oxford.

Lesniewska, E., Szynkowska, M.I. and Paryjczak, T. (2009) Main sources of mercury in human organisms not exposed professionally. *Rocznik Ochrona Srodowiska* 11, 403–419.

Lindsay, D.M. and Sanders, J.G. (1990) Arsenic uptake and transfer in a simplified estuarine food chain. *Environmental Toxicology and Chemistry* 9, 391–395.

Lionetto, M.G., Maffia, M., Cappello, M.S., Giordano, M.E., Storelli, C. and Schettino, T. (1998) Effect of cadmium on carbonic anhydrase and Na+-K+-ATPase in eel, Anguilla anguilla, intestine and gills. *Comparative Biochemistry and Physiology Part A: Molecular & Integrative Physiology* 120, 89–91.

Maes, G.E., Raeymaekers, J.A.M., Pampoulie, C., Seynaeve, A., Goemans, G., Belpaire, C. and Volckaert, F.A.M. (2005) The catadromous European eel *Anguilla anguilla* (L.) as a model for freshwater evolutionary ecotoxicology: relationship between heavy metal bioaccumulation, condition and genetic variability. *Aquatic Toxicology* 73, 99–114.

Mordenti, M., Di Biase, A., Sirri, R., Modugno, S. and Tasselli, A. (2012) Induction of sexual maturation in wild female European eels (*Anguilla anguilla*) in darkness and light. *The Open Access Israeli Journal of Aquaculture – Bamidgeh* 64, 1–9.

Mordenti, O., Emmanuele, P., Casalini, A., Lokman, P.M., Zaccaroni, A., Di Biase, A. and Parmeggiani, A. (2018) Effect of aromatable androgen (17-methyltestosterone) on induced maturation of silver European eels (Anguilla Anguilla): oocyte performance and synchronization. *Aquaculture Research* 49, 442–448.

Nowosad, J., Kucharczyk, D. and Łuczyńska, J. (2018) Changes in mercury concentration in muscles, ovaries and eggs of European eel during maturation under controlled conditions. *Ecotoxicology and Environmental Safety* 148, 857–861.

Palstra, A.P., Van Ginneken, V.J.T., Murk, A.J. and Van Den Thillart, G.E.E.J.M. (2006) Are dioxin-like contaminants responsible for the eel (Anguilla anguilla) drama? *Naturwissenschaften* 93, 145–148.

Pannetier, P., Caron, A., Campbell, P.G.C., Pierron, F., Baudrimont, M. and Couture, P. (2016) A comparison of metal concentrations in the tissues of yellow American eel (Anguilla rostrata) and European eel (Anguilla anguilla). *Science of the Total Environment* 569–570, 1435–1445.

Penglase, S., Hamre, K. and Ellingsen, S. (2014) Selenium and mercury have a synergistic negative effect on fish reproduction. *Aquatic Toxicology* 149, 16–24.

Pierron, F., Baudrimont, M., Bossy, A., Bourdineaud, J.-P., Brèthes, D., Elie, P. and Massabuau, J.-C. (2007a) Impairment of lipid storage by cadmium in the European eel (Anguilla anguilla). *Aquatic Toxicology* 81, 304–311.

Pierron, F., Baudrimont, M., Gonzalez, P., Bourdineaud, J.-P., Elie, P. and Massabuau, J.-C. (2007b) Common pattern of gene expression in response to hypoxia or cadmium in the gills of the European Glass eel (Anguilla anguilla). *Environmental Science & Technology* 41, 3005–3011.

Pierron, F., Baudrimont, M., Dufour, S., Elie, P., Bossy, A., Baloche, S., Mesmer-Dudons, N., Gonzalez, P., Bourdineaud, J.-P. and Massabuau, J.-C. (2008) How cadmium could compromise the completion of the European Eel's reproductive migration. *Environmental Science & Technology* 42, 4607–4612.

Saeki, K., Sakakibara, H., Sakai, H., Kunito, T. and Tanabe, S. (2000) Arsenic accumulation in three species of sea turtles. *Biometals* 13, 241–250.

Sanchez-Galan, S., Linde, A.R., Ayllon, F. and Garcia-Vazquez, E. (2001) Induction of micronuclei in eel (Anguilla anguilla L.) by heavy metals. *Ecotoxicology and Environmental Safety* 49, 139–143.

Sangalang, G.B. and O'Halloran, M.J. (1972) Cadmium-induced testicular injury and alterations of androgen synthesis in brook trout. *Nature* 240, 470–471.

Sangalang, G.B. and O'Halloran, M.J. (1973) Adverse effects of cadmium on brook trout testis and on in vitro testicular androgen synthesis. *Biology of Reproduction* 9, 394–403.

Santos, M.A. and Hall, A. (1990) Influence of inorganic lead on the biochemical blood composition of the eel, Anguilla anguilla L. *Ecotoxicology and Environmental Safety* 20, 7–9.

Scheuhammer, A., Braune, B., Chan, H.M., Frouin, H., Krey, A., Letcher, R., Loseto, L., Noël, M., Ostertag, S. and Ross, P. (2015) Recent progress on our understanding of the biological effects of mercury in fish and wildlife in the Canadian Arctic. *Science of the Total Environment* 509, 91–103.

Shukla, J.P. and Pandey, K. (1984) Impaired spermatogenesis in arsenic treated freshwater fish, Colisa fasciatus (Bl. and Sch.). *Toxicology Letters* 21, 191–195.

Srivastava, A. (1987) Changes induced by lead in fish testis. *Journal of Environmental Biology* 8, 329–332.

Székely, C., Molnár, K., Müller, T., Szabó, A., Romvári, R., Hancz, C. and Bercsényi, M. (2004) Comparative study of X-ray computerised tomography and conventional X-ray methods in diagnosis of swimbladder infection in eels caused by Anguillicola crassus. *Diseases of Aquatic Organisms* 58, 157–164.

Székely, C., Molnár, K. and Rácz, O.Z. (2005) Radiodiagnostic method for studying the dynamics of Anguillicola crassus (Nematoda: Dracunculoidea) infection and pathological status of the swimbladder in Lake Balaton eels. *Diseases of Aquatic Organisms* 64, 53–61.

Tulonen, J. and Vuorinen, P.J. (1996) Concentrations of PCBs and other organochlorine compounds in eels (Anguilla anguilla, L.) of the Vanajavesi watercourse in southern Finland, 1990–1993. *Science of the Total Environment* 187, 11–18.

Uthus, E.O. (1992) Evidence for arsenic essentiality. *Environmental Geochemistry and Health* 14, 55–58.

Uthus, E.O. (2003) Arsenic essentiality: A role affecting methionine metabolism. *The Journal of Trace Elements in Experimental Medicine* 16, 345–355.

Van Den Thillart, G. and Dufour, S. (2009) How to estimate the reproductive success of European silver eels. In: van den Thillart G., Dufour S. and Rankin J.C. (eds) *Spawning Migration of the European Eel.* Springer, Dordrecht, pp. 3–9.

Van Ginneken, V., Bruijs, M., Murk, T., Palstra, A. and Van Den Thillart, G. (2009) The effect of PCBs on the spawning migration of European silver eel (Anguilla anguilla L.). In: Van Den Thillart, G., Dufour, S. and Rankin, J.C. (eds) *Spawning Migration of the European Eel: Reproduction Index, a Useful Tool for Conservation Management.* Springer, Dordrecht.

Vøllestad, L.A. (1992) Geographic variation in age and length at metamorphosis of maturing European eel: environmental effects and phenotypic plasticity. *Journal of Animal Ecology* 61, 41–48.

CHAPTER 19

Parasite assemblages of European eels in Irish freshwater, mixohaline and marine habitats

T. Kieran McCarthy, Michelle Morrissey, Karen Creed, John Conneely and Eamonn S. Lenihan

Zoology, Ryan Institute, National University of Ireland Galway, Ireland

Introduction

Interest in the parasitology of the European eel (*Anguilla anguilla*) has increased in recent decades because of concerns about the adverse effects pathogenic invasive species may have on the dynamics of host populations. Surveys of eel parasites are often concerned only with documenting the spread of such non-indigenous species or identifying the pathological effects that such species have on individual infected eels or on eel population ecology. The Asian nematode *Anguillicola crassus*, that occurs in swimbladders of infected eels, was introduced accidentally to Europe in the 1980s. It is now widely distributed in Ireland (Becerra-Jurado et al., 2014) and throughout much of the continental range of the European eel, including parts of northern Africa (Jakob et al., 2016). Infection with *A. crassus* has been shown to affect the capacity for eels to undertake spawning migrations (Barry et al., 2014; Newbold et al., 2015; Palstra et al., 2007; Sjöberg et al., 2009). Other studies on eel parasite assemblages are not focused on damage caused by invasive parasites, and these can also provide insights concerning many aspects of host behaviour and ecology.

Development of parasitology, during the 20th century, extended appreciation of the ecological nature of parasitism (Dogiel et al., 1961; Esch et al., 1990). New terminology helped identify the paradigm shift (Esch et al., 1990). In Britain, the new ecological approach to parasitology of fish was pioneered by Kennedy (1975) and by other fish parasitologists elsewhere in Europe and in North America.

Studies on Irish eel parasites, which we conducted over four decades, provided insights concerning their biogeography, host habitat utilization, local movements of eels and their trophic interactions. These studies have facilitated the compilation of a new checklist of Irish eel parasites, which is presented and discussed in this brief review of eel parasitology research in Ireland. It is hoped that this may help focus attention on the need for further taxonomic, biogeographical and life-cycle research on parasites of European eel and those of other Anguillid eels.

Methodology

Figure 19.1 provides the locations of the sites (*N* = 18) at which eels were sampled during a series of postgraduate research projects by Conneely (1987), Callaghan (1993), Purcell (1994), Creed (1999) and Morrissey (2007). Eels were captured through a variety of sampling methods (electrofishing, fyke-netting, long-lining) and were narcotized, euthanized and frozen for laboratory analysis. After removal of all metazoan external parasites, eels were carefully dissected and the alimentary tract and major organ systems examined. Parasites were removed, preserved, identified and enumerated. Further details are given by Conneely and McCarthy (1986), Callaghan and McCarthy (1996), Morrissey and McCarthy (2007).

A seriation analysis was carried out using a presence–absence data matrix of parasite species and sampling sites. The seriation algorithm attempts to reorganize the data according to some

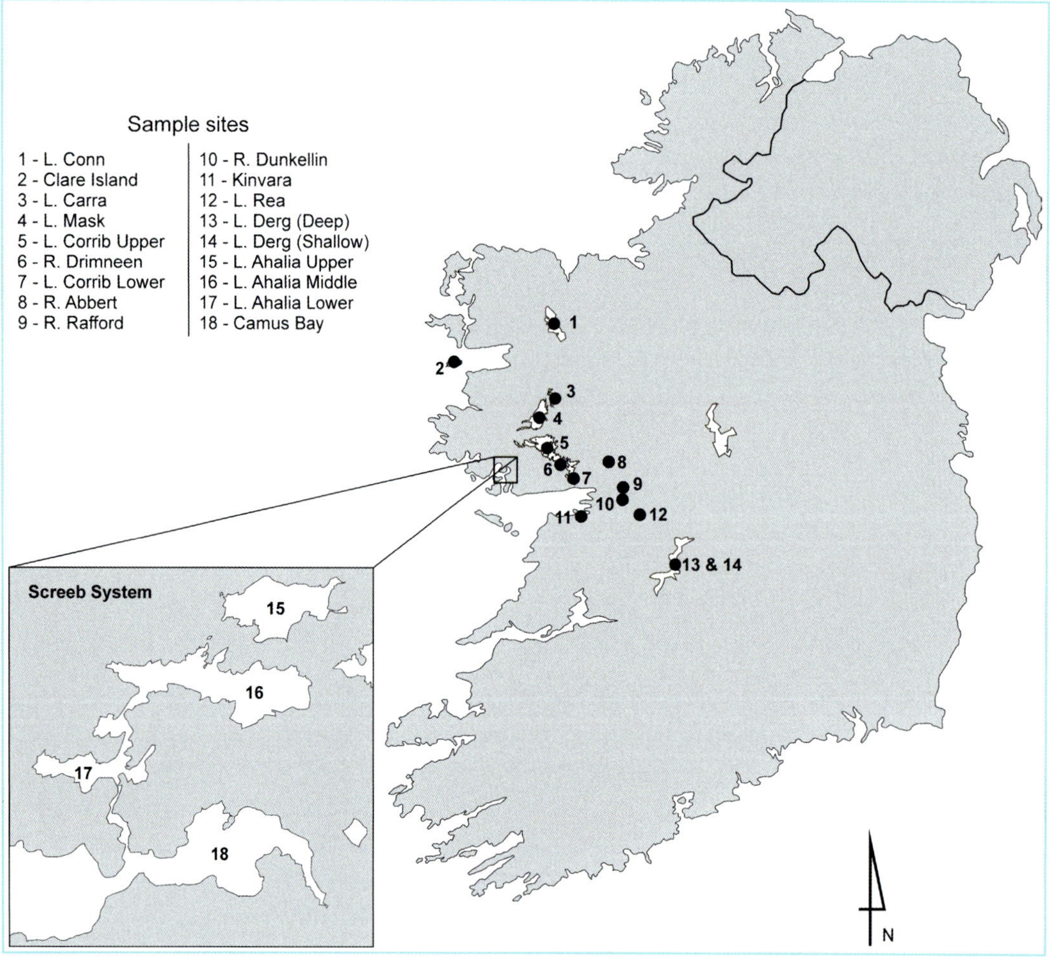

Figure 19.1 Map of Ireland with sampling sites indicated.

intrinsic characteristic of the data, which often corresponds to an environmental gradient. A CA was also conducted using abundance data of parasite species by site. CAs are a useful ordination technique that reduces large multivariate ecological datasets, set along some environmental gradient, making interpretation easier. PAST software (v.3.2.0) was used for these analyses.

Results

A checklist (Table 19.1) of 44 species (or operational taxonomic units) of metazoan parasites recorded in European eels in Irish freshwater, mixohaline and marine waters shows that digenetic flukes (36%), nematodes (27%) and acanthocephalans (14%) were the most biodiverse groups. The list is comprehensive for freshwater habitats. The 11 (26%) additions to the list are mostly marine generalist parasites of inshore fishes. Most parasites listed (84%) are not eel specialists and are regularly found in other fish. Some have alternative preferred hosts but occur as accidental parasites in eel. Others use eels as intermediate or paratenic hosts. A small proportion (max. 16%) of the Irish eel parasites are allogenic forms (that is, use piscivorous birds or mammals as their definitive host). Eel specific parasites comprise 16% of those listed (Table 19.1). Many generalist parasites were introduced to Ireland during importation of live fish. Over one-half of the freshwater fish now in Irish freshwater were introduced and about 25% of the parasites of Irish eels are introduced. Three bioinvasive specialist eel parasites (*Pseudodactylogyrus anguillae*, *P. bini*, and *A. crassus*) have spread rapidly within Ireland (McCarthy et al., 2009).

Seriation analysis (Figure 19.2), using species presence–absence data for sites ($N = 18$), highlighted how patterns in parasite component community composition were related to habitat characteristics. Marine parasites dominated the eel parasite assemblages at Camus Bay and Kinvara but freshwater forms (for example, *Acanthocephalus lucii* and *Echinorhynchus truttae* at Kinvara) indicated movements to and from sub-littoral freshwater discharge points. In the three basins of Lough Ahalia, a complex mixohaline lagoon, the salinity gradient from the entirely freshwater upper basin to the strongly brackish lower basin was reflected in the shift in species composition from freshwater to predominantly marine species. The seriation analysis also illustrated differences between lakes and rivers.

A CA (Figure 19.3) was undertaken using a species abundance matrix. This ordination illustrated differences in the eel parasite component communities along the Lough Ahalia to Camus Bay salinity gradient and their relationship to the other marine site (Kinvara). Groupings of river and lake sites are also evident. The offshore Clare Island site, with extremely low parasite abundances is clearly separated from mainland sites.

Clare Island (16 km^2), located 6 km offshore from the west coast of Ireland has a very limited freshwater fish fauna (flounder, brown trout and eel [McCarthy et al., in press]) and these have very few parasites. Only 10 of the 150 eels examined were infected by any metazoan parasites. The cumulative number (134) of the six parasite species (*P. anguillae*, *P. bini*, *Diplostomum spathcaceum*, *Raphidascaris acus*, *Pomphorhynchus tenerrima*, *A. clavula*) was also very low. The dominant species was the acanthocephalan *A. clavula*, which occurred in five eels and comprised 75% of all eel parasites recorded. This species uses the freshwater isopod *Asellus meridianus*, which is the only *Asellus* species on Irish offshore islands, as its intermediate host. On the mainland *A. aquaticus* is the most widespread and abundant freshwater isopod and the

Table 19.1 Checklist of metazoan parasites recorded in this study.

Parasite species	Authority
Monogenea	
Pseudodactylogyrus anguillae	Yin & Sproston, 1948
Pseudodactylogyrus bini	Kikuchi, 1929
Digenea	
Helicometra fasciata	Rudolphi, 1819
Podocotyle atomon	Rudolphi, 1802
Crepidostomum farionis	Muller, 1784
Crepidostomum metoecus	Braun, 1900
Sphaerostoma bramae	Muller, 1776
Deropristis inflata	Molin, 1859
Phyllodistomum sp.	
Lecithochirium rufoviride	Rudolphi, 1819
Lecithochirium furcolabiatum	Jones, 1933
Diplostomum chromatophorum	Brown, 1931
Diplostomum gasterostei	Williams, 1966
Diplostomum paraspathaceum	Shigin, 1965
Diplostomum pseudobaeri	Razmashkin & Andrynk, 1978
Diplostomum spathaceum	Rudolphi, 1819
Hemiuris communis	Odhner, 1905
Brachyphallus crenatus	Odhner, 1905
Cestoda	
Bothriocephalus claviceps	Goeze, 1782
Proteocephalus macrocephalus	Creplin, 1825
Nematoda	
Anguillacola crassus	Kuwahar, Nimi & Itagaki, 1974
Camallanus lacustris	Zoega, 1776
Capillaria sp.	

distribution and abundance of *A. lucii,* which uses it as intermediate host, reflects this biogeographical pattern.

Eels occur in a wide range of Irish transitional waters (McCarthy et al., 2009) but their parasite assemblages have not been extensively studied. A series of sites in the River Screebe area includes a complex triple basin freshwater lake/mixohaline lagoon area discharging to Camus Bay (Harrod et al., 2005). The component parasite communities of these four contrasting but closely linked habitats were included in multivariate analyses (Figure 19.3). Limited evidence of exchange in residency between salinity zones was indicated by the eel parasite data. In the marine littoral habitat at Kinvara the parasite component community was dominated by marine fish parasites. However, a number of eels were found to have freshwater parasites. This suggested that at times they forage at points where subterranean streams discharge below tidal limits, from the karst limestone Burren hinterland, where freshwater macrocrustacean intermediate hosts of acanthocephalans (*A. truttae* and *A. lucii*) occur. Habitat gradients, such as river continua or the east to west variation in the lower basin of Lough Corrib also reveal interesting

Table 19.1 (Continued)

Parasite species	Authority
Cucullanus truttae	Fabricius, 1794
Daniconema anguilla	Moravec & Køie, 1987
Paraquimperia tenerrima	Linstow, 1878
Raphidascaris sp.	
Raphidascaris acus	Bloch, 1779
Rhabdochona sp.	
Encysted Larvae spp. Indet	
Contracaecum sp.	Raillet & Henry, 1912
Hysterothylacium aduncum	Rudolphi, 1802
Acanthocephala	
Acanthocephalus anguillae	Muller, 1780
Acanthocephalus clavula	Dujardin, 1845
Acanthocephalus lucii	Muller, 1780
Echinorhynchus gadi	Zoega, 1776
Echinorhynchus truttae	Schrank, 1788
Pomphorhynchus laevis	Muller, 1776
Hirudinea	
Piscicola geometra	Linnaeus, 1758
Hemiclepsis marginata	Muller, 1844
Mollusca	
Anodonta cygnea	Linnaeus, 1758
Crustacea	
Argulus foliaceus	Linnaeus, 1758
Ergasilus gibbus	von Nordmann, 1832
Gnathia maxillaris	Montagu, 1804

variations in parasite distributions. Thus, in the case of Lough Corrib it has been noted that *A. lucii is* replaced by *A. clavula* and that this paralleled a shift in the relative abundances of the two *Asellus* species they use as intermediate hosts across the lake basin (Conneely and McCarthy, 1986).

Discussion

The updated checklist of eel parasites in Ireland (Table 19.1) includes 44 taxa, which represents a significant increase on the numbers listed for Ireland previously (Holland and Kennedy, 1997; Jakob et al., 2016; Kane, 1966; Kennedy, 1975). Unlike some checklists, we have not dealt with protozoan parasites or Myxozoa because no systematic surveys of these taxonomic groups have taken place in Ireland. Jakob et al. (2016) listed 161 parasites of the European eel of which 133 were metazoan species/taxa. However, the records they catalogued covered a wide range

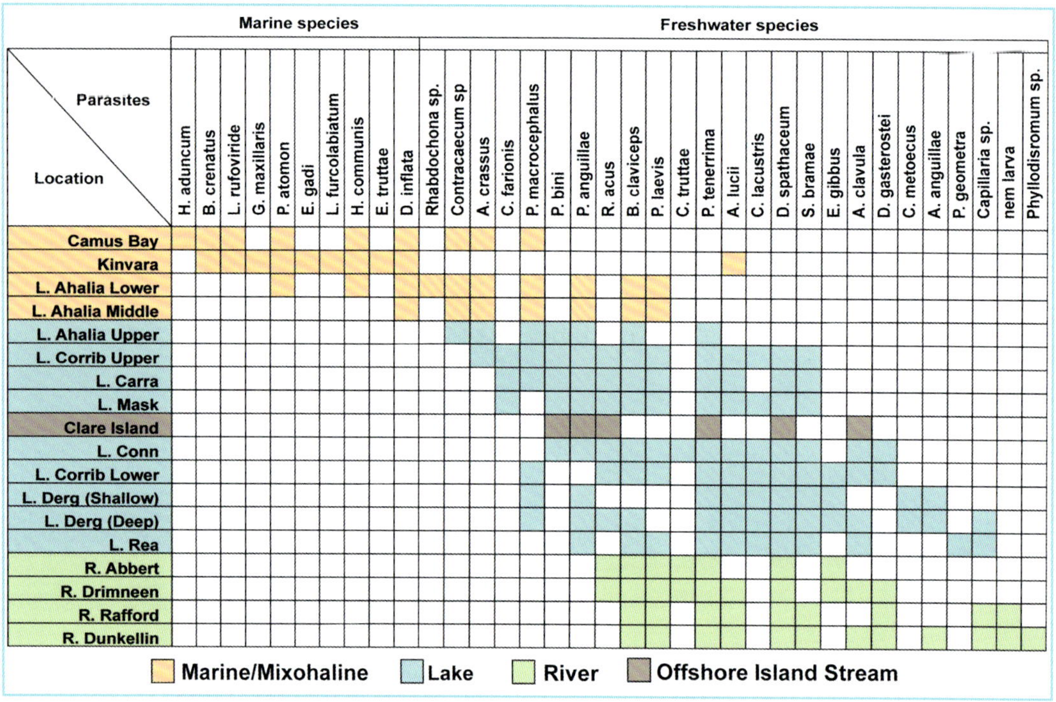

Figure 19.2 Seriation analysis of presence–absence parasite data for each site.

of continental European countries as well as Britain and Ireland. Brewster (2016) listed 34 eel parasites as having been recorded in Britain. The European eel is one of a relatively small number of the indigenous fish species found in Irish waters and as a euryhaline species it occurs in a wide range of mixohaline habitats as well as in all sorts of inland rivers and lakes. The biodiversity of eel parasites in Ireland reflects this range of habitats, with associated fish and invertebrate host assemblages. However, island biogeographical factors may also account for the apparent absence of some species that occur in Britain and for some unusual parasite distributional patterns in Ireland. The limited number of parasites affecting eels in Clare Island (site 2) is clearly linked to the isolation of the island and its limited freshwater fauna. The absence of the specialist nematode, *Spinetectus inermis*, in Ireland is still puzzling (Conneely and McCarthy, 1986). Likewise, the occurrence (Table 19.1) of *Ergasilus gibbus* in inland lakes in Ireland is not typical of the species elsewhere (Aguilar, et al., 2005; Di Cave et al., 2001; Hermida et al., 2008; Jakob et al., 2016; Maillo et al., 2005; Saraiva and Eiras, 1996; Soylu et al., 2013). The relatively recent records of marine fish parasites in eels (Table 19.1) has added considerably to our knowledge of the ecology of eels in marine and mixohaline waters and further records of generalist marine species can be anticipated in future. In Iceland, where eight eel parasites have been recorded, four marine species were recorded by Kristmundsson and Helgason (2007). Studies on eel parasites from coastal lagoons, estuaries and marine habitats include a wide range of taxa (Jakob et al., 2016).

The Irish fauna, like those of most European countries, has changed in recent decades and continues to change at an increasing rate and bioinvasions of non-indigenous ecologically

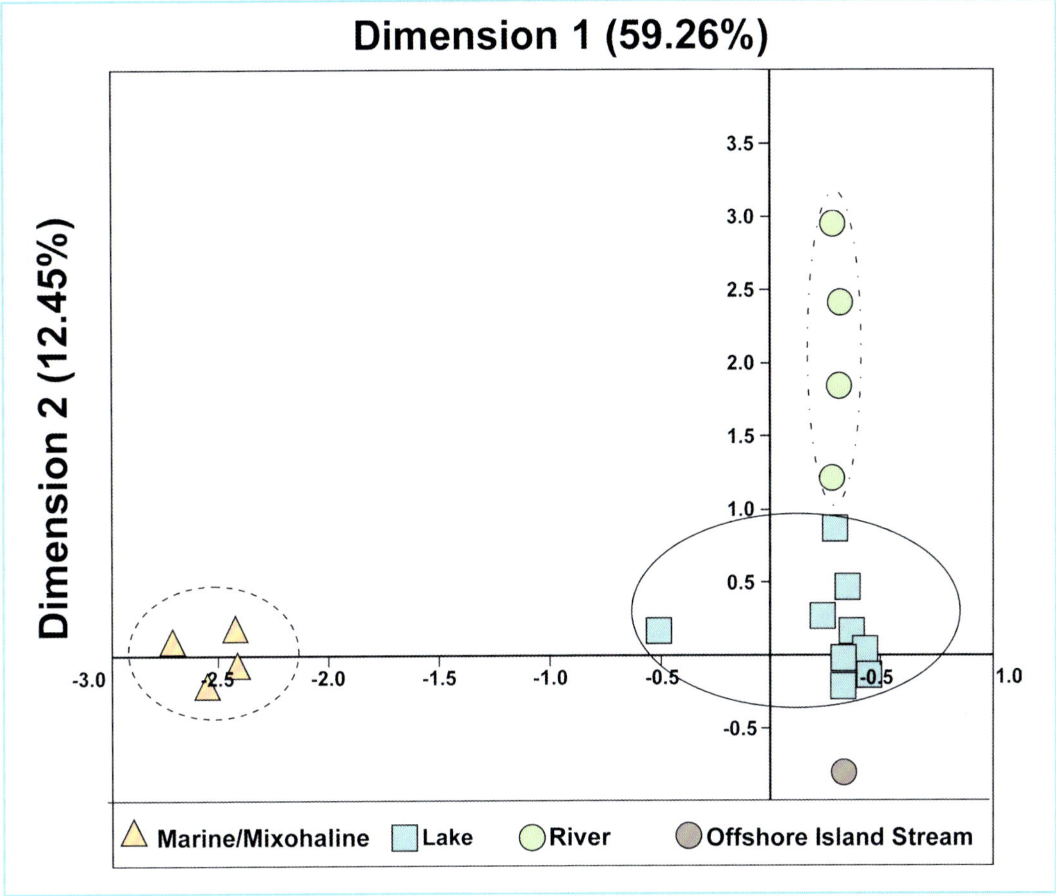

Figure 19.3 CA ordination of sampling sites based on parasite assemblages.

damaging taxa are among the many challenges affecting the natural communities of the island. Changes to the parasite communities of fish have been occurring in parallel with introductions and translocations of fish species. The composition of Ireland's eel parasite communities reflects this, though the eel is itself a native fish species. The changing composition of the islands freshwater invertebrate fauna (McCarthy, 1986; McCarthy et al., 2009), and especially macrocrustaceans, seems to be affecting the distributional patterns of some Acanthocephala. The restricted distribution of the North American *Paratenuisentis ambiguus* in Europe reflects the range of *Gammarus tigrinus*, its preferred intermediate host. This introduced North American amphipod is now long established and widely distributed in Ireland (Costello, 1993) but fortunately the parasite has not yet been introduced here. The introductions of *P. anguillae*, *P. bini*, and *A. crassus*, invasive and potentially pathogenic Asian parasites of eels, illustrates the risks associated with fish translocations. These species, though relatively recent introductions (Copley and McCarthy 2001; McCarthy and Rita, 1991; McCarthy et al., 2009) are now widely distributed in Ireland (Becerra-Jurado et al., 2014; McCarthy et al., 2009). The latest introductions of eel pathogens appear to be viral diseases (for example, Eel Virus European [EVE] and Eel Virus

European X [EVEX] [Anon., 2016; McConville et al., 2018]), and these are likely to have arrived via importations of live glass eels from continental Europe for stocking the commercial fishery in Lough Neagh in Northern Ireland. The impact of these pathogens on Irish eel stocks and the extent to which they affect spawner biomass escapement from major Irish rivers remains to be determined.

The Clare Island eel parasite survey revealed low species richness and parasite burdens in the island's eels. However, as is the case in mainland Ireland, the island freshwater habitats are already colonized by *Pseudodactylogyrus* species, only recently introduced to Ireland (McCarthy et al., 2009). There should be greater concern about the fact the natural composition and structure of the parasite assemblages of eels are being changed due to fish introductions and translocations. Non-equilibrium communities, such as those of islands, are susceptible to colonization by introduced species. Ireland, and its offshore islands, is increasingly being exposed to bioinvasions. The rapid spread of pathogenic species, such as *A. crassus*, is recognized as a matter of concern. However, other ecologically important changes may be occurring undetected.

Eel parasite assemblages reflect ecological patterns and can serve as indicators of home ranges, residency habitats and movements between different habitat types. Eel parasitology can complement research involving use of modern technologies, such as stable isotope analysis and otolith microchemistry (Arai et al., 2006; Harrod et al., 2005) in this respect. The results of the surveys of Irish eel parasite communities did not reveal much evidence of movement of eels between river and lake habitat but did confirm that in some mixohaline localities eels did sometimes move between saline and freshwater habitats. However, most remained resident as yellow eels in limited home ranges. The eel parasite assemblages at the four zones (sites 15, 16, 17 and 18) in the lower River Screebe, represented (Figure 19.1) by three sections of Lough Ahalia and Camus Bay, illustrated (Figures 19.2 and 19.3) the variation in component composition along a salinity gradient from an oligotrophic freshwater lake basin to a pure seawater area. At the coastal (Figure 19.1, site 11) Kinvara site the parasites recorded were mostly marine species. However, though no surface stream discharges in or near the sampling area, the occurrence of the freshwater acanthocephalan *E. truttae* indicated that eels may have been foraging at sub-littoral areas where subterranean rivers were entering Galway Bay from the karst limestone landscape around Kinvara.

The relatively infrequent observations of ectoparasites, other than gill parasites, on European eel has often been attributed to their natural protection in the form of skin mucous. Occasionally we have observed *Argulus foliacius* on eels but not listed them because they occurred only in fyke net catches arising from exchange from coarse fish by-catch or in eels being stored for excessively long times in lake storage boxes by fishermen (Evans and Matthews, 2000). Likewise, very few leeches are observed on eels in Irish freshwaters where two fish parasitic species, *Piscicola geometra* and *Hemiclepsis marginata*, are widely distributed (McCarthy, 1975). However, it is possible that the benthic habits and diurnal inactivity of eels enables these transient haematophagous ectoparasites to attach to eels. Zintl et al. (1997, 2000) investigated naturally occurring *Trypanosoma granulosum* in the River Burrishoole catchment in western Ireland. They observed that leeches (*H. marginata*), which they recorded on both yellow and silver eels, were effective vectors of trypanosomiasis in the eel population of the catchment. It is possible that such ectoparasites may also act as vectors of other eel diseases, including introduced viral pathogens.

The need for ongoing research on Irish eel parasites was highlighted by the comparatively recent discovery of *Daniconema anguillae* (Morrissey and McCarthy, 2008). This chance discovery, during a study (Morrissey and McCarthy, 2007) on *A. crassus*, suggests there may still be other species to be detected in Irish freshwater habitats. Though Bacerra-Jurado et al. (2014) surveyed 234 sites in their survey of *A. crassus* in Ireland they did not detect *D. anguillae*. However, this may have been due to the fact that they only recorded adult *A. crassus* and did not therefore look at the swimbladder walls of the eels they examined. *D. anguillae* may be more widely distributed in Ireland but detecting it will require more careful examination of the fins and beneath the skins of eels using protocols described by Molnár and Székely (1998). The original discovery of the species in Europe (Moravec and Køie, 1987) involved a range of swimbladder infections of larvae and adults in Lake Esrum, Denmark. However, subsequently Molnár and Székely (1998) recorded it in much higher numbers in fins and skin of eels in Hungary. They suggested that it may be possible that *A. foliaceus* may be involved as an intermediate host for this and other related nematode parasites of freshwater fish.

Analysis of variation in the component parasite communities of eels can provide biogeographical information that is especially interesting in respect of Ireland. However, the rather traditional approach adopted to-date could usefully be supported by molecular phylogeography and studies on population genetics of eel parasites in Ireland.

There is also a need for greater use of genetics to infer the biogeographical origin of Irish eel parasites, using the approaches recommended by Carlsson et al. (2014) for analysis of the colonization history of Ireland's fauna. Use of molecular taxonomy could also resolve questions concerning some of the taxa included (Table 19.1) in the Irish eel parasite checklist.

Acknowledgements

This research was partly funded during the HEA-PTRLI Cycle 3 project Population Biology of Eels in Marine and Mixohaline Waters. The Electricity Supply Board (ESB) Fisheries Conservation also contributed to funding as part of our investigations on eel populations of the Rivers Shannon and Erne. We also want to acknowledge the contributions made by postgraduate students and postdoctoral research assistants to the work (especially R. Callaghan, L. Copley, D. Hassett, O. Naughton and P. Purcell).

This paper is dedicated to Drs T. Burke, F. El Tuomi and P. Reilly, deceased former members of our Freshwater Ecology Group.

Bibliography

Aguilar, A., Álvarez, M.F., Leiro, L. and Sanmartin, M.L. (2005) Parasite populations of the European eel (*Anguilla anguilla* L.) in the Rivers Ulla and Tea (Galicia, northwest Spain). *Aquaculture* 249 85–94.

Anon. (2016) Examination of the presence and prevalence of European eel viruses in the Lough Neagh European eel population (*Anguilla anguilla*). Project Report – Invest NI Innovation voucher REF: IV1015007. AFBI, Belfast.

Arai T., Kotake, A. and McCarthy, T.K. (2006) Habitat use of the European eel *Anguilla anguilla* in Irish waters. *Estuarine, Coastal and Shelf Science* 67, 569–578.

Barry, J., McLeish, J., Dodd, J.A., Turnbull, J.F., Boylan, P. and Adams, C.E. (2014) Introduced parasite *Anguillicola crassus* infection significantly impedes swim bladder function in the European eel *Anguilla anguilla* (L.). *Journal of Fish Diseases* 37, 921–924.

Becerra-Jurado, G., Cruickshanks, R., O'Leary, C., Kelly, F., Poole, R. and Gargan, P. (2014) Distribution, prevalence and intensity of *Anguillicola crassus* (Nematoda) infection in *Anguilla anguilla* in the Republic of Ireland. *Journal of Fish Biology* 84, 1046–1062.

Brewster, B. (2016) Aquatic parasite information – a database on parasites of freshwater and brackish fish in the United Kingdom. PhD thesis, Kingston University, London

Callaghan, R. (1993) Fish species assemblages of the Dunkellin River system with particular reference to the ecology of the European eel, *Anguilla anguilla* L. PhD thesis, National University of Ireland, Galway.

Callaghan, R. and McCarthy, T.K. (1996) Metazoan parasite assemblages of eels in the Dunkellin catchment, Western Ireland. *Archives of Polish Fisheries* 4, 147–174.

Carlsson, J., Cross, T.F., McGinnety, P., Prodohl, P.A. and McDevitt, A.D. (2014) The use of genetics to infer the origins of the terrestrial and freshwater elements of the Irish fauna. In: Sleeman, D.P., Carlsson, J. and Carlsson, J.E.L (eds.). Mind the Gap II: new insights into the Irish postglacial. *Irish Naturalists' Journal* 33, 8–27.

Conneely, J.J. (1987) Ecological studies on the metazoan parasites of freshwater fishes from the Corrib catchment area, Western Ireland. PhD thesis, National University of Ireland, Galway.

Conneely, J.J. and McCarthy, T.K. (1986) Ecological factors influencing the composition of the parasite fauna of the European eel, *Anguilla anguilla* (L), in Ireland. *Journal of Fish Biology* 28, 207–219.

Copley, L. and McCarthy, T.K. (2001) The first record of the monogenean gill fluke *Pseudodactlylogyrus bini* (Kikuchi 1929) with observations on other ectoparasites of eels, *Anguilla anguilla* (L.) *Irish Naturalists' Journal* 26, 405–413.

Costello, M.J. (1993) Biogeography of alien amphipods occurring in Ireland and interactions with native species. *Crustaceana* 65, 287–299.

Creed, K. (1999) Studies on the Metazoan Parasites of *Salmo salar* (L.), *Anguilla anguilla* (L.) and *Platichthys flesus* (L.) in the West of Ireland with Reference to Clare Island. PhD thesis, National University of Ireland, Galway.

Di Cave, D., Berrilli, F., De Liberato, C., Orecchia, P. and Kennedy, C.R. (2001) Helminth communities in eels *Anguilla anguilla* from Adriatic coastal lagoons in Italy. *Journal of Helmithology* 75, 7–13.

Dogiel, V.A., Petrushevski, G.K. and Polyanski Y.I. (1961) *Parasitology of Fishes*. Oliver and Boyd, Edinburgh.

Esch, C.W., Shostak, A.W., Margogliese, D.J. and Goater, T.M. (1990) Patterns and processes in helminth parasite communities: an overview. In: Esch, G.W., Bush, A. and Aho, J. (eds) *Parasite Communities: Patterns and Processes*. Chapman Hall, London, pp. 1–19.

Evans, D. and Matthews, M.A. (2000) First record of *Argulus foleaceus* on the European eel in the British Isles. *Journal of Fish Biology* 57, 529–530.

Harrod, C., Grey, J., McCarthy, T.K. and Morrissey, M. (2005) Stable isotope analyses provide new insights into ecological plasticity in a mixohaline population of European eel. *Oecologia* 144(4), 673–683.

Hermida, M., Saraiva, A. and Cruz, C. (2008) Metazoan parasite community of a European eel (*Anguilla anguilla*) population from an estuary in Portugal. *Bulletin of the European Association of Fish Pathologists* 28, 35–40.

Holland, C.V. and Kennedy, C.R. (1997) A checklist of the parasitic helminth and crustacean species recorded in freshwater fish in Ireland. *Biology and Environment: Proceedings of the Royal Irish Academy* 97B, 225–243.

Jakob, E., Walter, T. and Hanel, R. (2016) A checklist of the protozoan and metazoan parasites of European eel (*Anguilla anguilla*): Checklist of *Anguilla anguilla* parasites. *Journal of Applied Ichthyology* 32, 757–804.

Kane, M.B. (1966) Parasites of Irish freshwater fishes. *Scientific Proceedings of the Royal Dublin Society* 18B, 205–220.

Kennedy, C. (1975) *Ecological Animal Parasitology*. Blackwell Scientific Publications, Oxford.

Kristmundsson, A. and Helgason, S. (2007) Parasite communities of eels *Anguilla anguilla* in freshwater and marine habitats in Iceland in comparison with other parasites communities in Europe. *Folia Parasitologica* 54, 141–153.

Maillo, P.A., Vich, M.A., Salvado, H., Marques, A. and Gracia, M.P. (2005) Parasites of *Anguilla anguilla* (L.) from three coastal lagoons of the River Ebro delta (Western Mediterranean). *Acta Parasitologica* 50, 156–160.

McCarthy, T.K. (1975) Observations on the distribution of the freshwater leeches (Hirudinea) of Ireland. *Proceedings of the Royal Irish Academy* 75B, 401–451.

McCarthy, T.K. (1986) Biogeographical aspects of Ireland's invertebrate fauna. In: Sleeman, D.P., Devoy, R.J. and Woodman, P.C. (eds) *Occasional Publications of the Irish Bigeographical Society* 88, 67–81.

McCarthy, T.K. and Rita, S.D. (1991) The occurrence of the monogenean *Pseudodactylogyrus anguillae* (Yin and Sproston) on migrating silver eels in western Ireland. *Irish Naturalists' Journal* 23, 473–477.

McCarthy, T.K., Creed, K., Naughton, O., Cullen, P. and Copley, L. (2009) The metazoan parasites of eels in Ireland: zoogeographical, ecological ad fishery management perspectives. *American Fisheries Society Symposium* 58, 175–187.

McCarthy, T.K., Creed, K. and Lenihan, E.S. (in press) Freshwater fish and their parasites. In: McCarthy, T.K., Breen, J. and Lenihan, E.S. (eds), *New Clare Island Survey: Zoology*. Royal Irish Academy, Dublin.

McConville, J., Fringuelli, E., Evans, D. and Savage, P. (2018) First examination of the Lough Neagh European eel (Anguilla anguilla) population for eel virus European, eel virus European X and Anguillid Herpesvirus-1 infection by employing novel molecular techniques. *Journal of Fish Diseases* https://doi.org/10.1111/jfd.12885.

Molnár, L.K. and Székely, C. (1998) Occurrence of skrjabillanid nematodes in fishes of Hungary and in the intermediate host, *Argulus foliaceus*. *Acta Veterinaria Hungarica* 46, 451–464.

Moravec, F. and Køie, M. (1987) *Daniconema anguillae* gen. et sp. n., a new nematode of a new family Daniconematidae fam. n. parasitic in European eels. *Folia parasitologica* 34(4), 335–340.

Morrissey, M. (2007) Biology and parasitology of eels in Irish marine and mixohaline waters. Ph.D thesis, National University of Ireland, Galway.

Morrissey, M. and McCarthy, T.K. (2007) The occurrence of *Anguillicola crassus* (Kuwahar, Nimi and Hagaki,1974), an introduced nematode, in an unexploited western Irish eel population. *Biology and Environment-Proceedings of the Royal Irish Academy* 107B, 13–18.

Morrissey, M. and McCarthy, T.K. (2008) A first record of the parasitic nematode *Daniconema anguillae* Moravec et Køie 1987 (Spiruridae, Dracunculoidea: Daniconematidae) from European eel (*Anguilla anguilla*) in Ireland. *Irish Naturalists' Journal* 29, 99–101.

Newbold, L.R., Hockley, F.A., Williams, C.F., Cable, J., Reading, A.J., Auchterlonie, N, and Kemp, P. (2015) Relationship between European eel *Anguilla anguilla* infection with non-native parasites and swimming behaviour on encountering accelerating flow. *Journal of Fish Biology* 86(5), 1519–1533.

Palstra, A.P., Heppener, D.F.M, van Ginneken, V.J.T., Székely, C. and van den Thillart, G.E.J.M. (2007) Swimming performance of silver eels is severely impaired by the swim-bladder parasite *Anguillicola crassus*. *Journal du Counceil international pour l'Exploration de la Mer* 352, 244–256.

Purcell, P.W. (1994) Metazoan parasites of the European Eel (*Anguillla anguilla* L.) in the River Shannon, with particular reference to Lough Derg. MSc thesis, National University of Ireland, Galway.

Saraiva, A. and Eiras, J.C. (1996) Parasite community of the European eel, *Angulla Anguilla* (L.) in the River Este, Northern Portugal. *Research and Reviews in Parasitology* 56, 179–183.

Sjöberg, N.B., Petersson, E., Wickström, H. and Hansson, S. (2009) Effects of the swim-bladder parasite *Anguillicola crassus* on the migration of European silver eels *Anguilla anguilla* in the Baltic Sea. *Journal of Fish Biology* 74, 2158–2170.

Soylu, E., Özesen, C., Erdogan, F, Erdogan, M. and Tektas, N. (2013) Microhabitat distribution of *Pseudodactylogyrus anguillae* (Monogenea), *Ergasilus gibbus* and *Ergasilus lizae* (Copepoda) on the gills of European eels (*Anguilla anguilla*, L.). *Acta Zoologica Bulgarica* 65, 251–257.

Zintl, A., Poole, R.R, Voorheis, H.P. and Holland, C. (1997) Naturally occurring *Trypanosoma granulosum* in the European eel, *Anguilla anguilla,* from the west coast of Ireland. *Journal of Fish Diseases* 20, 333–341.

Zintl, A., Voorheis, H.P. and Holland, C.V. (2000) Experimental infections of farmed eels with different *Trypanosoma granulosum* life-cycle stages and investigation of pleomorphism. *Journal of Parasitology* 86, 56–59.

CHAPTER 20

Silver service: a standard protocol for eel health examinations

Neil C. Lewin[1], Joanna James[1], Amy J. Reading[1], Jo Cable[2], Randolph Richards[3], Gareth D. Davies[1], Derek Evans[4], Claude Belpaire[5], Olga L. M. Haenen[6], Jon Bolland[7], Alan M. Walker[8] and Chris F. Williams[1]

[1]National Fisheries Laboratory, Environment Agency, UK; [2]School of Biosciences, Cardiff University, UK; [3]Fish Pharm Supplies Ltd, Alloa, UK; [4]Fisheries and Aquatic Ecosystems Branch, Agri-Food and Biosciences Institute, Belfast, UK; [5]INBO, Belgium; [6]Wageningen Bioveterinary Research, National Reference Laboratory for Fish, Shellfish and Crustacean Diseases, Lelystad, the Netherlands; [7]Hull International Fisheries Institute, University of Hull, UK; [8]Centre for Environment Fisheries and Aquaculture Science, UK

Introduction

The importance of eel health and spawner quality

Multiple environmental and anthropogenic pressures are thought to have contributed to the global decline of the critically endangered European eel (*Anguilla anguilla*) over recent decades (Feunteun, 2002). National management strategies and international regulatory frameworks to protect eels and help recovery of the species have largely focused on exploitation, barriers to migration, entrainment and habitat loss (Anon., 2009; ICES, 2014), with stock assessments being based largely on population size. However, there is increasing awareness of the importance of overall health and spawner quality in the recovery of European eel stocks.

European eels have a long and complex life cycle (Figure 20.1), every aspect of which will be influenced by health. Given their longevity (yellow eels can spend over 30 years in freshwater before migrating to sea [ICES, 2014]) eels are exposed to a wide range of pathogens and environmental contaminants, which can readily accumulate. Some of these have significant effects on fitness, behaviour, survival and reproductive success (Belpaire et al., 2009; Maes et al., 2005). For example, pathogens such as the swimbladder nematode *Anguillicoloides crassus* and the virus Eel Virus European X (EVEX) can cause fatigue, condition loss and mortality during the eels' long marine migration (Newbold et al., 2015; van Beurden et al., 2012). Similarly, mobilization of certain contaminants, such as cadmium, could also reduce fecundity and larval survival (Pierron

Figure 20.1 Different life stages of the European eel, all subject to different health pressures. Source: National Fisheries Laboratory, Environment Agency.

et al., 2008; Geeraerts and Belpaire, 2010). As such, considerations of health and spawner quality should be integral to management approaches for the recovery of European eels.

Drivers and recognition at a European level

Integrating health assessments into eel monitoring programs is now a pan-European initiative. Indeed, the Joint EIFAAC (European Inland Fisheries and Aquaculture Advisory Commission)/ICES (International Council for the Exploration of the Sea)/GFCM (General Fisheries Commission for the Mediterranean) Working Group on Eels (WGEEL), are united in recognizing the urgent need for member states to conduct routine monitoring to improve understanding of eel distribution and abundance, and assess the impact of contaminants, pathogens and diseases on overall health and reproductive success (ICES, 2012). These data will inform international eel stock monitoring and provide a platform to better integrate health assessments into the future management of this critically endangered species.

According to ICES (2012), there is currently insufficient data available to estimate the proportion of eel populations that are successfully reproducing. The relative importance of different contaminants and diseases on eel survival and recruitment remains poorly understood. Moreover, there is an urgent need to standardize approaches to future eel health assessments to improve data comparability and maximize data retrieval. In working towards this, the European Eel Quality Database (EEQD) was set up in 2011 as a pan-European initiative for the collection of eel health data (Belpaire et al., 2011). In 2015, the Planning Group on the Monitoring of Eel Quality met for its first workshop (WKPGMEQ), with the aim to document standardized and harmonized protocols for evaluating spawner quality of the European eel, with particular attention on the bioaccumulation of contaminants and the presence of diseases (ICES, 2015; Figure 20.2). This initiative identified the need to find common solutions to, and produce guidelines for, eel sampling procedures, diagnostic approaches and reporting of data on all aspects of eel health. Recommendations were made for member states to apply standardized methods for eel spawner quality assessment and reporting, with routine monitoring of contaminants, diseases and lipid levels.

Figure 20.2 Delegates from the ICES Workshop of the Planning Group on the Monitoring of Eel Quality, meeting to assess the 'Development of standardized and harmonized protocols for the estimation of eel quality' in 2015. Photograph: INBO, Belgium.

Development of the protocol

To meet the need for standardized eel health examinations, a collaborative project is underway to develop a comprehensive necropsy protocol. The broader aims of this project are to:

- Produce a single document providing a comprehensive framework for examining eels of all life stages; showing clear, annotated guidance that will allow practitioners of all abilities and experience to conduct consistent assessments of health and spawner quality.
- Provide a network of eel experts and fish health specialists covering all major disciplines to promote knowledge transfer, tissue sharing and partnership working.
- Provide information on the equipment and consumables required to support both laboratory and field-based eel examinations.
- Introduce basic eel biology with descriptions of handling, maintenance, anesthesia/euthanasia, general anatomy and dissection.
- Give information on methods for determining sex, reproductive status, silvering stage, lipid storage, age and otolith microchemistry.
- Provide a detailed and annotated protocol for eel necropsy to support the detection of parasites/pathogens and contaminants.
- Inform on approaches for non-destructive assessments of eel health, diseases and spawner quality to reduce the need for sacrificial sampling.
- Outline methods for archiving tissues to enable longer term material storage and promote tissue sharing.
- Produce a glossary of terms covering all aspects of eel health and a comprehensive list of references and recommended further reading.

Proposed structure of the eel health examination protocol

All major disciplines for an eel health examination will be considered within the protocol, with the initial chapter focusing on considerations for working with eels, eel handling and anatomy (Figure 20.3). This is followed by chapters on ecology and morphometrics; parasites and

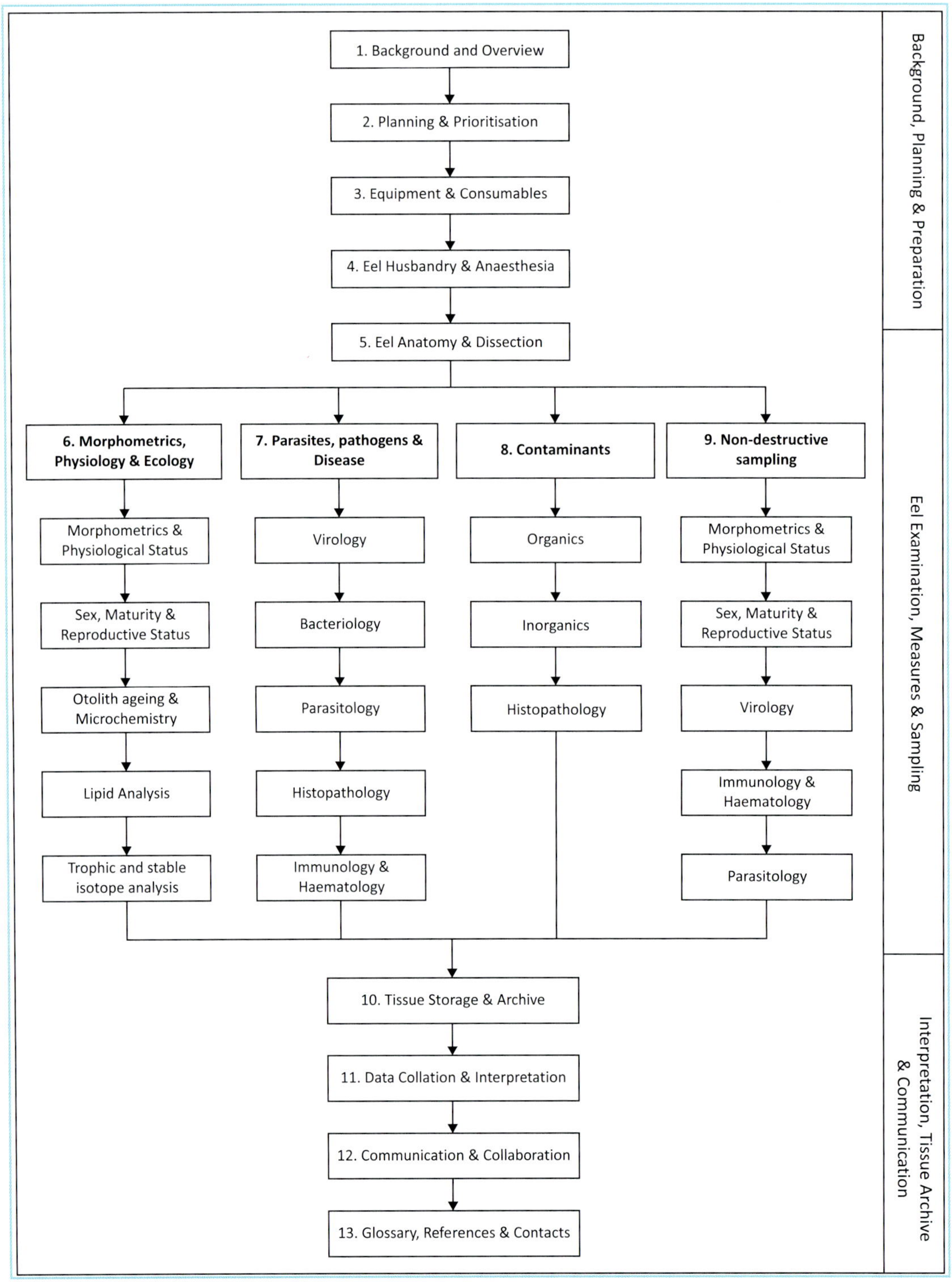

Figure 20.3 Proposed structure of the eel health examination protocol being developed through collaboration between international experts in the field. Source: National Fisheries Laboratory, Environment Agency.

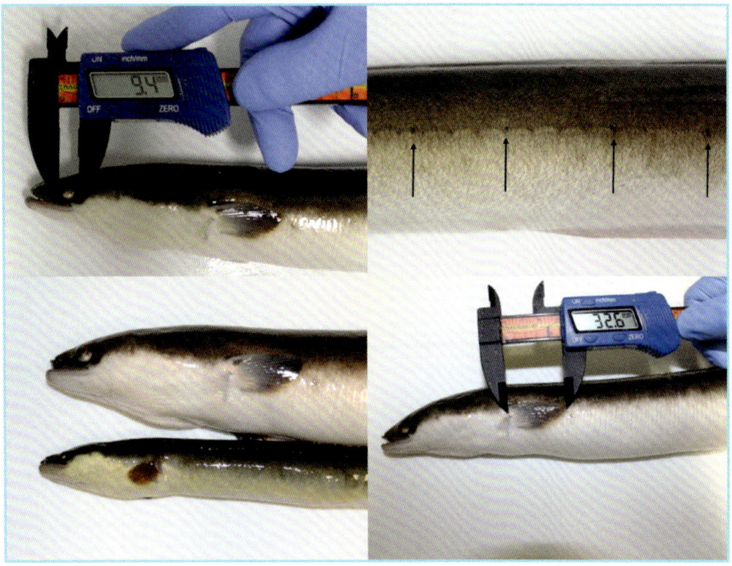

Figure 20.4 An example of the types of supporting images that will be provided throughout the eel health examination protocol in development. In this instance the images demonstrate the measurements that need to be taken to assess eel silvering stage. Source: National Fisheries Laboratory, Environment Agency.

pathogens; contaminants; and non-destructive measures of health (Figure 20.3). The eel health protocol will conclude with a discussion on interpretation of findings, partnership working and tissue archiving (Figure 20.3).

All of the chapters of the eel health examination protocol are being developed alongside experts in their respective field and will provide background information supported by published literature. Information will be presented in a step-by-step manner, with annotated images (Figure 20.4) to allow easy understanding of approaches and methods for sample collection.

Case studies

Although the eel health examination protocol is still in development, it has already been used by the National Fisheries Laboratory (Environment Agency, England) as a framework for conducting eel health assessments. Below are two case studies demonstrating how the protocol can be utilised.

Case study 1 – Anguillid herpesvirus distribution

In 2009, the Environment Agency investigated an eel specific mortality from a freshwater fishery. Post mortem analysis of eels sampled from this site revealed the presence of Anguillid herpesvirus (AngHV1). This was the first confirmed outbreak of this virus in wild eels in the UK. The lack of existing baseline data available on the distribution of AngHV1, makes it difficult to evaluate the relative importance of these outbreaks in determining eel population health at a national level. In response to this the Environment Agency, in collaboration with Cefas, Weymouth, undertook a national survey for this virus between 2011 and 2013. The primary objective of the project was to collect blood samples from live eels (Figure 20.5) and determine viral prevalence through development of an ELISA (enzyme-linked immunosorbent assay) test for AngHV1 antibodies. This

Figure 20.5 Blood sampling wild eels to improve understanding of the AngHV1 prevalence in rivers across England and Wales. Photograph: National Fisheries Services, Environment Agency.

study also required an assessment of eel life stage, general condition and signs of clinical disease. The eel health protocol was used to maximize data retrieval from this non-destructive sampling. Attention was therefore given to sections of the eel health examination protocol on project planning, field-based equipment, eel handling and anaesthesia, and then the suite of non-destructive measures to include morphometrics, evidence of clinical disease signs and blood sampling.

In total, blood samples for AngHv1 screening were collected from 687 eels from 36 rivers across 11 river basin districts in England and Wales. None of these eels were sacrificed and all of the samples were archived for potential future study. While this represented a relatively simple research question, the protocol provided a quick and easy way of standardizing the approaches needed and communicating these to all involved in sampling of eels. Work is underway to expand the range of non-destructive approaches available for eel health assessments as this represents a priority for future monitoring.

Case study 2 – Pumping station entrainment

An important consideration to the successful migration of European eels, is that of physical barriers within waterways, such as pumping stations and hydropower, and the ability of the eels to pass through such facilities. This is a Europe-wide issue, but little information is available on the effect of pumping station passage on eel health, particularly with regard to less visible sub-lethal changes or internal damage that could compromise survival and migration success. To address this question, a partnership project was set up between the Environment Agency and Hull Inland Fisheries Institute to assess the health of eels migrating through different types of pumping station. This raised a number of challenges and required a robust and consistent approach that could be applied to all sites. It also raised the need for both primary sampling to directly assess health and secondary sampling to support other disciplines (for example, ageing) that could influence behaviour during passage. With such destructive sampling, there was also a requirement to consider opportunities for tissue archiving or sampling to support wider initiatives and partnership work (for example, contaminant analysis and additional information

Figure 20.6 Pumping stations (top left), have been known to cause significant damage and mortality to migrating silver eels. A preliminary draft of the eel health examination protocol was used to design and conduct a comprehensive assessment of eel health, pre and post migration, through different pumping stations. Photographs: National Fisheries Laboratory, Environment Agency.

on virus prevalence). Assessments were conducted on tagged eels before and after entrainment through five different pumping stations in England (Figure 20.6).

Using the preliminary eel heath examination protocol as a framework, assessments of eel life stage, silvering and external damage were made in the field. Again using this version of the protocol as a guide, detailed dissections were conducted in the laboratory, and samples taken for histopathology testing. To maximise data retrieval from the eels sacrificed samples for virology and contaminant tests we also collected, and otoliths removed for age analysis. The project is ongoing but the results will be used to evaluate the effects of different pumping stations on eels being entrained. It is hoped that the methods used could be applied in future assessments of eel health.

Summary

Health is a central component of eel recovery and is integral to the success of measures to manage stocks across the geographical range of the species. The aim of the eel health examination protocol described here is to improve health monitoring and to help practitioners maximise use of post-mortem samples, thereby reducing the number of eels sacrificed. Also, to provide a tool to enable students, practitioners and research scientists to adopt a standardized approach for all future work conducted on eels irrespective of the underlying questions.

Adopting a standardized approach will facilitate the collection of better and more comparable data that can be better integrated into the management of eels. These data will help improve our understanding of current and future pressures upon the eel stock and help us manage these to promote eel recovery. Above all, the application of a single, agreed approach to understanding and promoting health, will increase the sharing of tissues, support more partnership working and facilitate knowledge transfer at local, national and international levels. Overall, it is hoped this protocol will help support the delivery of European initiatives (such as the EEQD) and, through informing management strategies, assist eel population recovery programmes.

Going forward the intention is to continue this collaborative work with international experts to ensure that the most appropriate and current methodologies for addressing different management and research questions are recommended in the protocol. The next priority will be the completion of individual chapters, with considerations being made for the future release of the complete protocol.

Acknowledgements

Thank you to all those who have assisted with this ongoing framework, particularly Environment Agency colleagues at National Fisheries Laboratory Brampton.

Bibliography

Anon. (2009) The Eels (England and Wales) Regulations. Available at: http://www.legislation.gov.uk/uksi/2009/3344/contents/made

Belpaire, C.G.J., Goemans, G., Geeraerts, C., Quataert, P., Parmentier, K., Hagel, P. and De Boer, J., (2009) Decreasing eel stocks: survival of the fattest? *Ecology of Freshwater Fish* 18(2), 197–214.

Belpaire, C., Geeraerts, C., Evans, D., Ciccotti, E. and Poole, R., (2011) The European eel quality database: towards a pan-European monitoring of eel quality. *Environmental Monitoring and Assessment* 183(1–4), 273–284.

Feunteun, E. (2002) Management and restoration of European eel population *Anguilla anguilla*: an impossible bargain. *Ecological Engineering* 18, 575–591.

Geeraerts, C., and Belpaire, C. (2010). The effects of contaminants in European eel: a review. *Ecotoxicology* 19, 239–266.

ICES (2012) Report of the 2012 Session of the Joint EIFAAC/ICES Working Group on Eels, Copenhagen, Denmark, 3–9 September 2012; ICES CM 2012/ACOM:18, 824 pp. EIFAAC Occasional Paper 49.

ICES (2014) Report of the Joint EIFAAC/ICES/GFCM Working Group on Eel, 3–7 November 2014, Rome, Italy. ICES CM 2014/ACOM.

ICES (2015) Report of the workshop of a planning group on the monitoring of eel. Development of standardized and harmonized protocols for the estimation of eel quality (WKPGMEQ), 20–22 January 2015, Brussels, Belgium. ICES CM 2014/SSGEF:14.

Maes, G.E., Raeymaekers, J.A.M., Pampoulie, C., Seynaeve, A., Goemans, G., Belpaire, C. and Volckaert, F.A.M.J. (2005) The catadromous European eel Anguilla anguilla (L.) as a model for freshwater evolutionary ecotoxicology: Relationship between heavy metal bioaccumulation, condition and genetic variability. *Aquatic Toxicology* 73(1), 99–114

Newbold, L.R., Hockley, F.A., Williams, C.F., Cable, J., Reading, A.J., Auchterlonie, N. and Kemp, P.S. (2015) Relationship between European eel Anguilla anguilla infection with non-native parasites and swimming behaviour on encountering accelerating flow. *Journal of Fish Biology* 86(5), 1519–1533.

Pierron, F., Baudrimont, M., Lucia, M., Durrieu, G., Massabuau, J.C. and Elie, P. (2008) Cadmium uptake by the European eel: Trophic transfer in field and experimental investigations. *Ecotoxicology and Environmental Safety* 70, 10–19.

Van Beurden, S.J., Engelsma, M.Y., Roozenburg, I., Voorbergen-Laarman, M.A., van Tulden, P.W., Kerkhoff, S., van Nieuwstadt, A.P., Davidse, A. and Haenen, O.L., (2012) Viral diseases of wild and farmed European eel Anguilla anguilla with particular reference to the Netherlands. *Diseases of Aquatic Organisms* 101(1), 69–86.

PART V

Aquaculture

CHAPTER 21

Captive breeding of the European eel: Italian review

Oliviero Mordenti, Antonio Casalini, Albamaria Parmeggiani, Pietro Emmanuele and Annalisa Zaccaroni

Department of Veterinary Medical Sciences, University of Bologna, Bologna, Italy

Introduction

The European eel (*Anguilla anguilla*) is a catadromous and semelparous species that spends most of its life as yellow eel in freshwater, brackish and coastal habitats. Upon reaching sexual maturity, it returns to the Sargasso Sea to spawn and die (Aschonitis et al., 2017). Before the seawater spawning migration, eels undergo significant morphological and physiological changes, a process known as 'silvering' (van den Thillart and Dufour, 2009, Sorteni et al., 2014). Thus, silvering is a requirement for downstream migration and reproduction and marks the end of the growth phase and the onset of sexual maturation (Durif et al., 2009).

Long-term records of eel species (*Anguilla* spp.) over the last four decades indicate an extensive worldwide reduction in numbers. European eel, American eel (*A. rostrata*) and Japanese eel (*A. japonica*) have decreased markedly (Dekker, 2003, Casselman, 2003) due to overfishing, environment destruction, oceanographic/climatic changes, parasitism, increased migration barriers, reduction of available prey in freshwater habitats and exotic fish invasions (EELREP, 2006, van Ginneken and Maes, 2005). The critical levels of the eel populations in Europe led to the application of measures for stock recovery based on European Council Regulation 1100/2007 (Regulation, 2007) and management plans for eel fisheries (Aschonitis et al., 2017).

One effective way to preserve these species is through the establishment of techniques for mass production of glass eels to reduce the fishing pressure on wild glass eel stocks (Okamura et al., 2014).

With reference to the European eel, research has focused on (1) successful artificial propagation protocols in terms of hormone dose and timing and (2) on the identification of optimal environmental parameters (water temperature, water salinity and photoperiod) (Durif et al., 2006, Mordenti et al., 2013, Mordenti et al., 2012, Parmeggiani et al., 2015, Di Biase et al., 2016, Di Biase et al., 2017). Current artificial induction methods for maturation are largely based on the repeated – typically weekly – use of human chorionic gonadotropin in male eels and of pituitary homogenates in female eels; fertilized eggs are successfully obtained by spontaneous spawning (Mordenti et al., 2018, Di Biase et al., 2017, Di Biase et al., 2016, Mordenti et al., 2014) or after gametes collection by stripping-insemination (Tanaka et al., 2001, Pedersen et al., 2003, Pedersen, 2004, Okamura et al., 2009, Palstra and van den Thillart, 2009, Burgerhout et al., 2011, Oliveira and Hable, 2010).

However, eels receiving the same treatment showed high variability in their maturation response, thus it is possible that these responses reflect not only individual reproductive capacities (Durif et al., 2006, Mordenti et al., 2013) but also the different habitats of each eel population. Consequently, one of the major problems in gamete production remains identification of broodstock with highly performing reproductive characteristics (Mordenti et al., 2013).

In Italy, extensive cultures are managed in coastal lagoons. The North Adriatic lagoons constitute nearly one-half of the total exploited Italian brackish lagoon environments. In this area, European eels are reared extensively via the traditional *Vallicoltura* (Mordenti et al., 2013, Mordenti et al., 2016, Ciccotti, 1997), carried out in the *Valli*, a sector of a lagoon or enclosed earthen pond (Ciccotti, 1997). The eel fishery of the Comacchio Lagoon in north-eastern Italy was one of the most important sites in terms of production. The lagoon has never been restocked, so the eel population is based only on natural recruitment.

In this chapter, we summarize the research carried to date on captive eel reproduction in Italy, focusing on the work of the University of Bologna.

Morpho-physiological features for the selection of broodstocks for artificial reproduction protocol

European eels have a complex life cycle. After a long growth period in continental waters (4–20 years), eels undertake a 5–6000 km spawning migration from Europe to the Sargasso Sea (Aarestrup et al., 2009). Before the seawater spawning migration, eels undergo significant morphological and physiological changes, in a process known as 'silvering' (van den Thillart and Dufour, 2009). Thus, silvering is a requirement for downstream migration and reproduction and marks the end of the growth phase and the onset of sexual maturation (Durif et al., 2009).

The morphological modifications include a change in belly colour from yellow to silver/bronze and in back and pectoral fins from white/grey to black and increased eye size (Fontaine, 1994, Han et al., 2003b, EELREP, 2006, Okamura et al., 2007). The physiological changes include degeneration of the digestive tract (Fontaine, 1994, Han et al., 2003a, Han et al., 2003b), changes of visual pigments, a more developed swim bladder and higher density of branchial chloride cells (Fontaine, 1994). It is suggested that these modifications in silver eels are pre-adaptation for the oceanic migration back to the Sargasso Sea (Han et al., 2003a, Durif et al., 2005).

Also, silvering marks the start of lipid mobilization and sexual maturation. It appears that silvering is not a true metamorphosis but a mere initiation of maturation (Palstra and van den Thillart, 2010). Silvering is more flexible than generally presumed (Svedäng and Wickström, 1997) and can be influenced by several trophic and environmental factors (Melià et al., 2006, Van Ginneken et al., 2007, van Ginneken and Maes, 2005, Durif et al., 2005, van den Thillart and Dufour, 2009).

At present, the only way to obtain sexually mature eels is to artificially induce sexual maturation in migratory individuals caught in brackish and freshwater environments, or cultured silver eels, using repeated injection of carp (CPE) or salmon (SPE) pituitary extract and a final injection of 17, 20 β-dihydroxy-4-pregnen-3-one (DHP) (Ohta et al., 1996, Palstra et al., 2005, Burgerhout et al., 2011). Studies have focused, particularly, on successful protocols based on dose and timing of hormone injection and on definition of environmental optimal parameters (water temperature, water salinity, photoperiod) (Durif et al., 2006). However, eels receiving the same treatment showed high variability in their maturation response, thus it is possible that these responses reflect not only individual reproductive capacities (Durif et al., 2006) but also the different habitats of each eel population.

Our research group wanted to describe the variability in the response of female silver eels in terms of gonad development and egg production to a standardized gonadotropic treatment (CPE) (Mordenti et al., 2013), and to relate this variability to population characteristics. For this purpose, sexual maturation and reproduction was induced in two populations coming from different locations (Comacchio [CM] and Marano-Grado [MG] lagoons) in the Adriatic Sea (Figure 21.1). Their reproductive capacity was assessed throughout the experimental periods using morphological and physiological indicators and related to their initial physiological and morphological characteristics (Silver Index – SI).

The SI encompasses many morphological and physiological parameters of wild female silver eels, and is therefore a better predictive value with respect to real silvering stage (Dufour and Van den Thillart, 2009). The results suggested that the MG and CM lagoons, in the northern Adriatic Sea, are environments with outstanding characteristics in which it is still possible to find female eels with a high SI to be used to start a programme of artificial reproduction and where it is possible to restore the conditions to host and breed eels. Indeed, about 90% of females treated with CPE reached ovulation. However, all MG eels were at maximum migrating stage (V) while CM eels, even though showing a good SI level, included females in pre-migrating stage (III) (Table 21.1). Morphometric parameters showed that the CM group presented highest values of body weight (BW), body length (BL) and condition factor (K), while the MG group presented

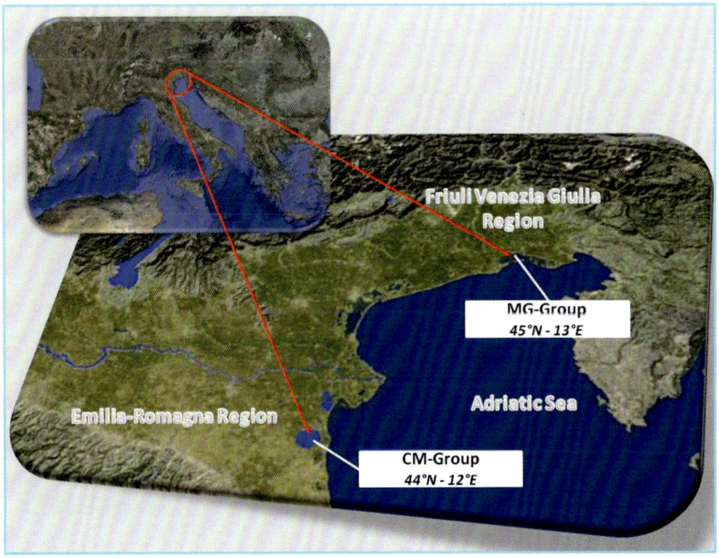

Figure 21.1 Map showing origin of eels.

Table 21.1 Initial characteristics of silver eel adopted for experimental procedure; *: statistical differences between CM and MG eels, $P < 0.05$.

		CM	MG
Silver index (SI) III	%	8.6	–
IV		77.1	–
V		14.3	100
Body Weight (BW)	g	1353±241*	358±94
Body Length (BL)	cm	85.62±4.98*	59.16±4.78
Condition factor (K)		2.13±0.22*	1.70±0.13
Eye index (EI)		10.12±1.35	10.80±2.19
Pectoral fin length index (PFLI)		4.68±0.40	5.55±0.37*
Gonadosomatic index (GSI)		1.61±0.26	2.17±0.36*
Age	years	5 (40.0%)	5 (54.3%)
		6 (48.6%)	6 (45.7%)
		7 (11.4%)	

the highest pectoral fin length index (PFLI) and gonadosomatic index (GSI) values. With regard to lipid content, this research underlined a higher energetic reserve in CM eels, persisting for the whole study length. Highest lipid storage in CM eels did not lead to a real advantage in terms of reproduction: indeed no higher lipid mobilization from muscle to gonads (eggs) was observed and egg production was lower with respect to MG eels. Our hypothesis is that the higher energetic storage was not used by eels for reproduction, as triggering of lipid mobilization and sexual maturation, but was required for swimming activity, absent in our study (recirculation system did not create any flow). In summary, the main lipid storage led to clear advantages in the wild, as only eels with sufficient lipid stores will start their reproductive migration to the spawning grounds in the Sargasso (Palstra and van de Thillart, 2010), while in captivity it seems to have no advantageous aspect. On the contrary, MG eels, characterized by a smaller size but a higher SI, showed a more regular gonadic development leading, although after a longer time, to a spontaneous egg emission.

With reference to zootechnical performance (Table 21.2), MG eels were statistically more productive (eggs/kg female) than the CM group. However, eels from CM group showed an earlier response to hormonal treatment: CM group females ovulated especially between 19th and 22nd weeks (77.8% spawned eels) while in MG eels the ovulation started at the 24th week and finished at the 28th week (100% spawned eels). Fertilization and hatching rates of eggs obtained from CM and MG eels suggested that the silvering states of females, as noted by Okamura et al. (2008), does not affect egg quality when maturation is induced by hormonal treatment.

In conclusion, the results of work suggested that the morpho-physiological features in European eel can affect ovarian development, being an important parameter in artificial reproduction protocols: bigger dimensions, highest condition factor and higher lipid storage in CM eels seems to hasten gonad maturation without positively increasing reproductive performance, while MG females, of a smaller size but a higher SI, showed instead a more regular gonad development, leading eels to spontaneous spawning.

Table 21.2 Zootechnical performances obtained by female eels at the end of the experimental protocol; *: statistical differences between CM and MG eels, $P < 0.05$.

		CM	MG
BWI increase	%	113.53±3.68	120.63±4.16*
Spawned eels	n/total	9/10	8/10
Spawned	%	88.1	–
Stripping Natural		11.1	100
Spawned eggs	% BW	25.16±7.88	40.10±6.33*
Stripping		27.13±5.59	–
Natural		9.4	40.10±6.33

Comparison of performances between spontaneous spawning and stripping-insemination approaches

The European eel is a species with a high potential market in freshwater aquaculture. To date, all the seedlings for cultivation are wild glass eels or elvers collected in estuarine waters. However, the European eel was recently included in the Red List of the IUCN as a Critically Endangered species; as a consequence, a short fishing season, a minimum capture size, larvae protection and an implemented trade regulation have been imposed to protect this species. Unfortunately, the application of the measures has not reduced the risk of extinction (Mordenti et al., 2012, Mordenti et al., 2013).

One effective solution to the issue would be to set artificial reproduction techniques for the production of seedlings for aquaculture so as to reduce the demand for wild glass eels. At present, studies have focused on the successful protocols based on hormone injection (carp or salmon pituitary extract) dose and timing and on the definition of optimal environmental parameters (water temperature, water salinity and photoperiod) (Mordenti et al., 2012, Durif et al., 2006) in order to obtain gametes by stripping for artificial fertilization; while laboratory experiments that have shown spontaneous spawning of artificially matured European eels in captivity remain rare.

Our research group focused its activity on the construction of a new closed recirculating aquaculture system including spawning and incubation chambers with the objective to obtain spontaneous reproduction in captivity of the European eel by using wild female silver eels.

The plant was designed to obtain spontaneous spawning in European eels in captivity as has been done for many years in the most important European marine farmed fish (*Dicentrarchus labrax*, *Sparus aurata*, *Scophthalmus maximus*, *Solea solea*, *Gadus morhua* and *Hippoglossus hippoglossus*) (Brown et al., 2006). Currently the reproduction of fish in captivity is mainly controlled by environmental manipulations, such as photoperiod, water temperature and salinity, or spawning substrate (Kelley et al., 1991, Bromage, 1993, Carrillo, 1995, Brooks et al., 1997, Anguis and Canavate, 2005, Mylonas et al., 2010). However, other factors such as tank size, water volume and/or depth, hydrodynamic conditions, water flow rate and stocking density were shown to influence reproductive success in some cultured fishes (Tsadik and Bart, 2007, Mylonas et al., 2010, Oca and Masalo, 2013).

Furthermore, spontaneous reproduction of fish in captivity removes the need to strip ovulating females and spermiating males for gamete collection and for egg manipulation during the artificial fertilization and their transfer to the hatcheries.

It has been widely demonstrated that the stress induced by frequent manipulation to obtain gamete emission can negatively affect the breeder performance and lead to death (Campbell et al., 1992, Bromage, 1995, Brooks et al., 1997, Schreck et al., 2001). In particular, in aquaculture broodstock and egg handling may determine deterioration in egg quality (Hilomen-Garcia, 1998, Kjørsvik et al., 1990, Brooks et al., 1997, Mylonas et al., 2004) and, in the case of species that do not spawn spontaneously, it is difficult to determine the optimal latency period for the collection of the gametes. It was demonstrated, in fact, that an early or delayed stripping is directly related to a loss in egg quality (Bromage et al., 1994, Brooks et al., 1997). Moreover, spontaneous spawning is preferable since it has the advantage of eliminating the time lapse involved in the artificial fertilization. In eels, in particular, insemination soon after ovulation is a prerequisite of success, for egg quality degenerates relatively soon after ovulation (Abe et al., 2010, Ijiri et al., 2011).

A new closed recirculating system is shown in Figure 21.2a and b. This system consisted of two fish-rearing tanks (0.47 m^3/tank; water volume 0.43 m^3/tank; water surface area 0.93 m^2/tank), a foam separation tank (0.05 m^3) (protein skimmer) and a biological filter (0.21 m^3) containing plastic porous balls (0.15 m^3) (Figure 21.2a). The total amount of water in the system was 1.12 m^3, which was transported by a circulating pump (1.1 kW; max delivery 16,000 l/hour). The rearing water reached the biological filter and the protein skimmer with an up-flow style (natural falling water) and the treated rearing water was returned to the rearing tank by a circulating pump. The system was also provided with a thermal regulation system (compact cooling equipment, 1.4 kW) to adjust the rearing water conditions (20±0.5°C), a UV-sterilizer lamp (max delivery 500 l/hour, 36W), an ozonizer (250 mg O_3/hour) and an aerator (electromagnetic air compressor; delivery 170 l/minute, pressure 34 kPa, 150 W) (Figure 21.2a).

The core of the system was the reproduction tank, which was made of four components: one spawning chamber (240 l), two incubation chambers (52 l/cad) and one outlet chamber (90 l).

The spawning chamber is connected to the incubation chambers through two 5 mm lengthwise splits located on the top side of the dividing panel. Two pipes allow the water to enter from the base of the spawning chamber (inlet tubes 1) in order to guarantee the water exchange and promote, once the gametes are released, the entrance of the eggs into the incubation chambers (Figure 21.2b). The water exchange rate is regulated by a valve (valve 1) (Figure 21.2b).

The incubation chamber has a cylindrical base and a tube on the top (inlet tube 2) and it is provided with inlet jets that produce a circular revolving current: an outlet mesh screen (200 × 200 mm; 300 μm diameter and exchangeable) is located on the dividing panel between the incubation and the outlet chambers; inlet jets push water across this mesh screen (Figure 21.2b). The water flow and the current speed in the incubation chamber are calibrated by the inlet tube valve of the spawning chamber (valve 1) and by the hatchery water pipe valve (valve 2) (Figure 21.2b).

The system has a vertical configuration and is inspired by the one originally developed by Greve (1975) (called 'planktonkreisel') for the maintenance of planktonic animals, later modified by various researchers for larval *Palinurus japonicas* culture (Matsuda and Takenouchi, 2007) and for rearing eel leptocephali (Okamura et al., 2007).

The overall tank water level is determined in the outlet chamber by adjusting the outlet tube height; at this point the water is discharged into the filtration system therefore adjusting the environmental parameters (Figure 21.2b). The water outflow takes place by overflowing, into a collecting tray (440 × 100 mm) positioned at the top of the outlet tube, in order to avoid water whirls in the outlet chamber.

Finally, three covers positioned on the tank had the purpose of maintaining darkness and trapping the breeders inside the tank.

The system was tested with 12 silver eel females which were induced with increasing doses of CPE (10, 20, 30 and 40 mg CPE/kg BW). Twenty-four hours after the last CPE injection, each female ovulation was induced by an injection of a DHP-solution. After which they were transferred to a new closed recirculating aquaculture system, where they were maintained for 16 hours with spermiating males (sex ratio 4 : 1) in order to achieve spontaneous reproduction. The reproduction was tested with six females in low-flow rate conditions and six females in high-flow rate conditions.

The results showed that the designed closed-loop system made it possible to carry out a more spontaneous reproduction for more than 80% of the females that underwent standardized gonadotropic treatment and favoured the automatic and complete transfer of the eggs to the hatchery.

This work suggests that it is possible to obtain spontaneous reproduction in captivity with European eels as already recorded in some work on Japanese eels (Dou et al., 2007, Dou et al., 2008, Horie et al., 2008, Okamura et al., 2009, Ahn et al., 2012). It is also important to highlight that spontaneous reproduction was possible with wild silver eels and not with farmed females or feminized eels (Ijiri et al., 2011) that were already adapted to life in captivity. The results also point out that high or low water current conditions in the tank do not hinder the mating and the emission of gametes by the breeders. The designed system, in short, seems to meet all the requirements for spontaneous reproduction in the European eel, thus eliminating all the issues related to 'stripping' and egg handling. The absence of mortality during the experiment is another extremely positive result considering that several studies with European eels reported high mortality rates (EELREP, 2006, van Ginneken et al., 2005, Pedersen, 2003, Pedersen, 2004, Palstra et al., 2005, Palstra and van den Thillart, 2010).

However, the experiment showed that a high flow rate in the incubation chamber causes a steady loss of viable eggs over time. Probably the high flow rate causes mechanical shocks within the incubation chamber that can lead to embryo death linked to the high fragility of eel eggs. However, mechanical shock sensitivity of fertilized eggs has been reported to vary among species (Hilomen-Garcia, 1998, Alderdice et al., 1988, Jensen and Alderdice, 1989, Thorsen et al., 2003).

Regarding the survival rate at the end of incubation (24 hours), a reduced number of fertile eggs was shown with increasing stocking density. There is no coincidence that the best results were obtained with those eels characterized by lower body weight and reduced egg emission rates, which showed a stocking density of less than 3000 eggs/l while the worst was recorded with eels that were close to 6000 eggs/l.

It is clear, at this point, that the recirculation system must be reformulated by increasing the volume of the two incubation chambers without, however, altering the circular revolving water current. A second solution would be to remove the breeder once it reaches a determined incubation density that should not exceed 2000 eggs/l, according to observations led in this study. currently the literature does not provide precise information about stocking density of incubated eel eggs and therefore it would be better to adopt prudent levels compared to those recorded in the experiment.

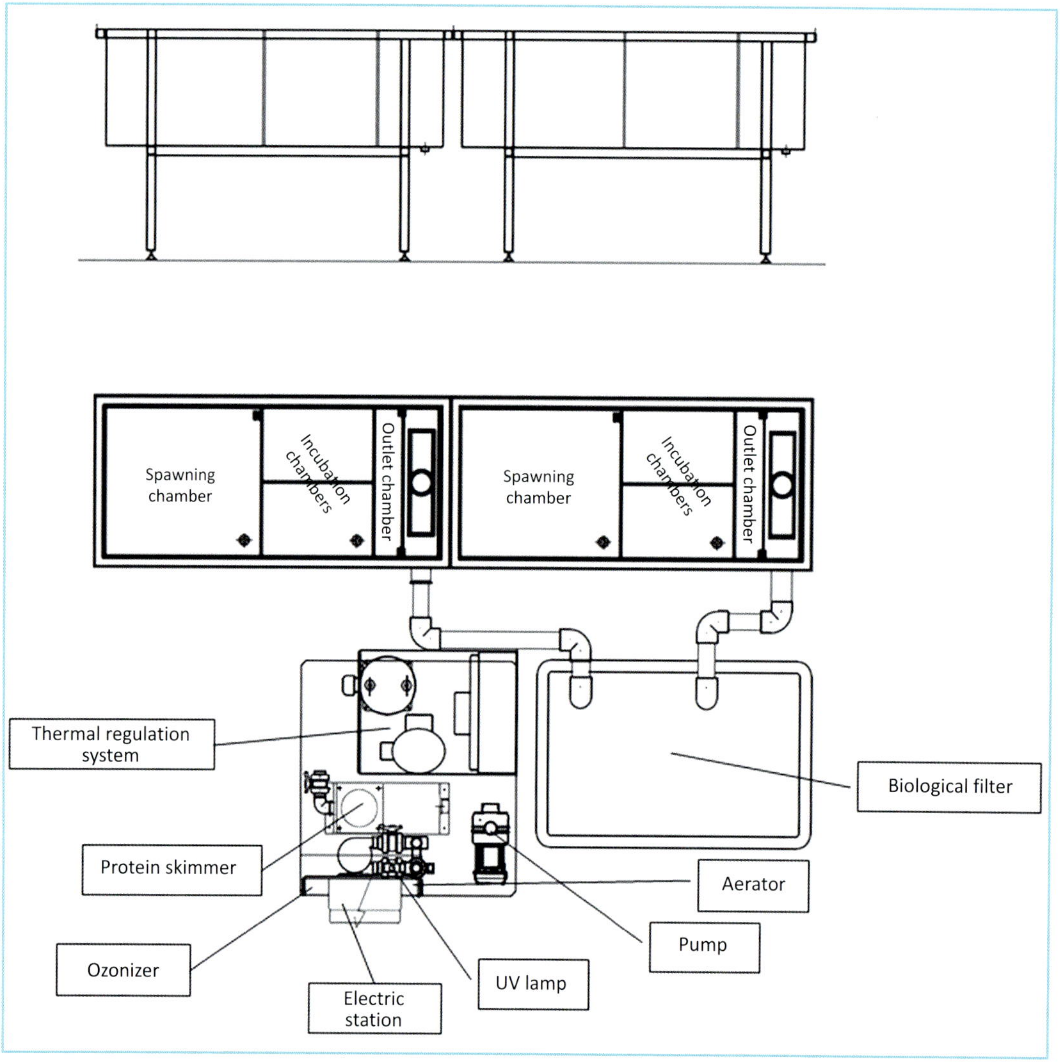

Figure 21.2a and b Schematic illustration of the closed recirculating aquaculture system for artificial seed production of *A. anguilla*.

The plant, which has been designed to avoid any egg manipulation from release to hatching, does not allow for the separation of viable eggs from the dead ones as happens for some farmed species, this may lead to a health risk. After fertilization, in fact, dying or dead eggs are colonized with bacteria/fungus, and if they are not quickly removed from the incubation trays, viable eggs may also be colonized (Brooks et al., 1997). During our experiments, however, no health issues were reported probably due to the rather short time of incubation (38±2.5 degrees day) (Mordenti et al., 2013) and to the fact that once hatching was reached, it was enough to block the water flow for a few minutes to achieve separation of the dying/dead eggs from the larvae: eggs, which were positioned on the bottom were easily removed by siphoning while larvae remained motionless on the water column. It is clear, finally, that

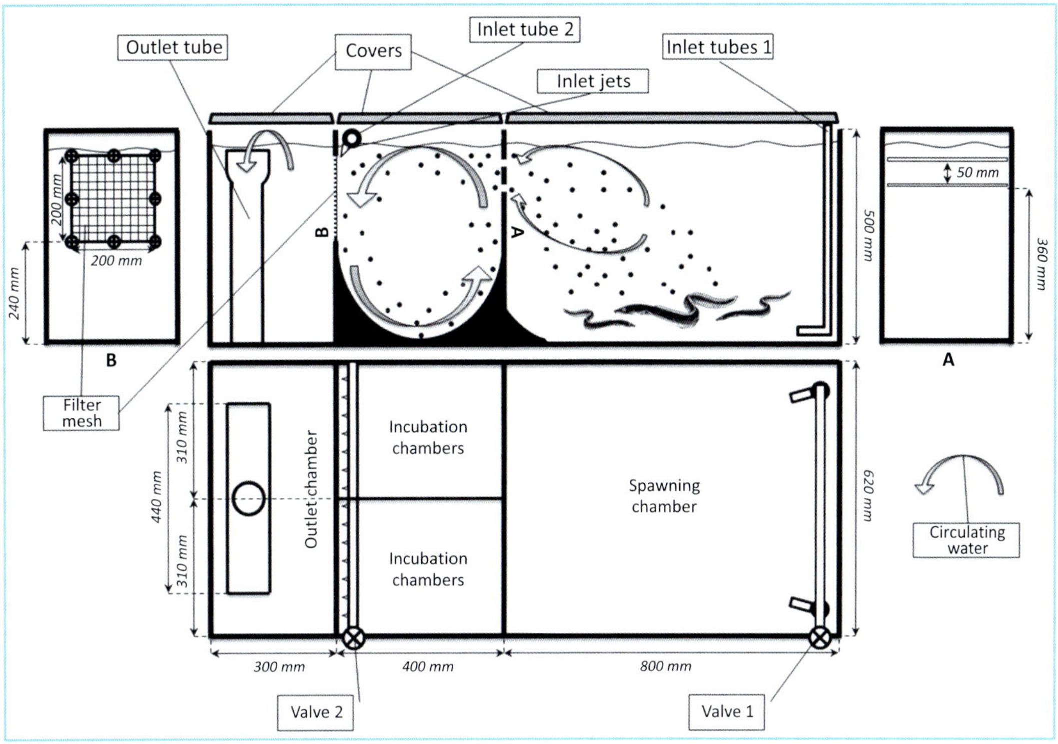

Figure 21.2a and b (Continued)

the designed plant can be used for larvae storage up to the yolk sac resorption and has all the characteristics to perform a first larval weaning phase without altering the conditions of rearing water.

The new closed recirculating aquaculture system facilitated spontaneous reproduction in captivity of the European eel, a fundamental requirement if we are to achieve significant aquacultural results in marine farming fish and to sort out the issues related to stripping. This result indeed allowed the research group to focus subsequent studies on the quality and fertility of eggs obtained instead of on reproductive success obtained by setting a specific artificial propagation protocols (hormone dose and timing and optimal environmental parameters) (Durif et al., 2006, Mordenti et al., 2013, Di Biase et al., 2016).

Indeed, one of the major problems facing seed production remains the constant availability of high-quality eggs. Egg quality, in terms of resulting fertilization, hatching and survival rates, is highly variable between the eggs from different individuals (Chai et al., 2010) or between different egg retrieval methods (Mordenti et al., 2014). Thus, it is necessary to improve the production of high-quality eggs in order to make mass production of eel larvae possible. The use of 15–25 weekly injections of salmon or carp pituitary extract can advance oocytes to the migratory nucleus stage after which final oocyte maturation and ovulation are typically induced by 17,20β-dihydroxy-4-pregnen-3-one (DHP). However, for this routine method, the relationship between timing of the DHP-induced final stage and the resulting egg quality remains to be explored. The objective of the present study was to investigate the quality and quantity of eggs produced by artificially matured European eels that were allowed to spawn spontaneously or that were subjected to stripping-insemination, and to relate these fertility parameters to the timing of DHP administration.

For this purpose, 18 wild female European eels captured in Bonello lagoon (North Adriatic Sea) were injected with CPE to undergo sexual maturation and ovulation; a final injection of DHP was administered when at least 30% of the oocytes were fully transparent. After the DHP-injection, nine eels were transferred to a new closed recirculating aquaculture system, where they were housed with spermiating males (sex ratio 4 : 1) to allow spontaneous spawning (SPT-group); the remaining nine eels were transferred to a 250 l tank and ovulation was checked at hourly intervals by manual stripping (STR-group) (Di Biase et al., 2016).

The results underlined that females that spawned spontaneously (SPT) were statistically more productive than those stripped manually (STR) (Table 21.3) and the total fertilization rate of eggs of SPT females was higher than that from eels that were spawned by stripping-insemination

Table 21.3 Reproductive performance of artificially matured female European eels that were strip-spawned (STR-group) or that spawned spontaneously (SPT-group) after hormone treatment (see text for details).

	BWI (% BW)	Oocytes (% FTO)	Ovulation (Yes/No)	Spawned eggs (% BW)
		STR-group		
1	122.2	61.4	Yes	32.40
2	108.3	40.4	Yes	24.50
3	115.5	37.5	Yes	12.70
4	128.5	68.4	Yes	38.40
5	112.8	63.2	Yes	30.80
6	118.1	58.3	Yes	33.80
7	131.3	38.5	Yes	25.40
8	123.7	33.5	No	–
9	118.5	41.5	Yes	23.60
Mean	119.88±7.36	49.90±13.37		27.70±7.93
		SPT-group		
1	117.7	64.5	Yes	42.50
2	116.7	62.3	Yes	47.50
3	118.5	35.8	Yes	39.60
4	124.9	70.5	Yes	52.80
5	134.1	32.4	No	–
6	107.9	59.2	Yes	44.50
7	125.8	38.9	Yes	40.40
8	126.2	40.2	Yes	37.60
9	120.4	72.3	Yes	50.20
Mean	121.35±7.44	52.90±15.88		44.39±5.38*

Notes: *Significant difference ($p < 0.01$) between eels in the STR-group and the SPT-group. BWI: body weight index (increase in relative body weight after injection of fish with 17,20 β -dihydroxy-4-pregnen-3-one); FTO: fully transparent oocytes at the time of treatment with 17,20 β-dihydroxy-4-pregnen-3-one).

(Table 21.4). Additionally, in both groups, a positive correlation was observed between the fully transparent oocytes (nucleus at the periphery and containing few large droplets- FTO), and the relative weight of spawned eggs: in fact, the female that retained the eggs displayed the lowest percentage of FTO at the time of DHP injection likewise the most productive females had the highest percentage of FTO (Table 21.3).

The research identified that among zootechnical performance indicators, body weight increase (BWI) at DHP injection was highly variable and it did not appear to be an adequate predictor of the final phase of ovarian maturation, as also reported by Palstra et al. (2005) and Mordenti et al. (2012), and for Japanese eels by many authors (Seoka et al., 2003, Dou et al., 2008, Chai et al., 2010, Ijiri et al., 2011, Unuma et al., 2011, Unuma et al., 2012). The lack of a correlation between the BWI and the percentage of FTOs in the gonads in study further reinforces the unsuitability of the BWI as a predictor of the right time for induction of the final maturation in the European eel. Having said that, the findings do not concur with those of Palstra et al. (2005), who showed a relationship between the number of oocytes with single fat droplets (over-ripe and not fertilizable oocytes) and higher BWIs.

Table 21.4 Fertilization rates of eggs from artificially matured female European eels that were strip-spawned (STR group) or that spawned spontaneously (SPT group) after hormone treatment (see text for details).

	Fertilization		
	(Yes/No)	Total (%)	Floating (%)
	STR-Group		
1	Yes	8.20	21.50
2	Yes	3.90	10.80
3	Yes	2.90	18.70
4	Yes	9.70	13.50
5	Yes	11.10	27.80
6	Yes	8.60	16.70
7	Yes	7.40	20.20
8	–	–	–
9	Yes	9.80	15.40
Mean		7.70±2.89	18.07±5.27
	SPT-Group		
1	Yes	64.20	93.20
2	Yes	52.30	74.50
3	No	0.00	0.00
4	Yes	57.30	78.60
5	–	–	–
6	Yes	44.70	81.20
7	No	0.00	0.00
8	Yes	38.90	68.40
9	Yes	48.60	70.70
Mean		38.25±24.81*	58.32±36.78*

Note: *Significance difference ($p < 0.01$) between eels in the STR-group and the SPT-group.

Ovarian biopsies obtained after the priming injection (the additional CPE administration following regular weekly injections) showed good synchronous development of oocytes, evidenced by 50% of oocytes being fully transparent. The stage synchrony of oocytes probably results from the low-dosage of hormone initially administered to the eels (Mordenti et al., 2013). Synchronous maturation in the final phase is very important as it enhances the quantity of spawned eggs; it is not coincidental that in this study the animals that provided the best yields (in terms of spawned eggs per %BW) had the highest percentage of FTOs.

Furthermore, the use of high-dose CPE injections (40 mg/kg BW) from the 10th week onwards led to an acceleration in the final maturation phase, as observed by Chai et al. (2010) in Japanese eels and, probably, to a better quality of eggs in terms of fertilization rate. However, there is a risk that migratory nucleus-stage oocytes develop too fast after the priming injection with CPE, and that a high proportion of oocytes contain a single fat droplet (over-ripe) prior to treatment with DHP (Chai et al., 2010). Unfortunately, over-ripening in eels progresses faster than in other teleosts (Ohta et al., 1996, Unuma et al., 2005).

Inappropriate timing of hormone administrations to pre-spawning eel broodstock may be a cause of poor egg quality. (Unuma et al., 2011) suggested that proper timing of ovulation induction is essential, but that additional factors are also important for the acquisition of good quality eggs. To predict the best time of ovulation induction, the developmental stages of the oocytes in the ovary during final maturation should be evaluated. The present study highlights that the best results in terms of spawned egg quantity and fertilization rate can be obtained when the proportion of FTO is at least 50% at the time of DHP injection. Indeed, it seems likely that the recent promising findings on spontaneous spawning of eels (Mordenti et al., 2014) reported by our team may be due to fine-tuning of the timing of DHP administration.

The present study convincingly illustrates that the spontaneous spawning method produces higher quality eggs than can be obtained by the stripping-insemination method, probably because the timing of spawning and fertilization is optimized by the parent eels themselves (Okamura et al., 2014). In addition, the high incidence of spontaneous spawning in European eels in captivity was striking. The presence of good-quality males in the tank is a decisive factor for spawning success: only a few minutes after a female was placed in the tank, typical mating behaviour was observed in males (approaching the head and urogenital region of the female) (van Ginneken et al., 2005). In sharp contrast, only females with unfertilized eggs were not approached by males displaying courtship behaviour. This could be caused by a lack of pheromone release by the female, a typical extrasomatic signal for the males that could chase the female and sniff at the urogenital pore or opercula (van Ginneken et al., 2005, Okamura et al., 2014). The better performance of the SPT group compared to the STR group is also reflected in the higher fertilization rate in the former group: for example, among the floating eggs in the SPT eels, five out of six females had a fertilization rate exceeding 70%. Retrieving floating eggs from all spawned females contrasts notably with the report on Japanese eels by (Seoka et al., 2003), in which floating eggs were obtained from 4 out of 31 females. The buoyancy of eggs, important for oceanic survival and dispersal as well as for the initiation of early embryogenesis (Kagawa et al., 2011), has often been used as an indicator in the assessment of egg quality (Unuma et al., 2005). While buoyancy is not universally indicative of egg quality in teleost fish ((Kohn and Symonds, 2012), the ratio of floating eggs to total spawned eggs at least correlates positively with egg hatchability in eels (Unuma et al., 2005).

In conclusion, this work showed that in European eels the spontaneous-spawning method can yield better egg production, both in terms of quality and quantity, than the stripping-insemination method. Furthermore, BWI is not itself a proper predictor to guide the timing of DHP injection; rather, the presence of a high proportion of fully transparent oocytes can be used to time DHP injection in order to obtain eggs of good fertility.

Effect of aromatable androgen (17-methyltestosterone) on induced maturation of silver European eels

At present, the only way to obtain sexually mature eels is to artificially induce sexual maturation in migratory individuals caught in brackish and freshwater environments or cultured silver eels using repeated injection of CPE or SPE and a final injection of DHP (Oliveira and Hable, 2010, Ijiri et al., 2011, Parmeggiani et al., 2015). Nevertheless, this method is not yet standardized and presents high variability between batches from different individuals (Chai et al., 2010, Okamura et al., 2014). Thus, it is necessary to improve techniques for the production of high-quality eggs to make mass production of larvae possible (Di Biase et al., 2016). Indeed, if this target could be reached, research could concentrate on larval weaning techniques, as first successful early feeding of European eel larvae has already been obtained (Di Biase et al., 2016, Butts et al., 2016).

More recently alternative induced spawning protocols have also been developed, and arguably, the most notable outcomes have been achieved in response to treatment with androgens [11-K testosterone (11-K)] of female silver eels (Lokman et al., 2015, Matsubara et al., 2005). For several years, the effect of androgen on eel reproductive activity was investigated. (Sudo and Tsukamoto, 2015) regarding Japanese eels, report that androgen is involved in silvering and also in migratory behaviour at the onset of the spawning migration (higher locomotor activity and increasing behavioural drive in the spawning migration). Furthermore, Sudo et al. (2012) suggested that 11-KT synchronously accelerates early development of the ovaries and morphological changes in the Japanese eel. Androgen treatment of eels resulted in morphological and physiological changes (silvering), such as increases in eye diameter and degeneration of the digestive tract (Rohr et al., 2001, Sudo et al., 2012). Wang and Lou (2007) showed that 17-α methyltestosterone (17-MT) provides a potential usefulness to precipitate the process of vitellogenesis when it combines with SPE treatments in Japanese eel. Furthermore, the synergistic effect actually enhances the growth and synchronous development of ovarian follicle throughout the reproductive process. Lokman et al. (2015) demonstrated that incorporation of androgen in female silver short-finned eels accelerated the gonadal maturation, increased lipid accumulation and oocyte size all without any apparent detrimental effect on egg quality.

The aim of the study was to verify the reproductive performances of European silver eel in terms of gonad development and egg production employing slow-release implants with the androgen 17-MT in combination with traditional weekly injection of CPE.

Eels were intramuscularly injected (i.m.) with 5 mg/kg BW (1st–4th week), 15 mg/kg BW (5th–8th week) and 30 mg/kg BW (9th week–final maturation) CPE following the originally developed protocol by Mordenti et al. (2012). Injections were repeated every 7 days until increases in body weight and a gravidic appearance indicated that oocyte hydration was likely to have started. At the first CPE injection, 50% of eels were implanted intraperitoneally (i.p.) with a slow-release implant containing 1 mg of 17-MT and at the fifth CPE injection they received the second and last implant (Y group), while the remaining eels were treated with CPE but received no implant (N group) (Mordenti et al., 2018).

Following standardized protocol (Di Biase et al., 2016, Mordenti et al., 2014, Mordenti et al., 2018), after DHP injection, each female was transferred to a new closed RAS and maintained for 20 hours with spermiating males (*sex ratio* 4M : 1F) in seawater at a temperature of 20±0.5°C to produce a thermic shock (Mordenti et al., 2014, Dou et al., 2008), in order to facilitate spontaneous spawning.

With regard to the artificial induction of maturation in female silver eels, the study demonstrated that females respond to androgen exposure in terms of hormonal response time at CPE treatments and in terms of egg productivity (Table 21.5). The results suggest that 17-MT may

Table 21.5 Morphometric parameters and egg yields of treated eels groups (N group and Y group).

		Body weight (BW)		BWI			Spawned eggs	
		Initial	at last CPE	at last CPE	at DHP	CPE=>DHP		
		g	g	%	%	%	g	% BW
N-group	N-001	830.7	1121.0	134.9	140.5	5.6	150.0	18.1
	N-002	799.6	1042.0	130.3	143.6	13.3	391.7	49.0
	N-004	824.0	939.7	114.0	130.3	16.3	289.0	35.1
	N-008	1183.4	1282.0	108.3	125.8	17.5	403.6	34.1
	N-006	698.9	784.8	112.3	134.8	22.5	374.9	53.6
	N-007	722.9	893.5	123.6	131.4	7.8	299.2	41.4
	N-014	905.8	1269.2	140.1	147.3	7.2	392.0	43.3
	N-010	708.1	995.1	140.5	154.7	14.2	513.0	72.5
	N-009	627.0	807.8	128.8	139.1	10.3	303.1	48.3
	N-011	594.0	726.0	122.2	135.7	13.5	210.3	35.4
	N-015	823.2	944.3	114.7	120.4	5.7	145.0	17.6
	N-017	666.9	775.6	116.3	133.2	16.9	367.8	55.2
	Mean	**782.0±157.0**	**965.1±186.0**	**123.8±11.1**	**136.4±9.4**	**12.6±5.3**	**320.0±110.0**	**40.9±15.6**
Y-group	Y-136	665.7	832.0	125.0	136.2	11.2	377.8	56.8
	Y-074	639.1	764.8	119.7	135.2	15.5	360.0	56.33
	Y-055	846.1	955.3	112.9	133.4	20.5	344.0	40.7
	Y-054	1284.8	1448.0	112.7	130.6	17.9	625.0	48.7
	Y-059	961.8	1132.0	117.7	133.1	15.4	593.0	61.7
	Y-060	698.9	909.1	130.1	138.2	8.2	416.0	59.5
	Y-061	755.0	993.6	131.6	146.2	14.6	478.0	63.3
	Y-062	570.6	669.6	117.4	154.0	36.7	456.8	80.1
	Y-063	602.7	757.7	125.7	141.4	15.6	426.8	70.8
	Y-064	533.0	656.2	123.1	153.7	30.6	389.1	73.0
	Y-137	633.3	823.0	123.0	162.3	32.4	374.5	59.1
	Y-138	516.4	650.3	125.9	141.5	15.6	374.2	72.5
	Mean	**725.6±218.3**	**882.6±230.9**	**122.6±6.5**	**142.2±10.0**	**19.5±8.9***	**434.6±90.7**	**61.9±11.1***

Note: *Significance difference ($p < 0.01$) between Y-group and N-group eels.

be directly involved in the improvement of the artificial reproduction of female silver European eels reducing the time to complete gonadic maturation by several weeks (Figure 21.3). Indeed, in similar trials from our group (Di Biase et al., 2016, Mordenti et al., 2014, Mordenti et al., 2013) wild eels from the same area (Northern Adriatic Sea), subjected to hormonal treatment with CPE only, rarely reached full gonadal maturation before the 20th week.

Lin et al. (1991, Lin et al., 1998) demonstrated in Japanese eels that multiple injections and implantations of 17-MT were effective in stimulating an increase in ovarian development. Furthermore, Wang and Lou (2006) demonstrated that in Japanese eel, gonadal development was significantly induced by co-treatment with salmon pituitary homogenate and 17-MT and Wang and Lou (2007) suggested that the androgen was not only converted to oestrogens by aromatase, but also had a potential usefulness to the process of vitellogenesis and oogenesis. More recently, Lokman et al. (2015) in short-finned and Di Biase et al. (2017) in European eels demonstrated that female silver eels strongly respond to androgen exposure during artificial induction of maturation with CPE. In addition, a sustained-release androgen implant during artificial maturation by CPE injection requires less time in induction, without any negative effect on quality of obtained eggs. Additionally, the data demonstrated that it is possible to obtain the same results without hypophysation with CPE during the first 4–5 weeks (Di Biase et al., 2017). The same authors emphasize that the amount of time, and consequently, the amount handling, hormones and money, needed to reach the pre-ovulatory stage is notably decreased in female eels pre-treated with 17-MT (Table 21.6). Finally, in Japanese eels it has been pointed out that a reduction gonadic maturation times and of post-ovulatory aging lead to better quality of ovulated eggs and to an increase in larval survival (Nomura et al., 2013, Okamura et al., 2014). Indeed, the success rates of induced maturation and production of viable eggs in European eels *A. anguilla* were generally low compared with other 'Anguillan' species, caused by low responsiveness of their gonads to exogenous hormones (Okamura et al., 2014). In conclusion, the use of 17-MT in artificial reproduction of eel seems advisable not only for reduction of time to maturation and number of CPE injections, but also for the production of good quality eggs in terms of hatching and/or fertilization rates.

An exciting finding from the present study is the fact that the use of androgen implants in artificial reproduction of European eels seems to lead to an acceleration in female maturation as well as to a synchronization of gonadal maturation. This is a very important aspect, as the hormonal answer of wild eel is often heterogeneous (Mordenti et al., 2012, Mordenti et al., 2013) and usually 11–29 weeks are required before reaching the final maturation phase of oocytes by weekly injections of SPE or CPE (Palstra et al., 2005, Okamura et al., 2014, Di Biase et al., 2016). Furthermore, this synchronization process agrees with that observed by Wang and Lou (2007) in Japanese eel, where the methyltestosterone causes the phenomenon of synchronous development when it combines with SPH treatments. It should be also noticed that the synchronization in female maturation is desirable in artificial reproduction because the time window for high-quality eggs is very narrow (Ijiri et al., 2011). Finally, having defined times for gamete emission by breeders is a mandatory and successful prerequisite for embryonated egg management in aquaculture. The mechanism beneath this synchronization could be linked to the increased amount and/or the activation of androgen receptors at the teca and granulosa level, as shown by Tosaka et al. (2010) in Japanese eels. Another important effect of 17-MT could be the increase in vitellogenin (VTG) synthesis. Indeed, 17-MT was converted to 17-methyl oestradiol

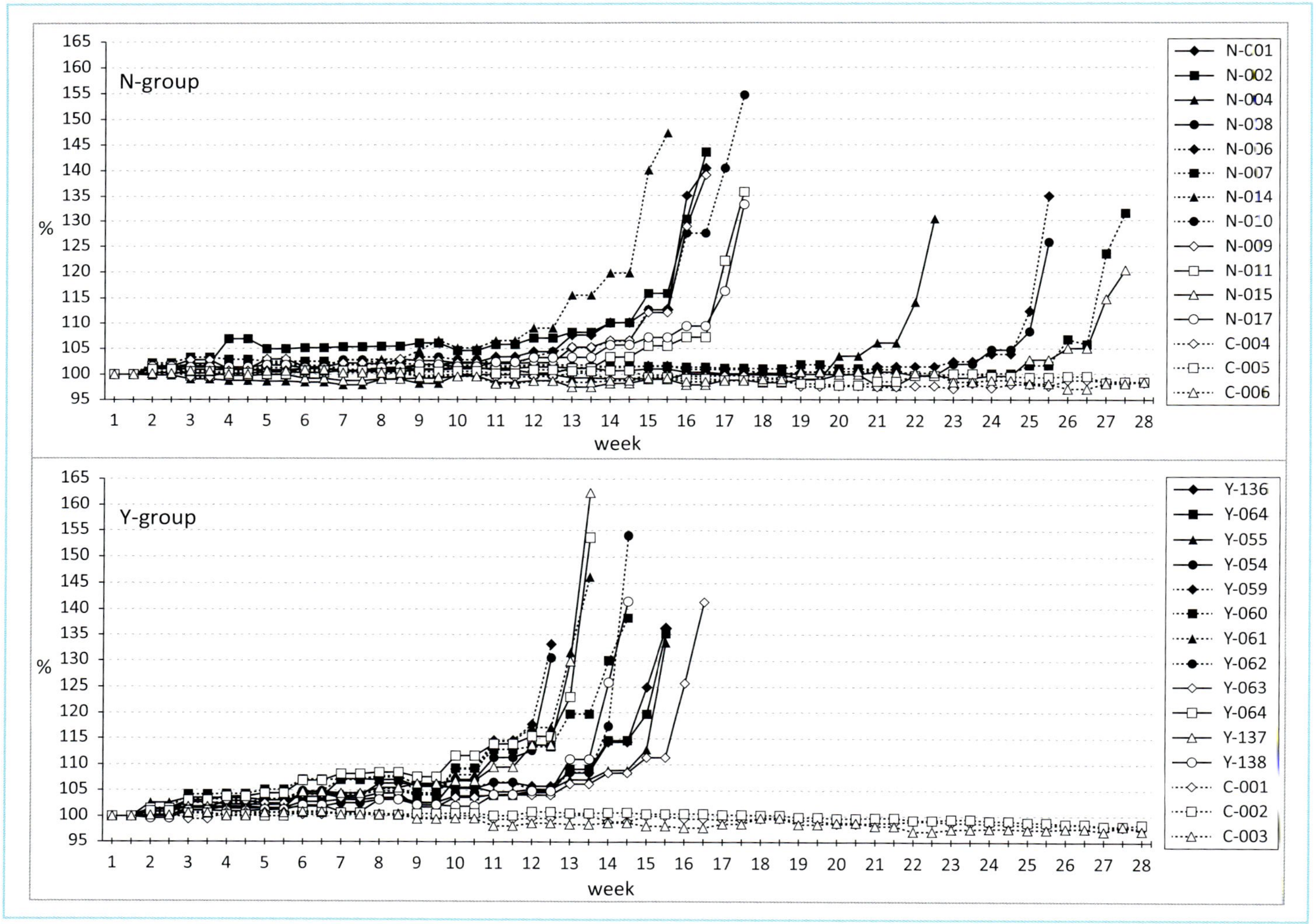

Figure 21.3 BWI trend recorded in female eel during reproductive cycle.

Table 21.6 CPE consumption and number of injections needed to induce spawning.

		CPE Consumption		CPE induction
		mg/female	mg/kg BW	n
N-group	N-001	285.6	343.9	16
	N-002	279.9	350.0	16
	N-004	415.6	504.4	22
	N-008	707.2	597.6	25
	N-006	420.5	601.7	25
	N-007	482.1	666.9	27
	N-014	291.5	321.9	15
	N-010	270.9	382.6	17
	N-009	213.3	340.2	16
	N-011	215.8	363.3	17
	N-015	535.5	650.6	27
	N-017	244.3	366.3	17
	Medium	363.5±151.7	457.4±135.9*	20.0±4.8*
Y-group	Y-136	209.3	314.3	15
	Y-074	197.8	309.5	15
	Y-055	261.2	308.7	15
	Y-054	272.1	211.8	12
	Y-059	208.3	216.6	12
	Y-060	203.7	291.5	14
	Y-061	192.9	255.5	13
	Y-062	158.9	278.5	14
	Y-063	206.2	342.1	16
	Y-064	136.5	256.2	13
	Y-137	160.0	252.6	13
	Y-138	142.7	276.3	14
	Medium	195.8±42.2	276.1±39.7	13.8±1.3

Note: *Significance difference ($p < 0.01$) between Y-group and N-group eels.

in fathead minnow, which activates the hepatic oestrogen receptors, leading to VTG production (Hornung et al., 2004). Although these two parameters have not been quantified in the present study, a combination of androgen receptor activation and an increase in vitellogenin synthesis cannot be excluded as basic mechanisms leading to synchronization of oocyte maturation and the improved quality of eggs.

One last noteworthy finding from this study is a better response to treatment with DHP of females given a sustained-release androgen implant, and a higher relative fertility of these same females.

Probably the treatment with 17-MT lead to a higher percentage of fully transparent oocytes (FTO) in pre-ovulatory phase (Di Biase et al., 2016, Palstra et al., 2005) and consequently to a higher production of spawned eggs. Indeed, Di Biase et al. (2016) observed a positive correlation in European eels between the FTO and the relative weight of spawned eggs. Furthermore, sustained-release androgen implants resulted in fully matured oocytes of Japanese eel (Lin et al.,

1998). Lokman et al. (2015) demonstrated that 17-MT implants in silver female shortfins eel induced a number of readily identifiable effects on body weight (increased), on GSI (increased), on oocyte diameters (increased). They are also likely to reflect their importance for a suite of physiological events, including oogenesis.

In summary in the study performed by the Italian group, silver European eels strongly respond to androgen exposure in terms of time of hormonal response to CPE treatments, in terms of synchronization of gonadal maturation and in terms of egg production. The data suggest that incorporation of androgen into the induced spawning protocol is an improvement that may help us get a step closer to completing the breeding of European eel in captivity.

Bibliography

Aarestrup, K., Økland, F., Hansen, M.M., Righton, D., Gargan, P., Castonguay, M., Bernatchez, L., Howey, P., Sparholt, H. and Pedersen, M.I. (2009) Oceanic spawning migration of the European eel (Anguilla anguilla). *Science* 325, 1660–1660.

Abe, T., Ijiri, S., Adachi, S. and Yamauchi, K. (2010) Development of an in vitro culture system for producing eel larvae from immature ovarian follicles in Japanese eel Anguilla japonica. *Fisheries Science* 76, 257–265.

Ahn, H., Yamada, Y., Okamura, A., Horie, N., Mikawa, N., Tanaka, S. and Tsukamoto, K. (2012) Effect of water temperature on embryonic development and hatching time of the Japanese eel Anguilla japonica. *Aquaculture* 330, 100–105.

Alderdice, D., Jensen, J. and Velse, N.F. (1988) Incubation of sablefish (Anoplopama fimbria) in a system designed for culture of fragile marine teleost eggs. *Aquaculture* 71, 271–283.

Anguis, V. and Canavate, J. P. (2005) Spawning of captive Senegal sole (Solea senegalensis) under a naturally fluctuating temperature regime. *Aquaculture* 243, 133–145.

Aschonitis, V., Castaldelli, G., Lanzoni, M., Rossi, R., Kennedy, C., Fano, E.A.J.A.C.M. and Ecosystems, F. (2017) Long-term records (1781–2013) of European eel (Anguilla anguilla L.) production in the Comacchio Lagoon (Italy): evaluation of local and global factors as causes of the population collapse. *Aquatic Conservation: Marine Freshwater Ecosystems* 27, 502–520.

Bromage, N. (1993) Environmental control of reproduction in salmonids. *Recent Advances Aquaculture*, 55–66.

Bromage, N. (1995) Broodstock management and seed quality-general considerations. *Brood-stock Management Egg Larval Quality*, 1–24.

Bromage, N., Bruce, M., Basavaraja, N., Rana, K., Shields, R., Young, C., Dye, J., Smith, P., Gillespie, M. and Gamble, J. (1994) Egg quality determinants in finfish the role of overripening with special reference to the timing of stripping in the Atlantic halibut Hippoglossus hippoglossus. *Journal of the World Aquaculture Society* 25, 13–21.

Brooks, S., Tyler, C.R. and Sumpter, J.P. (1997) Egg quality in fish: what makes a good egg? *Reviews in Fish Biology and fisheries*, 7, 387–416.

Brown, N., Shields, R. and Bromage, N. (2006) The influence of water temperature on spawning patterns and egg quality in the Atlantic halibut (Hippoglossus hippoglossus L.). *Aquaculture* 261, 993–1002.

Burgerhout, E., Brittijn, S.A., Kurwie, T., Decker, P., Dirks, R.P., Palstra, A.P., Spaink, H.P. and Van Den Thillart, G.E. (2011) First artificial hybrid of the eel species *Anguilla australis* and *Anguilla anguilla*. *BMC Developmental Biology* 11, 16.

Butts, I.A.E., Sørensen, S.R., Politis, S.N. and Tomkiewicz, J. (2016) First-feeding by European eel larvae: a step towards closing the life cycle in captivity. *Aquaculture* 464, 451–458.

Campbell, P., Pottinger, T. and Sumpter, J. (1992) Stress reduces the quality of gametes produced by rainbow trout. *Biology of Reproduction* 47, 1140–1150.

Carrillo, M. (1995) Sea bass (Dicentrachus labrax). *Broodstock Management Egg Larval Quality*, 138–168.

Casselman J.M. (2003) Dynamics of resources of the American eel, Anguilla rostrata: declining abundance in the 1990s. In: Aida, K., Tsukamoto, K. and Yamauchi, K. (eds) *Eel Biology*. Springer, Tokyo.

Chai, Y., Tosaka, R., Abe, T., Sago, K., Sago, Y., Hatanaka, E., Ijiri, S. and AdachI, S. (2010) The relationship between the developmental stage of oocytes in various seasons and the quality of the egg obtained by artificial maturation in the feminized Japanese eel *Anguilla japonica*. *Aquaculture Science* 58, 269–278.

Ciccotti, E. (1997) Italy *In:* Moriarty, C. and Dekker, W. (eds.) *Management of the European Eel.*

Dekker, W. (2003) Did lack of spawners cause the collapse of the European eel, Anguilla anguilla? *Fisheries Management Ecology* 10, 365–376.

Di Biase, A., Casalini, A., Emmanuele, P., Mandelli, M., Lokman, P.M. and Mordenti, O. (2016) Controlled reproduction in *Anguilla anguilla* (L.): comparison between spontaneous spawning and stripping-insemination approaches. *Aquaculture Research* 47, 3052–3060.

Di Biase, A., Lokman, P.M., Govoni, N., Casalini, A., Emmanuele, P., Parmeggiani, A. and Mordenti, O. (2017) Co-treatment with androgens during artificial induction of maturation in female eel, *Anguilla anguilla*: effects on egg production and early development. *Aquaculture* 479, 508–515.

Dou, S.-Z., Yamada, Y., Okamura, A., Shinoda, A., Tanaka, S. and Tsukamoto, K. (2008) Temperature influence on the spawning performance of artificially-matured Japanese eel, *Anguilla japonica*, in captivity. *Environmental Biology of Fishes* 82, 151–164.

Dou, S., Yamada, Y., Okamura, A., Tanaka, S., Shinoda, A. and Tsukamoto, K. (2007) Observations on the spawning behavior of artificially matured Japanese eels Anguilla japonica in captivity. *Aquaculture* 266, 117–129.

Dufour, S. and van den Thillart, G. (2009) Reproduction capacity of European eels. In: van den Thillart G., Dufour S. and Rankin, J.C. (eds) *Spawning Migration of the European Eel.* Springer, Dordrecht.

Durif, C., Dufour, S. and Elie, P. (2005) The silvering process of *Anguilla anguilla*: a new classification from the yellow resident to the silver migrating stage. *Journal of Fish Biology* 66, 1025–1043.

Durif, C.M., Dufour, S. and Elie, P. (2006) Impact of silvering stage, age, body size and condition on reproductive potential of the European eel. *Marine Ecology Progress Series* 327, 171–181.

Durif, C.M., Van Ginneken, V., Dufour, S., Müller, T. and Elie, P. (2009) Seasonal evolution and individual differences in silvering eels from different locations. In: van den Thillart G., Dufour S. and Rankin, J.C. (eds) *Spawning Migration of the European Eel.* Springer, Dordrecht.

EELREP (2006) Estimation of the Reproduction Capacity of European Eel. Final Report, EU Project, EELREP.

Fontaine, Y. (1994) L'argenture de l'anguille: métamorphose, anticipation, adaptation. *Bulletin Français de la Pe ̂che et de la Pisciculture* 335, 171–185.

Greve, W. (1975) The 'Meteor Planktonküvette': a device for the maintenance of macrozooplankton aboard ships. *Aquaculture* 6, 77–82.

Han, Y.-S., Liao, I.-C., Huang, Y.-S., He, J.-T., Chang, C.-W. and Tzeng, W.-N. (2003a) Synchronous changes of morphology and gonadal development of silvering Japanese eel Anguilla japonica. *Aquaculture* 219, 783–796.

Han, Y.-S., Liao, I.-C., Tzeng, W.-N., Huang, Y.-S. and Yu, J. Y.-L. (2003b) Serum estradiol-17β and testosterone levels during silvering in wild Japanese eel Anguilla japonica. *Comparative Biochemistry and Physiology – Part B: Biochemistry Molecular Biology* 136, 913–920.

Hilomen-Garcia, G.V. (1998) Sensitivity of fertilized milkfish (Chanos chanos Forsskal) eggs to mechanical shock and simulated transport. *Aquaculture* 159, 239–247.

Horie, N., Utoh, T., Mikawa, N., Yamada, Y., Okamura, A., Tanaka, S. and Tsukamoto, K. (2008) Influence of artificial fertilization methods of the hormone-treated Japanese eel Anguilla japonica upon the quality of eggs and larvae (Comparison between stripping-insemination and spontaneous spawning methods). *Nippon Suisan Gakkaishi* 74, 26–35.

Hornung, M.W., Jensen, K.M., Korte, J.J., Kahl, M.D., Durhan, E.J., Denny, J.S., Henry, T.R. and Ankley, G.T. (2004) Mechanistic basis for estrogenic effects in fathead minnow (*Pimephales promelas*) following exposure to the androgen 17α-methyltestosterone: conversion of 17α-methyltestosterone to 17α-methylestradiol. *Aquatic Toxicology* 66, 15–23.

Ijiri, S., Tsukamoto, K., Chow, S., Kurogi, H., Adachi, S. and Tanak, A.H. (2011) Controlled reproduction in the Japanese eel (*Anguilla japonica*), past and present. *Aquaculture Europe* 36, 13–17.

Jensen, J. and Alderdice, D. (1989) Comparison of mechanical shock sensitivity of eggs of five Pacific salmon (Oncorhynchus) species and steelhead trout (Salmo gairdneri). *Aquaculture* 78, 163–181.

Kagawa, H., Kishi, T., Gen, K., Kazeto, Y., Tosaka, R., Matsubara, H., Matsubara, T. and Sawaguchi, S. (2011) Expression and localization of aquaporin 1b during oocyte development in the Japanese eel (Anguilla japonica). *Reproductive Biology and Endocrinology* 9, 71.

Kelley, C., Tamaru, C., Lee, C., Moriwake, A. and Miyamoto, G. (1991) Effects of photoperiod and temperature on the annual ovarian cycle of the striped mullet, Mugil cephalus. In: Scott, A.P., Sumpter, J.P., Kime, D.E. and Rolfe, M.S. (eds) *Forth International Symposium on the Reproductive Physiology of Fish*. FishSymp91, 7–12 July 1991, Sheffield. Department of Animal and Plant Sicence, University of Sheffield, pp. 142–144.

Kjørsvik, E., Mangor-Jensen, A. and Holmefjord, I. (1990) Egg quality in fishes. *Advances in Marine Biology* 26, 71–113.

Kohn, Y.Y. and Symonds, J.E. (2012) Evaluation of egg quality parameters as predictors of hatching success and early larval survival in hapuku (Polyprion oxygeneios). *Aquaculture* 342, 42–47.

Lin, H.-R., Zhang, L.-H., Wang, X.-D. and CheN, L.-X. (1998) Artificial induction of gonadal maturation and ovulation in the Japanese eel (*Anguilla japonica* T. et S.). *Bulletin Français de la Pêche et de la Pisciculture* 163–176.

Lin, H.-R., Zhang, M.-L., Zhang, S.-M., Van Der Kraak, G. and Peter, R.E. (1991) Stimulation of pituitary gonadotropin and ovarian development by chronic administration of testosterone in female Japanese silver eel, *Anguilla japonica*. *Aquaculture* 96, 87–95.

Lokman, P., Wylie, M.J., Downes, M., Di Biase, A. and Damsteegt, E.L. (2015) Artificial induction of maturation in female silver eels, *Anguilla australis*: the benefits of androgen pre-treatment. *Aquaculture* 437, 111–119.

Matsubara, H., Lokman, P. M., Kazeto, Y., Adachi, S. and Yamauchi, K. (2005) Serum steroid profiles in artificially maturing female Japanese eel, Anguilla japonica. *Aquaculture* 243, 393–402.

Matsuda, H. and Takenouchi, T. (2007) Development of technology for larval Panulirus japonicus culture in Japan: a review. *Bulletin-Fisheries Research Agency Japan* 20, 77.

Melià, P., Bevacqua, D., Crivelli, A., Panfili, J., De Leo, G. and Gatto, M. (2006) Sex differentiation of the European eel in brackish and freshwater environments: a comparative analysis. *Journal of Fish Biology* 69, 1228–1235.

Mordenti, M., Di Biase, A., Sirri, R., Modugno, S. and Tasselli, A. (2012) Induction of sexual maturation in wild female European eels (*Anguilla anguilla*) in darkness and light. *The Open Access Israeli Journal of Aquaculture – Bamidgeh* 64, 1–9.

Mordenti, O., Casalini, A., Mandelli, M. and Di Biase, A. (2014) A closed recirculating aquaculture system for artificial seed production of the European eel (*Anguilla anguilla*): technology development for spontaneous spawning and eggs incubation. *Aquacultural Engineering* 58, 88–94.

Mordenti, O., Di Biase, A., Bastone, G., Sirri, R., Zaccaroni, A. and Parmeggiani, A. (2013) Controlled reproduction in the wild European eel (*Anguilla anguilla*): two populations compared. *Aquaculture International* 21, 1045–1063.

Mordenti, O., Di Biase, A., Casalini, A., Emmanuele, P., Melotti, P. and Roncarati, A. (2016) Growth performances and natural diet of European eel (*Anguilla anguilla* L.) reared in muddy and sandy ponds. *Aquatic Living Resources* 29, 105.

Mordenti, O., Emmanuele, P., Casalini, A., Lokman, P.M., Zaccaroni, A., Di Biase, A. and Parmeggiani, A. (2018) Effect of aromatable androgen (17-methyltestosterone) on induced maturation of silver European eels (Anguilla Anguilla): Oocyte performance and synchronization. *Aquaculture Research* 49, 442–448.

Mylonas, C.C., Fostier, A. and Zanuy, S. (2010) Broodstock management and hormonal manipulations of fish reproduction. *General Comparative Endocrinology* 165, 516–534.

Mylonas, C.C., Papandroulakis, N., Smboukis, A., Papadaki, M. and Divanach, P. (2004) Induction of spawning of cultured greater amberjack (Seriola dumerili) using GnRHa implants. *Aquaculture* 237, 141–154.

Nomura, K., Takeda, Y., Unuma, T., Morishima, K., Tanaka, H., Arai, K. and Ohta, H. (2013) Post-ovulatory oocyte aging induces spontaneous occurrence of polyploids and mosaics in artificial fertilization of Japanese eel, *Anguilla japonica. Aquaculture* 404, 15–21.

Oca, J. and Masalo, I. (2013) Flow pattern in aquaculture circular tanks: Influence of flow rate, water depth, and water inlet and outlet features. *Aquacultural Engineering* 52, 65–72.

Ohta, H., Kagawa, H., Tanaka, H., Okuzawa, K. and Hirose, K. (1996) Changes in fertilization and hatching rates with time after ovulation induced by 17, 20β-dihydroxy-4-pregnen-3-one in the Japanese eel, Anguilla japonica. *Aquaculture* 139, 291–301.

Okamura, A., Yamada, Y., Yokouchi, K., Horie, N., Mikawa, N., Utoh, T., Tanaka, S. and Tsukamoto, K. (2007) A silvering index for the Japanese eel Anguilla japonica. *Environmental Biology of Fishes* 80, 77–89.

Okamura, A., Yamada, Y., Mikawa, N., Horie, N., Utoh, T., Kaneko, T., Tanaka, S. and Tsukamoto, K. (2009) Growth and survival of eel leptocephali (Anguilla japonica) in low-salinity water. *Aquaculture* 296, 367–372.

Okamura, A., Horie, N., Mikawa, N., Yamada, Y. and Tsukamoto, K. (2014) Recent advances in artificial production of glass eels for conservation of anguillid eel populations. *Ecology of Freshwater Fish* 23, 95–110.

Oliveira, K. and Hable, W. (2010) Artificial maturation, fertilization, and early development of the American eel (*Anguilla rostrata*). *Canadian Journal of Zoology* 88, 1121–1128.

Palstra, A. and Van Den Thillart, G. (2009) Artificial maturation and reproduction of the European eel. *Spawning Migration of the European eel.* Springer, Dordrecht, pp. 297–322.

Palstra, A.P. and Van Den Thillart, G.E. (2010) Swimming physiology of European silver eels (Anguilla anguilla L.): energetic costs and effects on sexual maturation and reproduction. *Fish Physiology and Biochemistry* 36, 297–322.

Palstra, A., Cohen, E., Niemantsverdriet, P., Van Ginneken, V. and Van Den Thillart, G. (2005) Artificial maturation and reproduction of European silver eel: development of oocytes during final maturation. *Aquaculture* 249, 533–547.

Parmeggiani, A., Govoni, N., Zannoni, A., Di Biase, A., Sirri, R., Forni, M., Mandelli, M. and Mordenti, O. (2015) Effect of photoperiod on endocrine profiles and vitellogenin expression in European eels *Anguilla anguilla* during artificially induced ovarian development. *Theriogenology* 83, 478–484.

Pedersen, B.H. (2003) Induced sexual maturation of the European eel Anguilla anguilla and fertilisation of the eggs. *Aquaculture* 224, 323–338.

Pedersen, B.H. (2004) Fertilisation of eggs, rate of embryonic development and hatching following induced maturation of the European eel Anguilla anguilla. *Aquaculture* 237, 461–473.

Pedersen, K.H., Pedersen, S.N., Pedersen, K.L., Korsgaard, B. and Bjerregaard, P. (2003) Estrogenic effect of dietary 4-tert-octylphenol in rainbow trout (Oncorhynchus mykiss). *Aquatic Toxicology* 62, 295–303.

Regulation, C. J. O. J. O. T. E. U. L. (2007) No. 1100/2007 of 18 September 2007 establishing measures for the recovery of the stock of European eel. 248, 22.9.

Rohr, D.H., Lokman, P.M., Davie, P.S. and Young, G. (2001) 11-Ketotestosterone induces silvering-related changes in immature female short-finned eels, *Anguilla australis. Comparative Biochemistry and Physiology Part A: Molecular & Integrative Physiology,* 130, 701–714.

Schreck, C.B., Contreras-Sanchez, W. and Fitzpatrick, M.S. (2001) Effects of stress on fish reproduction, gamete quality, and progeny. *Aquaculture* 197, 3–24.

Seoka, M., Yamada, S., Iwata, Y., Yanagisawa, T., Nakagawa, T. and Kumai, H. (2003) Differences in the biochemical content of buoyant and non-buoyant eggs of the Japanese eel, Anguilla japonica. *Aquaculture* 216, 355–362.

Sorteni, C., Clavenzani, P., De Giorgio, R., Portnoy, O., Sirri, R., Mordenti, O., Di Biase, A., Parmeggiani, A., Menconi, V. and Chiocchetti, R.J. J.O.A. (2014) Enteric neuroplasticity in seawater-adapted European eel (Anguilla anguilla). *Journal of Anatomy* 224, 180–191.

Sudo, R. and Tsukamoto, K. (2015) Migratory restlessness and the role of androgen for increasing behavioral drive in the spawning migration of the Japanese eel. *Scientific Reports* 5.

Sudo, R., Tosaka, R., Ijiri, S., Adachi, S., Aoyama, J. and Tsukamoto, K. (2012) 11-ketotestosterone synchronously induces oocyte development and silvering-related changes in the Japanese eel, *Anguilla japonica. Zoological Science* 29, 254–259.

Svedäng, H. and Wickström, H. (1997) Low fat contents in female silver eels: indications of insufficient energetic stores for migration and gonadal development. *Journal of Fish Biology* 50, 475–486.

Tanaka, H., Kagawa, H. and Ohta, H. (2001) Production of leptocephali of Japanese eel (Anguilla japonica) in captivity. *Aquaculture* 201, 51–60.

Thorsen, A., Trippel, E.A. and Lambert, Y. (2003) Experimental methods to monitor the production and quality of eggs of captive marine fish. *Journal of Northwest Atlantic Fishery Science* 33, 55–70.

Tosaka, R., Todo, T., Kazeto, Y., Lokman, P.M., Ijiri, S., Adachi, S. and Yamauchi, K. (2010) Expression of androgen receptor mRNA in the ovary of Japanese eel, *Anguilla japonica*, during artificially induced ovarian development. *General and Comparative Endocrinology* 168, 424–430.

Tsadik, G.G. and Bart, A.N. (2007) Effects of feeding, stocking density and water-flow rate on fecundity, spawning frequency and egg quality of Nile tilapia, Oreochromis niloticus (L.). *Aquaculture* 272, 380–388.

Unuma, T., Hasegawa, N., Sawaguchi, S., Tanaka, T., Matsubara, T., Nomura, K. and Tanaka, H. (2011) Fusion of lipid droplets in Japanese eel oocytes: stage classification and its use as a biomarker for induction of final oocyte maturation and ovulation. *Aquaculture* 322, 142–148.

Unuma, T., Kondo, S., Tanaka, H., Kagawa, H., Nomura, K. and Oht, A.H. (2005) Relationship between egg specific gravity and egg quality in the Japanese eel, Anguilla japonica. *Aquaculture* 246, 493–500.

Unuma, T., Sawaguchi, S., Hasegawa, N., Tsuda, N., Tanaka, T., Nomura, K. and Tanaka, H. (2012) Optimum temperature of rearing water during artificial induction of ovulation in Japanese eel. *Aquaculture* 358, 216–223.

Van Den Thillart, G. and Dufour, S. (2009) How to estimate the reproductive success of European silver eels. In: van den Thillart G., Dufour S. and Rankin J.C. (eds) *Spawning Migration of the European Eel.* Springer, Dordrecht.

Van Ginneken, V.J. and Maes, G.E. (2005) The European eel (Anguilla anguilla, Linnaeus), its lifecycle, evolution and reproduction: a literature review. *Reviews in Fish Biology and Fisheries* 15, 367–398.

Van Ginneken, V., Durif, C., Balm, S.P., Boot, R., Verstegen, M.W., Antonissen, E. and Van Den Thillart, G. (2007) Silvering of European eel (Anguilla anguilla L.): seasonal changes of morphological and metabolic parameters. *Animal Biology* 57, 63–77.

Van Ginneken, V., Vianen, G., Muusze, B., Palstra, A., Verschoor, L., Lugten, O., Onderwater, M., Van Schie, S., Niemantsverdriet, P. and Van Heeswijk, R. (2005) Gonad development and spawning behaviour of artificially-matured European eel (Anguilla anguilla L.). *Animal Biology* 55, 203–218.

Wang, Y.-S. and Lou, S.-W. (2006) Structural and expression analysis of hepatic vitellogenin gene during ovarian maturation in *Anguilla japonica*. *The Journal of Steroid Biochemistry and Molecular Biology* 100, 193–201.

Wang, Y.-S. and Lou, S.-W. (2007) Influence of exogenous gonadotropin and sexual steroids on ovary development in Japanese eel *Anguilla japonica*. 臺灣水產學會刊 34, 261–273.

CHAPTER 22

European eel – an integrated approach to establish eel hatchery technology in Denmark

Jonna Tomkiewicz[1], *Sebastian N. Politis*[1], *Sune Riis Sørensen*[1,2], *Ian A.E. Butts*[1,3] *and Johanna S. Kottmann*[1]

[1]Technical University of Denmark, National Institute of Aquatic Resources, Lyngby, Denmark
[2]Billund Aquaculture, Billund, Denmark
[3]Auburn University, School of Fisheries, Aquaculture and Aquatic Sciences, Auburn, Alabama, USA

Introduction

As catadromous fishes, anguillid eels have a stage-structured life cycle involving continental juvenile stages followed by diverse oceanic reproductive and larval stages. Within the latter, silver eels and larvae (leptocephali) travel thousands of kilometres to complete their life cycle. For the European eel (*Anguilla anguilla*), knowledge about their natural reproduction and spawning habitats in the Sargasso Sea is negligible and insights into the ecophysiology of their offspring is scarce. Therefore, development of hatchery technology involves dedicated experimental work filling gaps in knowledge related to their complex reproductive biology, early life history ecology and physiological requirements throughout ontogeny. Such targeted research, steadily has advanced development of broodstock feeds and assisted reproduction methods for enhanced production of viable offspring as well as egg incubation and larval culture technology leading to first-feeding trials. The next challenge is to identify effective diets and to establish suitable culture conditions for growth of leptocephalus larvae. In Denmark, research on European eel reproductive biology, offspring developmental competence and larval ecophysiology has progressed during the past 10 years, using an integrated, multidisciplinary approach to generate new insights. Here, we summarize and discuss the implications of these results in relation to the establishment of innovative hatchery technology. Moreover, such new insights in eel reproduction and early life history may contribute to understanding the natural life cycle of European eel, thereby assisting management and conservation plans for this critically endangered species.

Background

Aquaculture plays a key role in sustainable market development by virtue of its potential to contribute to increased food production while helping to reduce pressure on natural fish populations as fishery resources (OECD, 2019). Based on its dynamic performance over the last 30 years, and with relatively stable catches from commercial fisheries, it is anticipated that the future growth of the fish and shellfish sector will mainly originate from aquaculture (FAO, 2016). This development relies on closing the life cycle for fish species in captivity, thereby replacing capture-based fish farming (Teletchea, 2015). Here, closed life cycle propagation promotes the success of aquaculture by ensuring high-quality, year-round hatchery production of offspring to sustain commercial farming and facilitate breeding programmes.

The future growth and sustainability of European eel aquaculture extensively depends on closing the life cycle in captivity. Here, the natural stock has declined and glass eel recruitment is at a historically low level, which has rendered the species critically endangered and led to trade restrictions (CITES, 2007; Jacoby and Gollock, 2014). Therefore, as eel farming continues to be capture based, hatchery production of glass eels is required in order to remove constraints on aquaculture production and markets for European eel. In addition, self-sustaining aquaculture production may support management and conservation measures by reducing fishery pressure on continental life history stages, including glass eels, yellow eels and silver eels. However, a background effort of applied research and technology development is imperative for overcoming obstacles related to the complex reproductive control mechanisms and inscrutable early life history stages of eel, which challenge development of suitable breeding and larval culture technologies.

Life cycle challenges for controlled propagation

As catadromous fishes, eels have a stage-structured life history, involving continental juvenile stages, followed by diverse oceanic reproductive, embryonic and larval stages (Figure 22.1a). Within the oceanic stages, silver eels and leptocephalus larvae travel thousands of kilometres to complete their life cycle. While the European eel spawning area is assumed to be the Sargasso Sea, due to presence of early larval stages, knowledge about their natural reproduction and spawning habitat is negligible and insights into the ecophysiology of their egg and early larval stages remain scarce (Tesch, 2003). First insights into the oceanic early life stages originate from leptocephalus larvae captured in the Sargasso Sea and along their migration route (Munk et al., 2010; Schmidt, 1923), followed by the well-known continental stages from glass eels via elvers to yellow eels and silver eels, that is, the spawning migration stage. The dotted line in Figure 22.1a indicates these undisclosed, enigmatic phases of the European eel natural life cycle.

This limitation of information challenges the development of hatchery innovation. As natural conditions cannot be consulted, the development of reliable culture technologies for different life history stages relies on targeted experimental research and analytical techniques, generating the knowledge base needed (Figure 22.1). This involves broodstock management, including dietary requirements, rearing conditions and protocols for gamete production, and fertilization. For the

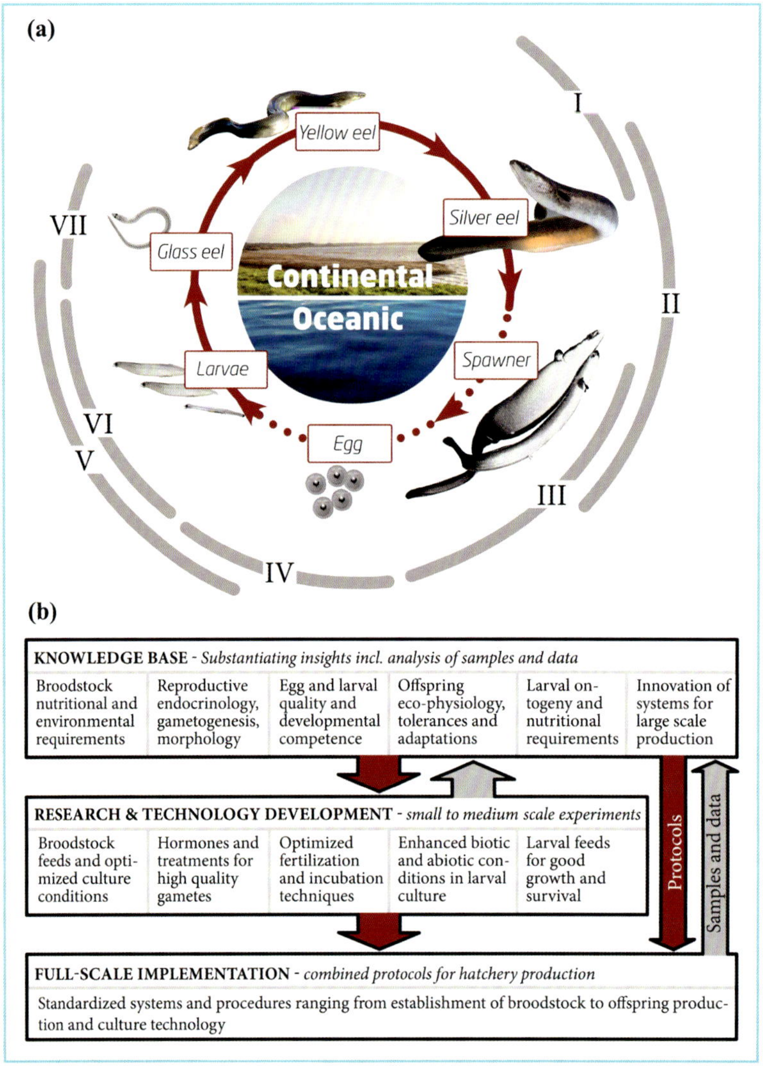

Figure 22.1 (a) Life cycle of European eel including life history stages related to the oceanic and continental phases. The solid line of the inner circle represents the known part in nature, while the dotted line shows the still unknown part. The outer ring illustrates phases of hatchery research and technology development: I Broodstock establishment: dietary requirements and conditioning; II Induction of gametogenesis and broodstock culture technology; III Follicular maturation, ovulation and sperm production; IV Fertilization techniques, egg incubation and embryonic development; V Bio-physical requirements and culture technology; VI Larval ontogeny and dietary requirements; and VII Metamorphosis and glass eel production. (b) Research concept using a systematic incremental approach linking basic science to application and technology development for implementation in full-scale production. Here, a knowledge base covering the reproductive phase and early life history stages is gradually substantiated through controlled experimental tests and dedicated analysis of samples and results. The insight gained is used to develop products, enhance protocols for application development, establish standardized full-scale production, and engineer culture systems. Photographs: Sune Riis Sørensen, DTU Aqua, and for Stage III Spawner, Inge Boëtius and Paul Juhlin, Danish Institute for Fisheries and Marine Research, now DTU Aqua.

offspring, it includes egg incubation and larval culture techniques and technology, insights into larval ontogeny and dietary requirements, and ultimately metamorphosis to obtain glass eels. In this context, a primary challenge for closing the life cycle of anguillid eels lies in an intricate hormonal control at the brain–pituitary level related to silvering and spawning migration, which inhibits sexual maturation and gamete production when inhabiting continental habitats (Dufour et al., 2003, 2005). Here, dopamine exerts an inhibitory control on gonadotropic function, that is, the release of follicle stimulating hormone (FSH) and luteinizing hormone (LH), thereby preventing gametogenesis in both sexes (Jolly et al., 2016; Vidal et al., 2004). In female eels, the follicular maturation and ovulation does not proceed, because signals that hold oocytes in meiotic arrest are not released (Yamauchi, 1990). Consequently, the development of breeding protocols involves assisted reproduction using hormonal treatments to induce gamete development and for females, additional administration of maturation inducing steroid (MIS) (Kagawa et al., 1995; Ohta et al., 1997; Yamauchi, 1990). Another major obstacle is lack of information about the ambient environment of spawners, gametes, and early life history stages, which challenges the development of larval culture technology. Additional challenges refer to the biology and ecology of the leaf-like leptocephalus larvae, unique to the Elopomorpha superorder, where limited knowledge about their natural diet and nutritional requirements hampers the establishment of efficient culture technology throughout the larval stage until their transformation into glass eels (Ayala et al., 2018; Miller et al., 2012; Okamura et al., 2018; Riemann et al., 2010; Tomoda et al., 2018).

In recent decades, intensified research addressing these challenges has substantiated insights throughout ontogeny and led to significant progress in terms of closing the life cycle of anguillid eels. Most advanced is the technology development related to the Japanese eel (*A. japonica*), where glass eel production has been accomplished (Kagawa et al., 2005; Tanaka et al., 2003) and now efficient upscaling for aquaculture is targeted (Masuda et al., 2012; Okamura et al., 2014; Tanaka, 2015). In the beginning of this century, breeding protocols described for the Japanese eel by Ohta et al. (1997) were implemented with modified methodologies for European eels resulting in embryonic development and larval hatch in a few cases (Palstra et al., 2005; Pedersen, 2003, 2004). Further refinement of methodologies and techniques has led to a steady production of viable offspring (Mordenti et al., 2014; Tomkiewicz, 2012) and progress in culture technology towards larval first feeding for this species (Butts et al., 2016; Di Biase et al., 2017; Politis et al., 2018a; Sørensen et al., 2016). Now, a prominent challenge includes identifying effective diets and establishing suitable culture conditions for growth of the leptocephalus larvae of European eel.

To respond to the goal of enabling a closed production cycle in a hatchery context, our research strategy uses a systematic, incremental approach linking basic science to application and technology development, for implementation in full-scale production (Figure 22.1b). Here, multidisciplinary experimental work and advanced analytical methodologies complement existing and otherwise emerging knowledge thereby helping to fill gaps in knowledge throughout their life history. In order to close the life cycle, we have addressed (1) broodstock establishment considering dietary requirements for high gamete quality, (2) assisted reproduction methodologies for induction of gametogenesis, follicular maturation, and gamete production, (3) efficient, standardized fertilization techniques and incubation technology sustaining embryonic development and (4) larval culture techniques and technology with present focus on larval growth and dietary requirements.

Insights advancing viable gamete production

The availability of high-quality broodstock and gametes is a prerequisite for closing the life cycle in captivity of any aquaculture species. The advantages of a closed production cycle include fertilized eggs and viable juveniles being produced on a year-round basis and ability of implementing selective breeding programmes. However, raising broodstock of farmed origin often requires nutritionally enhanced diets differing from on-growing feeds in order to meet the species' requirements for specific fatty acids, amino acids, vitamins and minerals ensuring high-quality gametes and viable offspring production (Izquierdo et al., 2001). In some cases, further acclimatization using environmental cues and/or hormonal induction of the gametogenesis and/or follicular maturation is required (Mylonas and Zohar, 2009; Mylonas et al., 2010). This applies to anguillid eels, where hormonal control of the entire reproductive development is required in order to obtain viable gametes.

Female broodstock origin and nutrition

Researching eel broodstock dietary requirements, in order to advance reproductive performance and offspring quality, is intricate, as eels cease feeding during the silvering process prior to spawning migration and gamete development in nature (Tesch, 2003). Moreover, the nutrient composition of eggs, embryos and yolk-sac larvae from the native spawning area cannot serve as a basis for diet development. Consequently, analyses of gametes produced by wild-caught broodstock using assisted reproduction serve as a benchmark for developing enhanced broodstock feeds for European eel. Studies comparing the egg composition of farmed and wild-caught female broodstock revealed differences in lipid content and fatty acid profiles, in particular arachidonic acid (20:4 n-6, ARA), eicosapentaenoic acid (20:5 n-3, EPA) and docosahexaenoic acid (22:6n-3, DHA). For instance, eggs from wild-caught silver eels were characterized by higher levels of ARA and lower levels of EPA in neutral lipids than levels in farmed eels (Heinsbroek et al., 2013; Støttrup et al., 2013). However, the content of essential fatty acids and their ratios in the eggs can be successfully adjusted via tailored feeds. For the European eel, increasing the dietary ARA level for farmed female broodstock to attain ~2–3% of the total fatty acids in the eggs led to higher fertilization rates as well as increased embryonic and larval survival (Kottmann et al., forthcoming; Støttrup et al., 2016). In contrast, studies on Japanese eels showed no marked relationship between egg quality and fatty acid levels of eggs, while fertilization success was negatively correlated with n-6 fatty acids in neutral lipids and with ARA in polar lipids (Furuita et al., 2003). A subsequent study by Furuita et al. (2007) furthermore suggested that a high ratio of n-6 to n-3 fatty acids negatively affected embryogenesis in Japanese eel. Moreover, optimal vitamin C and E levels in combination with highly unsaturated fatty acids (HUFA) improved egg and larval quality (Furuita et al., 2009a, 2009b) and high vitamin C level and vitamin E to HUFA ratio as well as optimized levels of vitamin A and E produced larvae with highest survival rates. These studies indicate that vitamin levels in Japanese eel play a major role defining high-quality eggs in combination with HUFA. In contrast, elevating vitamin C and E levels in broodstock diets for European eel did not affect the lipid peroxidation products in the eggs, indicating that applied vitamin levels were sufficient (Støttrup et al., 2016). An important finding for broodstock nutrition in European eel is that the feeding period needs to be long in order to change the female broodstock lipid profile prior to onset of maturation (Kottmann et al., forthcoming; Støttrup et al., 2016). At this time, gonads are still undeveloped, and as feeding ceases in readiness for silvering and the spawning migration, the main sources of lipids and protein for gamete development consist of muscle tissue and visceral fat (Støttrup et al., 2013).

Male broodstock origin and nutrition

Compared to female broodstock, dietary influences on male broodstock performance in terms of sperm volume and quality is little investigated, however, a few studies on male eel sexual maturation and viable gamete production exist. For example, Mazzeo et al. (2010) suggested that polyunsaturated fatty acids (PUFA), owing to their increase in testes and high concentration in milt, play an important role in sexual maturation of European eel. We studied the impact of varying levels of the essential dietary fatty acids, ARA, EPA and DHA, on muscle and liver lipid composition prior to induced maturation and quantified the resulting sperm composition and performance (that is, total volume and motility) (Butts et al., 2015). Results showed that optimal feeds for male European eel should contain high levels of DHA for production of high milt volumes, high EPA levels (even >7%) for higher motility, medium levels of ARA, while the n–3/n–6 ratio should be close to or >2. Likewise, Baeza et al. (2015a) found that dietary fatty acid composition impacted characteristics of sperm and milt production, while Baeza et al. (2015b) highlighted the important role of lipid metabolism affecting spermatogenesis. Moreover, the impact of male eel diets varying in amino acid composition were investigated (Butts et al., forthcoming). Here, diets supplemented with amino acids did not influence morphometric indices (body and organ-somatic indices: liver, testes, gonadosomatic index GSI) and testes morphology, while certain amino acids positively affected spermiation and sperm motility. Taken together, this new information may prove useful for developing male broodstock diets to improve sperm quality and subsequently, offspring production for this species.

While analyses of the composition of milt from wild-caught male silver eels have not yet been performed, fertilization success and offspring developmental competence have been compared among male broodstock, that is, wild-caught silver eels and farmed male broodstock. The comparison showed that male origin did not significantly affect early life history traits, that is, fertilization success, hatching success and larval deformities (Benini et al., 2018), despite some variability. Although male impact was less pronounced than female, the study demonstrated that some parental combination were highly successful producing viable offspring, while other combinations yielded significantly higher proportions of larvae with deformities. This indicates that male–female compatibility could be an issue of potential interest when defining the pool of males to be crossed with the female broodstock in order to improve the quality of offspring. Here, available genomic resources targeted for eel (Henkel et al., 2012; Jansen et al., 2017) can provide future opportunities creating tools and detecting genetic markers for the development of tests to predict female-male compatibility reducing the risk of deformities.

Gonadotropic function and induced gamete development

The complex mechanisms controlling anguillid eel gametogenesis (Dufour et al., 2003, 2005) are caused by dopaminergic control inhibiting the gonadotropic function, that is, the production and release of FSH and LH from the pituitary gland, while inhabiting continental habitats (Jolly et al., 2016; Vidal et al., 2004). This inhibition must be released, when the eels approach or reach the spawning area, but the cues and signalling pathways remain uncertain in spite of dedicated endocrine studies of the European and Japanese eel. Consequently, the development of breeding protocols presently involves assisted reproduction using hormonal treatments to induce gamete development (Mylonas et al., 2010).

As in all vertebrates, the brain–pituitary axis controls gonadal maturation, thus regulating the onset of reproduction through external and internal signals initiated at the brain level by activating gonadotropin-releasing hormone (GnRH). In response to GnRH stimulation, gonadotropic cells produce and release the gonadotropins, FSH and LH, which stimulate gonadal functions, that is, steroidogenesis and gametogenesis (Mylonas and Zohar, 2009; Rousseau et al., 2009). Anguillid eels in continental waters, however, remain juvenile influenced by a strong dopaminergic inhibitory effect on GnRH-induced gonadotropin synthesis and release (Dufour et al., 1988, 2003, 2005; Jolly et al., 2016; Vidal et al., 2004). Since the beginning of this century, studies of these intricate endocrine regulation mechanisms at the molecular level have significantly substantiated insights into function of the brain–pituitary–gonadal axis in European eel (Ager-Wick et al., 2013; Campo et al., 2018; Lafont et al., 2016; Minegishi et al., 2012; Morini et al., 2015; Pasquier et al., 2012; Rojo-Bartolomé et al., 2017).

The assembly and annotation of the European and Japanese eel genomes has accelerated acquisition of new knowledge, thereby providing a source of data to discover and identify new genes, characterize gene expression and identify genetic markers (Coppe et al., 2010; Henkel et al., 2012; Jansen et al., 2017). While continued endocrine research has the potential to develop novel methodologies that can release the dopaminergic inhibition, the present techniques for induction of gametogenesis typically rely on provision of gonadotropins using pituitary extracts (PE) from salmon or carp for females, while treatment for males includes human chorionic gonadotropin (hCG) (Mylonas and Zohar, 2009) or eel gonadotropins produced using recombinant technology (Kobayashi et al., 2010; Peñaranda et al., 2018).

Follicular development, maturation and oocyte quality

For female eel, the first protocols for induction of vitellogenesis that produced viable eggs used biweekly injection of carp pituitary extract (CPE) for the European eel (Bezdenezhnykh et al., 1983; Prokhorchik, 1986) and salmon pituitary extract (SPE) for the Japanese eel (Kagawa et al., 1995, 2003; Ohta et al., 1997). Subsequent experimental research on European eel typically has applied weekly injection, while varying in dosage and pituitary extract, that is, SPE (Da Silva et al., 2016; Pedersen, 2003; Tomkiewicz, 2012) and CPE (Mordenti et al., 2013; Palstra et al., 2005; Pérez et al., 2011), while research on Japanese eel primarily applies a constant dose of SPE (Okamura et al., 2014; Tanaka, 2015). During a period of 12–20 weeks of PE treatment at around 20 mg kg^{-1}, the follicles develop and ovaries increase in size (Figure 22.2a). At lower or variable dose, the developmental period is longer, however, oocyte stages and ovarian

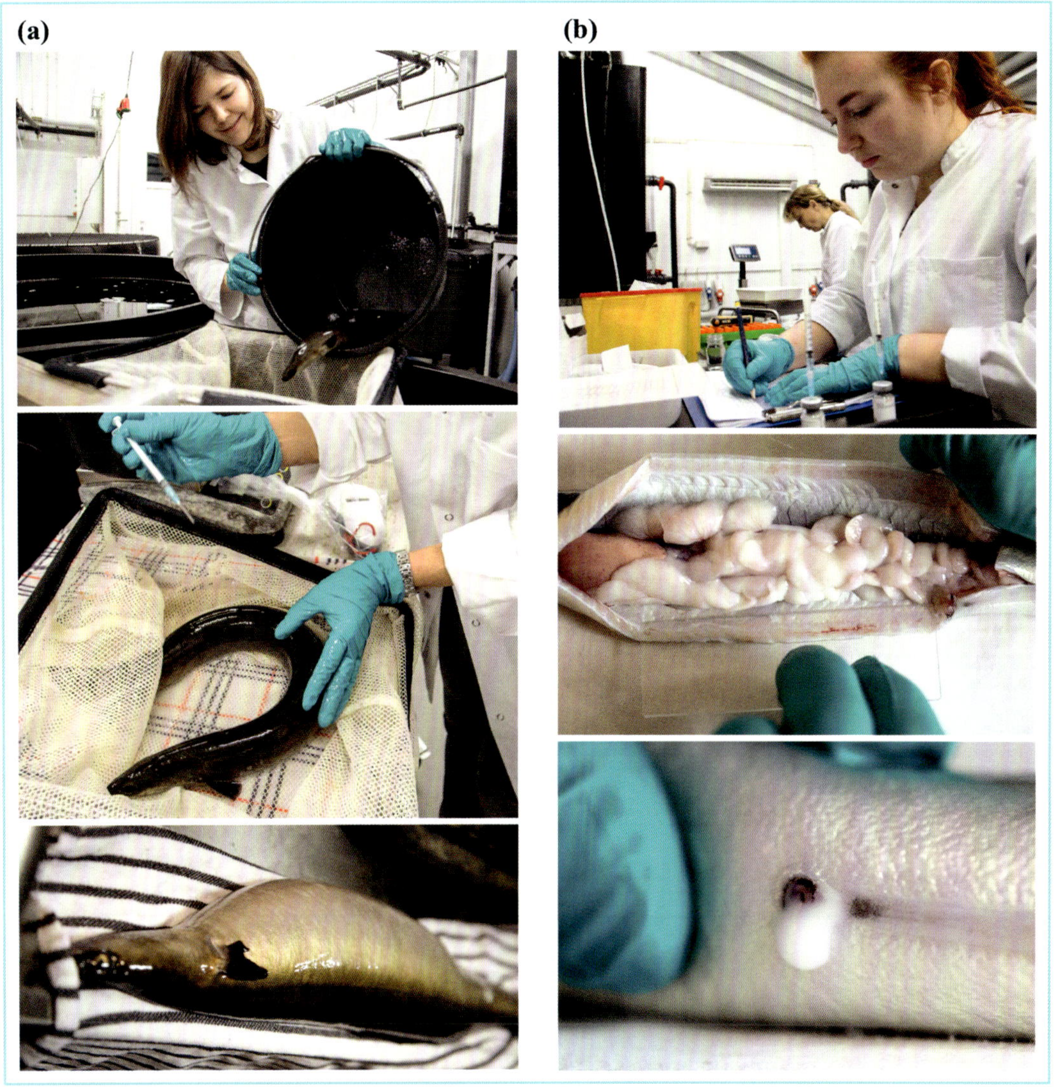

Figure 22.2 European eel broodstock reared in custom-designed recirculating aquaculture systems at the EEL-HATCH facility. Column (a) female broodstock handling, hormonal induction of ovarian development, and anaesthetized female in final maturation stage. Column (b) induced maturation in male broodstock, developed testis in abdomen, and milt production. Photographs: Sune Riis Sørensen, DTU Aqua.

development are similar (Da Silva et al., 2016; Tomkiewicz and Jarlbæk, 2008). However, responsiveness to treatment varies, which for European eel has led to tests of supplementary stimulation of sexual maturation and ovarian development including environmental conditions, for example, temperature, light and swimming (Mazzeo et al., 2014; Mordenti et al., 2012; Palstra et al., 2009; Pérez et al., 2011) as well as administration of androgens (Di Biase et al., 2017). For the Japanese eel and New Zealand freshwater eels, treatment using androgens has shown a positive effect on ovarian synchronization (Lokman et al., 2015; Matsubara et al., 2003), while feminization using estradiol is well established in propagation of Japanese eel, promoting growth, responsiveness, and ovarian development (Okamura et al., 2014; Tanaka, 2015).

While pituitary extracts enable oocyte growth, follicular maturation and ovulation require administration of a MIS, often preceded by a priming dose of pituitary extract (Ohta et al., 1997; Yamauchi, 1990). While an increase in female body weight can be used as an initial indicator, oocyte appearance, staging development according to germinal vesicle migration, coalesces of oil droplets and germinal vesicle breakdown can be used as indicators for optimal timing of MIS provision (Kagawa et al., 1995; Palstra et al., 2005). Further optimization of indicators include estimation of oocyte size and oil droplet diameter in preparation to priming and provision of MIS (da Silva et al., 2018a, 2018b; Unuma et al., 2011). After oocyte maturation, ovulation takes place and meiosis is reactivated and completed upon fertilization.

The induction of spawning must be precisely synchronized with the acquisition of follicular maturational and ovulatory competence (Mylonas and Zohar, 2009). Fish that are treated too early or too late will either not ovulate/spawn or produce eggs of low quality or poor fertilization capacity. Among various factors interfering with the induction of follicular maturation and subsequent gamete quality, is the heterogeneity in the development of the oocytes in the ovary. This asynchronous or group-synchronous development indicates batch spawning, which agrees with the ability of female eels to spawn several times and produce fertilizable eggs (Bezdenezhnykh and Prokhorchik, 1984; Da Silva et al., 2018a, b; Lokman and Young, 2000; Pedersen, 2003; Tomkiewicz and Jarlbæk, 2008). Moreover, the batch fecundity is high ranging from 0.7 to 2.6 million eggs per female (Boëtius and Boëtius, 1980), averaging approximately 80,000 per 100 g initial body weight with approximately the same batch size in the first and second batches (Tomkiewicz et al., 2013), which is promising for hatchery production.

Milt production and sperm quality

For male eel, spermatogenesis can be induced by administration of one or two injections of hCG (Miura et al., 1991; Ohta et al., 1996). However, in order to sustain spermatogenesis, spermiogenesis, and spermiation, weekly hormonal treatment with hCG or gonadotropins is common practice in the reproduction of European eel. In males that are given weekly hCG injections, spermiation starts in the first males around the fifth week of treatment (Asturiano et al., 2005, 2006; Pérez et al., 2000). Temperature has been shown to profoundly affect the progression of spermiation in European eel, reaching spermatogenesis earlier at higher temperatures (Baeza et al., 2014; Peñaranda et al., 2016). Figure 22.2b shows testes development and spermiation in European eel. Treatment schemes vary in dosage, but generally yield high-quality sperm. Here, stripping 24 hours after a priming injection shows optimal sperm motility (Asturiano et al., 2005). Furthermore, recombinant DNA technology has been used to produce homologous reFSH and reLH for Japanese eel (Kobayashi et al., 2010), as well as for European eel (Herranz-Jusdado et al., 2019; Peñaranda et al., 2018). These recombinant gonadotropins were shown to stimulate gonadotropin receptors in vitro and in vivo and appeared generally efficient to induce spermiation in European eel. As for female eels, different environmental scenarios to enhance spermatogenesis have been tested to optimize treatment, for example, temperature in combination with recombinant hCG (Gallego et al., 2012). The duration of spermatogenesis and the spermiation period depends on the treatment (that is, hormone dosage and injection frequency) (Asturiano et al., 2005; Peñaranda et al., 2010), with a continuous generation of spermatocytes and spermatozoa for several months as long as hCG is administered (Tomkiewicz et al., 2011). Here, the composition of different germ cell stages in the testis tissue remained constant with limited fluctuations for a period of 18 weeks until the end of the experiment. Such a reproductive strategy would match a batch spawning strategy of female eel.

Eel sperm quality is assessed widely using motility and velocity, by means of computer assisted analysis, CASA (Gallego et al., 2018; Mylonas et al., 2017; Vílchez et al., 2016, 2017), which serves as an objective measure of sperm swimming patterns. Alternatively, this can be evaluated by light microscopy, where motility is characterized using an arbitrary scale where 0: represents no motile sperm: I <25%; II 25–50%; III 50–75%; IV 75–90%; and V 90–100% of motile spermatozoa (Pérez et al., 2009). Using this method, sperm motility values of >IV are typically obtained and used for experimentation (Butts et al., 2014; Sørensen et al., 2013, 2015; Tomkiewicz, 2012).

Additionally, sperm density is quantified in order to standardize the sperm to egg ratio. Here, sperm cells are counted using a Neubauer hemocytometer, a method that is classified by the World Health Organization as the 'gold standard' for sperm density quantification (WHO, 1999). Faster counting methods (Fauvel et al., 2010), which correlate with Neubauer hemocytometer counts, have also been tested for European eel (Sørensen et al., 2013). Here, spermatocrit, CASA, and flow cytometry were positively correlated to haemocytometer counts. Additionally, it was shown that European eel sperm could be rapidly and accurately measured using a spectrophotometer (Butts et al., 2014).

Fertilization techniques, egg incubation and embryonic development

In common with most fishes, eels are external fertilizers, where gametes are released into the aquatic environment. As catadromous fishes, they naturally spawn in the marine environment, where the eggs are activated under hyper-osmotic conditions (Browne et al., 2015; Cosson et al., 2008). In hatcheries, common fertilization techniques include tank spawning with subsequent collection of fertilized eggs or manual stripping of gametes followed by in vitro fertilization (Mylonas and Zohar, 2009; Mylonas et al., 2010). Both procedures have been applied to European eel resulting in fertilized egg and larval production with increasing success over time (Butts et al., 2014; Di Biase et al., 2016; Mordenti et al., 2014; Palstra et al., 2005; Pedersen, 2003, 2004; Tomkiewicz, 2012). We utilize manual stripping followed by in vitro fertilization as these procedures can be controlled and standardized, thereby reducing noise in experimental results, allowing accurate determination of the floating egg layer and fertilization success (Butts et al., 2014; Sørensen et al., 2015; Tomkiewicz, 2012).

The in vitro fertilization protocol aims to optimize the sperm to egg ratio, considering sperm quality (motility, density) and post stripping ageing of eggs (Butts et al., 2014). An immobilizing medium mimicking seminal plasma (Asturiano et al., 2004; Peñaranda et al., 2010) is used for dilution and maintaining sperm cells in a quiescent state until in vitro fertilization (Butts et al., 2014). The pre-diluted milt enables a standardized sperm to egg ratio when mixing the gametes, while addition of artificial seawater at a volume dependent ratio enables a reproducible osmotic environment during gamete activation. Moreover, timing of fertilization has proven crucial, as egg quality deteriorates rapidly after ovulation (Butts et al., 2014), which also has been shown for eggs of Japanese eel (Ohta et al., 1996). After fertilization, buoyant eggs are incubated in artificial or treated natural seawater (Sørensen et al., 2015). This in vitro fertilization technique reliably yields fertilization rates up to 80–100% (Butts et al., 2014).

During the first hours after fertilization, a perivitelline space (PVS) forms as the chorion separates from the plasma membrane and the egg swells thereby enhancing the egg buoyancy (Figure 22.3) (Sørensen et al., 2016). Embryonic developmental success positively relates to the degree of swelling with failure of PVS development leading to no hatch (Tomkiewicz and Jarlbæk, 2008). During embryonic development, as well as in the fertilization procedure, both salt type and salinity significantly influence egg swelling and subsequent hatch (Sørensen et al., 2015). For instance, the use of treated natural sea water consistently produces larger eggs with diameters ranging up to 1.6–1.8 mm, which is comparable to wild-caught Japanese eel eggs (Tsukamoto et al., 2011; Yoshinaga et al., 2011). Buoyant fertilized eggs typically contain several smaller oil droplets that slowly fuse to form one large oil droplet (Figure 22.3). At 20°C, the first cell cleavages are observed at ~1 hour post fertilization (hpf), while the 16-cell stage is reached within ~4 hpf (Sørensen et al., 2016). At this time, fertilization success can be accurately assessed (Butts et al., 2014; Sørensen et al., 2015). The first somites become visible at ~24 hpf. At ~32 hpf, the embryo evolves two noticeable eye capsules and several somites. Thereafter, the tail bud forms, the yolk-sac becomes ellipsoid and hatching occurs at ~48 hpf (Sørensen et al., 2016). Rearing temperature significantly affects the duration of embryonic development and survival with detrimental effects above 22°C (Politis et al., 2017) as described below.

Embryonic development is influenced by gamete quality involving nutrition and cytoplasmatic factors, such as maternal RNA and activation of the embryo's genes (Lubzens et al., 2017). Differences in gamete quality among wild-caught and farmed broodstock receive much attention in aquaculture breeding and are of consequence also for eel, for example, in terms of essential fatty acids (Kottmann et al., forthcoming; Støttrup et al., 2013, 2016). While male broodstock origin, either farmed or wild-caught silver eels, did not affect fertilization and hatching success in a study by Benini et al. (2018), Kottmann et al. (forthcoming) found that enhancing essential fatty acid contents in eggs from farmed broodstock significantly improved fertilization success and embryonic survival, although survival and hatching success was lower than for wild-caught female broodstock. Here, the critical period for embryonic survival appeared to be between 8 and 16 hpf. This concurs with the timing of the expression of specific mRNA transcripts during embryonic development, that is, the transition from reliance of maternal mRNA to expression of the embryo's own genome (Lubzens et al., 2017). A study by Rozenfeld et al. (2016) showed that egg batches of farmed female European eel with either high or low hatching success did not differ in mRNA abundance between groups at 0 and 5 hpf, while they differed significantly at 30 hpf. At this point, embryos of the high hatch group showed higher abundance of five genes analysed (CPT1a, CPT1b, β-Tubulin, PHB2, and PIGF5 transcripts) than the low hatch group, suggesting that a general up-regulation of embryonic genes after the mid-blastula transition is needed to sustain embryonic development and hatching success.

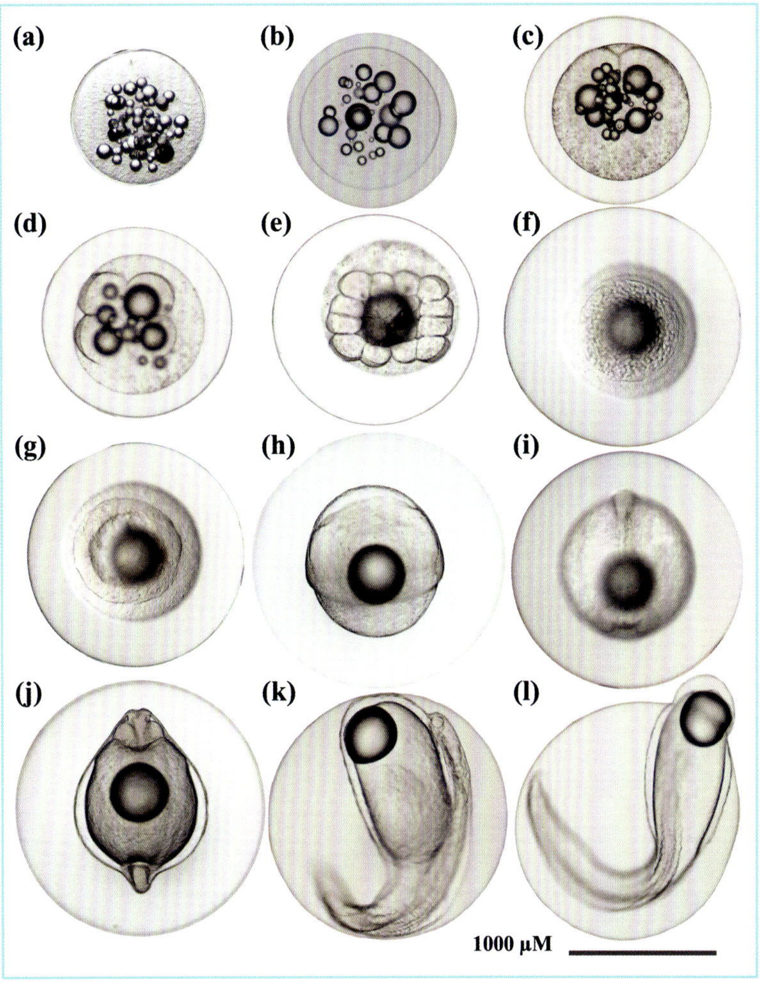

Figure 22.3 Embryonic development of European eel at 20°C. (a) Egg 5 min after extrusion by stripping, chorion not distended, multiple smaller oil droplets. (b) 0 h 25 min after fertilization, chorion has distended slightly. (c) 0 h 50 min, first cleavage and partial fusion of oil droplets. (d) 1 h 45 min, second cell cleavage and variable degree of oil droplet fusion (occasional full fusion of droplets at this stage). (e) 3 h, 16-cell stage and regular full fusion of oil droplet. (f) 8 h 30 min, early blastula stage, 100% fused oil droplets. (g) 13 h 0 min, germ ring formed. (h) 15 h, 1/2 epiboly. (i) 22 h 30 min, blastopore closure and first somites. (j) 32 h 15 min, embryo formed, eye vesicle visible and somites along most of the embryo, tail bud positioned downwards in figure. (k) 40 h, embryo has clearly distinguishable head, eyes and somites (> 20), occasional movements and onset of heart beat. (l) 45–48 h hatching starts by intense wriggling and repeated punch of head tip against chorion wall. Scale bar: 1 mm. Source: Sørensen et al. (2016).

Ontogeny of European eel pre-leptocephalus larvae

Russian scientists were the first to obtain and describe European eel embryos and larvae living up to 3.5 days (Bezdenezhnykh et al., 1983; Prokhorchik, 1986). Later, Pedersen (2004) reported on

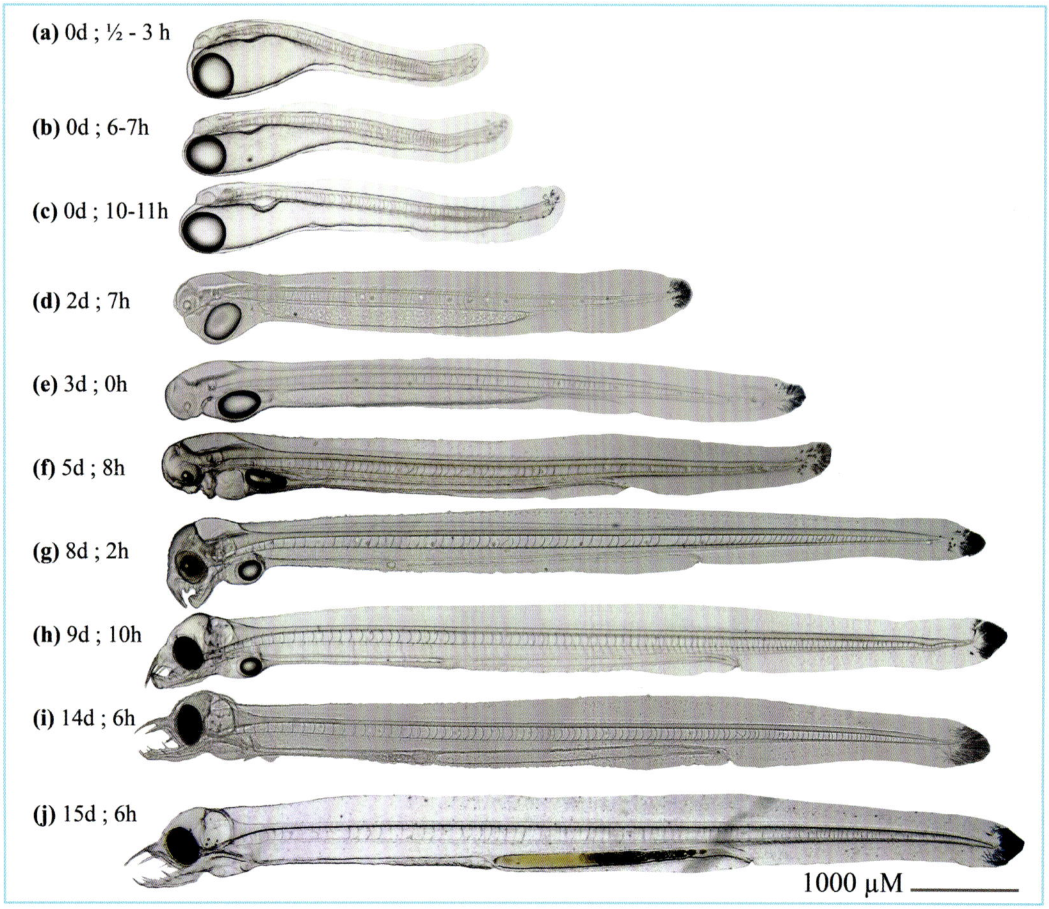

Figure 22.4 Development of European eel larvae from newly hatched to first-feeding stage [in hours post hatch (hph) and days post hatch (dph)] at 20°C. (a) 1/2–3 hph, larva still undeveloped, large yolk-sac with significant oil globule. (b) 6–7 hph, eye primordium apparent. (c) 10–11 hph prolonged tail, abdomen extends into yolk-sac, caudal pigments develop. (d) 2 dph, larval head protrudes anteriorly of yolk-sac, broad primordial fin apparent. (e) 3 dph, heart and brain clearly visible. (f) 5 dph, mouth starts to develop and eyes become pigmented. (g) 7 dph, upper and lower jaw distinguishable. (h) 9 dph, head angle changes, mouth points forward. (i) 14 dph, eyes clearly pigmented, gastro-intestinal tract enlarged, first-feeding stage. (j) 15 dph, first-feeding larva with food (rotifer paste) in the intestine. Scale bar: 1 mm. Adapted from Sørensen et al. (2016).

the chronology of embryological stages of European eel, while Sørensen et al. (2016) described early ontogeny and larval development throughout the yolk-sac stage, lasting 2 weeks when reared at 20°C. These results based on enhanced embryonic survival and mass hatching in the EU-FP7 project PRO-EEL outlined and illustrated for the first time the development throughout the pre-leptocephalus stage (Figure 22.4). At hatch, European eel larvae are little developed with a prominent yolk-sac encompassing a large oil droplet. During the first days, the larvae allocate resources to increase in length, develop optical capsules, a visible hindbrain and pericardium, as well as a wide primordial fin with a well-defined, rounded and pigmented tail. Towards

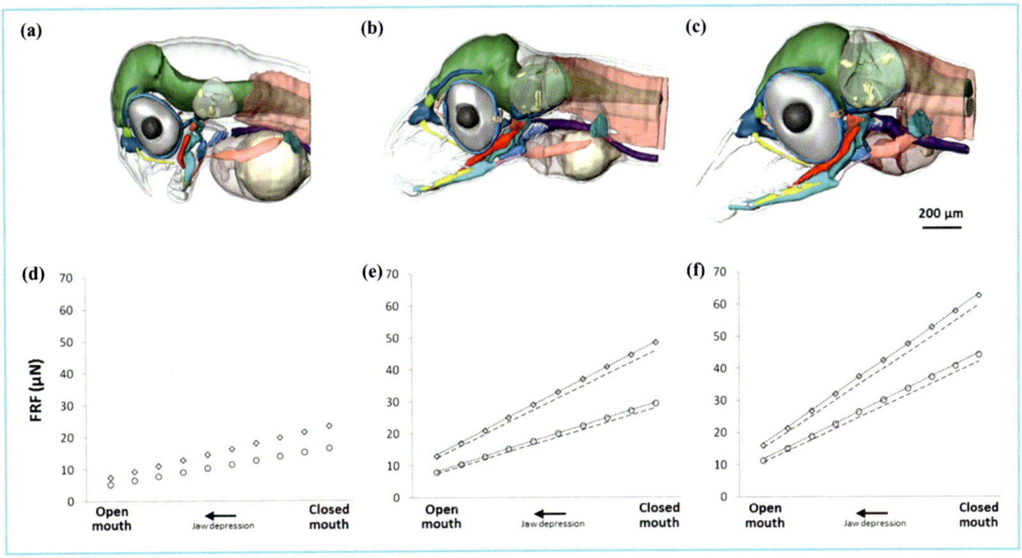

Figure 22.5 European eel larval head morphology: left lateral view on original 3D reconstructions. (a) 9 days post fertilization (dpf), (b) 12 dpf and (c) 15 dpf larvae of European eel. Cartilage is represented in blue, bone in yellow and muscles in red/pink. Graphical representation of food reaction forces (FRFs) for the (d) 9 dpf, (e) 12 dpf and (f) 15 dpf larvae. Circles and diamonds represent the force estimations at the level of first and third dentary teeth, respectively. All forces are calculated over a range of depression angles (15°, 30° and 40°) for the three investigated larval stages (9, 12 and 15 dpf) between a completely opened and closed mouth of the larvae. Modified from Bouilliart et al. (2015).

5 days post hatch (dph), the oral opening is observed as a small channel posterior to the eyes. At ~8 dph, the eyes start to pigment, upper and lower jaws emerge, and developing teeth appear. Towards 14 dph, yolk reserves appear utilized at 20°C (Sørensen et al., 2016) (Figure 22.4). During development, the angle between the head and trunk increases to enable the jaws pointing forward, finally resulting in the formation of the feeding apparatus with the characteristic, protruding teeth (Figure 22.5) (Bouilliart et al., 2015).

Early life history has been outlined for other anguillid eels including the Japanese eel (Ahn et al., 2012; Okamura et al., 2007; Yamamoto, 1981), American eel (*A. rostrata*) (Oliveira and Hable, 2010) as well as for hybrids of *A. dieffenbachii* ♀ × *A. australis* ♂ (Lokman and Young, 2000), *A. anguilla* and *A. japonica* (Okamura et al., 2004), and *A. australis* ♀ × *anguilla* ♂ (Burgerhout et al., 2011). Common to these species and their hybrids is that the larvae hatch at a relatively undifferentiated state, for example, little developed head morphology and digestive system compared to larvae of model species such as the zebrafish, *Danio rerio* (Kimmel et al., 1995). A characteristic that also applies to other members of Anguilliformes such as Japanese conger eel (*Conger myriaster*), daggertooth pike conger (*Muraenesox cinereus*) and the family of Sawtooth eels (*Serrivomeridae*) (Miller, 2009). Owing to species-specific temperature optima and rearing temperatures in embryonic and larval culture, the duration of the development phases differ. Thus, comparing morphological stages using degree day units (°D) as a measure of physiological time, starting at fertilization (Sørensen et al., 2016) shows that hatching time occurs at 38.3–40 °D for *A. anguilla*, 26.7–35.8 °D for *A. rostrata*, 37.6–39.4 °D for *A. australis* ♀ × *A. anguilla* ♂; 32.6–40.7 °D for *A. dieffenbachii/australis*, and 39.4–42.2 °D for *A. japonica*. Likewise, eye pigmentation in yolk-sac larvae of *A. anguilla* and hybrids *A. australis* ♀ × *A. anguilla* ♂ occurs at 220 °D, while at 196 °D for *A. japonica*.

Bio-physical requirements and culture technology

In almost all species, mortality is highest during early life, where subtle differences in survivorship can cause large variability in offspring production (Houde, 2008). In fish, the bio-physical conditions in the ambient environment during early life history stages are vital for their survival. Similarly, in hatchery practice, the bio-physical requirements need to be determined and suitable rearing conditions established (Conceição and Tandler, 2018). The choice of species-specific rearing conditions and prey/diets used for larval culture is typically based on knowledge originating from ambient conditions and food availability in their natural environment, supplemented by insights obtained from experiments. However, in the case of European eel where the natural environmental regimes of embryos and yolk-sac larvae remain unclear, focus is on assessing tolerances and optima using experimental studies of captive-produced offspring (Figure 22.1). Our research has focused on the impact of microbial activity, temperature, light, and salinity. Here, the goal has been to identify biotic and abiotic factors influencing larval survival and welfare in order to improve rearing conditions and increase numbers of survivors for first-feeding larval production, exploring first-feeding regimes enabling a successful transition to exogenous feeding.

Microbial activity

A major factor influencing hatchery production of offspring is the species' microbial sensitivity (Attramadal et al., 2012; Vadstein et al., 2013, 2018) that drives mortality and suboptimal development during early life stages (Skjermo and Vadstein, 1999). During the embryonic stage, the chorion provides protection against the external microbial environment (Finn, 2007; Laale, 1980). However, microbial activity on the egg tends to impede embryonic development (Salvesen and Vadstein, 1995) possibly caused by insufficient gas exchange and waste product secretion across the chorion (Salvesen et al., 1997). Sørensen et al. (2014) studied the effects of microbial interference on embryonic survival and larval hatch in European eel using different levels of microbial control and egg disinfection treatments. Here, survival was related to the efficiency of antibiotics to impede microbial activity, with low microbial interference resulting in increased embryonic survival, hatching success, and larval survival, thereby extending larval longevity. These results emphasize the importance of a low-level, stable microbial community in eel offspring culture (Sørensen et al., 2014).

Temperature

Identifying temperature regimes is essential for optimization of rearing conditions in larval culture, as temperature controls fundamental biochemical processes in living organisms thereby influencing developmental rates and survival of fish larvae (O'Connor et al., 2007). In the closely related Japanese eel, early ontogeny was found to be strongly influenced by temperature, with optimum thermal rearing conditions (~25°C) similar to those found in their recently identified natural spawning area (Ahn et al., 2012; Tsukamoto et al., 2011). In this regard, European eel larvae originate in the Sargasso Sea, which is a water mass characterized by a relatively constant salinity of 36.5 practical salinity units (PSU), a seasonal thermocline (~18°C) at a depth of ~300 m, and an upper warm water layer of 20–28°C (Castonguay and McCleave 1987; Worthington, 1959). In order to assess the optimal rearing temperature for European eel larvae, we selected a thermal regime ranging from 16 to 24°C (Politis et al., 2017), representing the thermal environment occurring between 600 m depth and close to the surface of the Sargasso Sea.

Temperature influenced all traits investigated such as time to hatch, hatching success, larval morphometrics and deformities, providing important insights on thermal phenotypic sensitivity. Moreover, this included the underlying gene expression of the molecular mechanisms relating to stress (heat shock proteins) as well as growth and development (growth hormone and insulin-like growth factors) (Politis et al., 2017). As no larvae hatched at 24°C, temperatures beyond this point were considered to be deleterious. Moreover, 22°C signalled the upper and 16°C the lower thermal tolerance limits, as ~90 and ~50% of larvae were deformed when reared at these temperatures, respectively. An optimal intermediate (18–20°C) thermal environment was identified with efficient growth and fewer deformities as well as high growth hormone expression (indicating better growth) and low heat shock protein expression (indicating lower stress).

Similarly, the expression of genes relating to the thyroid hormone (TH) signalling pathway, known to regulate growth, development, and metabolism in vertebrates, was followed and the sensitivity of this mechanism to temperature elucidated. Generally, THs are key regulators of growth, development, and metabolism in vertebrates (Power et al., 2001; Tata, 2006; Warner and Mittag, 2012), while evidence increases that they play important roles during early life development and metamorphosis in fish (Campinho et al., 2010; Infante et al., 2008; Marchand et al., 2004; Walpita et al., 2007). TH is produced in the thyroid gland (or thyroid follicles) mainly as T4 (thyroxine), which is then metabolized by deiodinase (DIO) enzymes in peripheral tissues, whereas their action is mostly exerted by binding to a specific nuclear thyroid hormone receptor (THR). The Japanese conger eel was previously chosen as a model species to investigate the TH role on development and metamorphosis of Anguilliformes (Kawakami et al., 2003), while a subsequent investigation regarding *thrs* showed differentially regulated expression during development and metamorphosis in Japanese eel (Kawakami et al., 2013). In European eel larvae, Politis et al. (2018b) identified three deiodinases (*dio1*, *dio2* and *dio3*) as well as two isoforms of *thra* (*thraA* and *thraB*) and two isoforms of *thrβ* (*thrβA* and *thrβB*), which showed high similarity to other mammalian, bird, amphibian or fish species, but highest similarity to the closely related Japanese eel.

Generally, regarding gene expression, the warmer the temperature the earlier the expression response was observed for a specific target gene. Moreover, in real time, the expression profiles appeared very similar and only shifted with temperature, while in developmental time, expression of all genes differed across selected developmental stages, such as hatching, during teeth formation or at first feeding. As such, understanding the biological responses, limits and adaptabilities or preferences to extrinsic environmental factors, such as temperature, provides enhanced knowledge for the optimization of rearing techniques.

Light

Light influences development and behaviour during early life history of fish (Downing and Litvak, 2002). Most animals possess well-developed photoreceptors and neuronal networks to organize, identify, and interpret sensory information through the perception of light in order to react to environmental changes and adapt accordingly (Meissl et al., 1986). Before hatch, the retinae of most teleost larvae consist of only cone photoreceptors and development is not completed until metamorphosis (Blaxter and Staines, 1970). Primarily under the control of rods, scotopic vision seems impaired in early life stages (Powers and Raymond, 1990), but early developed pineal photoreceptors are generally well-suited for detecting luminance changes (Kusmic et al., 1992; Meissl and Ekström, 1988). We performed a study on the reaction of European eel embryos and larvae to photoperiod, light intensity and wavelength (Politis et al., 2014). Here, embryonic survival was highest when embryos were reared under a 12:12 hour light/dark photoperiod and low-intensity light regime, while low-intensity light also led to the highest hatching success. However, timing and development of colour sensitivity is not clearly understood, as eel embryos were not affected by spectral composition, probably due to the perceptive mechanism and corresponding receptors not being fully developed at this stage. On the contrary, European eel larvae were sensitive to all physical parameters of light. Here, larvae reared in low-intensity light had higher survival than those reared in high-intensity light, larvae reared in a 12:12 hour photoperiod had higher survival than those reared in constant light, and larvae reared in red light had higher survival than those reared in green or white light.

Furthermore, a high-degree of offspring survived and developed when reared under continuous darkness, similar to the best intensity-photoperiod-spectral composition light regime. As such, rearing in darkness or under a red, low-intensity and 12 hour photoperiod light regime seems to be of benefit for European eel yolk-sac stage pre-leptocephali (Politis et al., 2014). However, the incidence of first feeding and percentage of gut fullness increased with light intensity (Butts et al., 2016), clearly indicating a shift in light sensitivity and preference at later developmental stages.

Salinity

Most fish species are hyper-osmotic in freshwater, where plasma osmolality is higher than the environment and hypo-osmotic in seawater, where plasma osmolality is lower than the environment (Marshall and Grosell, 2005). Thus, in freshwater they need to actively take up ions to counteract the diffusive ion loss and osmotic water gain, while in seawater they need to maintain osmotic balance through a desalting process to counteract osmotic water loss (Evans, 2008). Eels are euryhaline species that have adapted to cope with both hyper- and hypo-osmotic environments, likely due to regular salinity changes in their habitats and migrations between freshwater and marine environments at different developmental stages (Tesch, 2003). Eel offspring naturally occur in a hyper-osmotic environment in the ocean (Castonguay and McCleave, 1987), and interestingly, Okamura et al. (2009) showed that reducing salinity during early life history rearing under culture conditions, results in better growth and survival of Japanese eel larvae.

We tested this proposition also for European eel by rearing yolk-sac larvae under different salinity reduction scenarios with the goal to ease osmotic 'stress' by adjusting salinity towards iso-osmotic levels (Politis et al., 2018c). The results showed that gradually reducing salinity towards iso-osmotic conditions, that is, from 36 to 18 PSU, improved growth and lead to a four-fold increase in survival. Thus, the larvae were able to keep energy metabolism at stable levels judged from the related gene expression (*atp6, cox1*). Moreover, the lower salinity allowed for an energy surplus by reducing metabolic demands associated with osmoregulation and stress revealed by lower expression of *$Na^+K^+2Cl^-$ cotransporters, aquaporins* and *heat shock proteins*. This likely facilitated efficient use of energy for conversion of endogenous resources into somatic development, reflecting improved survival, biometry, and growth efficiency. Additionally, salinity reduction decreased the amount of severe deformities such as spinal curvature and emaciation but tended to induce an oedematous state of the larval heart. Here, the most balanced survival/deformity ratio was obtained when salinity was decreased from 3 dph and onwards at 2 PSU per day.

Ontogeny of the immune system and feeding apparatus

Overall, the identification of bio-physical requirements and regulated underlying molecular processes during the embryonic and yolk-sac larval stage has advanced our understanding on larval physiology. Here, the new insights gained by morphologically and molecularly defined physiological tolerance limits and preferences have gradually improved protocols for incubation and pre-leptocephalus larval culture technology and resulted in high numbers of healthy larvae capable of reaching the first-feeding stages. This increase in larval survivorship, allowed us to initiate comprehensive experimental studies on the development of the larval immune system and feeding capabilities.

Larval immunity

Teleost fish were one of the first phylogenetic groups to develop an immune system that possesses both the innate and adaptive arm of the immune response, characteristic to higher vertebrates (Uribe et al., 2011). However, evidence has accumulated that newly hatched fish larvae are particularly sensitive to pathogens as their immune system is not fully developed throughout the first weeks of life. During this period, they solely rely on the innate arm of the immune system, while exposure to pathogens intensifies due to hatching, mouth opening and first feeding (Ferraresso et al., 2016; Magnadóttir et al., 2004; Vadstein et al., 2013). Moreover, temperature is a fundamental modulator of the immune system of fish (Bowden, 2008) and has been shown to affect immunity during fish early life history (Dios et al., 2010). Therefore, in the light of our previous results, showing notable microbial sensitivity and temperature influence (Politis et al., 2017; Sørensen et al., 2014), we studied the molecular ontogeny of the immune system considering temperature as an immunomodulatory factor (Miest et al., 2019).

The molecular ontogeny of both the innate and the adaptive arm of the immune system during yolk-sac larval development was tested experimentally at different temperatures to evaluate the sensitivity of the early larval stages to microbial interference (Miest et al., 2019). The expression patterns of 11 immune genes were analysed throughout development, from hatch until the first-feeding stage. Interactions with temperature and immune competency were reduced at the lower end of the thermal spectrum (16°C), while close to the upper thermal limit (22°C) larvae showed signs of thermal stress. At the larvae's thermal optimum (18°C), the pattern of immune gene expression revealed an immunocompromised phase between hatch and teeth development (0–8 dph), caused by a lag period between initial protection and development of inherent immune competence. These results, together with the previous studies, emphasize the importance of a low-level, stable microbial community in eel offspring culture (Sørensen et al., 2014) in order to reduce detrimental fish–microbe interactions that commonly cause high and unpredictable mortality in the first weeks after hatching of marine fish larvae (Vadstein et al., 2013).

Insights leading to first feeding larval trials

Generally, when initiating first feeding, larvae detect prey through a range of chemical, visual and physical stimuli (Rønnestad et al., 2013). For instance, natural or synthetic chemo-tactic stimulants attract larvae, foster orientation, and promote the initiation of prey capture and ingestion (Barroso et al., 2013; Kamstra and Heinsbroek, 1991; Reig et al., 2003). However, nutritional requirements of fish larvae are species specific and differ across developmental stages within a species mainly due to morphological and physiological changes during ontogeny (Zambonino-Infante and Cahu, 2001). As such, scientists have aimed at revealing the natural feeding sources of larval eel, but despite increasing insights relying on advanced analyses of stomach content, their natural diet is not fully uncovered (Ayala et al., 2018; Miller et al., 2012; Riemann et al., 2010; Tomoda et al., 2018). At the same time, increased scientific inquiry has been directed towards developing suitable first feeding diets for eel larvae. In the case of Japanese eel, larvae are attracted to and successfully feed on a slurry diet based on shark egg yolk (Tanaka et al., 2003), hen egg yolk, exoskeleton-free Antarctic krill (Okamura et al., 2013), or a protein hydrolysate-based diet (Masuda et al., 2016). Additionally, a minute rotifer (*Proales similis*) was suggested as an alternative diet closer to natural larval trophic levels (Wullur et al., 2013). Similarly, for European eel, suitable feeds for larval culture covering the nutritional requirements need to be tailored and tested. First insights on potential feed items are revealed from investigating the morphology of the feeding apparatus.

In order to estimate the biting force and size of ingestible particles, the musculoskeletal anatomy of the larval head was investigated by histological 3D-reconstruction of the somatic structure of the feeding apparatus (Bouilliart et al., 2015). All cartilaginous and bony elements, as well as ligaments, tendons, and muscles were traced and identified in larvae close to the first-feeding stage, that is, 9, 12 and 15 days post fertilization. Here, Figure 22.5 shows the teeth morphology and orientation, the presence of branchial arches (no gills), a non-rigid operculum, and the ligament between the hyoid and lower jaw. Bigger anatomical structures were also reconstructed including eyes, brain, heart and otic vesicle. Using this reconstruction, both the biting force and kinematic efficiency of the present hyoid four-bar mouth opening mechanism showed minimal biting force. Still though, the distinct attachment sites of the first pair of teeth, in both the upper and lower jaw, indicate that these teeth are involved in biting. As a combined result, a rather weak bite force (± 50 μN) and a relatively small maximum gape angle of the lower jaw (± 25°) suggests a preference for very soft and/or small food organisms and/or particles (Bouilliart et al., 2015).

First feeding in European eel larvae

The stable production of larvae reaching the feeding stage and indication of potential diets led to first feeding experiments, studying, for example, feeding behaviour, ingestion rate and mechanisms regulating ingestion, digestion, and growth (Figure 22.6). Here, we aimed at determining whether tailored diets, chemo-attractants, and light conditions impact the incidence of first feeding, gut fullness, and behaviour in the larvae. First results by Butts et al. (2016) showed that up to 50% of cultured larvae in the experiment ingested a diet composed of enriched rotifers (*Brachionus plicatilis*), concentrated and emulsified into a paste, with or without potential stimulants of feeding. Moreover, it was documented that first-feeding eel larvae are able to execute a complex goal-oriented motor response, where highly distinctive modes of swimming were observed from short-term bouts, slow steady-state cruising to quick lunges for either prey attacks or spontaneous escape behaviours. Swimming activity increased over ontogeny, co-varied with the frequency of attacks and increased in the presence of live rotifers or chemo-attractants, probably by increasing the awareness of food availability (Figure 22.7). Furthermore, improved ingestion was detected at higher light intensities, suggesting that these larvae are visual feeders, which probably explains their large eye globules. However, the observed chemo-attraction and successful food intake with and without light, indicates that this species is able, besides using visual cues, to utilize other stimuli (olfaction, taste) to detect prey (Butts et al., 2016). Although the diet did not enable larval survival and growth, the study significantly enhanced insights in the eel larval feeding biology and provided a benchmark diet for future feeding trials.

Molecular ontogeny of first-feeding European eel larvae

Based on the first-feeding trial, we focused on larval nutritional condition examining key physiological mechanisms that regulate ingestion, digestion, and growth (Figure 22.8). As such, eel larvae were reared with or without the presence of algae (*Nannochloropsis*, *Pavlova*, and

Figure 22.6 Captive-bred, feeding European eel larva (~15 dph) targeting small (<100 μm) food particles. Photograph: Sune Riis Sørensen, DTU Aqua.

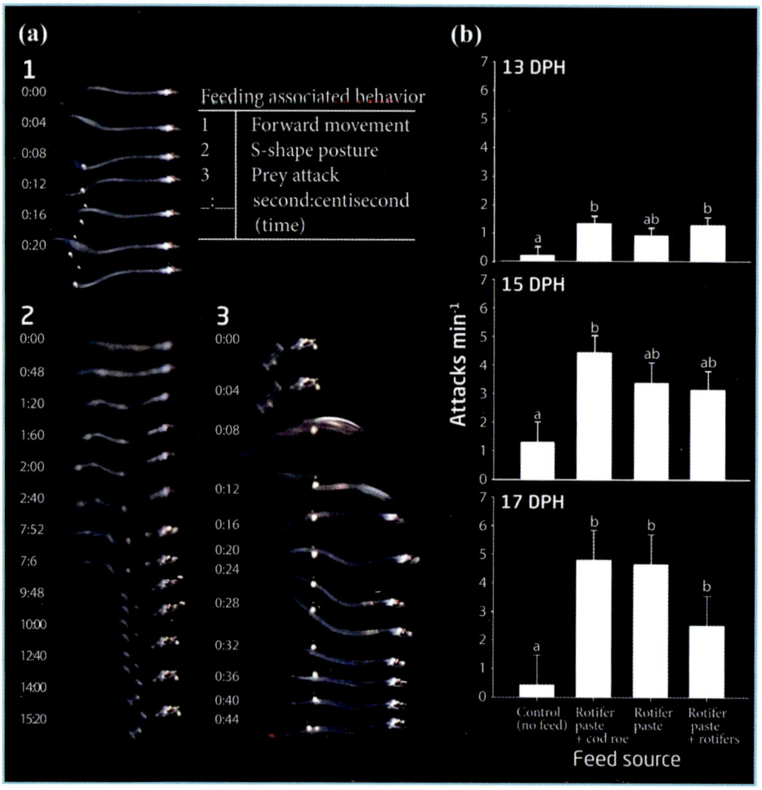

Figure 22.7 European eel larval feeding associated behaviours. (a) Illustrations of S-bending attacks: larva remains stationary, using frequent caudal beats, before exhibiting an s-shape posture and executing an explosive lunge. (b) Effect of feed source on the frequency of larval attacks at 13, 15 and 17 dph. Feed sources with different letters are significantly different ($p < 0.05$, least square means, ANOVA). Error bars represent least square means standard error. Source: Butts et al. (2016).

Tetraselmis) from 0 to 14 dph and with or without the presence of food (rotifer paste) from 15 to 24 dph (Politis et al., 2018a). Genes encoding appetite stimulators (*ghrelin*) and inhibitors (*cholecystokinin*) were expressed at basic levels on 4 dph, while their expression increased on 12 dph, indicating the beginning of the first-feeding window. Moreover, expression patterns of genes encoding important digestive enzymes relating to hydrolysis of protein (*trypsin*), lipid (*triclyceride lipase*) and carbohydrate (*amylase*) revealed that essential digestive ontogenetic processes occur from 14 to 20 dph (Figure 22.8). These ontogenetic molecular mechanisms were not affected by algal presence (green-water) and occurred irrespective of exogenous food ingestion signifying that they are linked to internal rhythms, which are under endocrine control.

The larvae that ingested a paste consisting of enriched rotifers as described in Butts et al. (2016) showed a molecular response to initiation of exogenous feeding in the expression pattern of genes relating to energy metabolism (*atp6*, *cox1*), food intake (*pomc*), growth (*igf1*) and thyroid metabolism (*thrαA*, *thrβB*). Moreover, the transcript levels of protein (*try*) digestion enzymes were higher than those of carbohydrate (*amyl*) and lipid (*tgl*) digestion enzymes, similar to the findings for pre-leptocephali and leptocephali larvae of Japanese eel (Hsu et al., 2015), indicating a nutritional predisposition for proteins during those life stages. This would be in accordance with their natural feeding regime, as they are assumed to feed on marine snow, primarily consisting of protein detritus and/or gelatinous plankton (Ayala et al., 2018; Miller et al., 2012; Tomoda et al., 2018). Likewise, as with Japanese eel (Hsu et al., 2015),

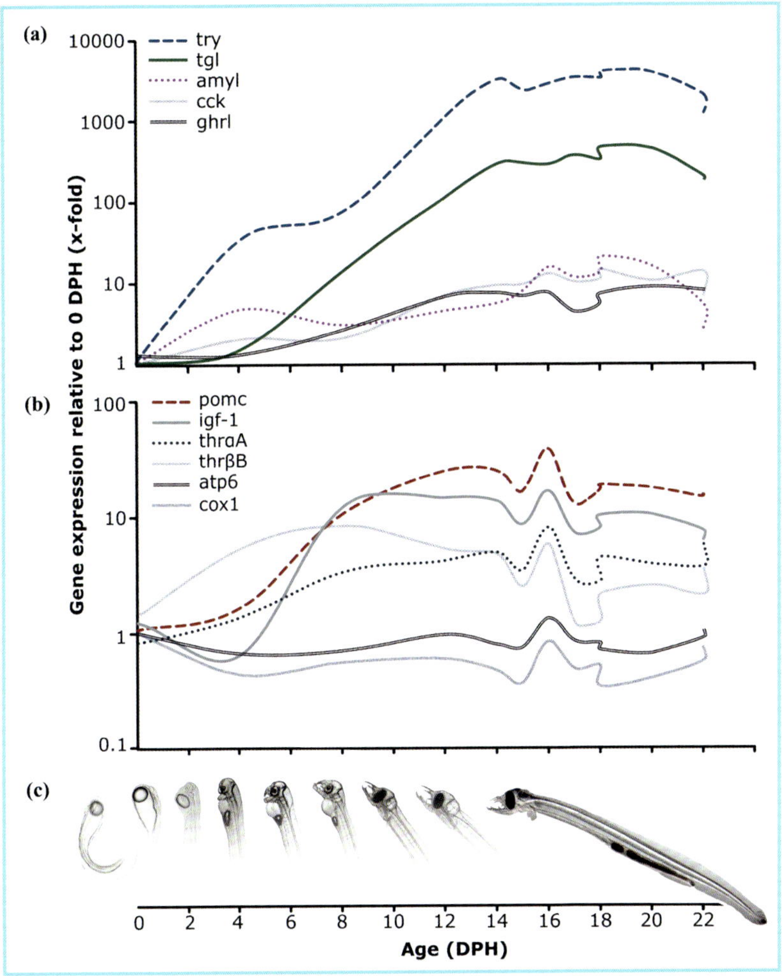

Figure 22.8 Conceptual overview of European eel larval gene expression in relation to age in days post hatch. Expression ($2^{-\Delta\Delta ct}$) calculated in relation to the average expression of each gene on day 0. (a) Relative expression for trypsin (*try*), triglyceride lipase (*tgl*), amylase (*amyl*), cholecystokinin (*cck*) and ghrelin (*ghrl*). (b) Relative expression for proopiomelanocortin (*pomc*), insulin-like growth factor (*igf1*), thyroid hormone receptors (*thrαA*, *thrβB*), ATP synthase F0 subunit 6 (*atp6*) and cytochrome-c-oxidase (*cox1*). (c) Comparable larval development from hatch into the feeding stage. Source: Politis et al. (2018a).

elevated expression levels of *amyl* (carbohydrate hydrolysis) were detected within the first-feeding window (Figure 22.8), which may reflect a primary mode of digestion (Zambonino-Infante and Cahu, 2001). Furthermore, higher RNA content in feeding compared to non-feeding larvae was observed, which indicates increased metabolic activity associated to protein synthesis. Additionally, increased DNA content in feeding larvae compared to non-feeding larvae was observed, which in combination with increased RNA content and greater body area, indicated a better growth pattern in the feeding larvae. However, RNA : DNA ratios still decreased from 12 dph onwards, indicating a decreasing larval condition, probably leading to the 'point-of-no-return' and subsequent mortality.

Thus, the study molecularly identified the first-feeding window in European eel, revealing that exogenous feeding success occurs concurrently with the onset of a broad array of genetically pre-programmed molecular factors, which regulate physiological functions of feeding. Here, feeding eel larvae sustained growth and condition at a higher level than non-feeding larvae. These results add to our understanding of the physiological changes occurring during the transition from newly hatched endogenous feeding pre-leptocephalus larvae to the exogenous leptocephalus stage. This new insight regarding a still undisclosed phase in the European eel life cycle may help developing efficient feeds and feeding strategies for European eel larvae.

Conclusion and perspectives

European aquaculture production has a substantial potential for expansion through species diversification and domestication, while at the same time reducing pressure on natural populations. In this regard, closing the life cycle of European eel in captivity would enable a sustainable and profitable aquaculture sector along with supporting conservation measures by replacing wild-caught glass eels with hatchery propagation of fry. Here, we have outlined a wide range of challenges that need to be overcome in order to develop closed-cycle propagation and mass production of glass eels. We also described the recent progress in establishing the scientific knowledge base needed to overcome these barriers for anguillid eel targeted for aquaculture. This research, encompassing the complex physiology, molecular adaptations and life history traits of eels, has since the 1990s, rapidly substantiated insights throughout ontogeny. At the same time, it has led to significant progress in advancing methodologies for captive breeding as well as larval culture technology leading to glass eel production for the Japanese eel and first-feeding larval trials for the European eel, bringing new attention to eel as a candidate species for commercial hatchery development.

Overall, this rapid substantiation of insights on eel reproduction and early life history, originate from combining extensive experimental work and field collection with advanced analytical methods and techniques. Particularly, the assembly and annotation of the European and Japanese eel genomes have accelerated acquisition of new knowledge, thereby providing a source of data for the discovery of genetic markers, and the comprehensive characterization of genes and their expression in different developmental stages and environments. Moreover, these advancements will allow the fast detection of biomarkers associated with eel physiological status that can guide future research in breeding for domestication. The application of these tools, with certainty, will increase our understanding of the endocrine reproductive control mechanisms and viable offspring production, uncover further aspects of parental influences on embryonic and larval developmental competence, and assist in identifying effective diets and establishing suitable culture conditions for growth of European eel leptocephali. Another aspect relates to integrating knowledge gained from experimental and field research in order to enhance our understanding of the European eel biology and ecology. For instance, the experimental identification of the optimal thermal environment for embryonic and early larval development may support new hypotheses regarding the spawning location in nature, while defining temperature tolerance limits may help predict environmental impacts of rising ocean temperatures. On the other hand, sampling during surveys enables analysis of larval biochemical composition and identification of the natural larval diet, which can help optimizing hatchery feeds for eel larvae. In future scenarios, integrating such multi- and interdisciplinary research may mutually benefit hatchery development and conservation of the European eel, and help preserving a fascinating, catadromous species of high ecological and economical importance.

Acknowledgements

The research presented has been promoted through international, collaborative research and technology development that received funding from the European Commission to the FP7 project 'Reproduction of European Eel: Toward a Self-sustained Aquaculture' (PRO-EEL), Grant Agreement no. 245257; Innovation Fund Denmark the projects 'Eel Hatchery Technology for a Sustainable Aquaculture' (EEL-HATCH), Grant no. 5184-00093B; and 'Improve technology and scale-up production of offspring for European eel aquaculture' (ITS-EEL) Grant no. 7076–00125B; and the Food and Agriculture COST Action FA1205: Assessing and improving the quality of aquatic animal gametes to enhance aquatic resources – The need to harmonize and standardize evolving methodologies, and improve transfer from academia to industry (AQUAGAMETE). Furthermore, the industrial partners: Billund Aquaculture A/S, The Danish Aquaculture Organization, STMI Aqua Systems, SPAROS Lda, Bioneer A/S, and the North Sea Science Park have contributed scientifically, technically and financially to the research presented.

Bibliography

Ager-Wick, E., Dirks, R.P., Burgerhout, E., Nourizadeh-Lillabadi, R., de Wijze, D.L., Spaink, H.P., van den Thillart, G.E., Tsukamoto, K., Dufour, S., Weltzien, F.A. and Henkel, C.V. (2013) The pituitary gland of the European eel reveals massive expression of genes involved in the melanocortin system. *PLoS ONE* 8(10), e77396.

Ahn, H., Yamada, Y., Okamura, A., Horie, N., Mikawa, N., Tanaka, S. and Tsukamoto, K. (2012) Effect of water temperature on embryonic development and hatching time of the Japanese eel *Anguilla japonica*. *Aquaculture* 330, 100–105.

Asturiano, J.F., Marco-Jiménez, F., Pérez, L., Balasch, S., Garzón, D.L., Peñaranda, D.S., Vicente, J.S., Viudes-de-Castro, M.P. and Jover, M. (2006) Effects of hCG as spermiation inducer on European eel semen quality. *Theriogenology* 66, 1012–1020.

Asturiano, J.F., Pérez, L., Garzón, D.L., Marco-Jiménez, F., Peñaranda, D.S., Vicente, J.S. and Jover, M. (2004) Physio-chemical characteristics of seminal plasma and development of media and methods for the cryopreservation of European eel sperm. *Fish Physiology and Biochemistry* 30, 283–293.

Asturiano, J.F., Pérez, L., Garzón, D.L., Peñaranda, D.S., Marco-Jiménez, F., Martínez-Llorens, S., Tomás, A. and Jover, M. (2005) Effect of different methods for the induction of spermiation on semen quality in European eel. *Aquaculture Research* 36, 1480–1487.

Attramadal, K.J.K., Salvesen, I., Xue, R., Øie, G., Størseth, T.R., Vadstein, O. and Olsen, Y. (2012) Recirculation as a possible microbial control strategy in the production of marine larvae. *Aquacultural Engineering* 46, 27–39.

Ayala, D.J., Munk, P., Lundgreen, R.B.C., Traving, S.J., Jaspers, C., Jørgensen, T.S., Hansen, L.H. and Riemann, L. (2018) Gelatinous plankton is important in the diet of European eel (*Anguilla anguilla*) larvae in the Sargasso Sea. *Scientific Reports* 8(1), 6156, DOI 10.1038/s41598-018-24388-x.

Baeza, R., Mazzeo, I., Vílchez, M.C., Gallego, V., Peñaranda, D.S., Pérez, L. and Asturiano, J.F. (2014) Effect of thermal regime on fatty acid dynamics in male European eels (*Anguilla anguilla*) during hormonally induced spermatogenesis. *Aquaculture* 430, 86–97.

Baeza, R., Mazzeo, I., Vílchez, M.C., Gallego, V., Peñaranda, D.S., Pérez, L. and Asturiano, J.F. (2015a) Relationship between sperm quality parameters and the fatty acid composition of the muscle, liver and testis of European eel. *Comparative Biochemistry and Physiology* 181, 79–86.

Baeza, R., Peñaranda, D.S., Vílchez, M.C., Tveiten, H., Pérez, L. and Asturiano, J.F. (2015b) Exploring correlations between sex steroids and fatty acids and their potential roles in the induced maturation of the male European eel. *Aquaculture* 435, 328–335.

Barroso, F.G., Rodiles, A., Vizcaino, A.J., Martínez, T.F. and Alarcón, F.J. (2013) Evaluation of feed attractants in juvenile Senegalese sole, *Solea senegalensis*. *Journal of the World Aquaculture Society* 44, 682–693.

Benini, E., Politis, S.N., Kottmann, J.S., Butts, I.A.E., Sørensen, S.R. and Tomkiewicz, J., (2018) Effect of parental origin on early life history traits of European eel. *Reproduction in Domestic Animals* 53(5), 1149–1158.

Bezdenezhnykh, V.A. and Prokhorchik, G.A. (1984) Period of oocyte maturation and assessment of egg quality of eel, *Anguilla anguilla* (Anguillidae), under stimulation of maturation with gonadotropic hormones. *Voprosy Ikhtyologii* 5, 814–821 (in Russian).

Bezdenezhnykh, V.A., Prokhorchik, G.A., Petrikov, A.M., Petukov, V.B. and Plyuta, M.V. (1983) Obtaining the larvae of European eel *Anguilla anguilla* L. (Pisces, Anguillidae) under experimental conditions. *Doklady Akademii nauk SSSR* 268(5), 1264–1266 (in Russian).

Blaxter, J.H.S. and Staines, M. (1970) Pure-cone retinae and retinomotor responses in larval teleosts. *Journal of the Marine Biological Association of the United Kingdom* 50, 449–460.

Boëtius, I. and Boëtius, J. (1980) Experimental maturation of female silver eels, *Anguilla anguilla*: Estimates of fecundity and energy reserves for migration and spawning. *Dana* 1, 1–28.

Bouilliart, M., Tomkiewicz, J., Lauesen, P., De Kegel, B. and Adriaens, D. (2015) Musculoskeletal anatomy and feeding performance of pre-feeding engyodontic larvae of the European eel (*Anguilla anguilla*). *Journal of Anatomy* 227, 325–340.

Bowden, T.J. (2008) Modulation of the immune system of fish by their environment. *Fish and Shellfish Immunology* 25, 373–383.

Browne, R.K., Kaurova, S.A., Uteshev, V.K., Shishova, N.V., McGinnity, D., Figiel, C.R., Mansour, N., Agnew, D., Wu, M., Gakhova, E.N., Dzyuba, B. and Cosson, J. (2015) Sperm motility of externally fertilizing fish and amphibians. *Theriogenology* 83, 1–13.

Burgerhout, E., Brittijn, S.A., Kurwie, T., Decker, P., Dirks, R.P., Palstra, A.P., Spaink, H.P., and Van den Thillart, G.E. (2011) First artificial hybrid of the eel species *Anguilla australis* and *Anguilla anguilla*. *BMC Developmental Biology* 11: 16, https://doi.org/10.1186/1471-213X-11-16.

Butts, I.A.E., Sørensen, S.R., Politis, S.N., Pitcher, T.E. and Tomkiewicz, J. (2014) Standardization of fertilization protocols for the European eel, *Anguilla anguilla*. *Aquaculture* 426, 9–13.

Butts, I.A.E., Baeza, R., Støttrup, J.G., Krüger-Johnsen, M., Jacobsen, C., Pérez, L., Asturiano, J.F. and Tomkiewicz, J. (2015) Impact of dietary fatty acids on muscle composition, liver lipids, milt composition and sperm performance in European eel. *Comparative Biochemistry and Physiology Part A: Molecular & Integrative Physiology* 183, 87–96.

Butts, I.A.E., Sørensen, S.R., Politis, S.N. and Tomkiewicz, J. (2016) First-feeding by European eel larvae: A step towards closing the life cycle in captivity. *Aquaculture* 464, 451–458.

Butts, I.A.E., Hilmarsdóttir, G.S., Gallego, V., Støttrup J.G, Jacobsen, C., Zadmajid, V., Krüger-Johnsen, M., Politis, S.N., Asturiano and J.F., Tomkiewicz. J. (forthcoming) Dietary lysine and arginine have an impact on sperm performance traits in European eel, *Anguilla anguilla*. *Aquaculture*.

Campinho, M.A., Galay-Burgos, M., Sweeney, G.E. and Power, D.M. (2010) Coordination of deiodinase and thyroid hormone receptor expression during the larval to juvenile transition in sea bream (*Sparus aurata*, Linnaeus). *General and Comparative Endocrinology* 165(2), 181–194.

Campo, A., Lafont, A.-G., Lefranc, B., Leprince, J., Tostivint, H., Kamech, N., Dufour, S. and Rousseau, K. (2018) Tachykinin-3 Genes and Peptides Characterized in a Basal Teleost, the European Eel: Evolutionary Perspective and Pituitary Role. *Frontiers in Endocrinology* 9, 304.

Castonguay, M. and McCleave, J.D. (1987) Vertical distributions, diel and ontogenetic vertical migrations and net avoidance of leptocephali of Anguilla and other common species in the Sargasso Sea. *Journal of Plankton Research* 9, 195–214.

CITES (2007) Consideration of proposals to amend the appendices I and II. CoP14. Proposal 18. Fourteenth meeting of the conference of the parties, The Hague.

Conceição, L.E.C. and Tandler, A. (eds) (2018) *Success Factors for Fish Larval Production. John Wiley & Sons*, Oxford, UK.

Coppe, A., Pujolar, J.M., Maes, G.E., Larsen, P.F., Hansen, M.M., Bernatchez, L., Zane, L. and Bortoluzzi, S. (2010) Sequencing, de novo annotation and analysis of the first *Anguilla anguilla* transcriptome: EeelBase opens new perspectives for the study of the critically endangered European eel. *BMC Genomics* 11: 635, https://doi.org/10.1186/1471-2164-11-635.

Cosson, J., Groison, A.-L., Suquet, M., Fauvel, C., Dreanno, C. and Billard, R. (2008) Marine fish spermatozoa: racing ephemeral swimmers. *Reproduction* 136, 277–294.

Da Silva, F.F.G., Støttrup, J.G., Kjørsvik, E., Tveiten, H. and Tomkiewicz, J. (2016) Interactive effects of dietary composition and hormonal treatment on reproductive development of cultured female European eel, *Anguilla anguilla. Animal Reproduction Science* 171, 17–26.

Da Silva, F.F.G., Jacobsen, C., Kjorsvik, E., Støttrup, J.G. and Tomkiewicz, J. (2018a) Oocyte and egg quality indicators in European eel: Lipid droplet coalescence and fatty acid composition. *Aquaculture* 496, 30–38.

Da Silva, F.F.G., Tveiten, H., Maugars, G., Lafont, A.-G., Dufour S., Støttrup, J.G., Kjørsvik, E. and Tomkiewicz, J. (2018b) Differential expression of gonadotropin and estrogen receptors and oocyte cytology during follicular maturation associated with egg viability in European eel (*Anguilla anguilla*). *Comparative Biochemistry and Physiology, Part A* 221, 44–54.

Di Biase, A., Casalini, A., Emmanuele, P., Mandelli, M., Lokman, P.M. and Mordenti, O. (2016) Controlled reproduction in *Anguilla anguilla (*L.): comparison between spontaneous spawning and stripping-insemination approaches, *Aquaculture Research* 47, 3052–3060.

Di Biase, A., Lokman, P.M., Govonia, N., Casalinia, A., Emmanuelea, P., Parmeggiania, A. and Mordenti, O. (2017) Co-treatment with androgens during artificial induction of maturation in female eel, *Anguilla anguilla*: Effects on egg production and early development. *Aquaculture* 479, 508–515.

Dios, S., Romero, A., Chamorro, R., Figueras, A. and Novoa, B. (2010) Effect of the temperature during antiviral immune response ontogeny in teleosts. *Fish and Shellfish Immunology* 29, 1019–1027.

Downing, G. and Litvak, M.K. (2002) Effects of light intensity, spectral composition and photoperiod on development and hatching of haddock (*Melanogrammus aeglefinus*) embryos. *Aquaculture* 213, 265–278.

Dufour, S., Lopez, E., Le Menn, F., Le Belle, N., Baloche, S., Fontaine, Y.A. (1988) Stimulation of gonadotropin release and of ovarian development, by the administration of a gonadoliberin agonist and of dopamine antagonists, in female silver eel pretreated with estradiol. *General and Comparative Endocrinology* 70, 20–30.

Dufour, S., Burzawa-Gerard, E., Le Belle, N., Shaihi, M. and Vidal, B. (2003) Reproductive endocrinology of the European eel, *Anguilla anguilla*. In: Aida, K., Tsukamoto, K. and Yamauchi, K. (eds), *Eel Biology*. Springer-Verlag, Tokyo, pp. 373–383.

Dufour, S., Weltzien, F,M.E.S, Le Belle, N., Vidal, B., Vernier, P. and Pasqualini, C. (2005) Dopaminergic inhibition of reproduction in teleost fishes, *Annals of the New York Academy of Science* 1040, 9–21.

Evans, D.H. (2008) Teleost fish osmoregulation: what have we learned since August Krogh, Homer Smith, Ancel Keys. *American Journal of Physiology-Regulatory, Integrative and Comparative Physiology* 295, 704–713.

FAO (2018) Global Anguilla spp. production 2015. FISHSTAT Global Capture and Aquaculture Production Databases. Available at: www.fao.org/fishery/species/2203/en (accessed 4 March 2019).

Fauvel, C., Suquet, M. and Cosson, J. (2010) Evaluation of fish sperm quality. *Journal of Applied Ichthyology* 26, 636–643.

Ferraresso, S., Bonaldo, A., Parma, L., Buonocore, F., Scapigliati, G., Gatta, P.P. and Bargelloni, L. (2016) Ontogenetic onset of immune-relevant genes in the common sole (*Solea solea*). *Fish and Shellfish Immunology* 57, 278–292.

Finn, R.N. (2007) The physiology and toxicology of salmonid eggs and larvae in relation to water quality criteria. *Aquatic Toxicology* 81, 337–354.

Furuita, H., Hori, K., Suzuki, N., Sugita, T. and Yamamoto, T. (2007) Effect of n-3 and n-6 fatty acids in broodstock diet on reproduction and fatty acid composition of broodstock and eggs in the Japanese eel *Anguilla japonica. Aquaculture* 267, 55–61.

Furuita, H., Ohta, H., Unuma, T., Tanaka, H., Kagawa, H., Suzuki, N. and Yamamoto, T. (2003) Biochemical composition of eggs in relation to egg quality in the Japanese eel, *Anguilla japonica. Fish Physiology and Biochemistry* 29, 37–46.

Furuita, H., Ishida, T., Suzuki, T.;, Unuma, T., Kurokawa, T., Sugita, T. and Yamamoto, T. (2009a) Vitamin contents of eggs that produce larvae showing a high survival rate in the Japanese eel *Anguilla japonica. Aquaculture Research* 40, 1270–1278.

Furuita, H., Ishida, T., Suzuki, T., Unuma, T., Kurokawa, T., Sugita, T. and Yamamoto, T. (2009b) Vitamin content and quality of eggs produced by broodstock injected with vitamins C and E during artificial maturation in Japanese eel *Anguilla japonica. Aquaculture* 289(3–4), 334–339.

Gallego, V., Mazzeo, I., Vílchez, M.C., Peñaranda, D.S., Carneiro, P.C.F., Pérez, L. and Asturiano J.F. (2012) Study of the effects of thermal regime and alternative hormonal treatments on the reproductive performance of European eel males (*Anguilla anguilla*) during induced sexual maturation. *Aquaculture* 354–355, 7–16.

Gallego, V., Herranz-Jusdado, J.G., Rozenfeld, C., Pérez, L. and Asturiano, J.F. (2018) Subjective and objective assessment of fish sperm motility: when the technique and technicians matter. *Fish Physiology and Biochemistry* 44, 1457–1467.

Heinsbroek, L., Støttrup, J., Jacobsen, C., Corraze, G., Kraiem, M., Holst, L.K., Tomkiewicz, J. and Kaushik, S. (2013) A review on broodstock nutrition of marine pelagic spawners: the curious case of the freshwater eels (*Anguilla spp.). Aquaculture Nutrition* 19, 1–24.

Henkel, C.V., Dirks, R.P., de Wijze, D.L., Minegishi, Y., Aoyama, J., Jansen, H.J., Turner, B.;, Knudsen, B., Bundgaard, M., Hvam, K.L., Boetzer, M., Pirovano, W., Weltzien, F.-A., Dufour, S., Tsukamoto, K., Spaink, H.P., van den Thillart, G.E.E.J.M. (2012) First draft genome sequence of the Japanese eel, *Anguilla japonica. Gene* 511, 195–201.

Herranz-Jusdado, J.G., Rozenfeld, C., Morini, M., Perez, L., Asturiano, J.F. and Gallego, V. (2019) Recombinant vs purified mammal gonadotropins as maturation hormonal treatments of European eel males. *Aquaculture* 501, 527–536.

Houde, E.D. (2008) Emerging from Hjort's Shadow. *Journal of Northwest Atlantic Fishery Science* 41, 53–570.

Hsu, H.-Y., Chen, S.-H., Cha, Y.-R., Tsukamoto, K., Lin, C.-Y. and Han, Y.-S. (2015) De Novo Assembly of the Whole Transcriptome of the Wild Embryo, Preleptocephalus, Leptocephalus, and Glass Eel of *Anguilla japonica* and Deciphering the Digestive and Absorptive Capacities during Early Development. *PLOS ONE* 10(9), e0139105.

Infante, C., Asensio, E., Cañavate, J.P. and Manchado, M. (2008) Molecular characterization and expression analysis of five different elongation factor 1 alpha genes in the flatfish Senegalese sole (*Solea senegalensis* Kaup): differential gene expression and thyroid hormones dependence during metamorphosis. *BMC Molecular Biology* 9: 19, https://doi.org/10.1186/1471-2199-9-19.

Izquierdo, M.S., Fernández-Palacios, H. and Tacon, A.G.J. (2001) Effect of broodstock nutrition on reproductive performance of fish. *Aquaculture* 197, 25–42.

Jacoby, D and Gollock, M. (2014) *Anguilla anguilla*. The IUCN Red List of Threatened Species 2014: e.T60344A45833138.

Jansen, H.J., Liem, M., Jong-Raadsen, S.A., Dufour, S., Weltzien, F.A., Swinkels, W., Koelewijn, A., Palstra, A.P., Pelster, B., Spaink, H.P., Thillart, G., Dirks, R.P. and Henkel, C.V. (2017) Rapid de novo assembly of the European eel genome from nanopore sequencing reads. *Scientific Reports* 7(1), 7213, 10.1038/s41598-017-07650-6.

Jolly, C., Rousseau, K., Prézeau, L., Tomkiewicz, J., Dufour, S., Pasqualini, C. (2016) Functional characterization of eel dopamine D2 receptors and involvement in the direct inhibition of pituitary gonadotropins. *Journal of Neuroendocrinology* 28(9), 10.1111/jne.12411.

Kagawa, H. (2003) Artificial induction of oocyte maturation and ovulation. In: Aida, K., Tsukamoto, K. and Yamauchi, Y. (eds) *Eel Biology*. Springer-Verlag, Tokyo, pp. 401–414.

Kagawa, H., Tanaka, H., Ohta, H., Okuzawa, K. and Hirose, K. (1995) In vitro effects of 17α-hydroxyprogesterone and 17α,20β-dihydroxy-4-pregnen-3-one on final maturation of oocytes at various developmental stages in artificially matured Japanese eel *Anguilla japonica*. *Fisheries Science* 61(6), 1012–1015.

Kagawa, H., Tanaka, H., Ohta, H., Unuma T. and Nomural. K. (2005) The first success of glass eel production in the world: basic biology on fish reproduction advances new applied technology in aquaculture. *Fish Physiology and Biochemistry* 31, 193.

Kamstra, A. and Heinsbroek, L.T.N. (1991) Effects of attractants on start of feeding of glass eel, *Anguilla anguilla* L. *Aquaculture and Fisheries Management* 22, 47–56.

Kawakami, Y., Tanda, M., Adachi, S. and Yamauchi, K. (2003) cDNA cloning of thyroid hormone receptor betas from the conger eel, *Conger myriaster*. *General and Comparative Endocrinology* 131, 232–240.

Kawakami, Y., Nomura, K., Ohta, H. and Tanaka, H. (2013) Characterization of thyroid hormone receptors during early development of the Japanese eel (*Anguilla japonica*). *General and Comparative Endocrinology* 194, 300–310.

Kimmel, C.B., Ballard, W.W., Kimmel, S.R., Ullmann, B. and Schilling, T.F. (1995) Stages of embryonic development of the zebrafish. *Developmental dynamics* 203, 253–310.

Kobayashi, M., Hayakawa, Y., Park, W., Banba, A., Yoshizaki, G., Kumamaru, K., Kagawa, H., Kaki, H., Nagaya, H. and Sohn, Y.C. (2010) Production of recombinant Japanese eel gonadotropins by baculovirus in silkworm larvae. *General and Comparative Endocrinology* 167, 379–386.

Kottmann, J.S., Butts, I.A.E., Støttrup, J.G., Lund, I., Holst, L., Jacobsen, C. and Tomkiewicz, J. (forthcoming) Fatty acid composition and quality of eggs, embryos, and larvae of European eel broodstock fed formulated diets under hatchery conditions compared to those of wild-caught broodstock. Forthcoming.

Kusmic, C., Marchiafava, P.L., Strettoi, E. (1992) Photoresponses and light adaptation of pineal photoreceptors in the trout. *Proceedings of the Royal Society B* 248(1322), 149–157.

Laale, H.W. (1980) The perivitelline space and egg envelopes of bony fishes: a review. *Copeia* 2, 210–226.

Lafont, A.G., Rousseau, K., Tomkiewicz, J. and Dufour S. (2016) Three nuclear and two membrane estrogen receptors in basal teleosts, *Anguilla sp.*: identification, evolutionary history and differential expression regulation. *General and Comparative Endocrinology* 235, 177–191.

Lokman, P. and Young, G. (2000) Induced spawning and early ontogeny of New Zealand freshwater eels (*Anguilla dieffenbachii* and *A. australis*). *New Zealand Journal of Marine and Freshwater Research* 34, 135–145.

Lokman, P.M., Wylie, M.J., Downes, M., Di Biase, A. and Damsteegt, E.L. (2015) Artificial induction of maturation in female silver eels, An*guilla australis*: the benefits of androgen pre-treatment. *Aquaculture* 437, 111–119.

Lubzens, E., Bobe, J., Young, G. and Sullivan, C.V. (2017) Maternal investment in fish oocytes and eggs: The molecular cargo and its contributions to fertility and early development. *Aquaculture* 472, 107–143.

Magnadóttir, B., Lange, S., Steinarsson, A. and Gudmundsdóttir, S. (2004) The ontogenic development of innate immune parameters of cod (*Gadus morhua* L.). *Comparative Biochemistry and Physiology Part B: Biochemistry and Molecular Biology* 139, 217–224.

Marchand, O., Duffraisse, M., Triqueneaux, G., Safi, R. and Laudet, V. (2004) Molecular cloning and developmental expression patterns of thyroid hormone receptors and T3 target genes in the turbot (*Scophtalmus maximus*) during postembryonic development. *General and Comparative Endocrinology* 135, 345–357.

Marshall, W.S. and Grosell, M. (2005) Ion transport, osmoregulation, and acid-base balance In: Evans D.H. and Claiborne J.B. (Eds.) The Physiology of Fishes, 3rd edn. CRC Press, New York, pp. 177–230.

Masuda, Y., Imaizumi, H., Oda, K., Hashimoto, H., Usiki. H. and Teruya K. (2012) artificial completion of the Japanese Eel, *Anguilla japonica*, life cycle: challenge to mass production. *Bulletin of Fisheries Research Agency* 35, 111–117.

Masuda, Y., Yatabe, T., Matsunari, H., Furuita, H., Kamoshida, M., Shima, Y. and Kuwada, H. (2016) Rearing of larvae of Japanese eel *Anguilla japonica* to metamorphosis into glass eel by feeding with fish protein hydrolysate-based diets. *Nippon Suisan Gakkaishi* 82(2), 131–133.

Matsubara, H., Lokman, P.M., Senaha, A., Kazeto, Y., Ijiri, S., Kambegawa, A., Hirai, T., Young, G., Todo, T., Adachi, S. and Yamauchi, K. (2003) Synthesis and possible function of 11-ketotestosterone during oogenesis in eel (*Anguilla* spp.). *Fish Physiology and Biochemistry* 28, 353–354.

Mazzeo, I., Gallego, V., Pérez, L., Peñaranda, D.S., Jover, M. and Asturiano, J.F. (2010) Variations in fatty acids composition in different tissues of the European eel (*Anguilla anguilla* L.) males during induced sexual maturation. *Journal of Applied Ichthyology* 26, 763–774.

Mazzeo, I., Peñaranda, D.S., Gallego, V., Balochec, S., Nourizadeh-Lillabadi, R., Tveiten, H., Dufour, S., Asturiano, J.F., Weltzien, F.-A. and Pérez, L. (2014) Temperature modulates the progression of vitellogenesis in the European eel. *Aquaculture* 434, 38–47.

Meissl, H. and Ekström, P. (1988) Dark and light adaptation of pineal photoreceptors. Vision Research 28(1), 49–56.

Meissl, H., Nakamura, T. and Thiele, G. (1986) Neural response mechanisms in the photoreceptive pineal organ of goldfish. *Comparative Biochemistry and Physiology – Part A: Molecular & Integrative Physiology* 84(3), 467–473.

Miest, J.J., Politis, S.N., Adamek, M., Tomkiewicz, J. and Butts, I.A.E. (2019) Molecular ontogeny of larval immunity in European eel at variable temperatures. *Fish and Shellfish Immunology* 87, 105–119.

Miller, M.J. (2009) Ecology of Anguilliform leptocephali: Remarkable transparent fish larvae of the ocean surface layer. *Aqua-Bio Science Monographs* 2, 1–94.

Miller, M.J., Chikaraishi, Y., Ogawa, N.O., Yamada, Y., Tsukamoto, K., and Ohkouchi, N. (2012) A low trophic position of Japanese eel larvae indicates feeding on marine snow. *Biology Letters* 9: 20120826.

Minegishi, Y., Henkel, C.V., Dirks, R.P. and van den Thillart, G.E. (2012) Genomics in eels – towards aquaculture and biology. *Marine Biotechnology* (NY) 14, 583–590.

Miura, T., Yamauchi, K., Takahashi, H. and Nagahama, Y. (1991) Hormonal induction of all stages of spermatogenesis in vitro in the male Japanese eel (*Anguilla japonica*). *Proceedings of the National Academy of Sciences* 88, 5774–5778.

Mordenti, O., Di Biase, A., Sirri, R., Modugno, S. and Tasselli A. (2012) Induction of sexual maturation in wild female European eels (*Anguilla anguilla*) in darkness and light. *The Israeli Journal of Aquaculture – Bamidgeh, IJA*:64.2012.674, 6 pages.

Mordenti, O., Di Biase A., Bastone G., Sirri R., Zaccaroni A. and Parmeggiani A. (2013) Controlled reproduction in the wild European eel (*Anguilla anguilla*): two populations compared. *Aquaculture International* 21, 1045–1063.

Mordenti, O., Casalini, A., Mandelli, M. and Di Biase, A. (2014) A closed recirculating aquaculture system for artificial seed production of the European eel (*Anguilla anguilla*): technology development for spontaneous spawning and egg incubation. *Aquacultural Engineering* 58, 88–94.

Morini, M., Pasquier, J., Dirks, R., Thillart, G., Tomkiewicz, J., Rousseau, J., Dufour, S. and Lafont A.-G. (2015). Duplicated leptin receptors in two species of eel bring new insights into the evolution of the leptin system in vertebrates. *PLoS ONE* 10(5), e0126008.

Munk, P., Hansen, M.M., Maes, G.E., Nielsen, T.G., Castonguay, M., Riemann, L., Sparholt, H., Als, T.D., Aarestrup, K., Andersen, N.G. and Bachler, M. (2010). Oceanic fronts in the Sargasso Sea control the early life and drift of Atlantic eels. *Proceedings of the Royal Society B Biological Sciences* 277, 3593–3599.

Mylonas, C. and Zohar, Y. (2009) Controlling fish reproduction in aquaculture. In: New Technologies in Aquaculture Improving Production Efficiency, Quality and Environmental Management. *Woodhead Publishing Series in Food Science, Technology and Nutrition*, pp. 109–142.

Mylonas, C., Fostier, A. and Zanuy, S. (2010) Broodstock management and hormonal manipulations of fish reproduction. *General and Comparative Endocrinology* 165, 516–534.

Mylonas, C., Duncan, N.J. and Asturiano, J.F. (2017) Hormonal manipulations for the enhancement of sperm production in cultured fish and evaluation of sperm quality. *Aquaculture* 472, 21–44.

O'Connor, M.I., Bruno, J.F., Gaines, S.D., Halpern, B.S., Lester, S.E., Kinlan, B.P. and Weiss, J.M. (2007) Temperature control of larval dispersal and the implications for marine ecology, evolution, and conservation. *Proceedings of the National Academy of Sciences of the United States* 104(4), 1266–1271.

OECD (2019) Aquaculture production (indicator). doi: 10.1787/d00923d8-en (accessed 4 March 2019).

Ohta, H., Kagawa, H., Tanaka, H., Okuzawa, K. and Hirose, K. (1996) Changes in fertilization and hatching rates with time after ovulation induced by 17, 20b-dihydroxy-4- pregnen-3-one in the Japanese eel, *Anguilla japonica*. *Aquaculture* 139, 291–301.

Ohta, H., Kagawa, H., Tanaka, H., Okuzawa, K., Iinuma, N. and Hirose, K. (1997) Artificial induction of maturation and fertilization in the Japanese eel, *Anguilla japonica*. *Fish Physiology and Biochemistry* 17, 163.

Okamura, A., Zhang, H., Utoh, T., Akazawa, A., Yamada, Y., Horie, N., Mikawa, N., Tanaka, S. and Oka, H.P. (2004) Artificial hybrid between *Anguilla anguilla* and *A. japonica*. *Journal of Fish Biology* 64, 1450–1454.

Okamura, A., Yamada, Y., Horie N., Utoh, T., Mikawa, N., Tanaka, S. and Tsukamoto, K. (2007) Effects of water temperature on early development of Japanese eel *Anguilla japonica*. *Fisheries Science* 73, 1241–1248.

Okamura, A., Yamada Y., Mikawa, N., Horie, N., Utoh, T., Kaneko, T., Tanaka, S. and Tsukamoto. K. (2009) Growth and survival of eel leptocephali (*Anguilla japonica*) in low-salinity water. *Aquaculture* 296, 367–372.

Okamura, A., Yamada, Y., Horie, N., Mikawa, N., Tanaka, S., Kobayashi, H. and Tsukamoto, K. (2013) Hen egg yolk and skinned krill as possible foods for rearing leptocephalus larvae of *Anguilla japonica* Temminck & Schlegel. *Aquaculture Research* 44, 1531–1538.

Okamura, A., Horie, N., Mikawa, N., Yamada, Y. and Tsukamoto, K. (2014) Recent advances in artificial production of glass eels for conservation of anguillid eel populations. *Ecology of Freshwater Fish* 23, 95–110.

Okamura, A., Sakamoto, Y., Yamada, Y. and Tsukamoto, K. (2018) Accumulation of hyaluronan in reared Japanese eel *Anguilla japonica* during early ontogeny. *Aquaculture* 497, 220–225.

Oliveira, K. and Hable, W.E. (2010) Artificial maturation, fertilization, and early development of the American eel (*Anguilla rostrata*). *Canadian Journal of Zoology* 88, 1121–1128.

Palstra, A.P. and van den Thillart, G.E.E.J.M. (2009). Artificial maturation and reproduction of the European eel. In: van den Thillart, G.E.E.J.M., Dufour S. and Rankin, C. (eds) *Spawning Migration of the European Eel.* Springer, New York, pp. 309–332.

Palstra, A.P., Cohen, E.G.H., Niemantsverdriet, P.R.W., van Ginneken, V.J.T. and van den Thillart, G.E.E.J.M. (2005) Artificial maturation and reproduction of European silver eel: development of oocytes during final maturation. *Aquaculture* 249, 533–547.

Pasquier, J., Lafont, A.G., Jeng, S.R., Morini, M., Dirks, R.P., van den Thillart, G., Tomkiewicz, J., Chang, C.F., Rousseau, K. and Dufour, S. (2012) Multiple kisspeptin receptors in early osteichthyans provide new insights into the evolution of this receptor family. *PLoS One* 7(11):, e48931.

Pedersen, B.H. (2003). Induced sexual maturation of the European eel *Anguilla anguilla* and fertilization of the eggs. *Aquaculture* 224, 323–338.

Pedersen, B.H. (2004) Fertilization of eggs, rate of embryonic development and hatching following induced maturation of the European eel *Anguilla anguilla. Aquaculture* 237, 461–473.

Peñaranda, D.S., Pérez, L., Gallego, V., Barrera, R., Jover, M. and Asturiano, J.F. (2010) European Eel Sperm Diluent for Short-term Storage. *Reproduction in Domestic Animals* 45, 407–415.

Peñaranda, D.S., Morini, M., Tveiten, H., Vílchez, M.C., Gallego, V., Dirks, R.P., van den Thillart, G.E.E.J.M., Pérez, L. and Asturiano, J.F. (2016) Temperature modulates testis steroidogenesis in European eel. *Comparative Biochemistry and Physiology Part A: Molecular and Integrative Physiology 197*, 58–67.

Peñaranda, D.S., Gallego, V., Rozenfeld, C., Herranz-Jusdado, J.G., Pérez, L., Gómez, A., Giménez, I. and Asturiano, J.F. (2018) Using specific recombinant gonadotropins to induce spermatogenesis and spermiation in the European eel (*Anguilla anguilla*). *Theriogenology* 107, 6–20.

Pérez, L., Asturiano, J.F., Tomás, A., Zegrari, S., Barrera, R., Espinós, F.J., Navarro, J.C. and Jover, M. (2000) Induction of maturation and spermiation in the male European eel: assessment of sperm quality throughout treatment. *Journal of Fish Biology* 57, 1488–1504.

Pérez, L., Peñaranda, D., Gallego, V. and Asturiano, J. (2009) Testis development, sperm quality evaluation and cryopreservation in the European eel. In: Thillart, G., Dufour, S. and Rankin, J.C. (Eds.) Spawning Migration of the European Eel, Fish and Fisheries Series. Springer, Dordrecht, Netherlands, pp. 333–362.

Perez, L., Peñaranda, D.S., Dufour, S., Baloche, S. Palstra,, A.P., Van Den Thillart, G.E.E.J.M. and Asturiano J.F. (2011) Influence of temperature regime on endocrine parameters and vitellogenesis during experimental maturation of European eel (*Anguilla anguilla*) females. *General and Comparative Endocrinology* 174, 51–59.

Politis, S.N., Butts, I.A.E. and Tomkiewicz, J. (2014) Light impacts embryonic and early larval development of the European eel, *Anguilla anguilla*. *Journal of Experimental Marine Biology and Ecology* 461, 407–415.

Politis, S.N., Mazurais, D., Servili, A., Zambonino-Infante, J.-L., Miest, J.J., Sørensen, S.R., Tomkiewicz, J. and Butts, I.A. (2017) Temperature effects on gene expression and morphological development of European eel, *Anguilla anguilla* larvae. *PloS One* 12, e0182726.

Politis, S.N., Sørensen, S.R., Mazurais, D., Servili, A., Zambonino-Infante, J-L., Miest, J., Clemmesen, C., Tomkiewicz J. and Butts, I.A.E (2018a) Molecular ontogeny of first-feeding European eel larvae. *Frontiers in Physiology* 9,1477, doi: 10.3389/fphys.2018.01477.

Politis, S.N., Servili, A., Mazurais, D., Zambonino-Infante, J.-L., Miest, J.J., Tomkiewicz, J. and Butts, I.A.E. (2018b) Temperature induced variation in gene expression of thyroid hormone receptors and deiodinases of European eel (*Anguilla anguilla*) larvae. *General and Comparative Endocrinology* 259, 54–65.

Politis, S.N., Mazurais, D., Servili, A., Zambonino-Infante, J.-L., Miest, J.J., Tomkiewicz, J. and Butts, I.A. (2018c) Salinity reduction benefits European eel larvae: Insights at the morphological and molecular level. *PLoS ONE* 13(6), e0198294.

Power, D.M., Llewellyn, L., Faustino, M., Nowell, M.A., Bjornsson, B.T., Einarsdottir, I. E., Canario, A.V. and Sweeney, G.E. (2001) Thyroid hormones in growth and development of fish. *Comparative Biochemistry and Physiology Part C: Toxicology & Pharmacology* 130, 447–459.

Powers, M.K. and Raymond, P.A. (1990) Development of the visual system. In: Douglas, R.H. and Djamgoz, M.B.A. (Eds.) The Visual System of Fish. Chapman and Hall, London, pp. 419–442.

Prokhorchik, G.A. (1986) Postembryonic development of European eel, *Anguilla anguilla*, under experimental conditions. *Journal of Ichthyology* 26(6), 121–127.

Reig, L., Ginovart, M. and Flos, R. (2003) Modification of the feeding behaviour of sole (Solea solea) through the addition of a commercial flavour as an alternative to betaine. *Aquatic Living Resources* 16, 370–379.

Riemann, L., Alfredsson, H., Hansen, M.M., Als, T.D., Nielsen, T.N., Munk, P., Aarestrup, K., Maes, G.E., Sparholt, H., Petersen, M.I., Bachler, M. and Castonguay, M. (2010) Qualitative assessment of the diet of European eel larvae in the Sargasso Sea resolved by DNA barcoding. *Biology Letters* 6, 819–822.

Rønnestad, I., Yúfera, M., Ueberschär, B., Ribeiro, L., Sæle, Ø. and Boglione, C. (2013) Feeding behaviour and digestive physiology in larval fish: current knowledge, and gaps and bottlenecks in research. *Reviews in Aquaculture* 5 (Issue Supplement s1), 59–98.

Rojo-Bartolomé, I., Martínez-Miguel, L., Lafont, A. G., Vílchez, M. C., Asturiano, J. F., Pérez, L. and Cancio, I. (2017) Molecular markers of oocyte differentiation in European eel during hormonally induced oogenesis. *Comparative Biochemistry and Physiology Part A: Molecular & Integrative Physiology, 211*, 17–25.

Rousseau, K., Lafont, A.-G., Maugars, G., Jolly, C., Sébert, M.-E., 1 Aroua, S., Pasqualini, C. and Dufour, S. (2009) Advances in Eel Reproductive Physiology and Endocrinology. In: Trischitta, F., Takei,,Y. and Sébert, P. (Eds.) Eel Physiology. CRS Press, Boca Raton, Fl. USA, pp. 1–43.

Rozenfeld, C., Butts, I.A.E., Tomkiewicz, J., Zambonino-Infante, J.-L. and Mazurais, D. (2016) Abundance of specific mRNA transcripts impacts hatching success in European eel, *Anguilla anguilla* L. *Comparative Biochemistry and Physiology, Part A* 191, 59–65.

Salvesen, I. and Vadstein, O. (1995) Surface disinfection of eggs from marine fish: evaluation of four chemicals. *Aquaculture International* 3, 155–171.

Salvesen, I., Øie, G. and Vadstein, O. (1997) Surface disinfection of Atlantic halibut and turbot eggs with glutaraldehyde: Evaluation of concentrations and contact times. *Aquaculture International* 5, 249–258.

Schmidt, J. (1923) The Breeding Places of the Eel. *Philosophical Transactions of The Royal Society B Biological Sciences* 211, 179–208.

Skjermo, J. and Vadstein, O. (1999) Techniques for microbial control in the intensive rearing of marine larvae. *Aquaculture* 177, 333 343.

Sørensen, S., Gallego, V., Pérez, L., Butts, I., Tomkiewicz, J. and Asturiano, J. (2013) Evaluation of methods to determine sperm density for the European eel, *Anguilla anguilla. Reproduction in Domestic Animals* 48(6), 936–944.

Sørensen, S.R., Skov, P.V., Lauesen, P., Tomkiewicz, J., Bossier, P. and De Schryver, P. (2014) Microbial interference and potential control in culture of European eel (*Anguilla anguilla*) embryos and larvae. *Aquaculture* 426–427, 1–8.

Sørensen, S.R., Butts, I.A.E., Munk, P. and Tomkiewicz, J. (2015) Effects of salinity and sea salt type on egg activation, fertilization, buoyancy and early embryology of European eel, *Anguilla anguilla. Zygote,* 24, 121–138.

Sørensen, S.R., Tomkiewicz, J., Munk, P., Butts, I.A., Nielsen, A., Lauesen, P. and Graver, C. (2016) Ontogeny and growth of early life stages of captive-bred European eel. *Aquaculture* 456, 50–61.

Støttrup, J.G., Jacobsen, C., Tomkiewicz, J. and Jarlbæk, H. (2013) Modification of essential fatty acid composition in broodstock of cultured European eel *Anguilla anguilla* L. *Aquaculture Nutrition* 19, 172–185.

Støttrup, J.G., Tomkiewicz, J., Jacobsen C., Butts, I.A.E., Holst, L.K., Krüger-Johnsen, M., Graver, C., Lauesen, P., Fontagné-Dicharry, S., Corraze, G., Heinsbroek, L.T.N. and Kaushik, S. (2016) Development of broodstock diet to improve developmental competence of embryos in European eel, *Anguilla anguilla. Aquaculture Nutrition* 22, 725–737.

Tanaka, H. (2015) Progression in artificial seedling production of Japanese eel *Anguilla japonica. Fisheries Science* 81, 11–19.

Tanaka, H., Kagawa, H., Ohta, H., Unuma, T. and Nomura, K. (2003) The first production of glass eel in captivity: fish reproductive physiology facilitates great progress in aquaculture. Fish Physiology and Biochemistry 28, 493–497.

Tata, J.R. (2006) Amphibian metamorphosis as a model for the developmental actions of thyroid hormone. *Molecular and Cellular Endocrinology* 246(1–2), 10–20.

Teletchea, F. (2015) Domestication of Marine Fish Species: Update and Perspectives. *Journal of Marine Science and Engineering* 3, 1227–1243.

Tesch, F-W. (2003) The Eel. Fifth ed. Blackwell Publishing, Oxford, UK. 416p.

Tomkiewicz, J., (Ed.) (2012) Reproduction of European eel in aquaculture (REEL), Consolidation and new production methods. *DTU Aqua Report* 249–2012, DTU Aqua. 47 p.

Tomkiewicz, J. and Jarlbæk, H. (2008) Kunstig reproduktion af ål: ROE II og IIB., *DTU Aqua Rapport* 180–08. DTU Aqua. 79 pp. (in Danish).

Tomkiewicz, J., Kofoed, T.M.N. and Pedersen, J.S. (2011) Assessment of testes development during induced spermatogenesis in European eel *Anguilla anguilla. Marine and Coastal Fisheries: Dynamics, Management and Ecosystem Science* 3, 106–118.

Tomkiewicz, J., Støttrup, J.G., Corraze, G., Kausik, S., Holst, L.K., McEvoy, F., Dufour, S., Lafont, A.G., Asturiano, J., Sørensen, S.R., Tveiten, H., De Schryver, P., Butts, I.A., Munk, P., Zambonino-Infante, J., Politis, S.N., Krüger-Johnsen, M., Lauesen, P. (2013) Reproduction of European eel and larval culture: state of the art. *Communications in Agricultural and Applied Biological Sciences* 78(4), 455–6.

Tomoda, T., Chow, S., Kurogi, H., Okazaki, M., Ambe, D., Furuita, H., Matsunari, H., Nagai, S., Yokouchi, K.A., Sawayama, S., Nomura, K.A., Tanaka, H., Sudou, R., Hasegawa, D. and Inaba,

N. (2018) Observations of gut contents of anguilliform leptocephali collected in the western North Pacific. *Nippon Suisan Gakkaishi* 84, 32–44.

Tsukamoto, K., Chow, S., Otake, T., Kurogi, H., Mochioka, N., Miller, M.J., Aoyama, J., Kimura, S., Watanabe, S., Yoshinaga, T., Shinoda, A., Kuroki, M., Oya, M., Watanabe, T., Hata, K., Ijiri, S., Kazeto, Y., Nomura, K. and Tanaka, H. (2011) Oceanic spawning ecology of freshwater eels in the western North Pacific. *Nature Communications* 2, 179.

Unuma, T., Hasegawa, N., Sawaguchi, S., Tanaka, T., Matsubara, T., Nomura, K. and Tanaka, H. (2011) Fusion of lipid droplets in Japanese eel oocytes: stage classification and its use as a biomarker for induction of final oocyte maturation and ovulation. *Aquaculture* 322–323, 142–148.

Uribe, C, Folch, H., Enriquez, R. and Moran, G. (2011) Innate and adaptive immunity in teleost fish: a review. *Veterinární medicína (Praha)* 56, 486–503.

Vadstein, O., Bergh, Ø., Gatesoupe, F.-J., Galindo-Villegas, J., Mulero, V., Picchietti, S., Scapigliati, G., Makridis, P., Olsen, Y., Dierckens, K., Defoirdt, T., Boon, N., De Schryver, P. and Bossier, P. (2013) Microbiology and immunology of fish larvae. *Reviews in Aquaculture* 5, 1–25.

Vadstein, O., Attramadal, K.J.K., Bakke, I., Forberg, T., Olsen, Y., Verdegem, M., Giatsis, C., Skjermo, J., Aasen, I.M., F-J, Dierckens, K., Sorgeloos, P. and Bossier, P. (2018) Managing the Microbial Community of Marine Fish Larvae: A Holistic Perspective for Larviculture. *Frontiers in Microbiology* 9: 1820, DOI 10.3389/fmicb.2018.01820.

Vidal, B., Pasqualini, C, Le Belle, N, Holland, M.C.H., Sbaihi, M., Vernier, P., Zohar, Y. and Dufour, S. (2004) Dopamine inhibits luteinizing hormone synthesis and release in the juvenile European eel: a neuroendocrine lock of the onset of puberty. *Biology of Reproduction* 71, 1491–1500.

Vílchez, M.C., Morini, M., Peñaranda, D.S., Gallego, V., Asturiano, J.F. and Pérez, L. (2016) Sodium affects the sperm motility in the European eel. *Comparative Biochemistry and Physiology Part A: Molecular and Integrative Physiology* 198, 51–58.

Vílchez, M.C., Morini, M., Peñaranda, D.S., Gallego, V., Asturiano, J.F. and Pérez, L. (2017) Role of potassium and pH on the initiation of sperm motility in the European eel. *Comparative Biochemistry and Physiology Part A: Molecular and Integrative Physiology* 203, 210–219.

Walpita, C.N., Van der Geyten, S., Rurangwa, E. and Darras, V.M. (2007) The effect of 3, 5, 30-tri-iodothyronine supplementation on zebrafish (*Danio rerio*) embryonic development and expression of iodothyronine deiodinases and thyroid hormone receptors. *General and Comparative Endocrinology* 152, 206–214.

Warner, A. and Mittag, J. (2012) Thyroid hormone and the central control of homeostasis. *Journal of Molecular Endocrinology* 49(1), 29–35.

WHO (1999) WHO Laboratory Manual for the Examination of Human Semen and Semen Cervical Mucus Interaction, Cambridge University Press, Cambridge. 128 p.

Worthington, L.V. (1959) The 18° water in the Sargasso Sea. *Deep Sea Research* 5(2–4), 297–305.

Wullur, S., Yoshimatsu, T., Tanaka, H., Ohtani, M., Sakakura, Y., Kim, H.-J. and Hagiwara, A. (2013) Ingestion by Japanese Eel *Anguilla japonica* Larvae on Various Minute Zooplanktons. *Aquaculture Science* 61(4), 341–347.

Yamamoto, K. (1981) Cleavage pattern and formation of the blastocoel in the egg of the eel, *Anguilla japonica*. *Japanese Journal of Ichthyology* 27, 301–308.

Yamauchi, K. (1990) Studies of Gonadal Steroids Involved in Final Gonadial Maturation in the Japanese Eel, *Anguilla japonica*, a Review. *Internationale Revue der gesamten Hydrobiologie*. 76, 859–860.

Yoshinaga, T., Miller, M.J., Yokouchi, K., Otake, T., Kimura, S., Aoyama, J., Watanabe, S., Shinoda, A., Oya, M., Miyazaki, S., Zenimoto, K., Sudo, R., Takahashi, T., Ahn, H., Manabe, R., Hagihara, S., Morioka, H., Itakura, H., Machida, M., Ban, K., Shiozaki, M., Ai, B. and Tsukamoto, K. (2011)

Genetic identification and morphology of naturally spawned eggs of the Japanese eel *Anguilla japonica* collected in the western North Pacific. *Fisheries Science* 77, 983–992.

Zambonino-Infante, J.L. and Cahu, C.L. (2001) Ontogeny of the gastrointestinal tract of marine fish larvae. *Comparative Biochemistry and Physiology Part C* 130, 477–487.

Part VI

Exploitation

CHAPTER 23

Global exploitation of freshwater eels (genus *Anguilla*): fisheries, stock status and illegal trade

Kenzo Kaifu[1], Florian Stein[2,3], Willem Dekker[4], Nick Walker[5], C. Andrew Dolloff[6], Kristen Steele[7], A. Alonso Aguirre[5], Vincent Nijman[8], Penthai Siriwat[8] and Pierre Sasal[9]

[1]Chuo University, Japan; [2]Technische Universität Braunschweig, Germany; [3]Sustainable Eel Group (SEG), Belgium; [4]Swedish University of Agricultural Sciences, Stockholm, Sweden; [5]George Mason University, Fairfax, Virginia, USA; [6]USDA Forest Service, Southern Research Station, Fish and Wildlife Conservation, Virginia Tech University, Blacksburg, Virginia, USA; [7]University College of London, UK; [8]Oxford Brookes University, Oxford, UK; [9]PSL Research University: EPHE-UPVD-CNRS USR CNRS 3278 CRIOBE, Université de Perpignan Via Domitia, Perpignan, France

Introduction

Freshwater eels of the genus *Anguilla* are among the most widespread freshwater fish species and many of the 19 species are globally exploited for consumption. Historically, the Japanese eel (*A. japonica*) and the European eel (*A. anguilla*) were the primarily targeted species in Asia and in Europe. During the past century, small-scale fisheries, which originally provided for local consumption, have developed into internationally linked and economically important industries, centred in East Asia. Recent declines in recruitment of Japanese and European species have resulted in several side effects: (1) the European eel has been regulated under CITES since 2010 and its export outside the EU is prohibited; (2) the American eel (*A. rostrata*) and tropical *Anguilla* species have been targeted as replacements for European and Japanese eels in aquaculture; and (3) Sustained global demand triggers a black market for trade in young, wild-caught eels that raises concerns about the sustainable use of freshwater eel species in the future.

Japanese eel (*Anguilla japonica*)

Kenzo Kaifu

Japanese eel occurs as a native species in Japan, China, Taiwan, Province of China (PoC) and Korea. The range of this species extends from the southern Pacific coast of Japan and further south to Hainan Island covering large areas of mainland China, Taiwan and Korea (Jacoby and Gollock, 2014a). Japanese eel is one of the most economically important fisheries in East Asia, especially in Japan. As well as other anguillid species, Japanese eel farming is reliant on wild-caught juvenile eels (elvers or glass eels). Although artificial breeding in captivity has been successful (Tanaka et al., 2003), it is not yet commercially viable.

Eels are commercially and culturally important fish in Japan and are consumed almost exclusively as kabayaki (Figure 23.1a). Kabayaki is a grilled eel served with sweet soy sauce, usually on steamed rice. This cooking technique has been increasing in popularity in other East Asian countries (Figure 23.1b). Eels are also steamed, fried, and stewed in mainland China and Taiwan, PoC and Korean-style-barbecued in South Korea (Fig 23.1c, d). Outside Asia, kabayaki fillet slices are primarily used in Sushi dishes.

Consumption

East Asia has the highest demand for anguillid eels. According to Food and Agriculture Organization (FAO) of the United Nations data, Japan's consumption accounted for approximately 70% of global *Anguilla* production in 2000 and 2001, but less than 15% in 2013 (Shiraishi and Crook, 2015). However, according to production data from the 'Joint Statement of the Bureau of Fisheries of People's Republic of China, the Fisheries Agency of Japan, the Ministry of Oceans and Fisheries of the Republic of Korea and the Fisheries Agency of Taiwan, on International Cooperation for Conservation and Management of Japanese Eel Stock and Other Relevant Eel Species' ('Joint Statement', hereafter, [Anonymous, 2014a]) Japan's role as a global consumer does not appear to have declined as significantly, with Japan's consumption

Figure 23.1 Eel cuisine in East Asia. (a) Kabayaki on steamed rice; (b) Japanese-style kabayaki sold in a Chinese supermarket; (c) and (d) Korean-style barbecued eel.

estimated to account for 46 and 34% of global eel production in 2012 and 2013, respectively (Shiraishi and Crook, 2015). Japan is still a major consumer of eel, but its share of the global market has decreased from 2000 to 2013, regardless of the data source. Shiraishi and Crook (2015) reported that mainland China could be considered an increasingly significant eel consumer. With annual exports from mainland China being significantly less than annual production according to FAO, annual consumption was estimated at more than 150,000 tonnes of eels in 2012 and 2013. It should be noted, however, that it is difficult to estimate the true extent of eel consumption in mainland China, because different data sources provide very wide production estimates and there are many uncertainties surrounding the accuracy of these data (see Shiraishi and Crook [2015] for details).

According to the FAO, aquaculture production of Japanese eel has increased from 69,926 tonnes in 1980 to 236,344 tonnes in 2012 (Figure 23.2). In contrast, according to the Joint Statement, the sum of reported aquaculture production in mainland China, Japan, South Korea and Taiwan, PoC, the distribution area of the species, was only 62,271 tonnes in 2012. Shiraishi and Crook (2015) reported that aquaculture production data of the FAO and the Joint Statement are similar for Japan and Taiwan, PoC, but demonstrate significant differences for mainland China and South Korea. According to the Joint Statement (Anonymous, 2014a), the overall catch of glass eels (amount of glass eels used in aquaculture) in the four countries/territories was 37.8 and 40.8 tonnes in the 2014–15 and 2015–16 fishing seasons. Taking the average weight of a glass eel as 0.2 g, about 200 million individuals are estimated to be caught and farmed annually in these countries.

In addition to individuals caught as glass eel, naturally recruited wild Japanese eels ('wild eels' or 'wild individuals' hereafter) inhabiting rivers, lakes and coastal areas are also caught for consumption. Wild eels are preferred, and priced higher, than cultured eels in Japan. According to the Ministry of Agriculture, Forest and Fisheries, the catch of wild eels in rivers and lakes in Japan was about 3000 tonnes in the 1960s and steeply decreased to 70 tonnes by 2015 (Figure 23.3). Wild eels are also caught in coastal areas. Kaifu et al. (2018) reported eel catch data obtained from five fishery cooperatives in the coastal area of Okayama Prefecture, Japan. The annual eel catch of these cooperatives was approximately 1272 kg in 2002 and significantly decreased to 86 kg by 2015. Moreover, 67.7% of the annual eel catch was caught from October to November for the period 2002–15 (Figure 23.4). The large eel catches in fall and early winter imply dominance of silver eels in eel fisheries in coastal areas, because spawning migration occurs from fall to early winter in Japanese eels (Sudo et al., 2017). The eel catch from October to November in the coastal area of Okayama Prefecture was 977.7 kg in 2002, and significantly decreased to 18.6 kg in 2015. In Taiwan, PoC, catching eels larger than 8 cm is prohibited in some rivers by law. Information about wild eel consumption in mainland China and Korea is limited.

Population dynamics

To the best of our knowledge, only one study has assessed the stock of Japanese eel to date (Tanaka, 2014). That study estimated that the age 1+ stock of this species has been increasing since 1990 based on glass eel and exploitable eel (yellow and silver eels) fisheries data, including their biological characteristics such as survival rate, growth rate and sex ratio. This estimation

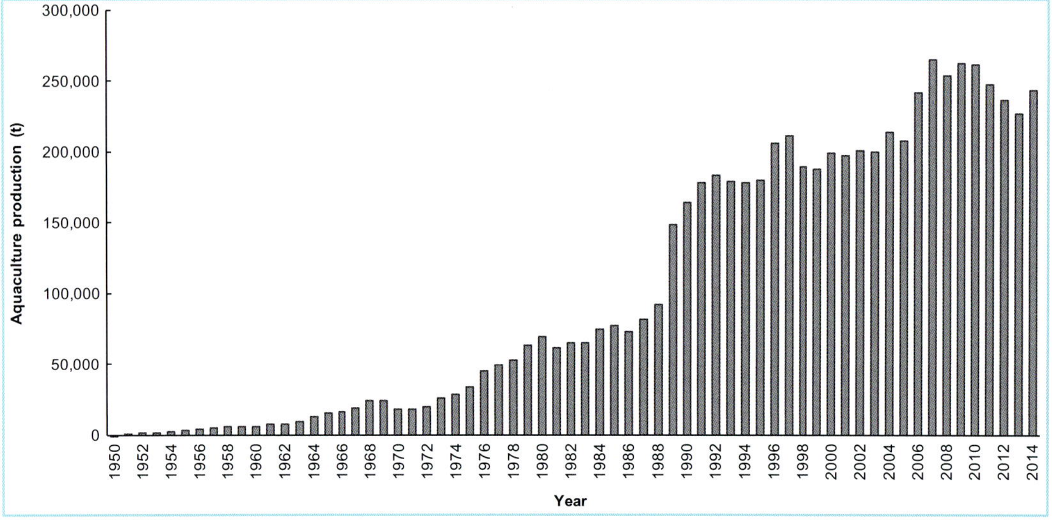

Figure 23.2 Annual aquaculture production of Japanese eel according to the FAO.

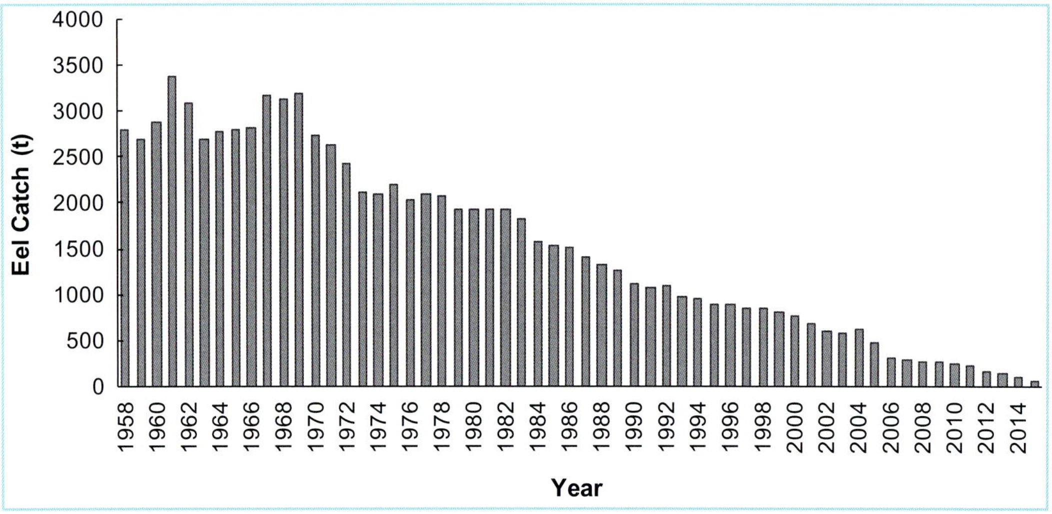

Figure 23.3 Annual catch of wild eels in Japanese rivers and lakes (Ministry of Agriculture, Forest and Fisheries, Japan).

(Tanaka, 2014) contradicts the Red List assessments of the IUCN, which classifies Japanese eel as Endangered, with declining populations (Jacoby and Gollock, 2014a).

One possible explanation of this contradiction is potential bias with respect to stocked individuals. In Japan, the Fisheries Act mandates to increase eel populations for inland water fishery cooperatives that catch eels in rivers and lakes and these cooperatives typically stock eels to fulfil this obligation (Kaifu et al., 2014). The Fisheries Agency of Japan financially supported eel farming associations that stock eels until 2015. The primary method of eel stocking in Japan is to

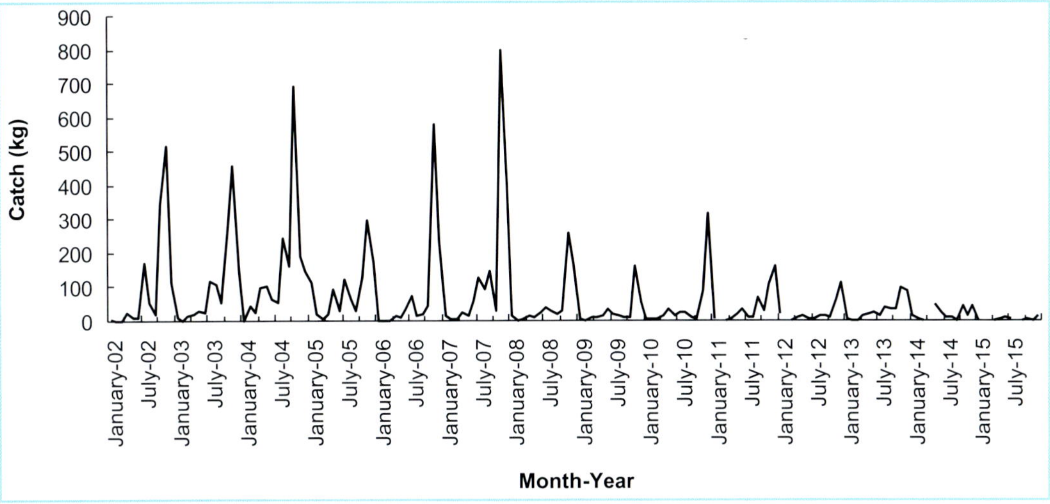

Figure 23.4 Monthly eel catches of five fisheries cooperatives in the coastal area of Okayama Prefecture, Japan (Kaifu et al., 2018).

release small yellow eels from eel farms into rivers and lakes (Kaifu et al., 2014). Tanaka (2014) estimated the stock dynamics of Japanese eels based on glass eel and exploitable eel (yellow and silver eels) fisheries data in Japan. However, the yellow and silver eel fisheries data was collected from water systems wherein eel stocking was conducted; thus, the yellow and silver fisheries data used by Tanaka (2014) may have been affected by additional stocked populations.

Kaifu et al. (2018) conducted a case study in Okayama Prefecture, Japan to estimate the population dynamics of wild Japanese eels. In this study, wild and stocked eels were classified using a recently developed method based on otolith stable isotopes (see Kaifu et al. [2017] for details). Of the 161 eels captured in freshwater areas where eels had been stocked, 98.1% were classified as stocked. In contrast, 82.8% of 128 eels were captured in coastal areas where eels were classified as wild. Longline and set-net eel catches per unit of effort (CPUE) in coastal areas where wild eels were dominant significantly decreased between 2003 and 2016, indicating ongoing depletion of wild Japanese eels in these waters (Figure 23.5). This study (Kaifu et al., 2018) collected fisheries data from a small number of fisheries in a single prefecture in Japan, which has a total of 47 prefectures, although Japanese eels are widely distributed across East Asia. However, it is possible that the dynamics of the local population studied in Okayama Prefecture accurately reflect those of the entire population, because Japanese eel has a single spawning population (Han et al., 2010; Minegishi et al., 2012).

At present, the dynamics of the Japanese eel population is not well-understood. To determine the population dynamics of Japanese eel accurately, fisheries data of wild eels in multiple areas across the entire distribution range of the species are necessary.

Management

The Japanese eel is a panmictic fish species (Han et al., 2010; Minegishi et al., 2012), and thus should be managed as a whole population. In the context of management of the whole

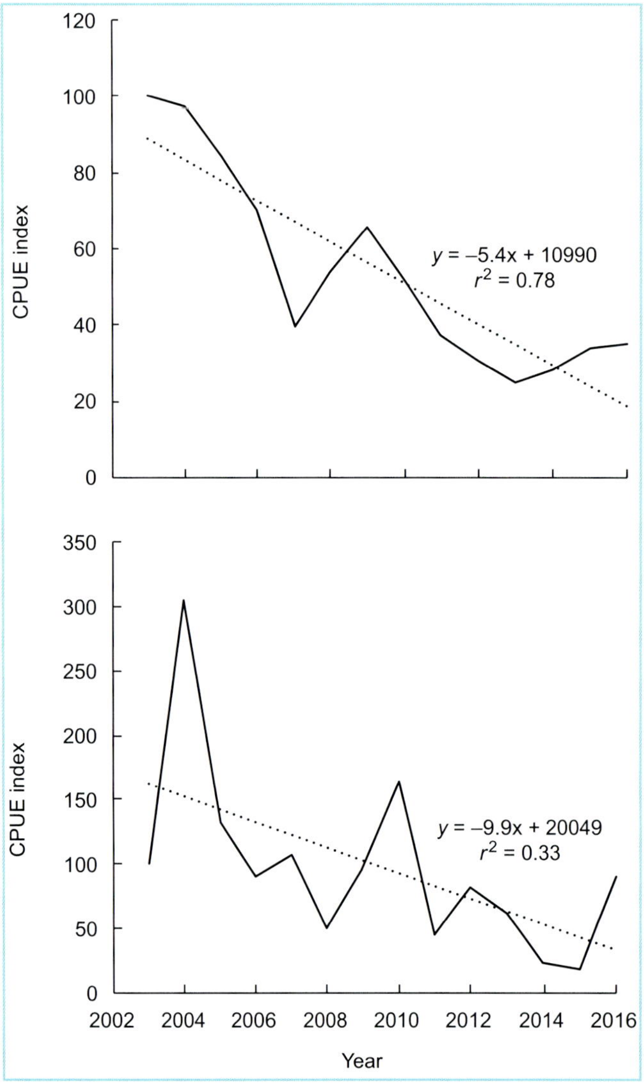

Figure 23.5 Standardized eel catch in the coastal area of Okayama Prefecture, Japan (Kaifu et al., 2018). Upper: longline eel catch per month; lower: eel catch per fixed net. Values are indicated as indices based on a catch per unit effort (CPUE) in 2003 of 100.

population, the Joint Statement placed a limit on the amount of Japanese eel that can be used for aquaculture. The annual input of Japanese eel seeds (glass eels) has been limited to 78.8 tonnes, which was 80% of those caught from 1 November 2013 to 31 October 2014, since the 2015 season (from 1 November 2014 to 31 October 2015) (Table 23.1) (Anonymous, 2014a). Although the limit is 78.8 tonnes, the actual amount used for aquaculture was 37.8 and 40.8 tonnes in the 2014–15 and 2015–16 seasons, respectively (Anonymous, 2014a). The limit of the Joint Statement is not considered to have decreased Japanese eel exploitation, because the ceiling is about twice as large as the actual glass eel input.

The Joint Statement limits the annual input of Japanese eel glass eels based on the glass eel input from 1 November 2013 to 31 October 2014 (Anonymous, 2014a). Glass eel inputs in China, Taiwan, PoC, Japan and Korea in this season were increased 6.4, 8.3, 1.3 and 4.6-fold

Table 23.1 Maximum versus actual inflow of Japanese glass eels in Japan, China, Korea, and Taiwan, PoC.

Country	Maximum inflow volume (tonnes)	Actual inflow during the 2015 fishing season (tonnes)	Actual inflow during the 2016 fishing season (tonnes)
Japan	21.7	18.3	19.7
Mainland China	36.0	9.3	8.2
Korea	11.1	7.4	9.3
Taiwan, PoC	10.0	2.8	3.6
Total	**78.8**	**37.8**	**40.8**

of those in previous season, respectively (Figure 23.6). In Japan, the proportions of the glass eel input to other values, such as eel aquaculture production and the amount of feed used in eel culture ponds, are larger compared to those in the other years. Table 23.2 shows the glass eel input to Japanese eel culture ponds according to two different sources: the Fisheries Agency of Japan and a trade paper for aquaculture in Japan (*Nihon Yoshoku Shinbun*). The difference between the glass eel input reported by the Fisheries Agency and the trade paper is significantly larger in 2014 than those in the other years (exceeding the upper 95% confidence interval). Similarly, the ratios of 2 year-averaged eel aquaculture production and feed used for eel culture to the glass eel

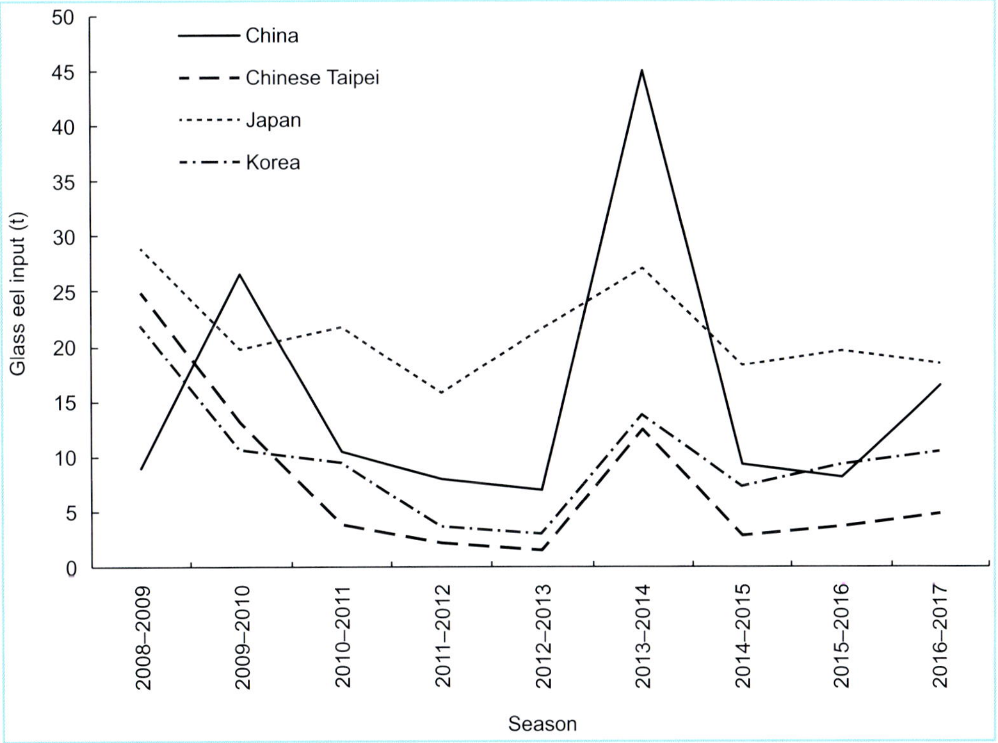

Figure 23.6 Glass eel input in mainland China, Taiwan, PoC, Japan and Korea reported by the Joint Statement (Anonymous, 2014a).

Table 23.2 Glass eel input to Japanese eel culture ponds based on two sources, the Fisheries Agency and a trade paper for aquaculture in Japan, and annual eel production data and the estimated amount of food used for eel aquaculture in Japan.

Year	A Glass eel input reported by Fisheries Agency of Japan	B Glass eel input estimated by trade paper	C A-B	D C/A	D Eel production	E Eel production (2 years average)	F E/A	G Food used for eel culture (April to next March)	H A/G
2003	26	NA	NA	NA	22,600	22,505	865.6	NA	NA
2004	24.7	NA	NA	NA	22,409	21,318	863.1	NA	NA
2005	18.8	NA	NA	NA	20,227	20,631	1097.4	NA	NA
2006	29.2	NA	NA	NA	21,035	21,984	752.9	NA	NA
2007	25.1	NA	NA	NA	22,933	22,078	879.6	NA	NA
2008	21.7	22.5	–0.8	–0.0369	21,222	21,946	1011.3	NA	NA
2009	28.9	29.5	–0.6	–0.0208	22,669	21,724	751.7	NA	NA
2010	19.9	20.2	–0.3	–0.0151	20,778	21,507	1080.7	NA	NA
2011	22	20.9	1.1	0.0500	22,235	19,889	904.0	NA	NA
2012	15.9	16.1	–0.2	–0.0126	17,542	15,941	1002.6	NA	NA
2013	12.6	12.9	–0.3	–0.0238	14,339	16,039	1272.9	21,038	1669.7
2014	27.1	25	2.1	0.0775	17,739	18,964	699.8	30,455	1123.8
2015	18.3	18	0.3	0.0164	20,189	19,565	1069.1	28,032	1531.8
2016	19.7	19.4	0.3	0.0152	18,941	NA	NA	29,222	1483.4
2017	19.6	19.5	0.1	0.0051	NA	NA	NA	25,630	1307.7
								95% confidence interval	
lower	19.31	17.1	–0.4491	–0.02003	18,922	18,998	842.8	22,259	1160
upper	24.63	23.7	0.7891	0.03105	21,772	21,614	1042	31,491	1686

Notes: Eel volumes are indicated as tonnes. NA indicates data were not available. Values of A and B are from 1 November of the previous year to 31 October of the current year. Values of B are based on the Japanese aquaculture trade paper 'Nihon Yoshoku Shinbun'. Values of G are from the Japan Aquaculture Food Organization, 'Nihon-Yoshoku-Shiryo-Kyokai'.

input reported by the Fisheries Agency were significantly smaller in 2014 than those in the other years. The glass eel input quota was distributed among eel farms in Japan according to the input in 2013–14 season. Thus, the higher the input reported by a given Japanese eel farm, the greater the proportion of the quota assigned to that farm. Although there is no direct evidence, the differences among the reported values in Japan indicate a possibility that the reported glass eel input in 2013–14 season in Japan exceeded the actual amount. It stands to reason that, if excess reporting can occur, so too can underreporting. To limit the input of glass eels into eel culture ponds, which could contribute to conservation and management of eel resources, it is necessary to develop an appropriate protocol to confirm reported glass eel inputs by eel farms.

According to the joint press release by the Bureau of Fisheries and Fisheries Law Enforcement under Ministry of Agriculture of People's Republic of China, the Fisheries Agency of Japan, the Ministry of Oceans and Fisheries of the Republic of Korea and the Fisheries Agency of Taiwan, PoC (Anonymous, 2014a), catch of glass eels is subject to licences to be issued by the provincial governments and duration of fishing season is limited from December to March in mainland

China. In Japan, eel fisheries are prohibited from catching eels smaller than about 20 cm by prefectural law, in all 47 prefectures, although the size limit varies among prefectures. Japanese eel glass eels approximately 6 cm in length (Leander et al., 2012) can be caught with special permission (*Tokubetsu Saiho Kyoka*) from prefectural governments for the purposes of scientific research or aquaculture. To our knowledge, there is no catch limit on yellow and silver eel fisheries in Japan, however, catching eels from fall to winter is prohibited by law in some prefectures to protect downstream migrating individuals. In Korea, eel fishing is prohibited from 1 October to 31 March, and the catch of eels between 15 cm and 45 cm is prohibited. In Taiwan, PoC, glass eel fishing is only permitted from November to February (the traditional fishing season for glass eel is from October to April). A licence system has also been introduced to vessels fishing for glass eel. Yilian County, the traditional major glass eel harvest region, has prohibited the yellow and silver eel fishing. In addition, it has prohibited the catch of yellow and silver eels in 19 rivers of other regions.

Illegal, unreported and unregulated (IUU) fisheries and illegal trading

The illegal catching and trade of glass eels is one of the major threats to Japanese eels as it would undermine effectiveness of conservation and management measures. In Japan, fishing for glass eels requires a special permit, and individuals with a permit are required to report their catch. During the 2014–15 fishing season, however, the reported catch was no more than 37% of the total volume of glass eels caught in Japan's domestic waters (Figure 23.7). The remaining 63% (equivalent to 9.6 tonnes of glass eels) was traded through illegal means, either because it was caught by individuals fishing without a licence, or because it exceeded the amount that licensed eel fishers actually reported (unreported catch). There were a further 3 tonnes of imported glass eels, which, according to the trade statistics of Japan's Ministry of Finance, had been imported from Hong Kong, while the glass eels that Japan imports from Hong Kong are considered to have been smuggled into the region from Taiwan, PoC, and other areas subject to export restrictions.

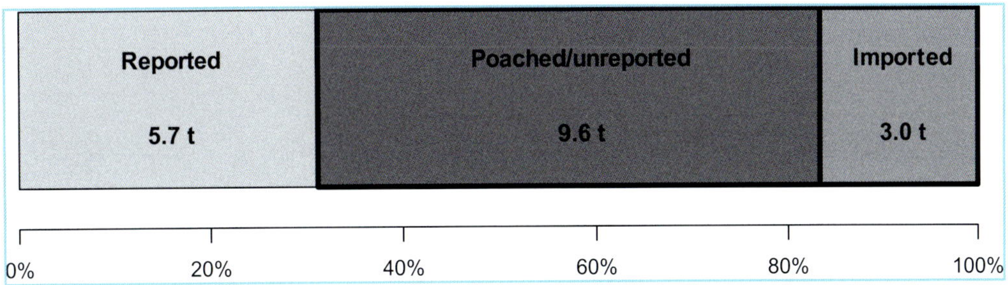

Figure 23.7 Sources of Japanese glass eels used in Japanese culture ponds in the 2014–2015 season (data from the Fishery Agency and Foreign Trade Statistics of Japan). Notes: The reported inflow of glass eels used by domestic eel farms during the 2014–15 season was 18.3 tonnes. Subtracting the 3.0 tonnes of imported glass eels gives a domestic catch of 15.3 tonnes. Of that, 5.7 tonnes were properly reported, leaving 9.6 tonnes of domestic glass eels caught through illegal means – via either poaching or underreporting. The 3.0 tonnes of glass eels were imported from Hong Kong, which lacks a glass eel fishing industry. It is therefore suspected to have been smuggled in from a different country.

Japan imports live and processed eels, including Japanese eel, American eel, European eel, shortfin eel (*A. bicolor*) and some other species. In 2015, 18,138 and 3275 tonnes of eels were imported to Japan from mainland China and Taiwan (Pramod et al., 2017), estimating that 45–75% and 22–35% of eel products exported to Japan from China and Taiwan, respectively, in 2015 were from IUU catches.

European eel (*Anguilla anguilla*)

Florian Stein and Willem Dekker

The exploitation of the European eel across Europe since the Second World War has become well documented in a number of publications, and recent reports of the Joint EIFAAC/ICES/GFCM Working Group on Eels (WGEEL). Therefore, we provide a brief summary with references to further literature. Glass eel trafficking, a rather new phenomenon, developed after the implementation of the zero quota for the trade in European eels from the European Union (EU) to elsewhere in 2010. Since this trafficking is relatively new, we present the information on this in detail. Exploitation of European eel outside Europe, but within its natural range, is much less well documented and is therefore largely ignored here.

Stock status and eel fishery

The European eel has been the most widespread single fish stock in Europe and the Mediterranean. The species comprises a single stock spanning coastal and inland waters from North Africa to the Barents Sea and throughout the Mediterranean Sea region (Dekker, 2003b). Eel fishing has been practised for millennia (Dekker, 2019), targeting all continental life stages: juveniles recruiting from the ocean into continental waters (glass eel), the growing stage (yellow eel) and the maturing eels on their return towards the ocean (silver eels); in continental waters and along the coast; in fresh, brackish and marine habitats (ICES, 2017).

The stock of the European eel is currently at a historical minimum. For more than half a century, stock abundance and fishing yield have declined by about 5% per year, to less than 10% of its historical level (Dekker, 2019; ICES, 2017). From 1980 to 2010, recruitment of glass eels from the ocean towards the continent dropped consistently by approximately 15% per year, to 1–10% of its former levels (Dekker, 2019; ICES, 2017). The causes of these downward trends are not clear and consequently mitigation measures are hard to design (Dekker, 2016). To address the bad state of the stock, the EU adopted a protection and recovery plan in 2007 and implemented Council Regulation (EC) No. 1100/2007 of 18 September 2007 establishing measures for the recovery of the stock of European eel (EU, 2007). This so-called 'EU Eel Regulation' instructed EU Member States to develop national Eel Management Plans by 2009, aiming at a common objective: to reduce human mortality in order to restore a spawner run of at least 40% of the notional pristine run. Additionally, the European eel was listed on Appendix II of CITES – since the end of 2010, trade of European eel to or from the EU has been effectively prohibited (Anonymous, 2010, 2015).

Until the late 1800s, most eel fisheries were small-scaled subsistence fisheries, catching far less than 1 tonne per fisher per year, in smaller streams and along the shores of lakes and lagoons; most trade was localized, with low prices. Around the turn of the century, however, a deliberate and effective development started to improve fishing techniques, to exploit larger waterbodies, to better valorize the catch by further processing; this gave rise to the now well-remembered eel fisheries of the mid-1900s (Dekker, 2019). In the 1950s, some 20,000 tonnes of catch were reported annually (Dekker, 2003a, Figure 1) across Europe (and probably a comparable quantity in the Mediterranean). In 1997, the number of people receiving an income from eel fishing in Europe was estimated at 25,000 (Moriarty and Dekker, 1997) but the socio-economic importance of eel may have been significantly reduced, due to the decline of the stock. Since 2009, only 2000 to 3000 tonnes per year have been reported (ICES, 2017). Still, freshwater eels (of the genus *Anguilla*) are among the most important commercial freshwater fish species in the world (Violi et al., 2015) and European eel sustains local artisanal and recreational fisheries in many areas of Europe.

Glass eel fisheries, like the yellow and silver eel fisheries, have existed since times immemorial, but they expanded considerably in the 1960s and later, from circa 200 tonnes to more than 2000 tonnes per year (Briand et al., 2008). In 1980, glass eel catches in Europe exceeded 3000 tonnes per year, but have since decreased to less than 100 tonnes in 2004–07, and below 60 tonnes annually in 2010–16 (ICES, 2017). Major markets were in France, Spain and Mexico (Briand et al., 2008).

Recreational fisheries for eel (of different life stages) occur all over Europe; their catches, however, are difficult to quantify since the landings are generally poorly reported. Attempts are being made to improve this situation (ICES, 2017).

In addition to the impact of exploitation, many other anthropogenic activities often have an impact on eel stocks. This includes habitat loss, blockage of migration routes (hydropower generation, water management) and pollution. Their combined impact is in the same order of magnitude as the impact of fishing (Figure 23.8; Dekker, 2016).

Restocking

Since the mid-1800s – and especially since 1950 – young eels have been transported from areas of highest abundance (France, Italy, parts of England, Spain) to more peripheral parts of the distribution area in central and northern Europe (Dekker and Beaulaton, 2016a), aiming to support the dwindling fisheries – a practice known as restocking. Restocking thus pre-dates the EU Eel Regulation (EU, 2007), but the Regulation has changed the objective, from 'support the dwindling fisheries' to 'increase the number of silver eel, escaping to the sea for reproduction'. The EU Eel Regulation applies restocking as a 'conservation measure' and many national eel management plans (EMP) rely on restocking to meet their protection targets. From 2011 to 2017, the demand for glass eels in national EMPs has never been fulfilled. On average, only 51% of the aggregated national targets were achieved between 2015–17 (Anonymous, 2018a).

Aquaculture and intra-European trade

Aquaculture of the European eel (that is: on-growing of wild-caught recruits) is practised in many countries, including the Netherlands, Denmark, Germany, Sweden, Estonia, Lithuania, Finland,

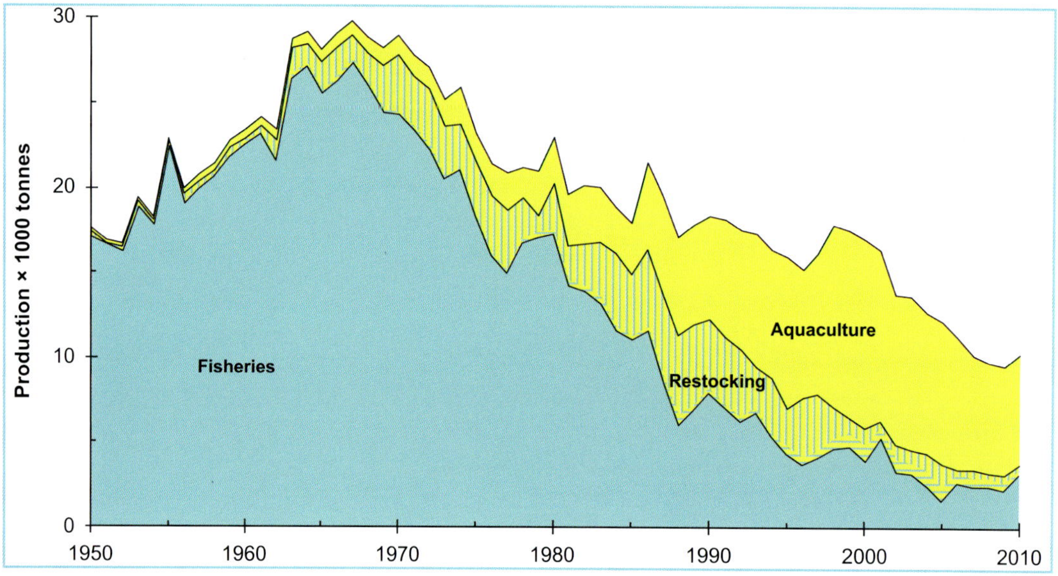

Figure 23.8 Trend in eel fisheries catch, restocking (reconstructed production) and aquaculture for the European eel inside the EU (from Dekker and Beaulaton, 2016b).

Spain, Greece and Italy. Records from 1984 onwards report highest production peaking at 8751 tonnes in 2005, then decreasing to 4935 tonnes within 10 years until 2016 (ICES, 2017). In 2014, there were 33 eel farms across Europe, with the majority located in the Netherlands (13) followed by Denmark and Germany (Anonymous, 2014c; Eding, 2014).

The intra-European trade in eel commodities is dominated by different life stages of live eels. Trade in live eels longer 20 cm accounts for 50% of the dispatches equivalent to exports within the EU, and the total of live eels of all sizes for 70% (CITES, 2018a).

Trade restrictions and the illegal trade

Information on illegal exploitation is scarce. Most likely, there is a widespread, non-commercial fishery (ICES, 2010) often with some links to legal commercial and/or recreational fisheries, on a small scale, used for private consumption or very localized trade. Being widespread, and small-scaled it is extremely hard to estimate their total impact. In addition, there is illegal trade in legally obtained catches. The global trade in glass eel is restricted in volume, but given the extreme prices paid, it is by far the most lucrative commodity. This illegal trade is primarily driven by demand from the eel farms in East Asia, and especially in mainland China. Given the reduced abundance of glass eels of the Japanese eel over the past 40 years, European glass eels have been increasingly used as a substitute in East Asian farms.

Since 2007, the European eel has been listed on Appendix II of the Convention on International Trade in Endangered Species of Wild Fauna and Flora (CITES). The Scientific Review Group (SRG) of the EU has assessed the status of the stock, and in 2010, they concluded that the eel does not meet the conditions for issuing a so-called Non-Detriment Finding (NDF) – a NDF is a requirement for issuing permits for commercial trade in species listed on CITES Appendix II. Consequently, no export or import is allowed from the EU since late 2010 (Anonymous, 2010, 2015). In most recent years, however, increasing evidence indicates that despite this export ban, substantial illegal trade occurs from Europe to Asia. This information being relatively fresh, and the illegal trade rapidly shifting in space, time and modus operandi, we are not able to provide a balanced and reliable picture of the total impact of this illegal trade. Noting the potentially precarious effect on the efforts to protect the stock, we discuss this illegal export in rather more detail here.

Trade statistics, customs data and scientific studies

Even though export of European eel across the outer borders of the EU is prohibited, records of successful exports appear in official trade statistics. EUROSTAT data contain records (customs code 03019210, live *Anguilla* glass eels, less than 12 cm) of 3200 kg exported from France to Hong Kong SAR in season 2013–14 and 1600 kg in season 2014–15 (CITES, 2018a).

According to CITES (2018a), even more records of live glass eel imports from European countries appear in Hong Kong customs data. For the period between 2012 and 2016, these data contain records on imports from Italy: 1132 kg; from France: 1680 kg; from the UK: 1784 kg and from Greece: 3124 kg. East Asian customs data furthermore reveal live eel imports (unspecified life stage) from Bulgaria accounting for 12,731 kg and 3676 kg from France (CITES, 2018a).

ICES-WGEEL and the Sustainable Eel Group (SEG) regularly survey the European glass eel market. SEG found that for up to 50% of the EU glass eel catch declared in season 2015/16 and 2016/17, the destination was unknown (Anonymous, 2018b; ICES, 2013, 2016).

Major enforcement operations

From 2011 onwards, the Spanish Nature Protection Service (SEPRONA) of the Guardia Civil initiated major enforcement operations combating the illegal export of European glass eels.

In the 2011–12 fishing season, law enforcement agencies arrested 14 people and seized 1.5 tonnes of glass eels, transported in cargo containers and a private jet. Besides citizens from the source countries Spain and Portugal, people from Romania and Bulgaria were involved in illegal trade activities detected by operation SUCULENTA (Anonymous, 2012b).

During operation SUCULENCIAS (2012–14), law enforcement agencies arrested 12 people and seized 0.75 tonnes of glass eels that originated from Spain and Portugal and were en route to Asia via cargo shipments (Anonymous, 2014b).

Operation BLACK GLASS was conducted in 2015–16 and resulted in the seizure of 2.5 tonnes of glass eels and arrests of 20 Chinese and Spanish citizens that operated from a house close to the Adolfo Suárez Madrid-Barajas Airport. Suitcases containing plastic bags filled with glass eels, water and oxygen were prepared and carried on regular flights to Hong Kong and other Asian destinations. Suitcases were intercepted on departure in Madrid as well as on arrival in Hong Kong (Anonymous, 2016b; CITES, 2017; Stein et al., 2016).

In the 2016–17 glass eel fishing season, SEPRONA supported by law enforcement agencies from Portugal, Italy, France, Greece, United Kingdom as well as Europol and EUROJUST dismantled an international organization that illegally exported European glass eels from Spain to China via Italy, Greece and Hong Kong. SEPRONA estimates that this single organization exported 7 tonnes (approximately 24.5 million fish) annually between 2012 and 2017. During so-called operation ABAIA, eight Spaniards and nine Greeks were arrested (Anonymous, 2017b).

In the 2017–18 glass eel fishing season, European law enforcements seized 5.8 tonnes of glass eels and arrested 90 suspected smugglers (Europol, 2018b, updated by pers. comm., José J. A. Alfaro Moreno, Europol, 6 November 2018). Operation LAKE was supported by the European Commission within the framework of the European Union Action Plan against Wildlife Trafficking (Anonymous, 2016a) and law enforcement agencies from France, Portugal, Spain and the UK, as well as EUROJUST. Italy, the Netherlands, Sweden and INTERPOL have joined in 2018. More operational actions will be planned under the guidance of EU CITES Enforcement Group of the European Commission (Directorate-General for Environment) (Europol, 2018a).

Environmental crimes – including illegal trade in endangered species – have been identified as one of ten priorities in the EU and was therefore added to the EU Policy Cycle for organized and serious international crime for the period 2018–2021 (Anonymous, 2017a).

Scale of trafficking

Based on the information from the enforcement operations presented above, it has been estimated that 100 tonnes of glass eels, equivalent to about 350 million fish, are trafficked from Europe to Asia for farming purposes and consumption annually as of April 2018 (Europol, 2018b). This comprises glass eels from (1) declared European catches and (2) fish that originate from IUU.

The trend in glass eel recruitment (the number of glass eels that enter continental waters) is one of the indicators used by the ICES-WGEEL to assess the stock status. Bornarel et al. (2017) applied a Bayesian model, the Glass Eel Recruitment Estimation Model (GEREM), to model the annual absolute recruitment. According to the model, the European eel recruitment was 10,825 tonnes in 1960 and 440 tonnes in 2015. For simplification, we considered 440 tonnes as the current annual recruitment. We visualized the proportion of the estimated 100 tonnes in relation to the annual recruitment model and the legal European annual demand for aquaculture (18 tonnes) and restocking (13 tonnes), both reported by the Sustainable Eel Group (Anonymous, 2018b) (Figure 23.9). Since 100 tonnes accounts for 23% of the annual European glass eel recruitment, the trafficking of glass eels from Europe to Asia is undermining conservation efforts across the species' natural range.

There are growing indications that European eel meat is re-exported from China into retail markets across the globe. German authorities detained 24 tonnes of eel meat imported from China in a shipping container (CITES, 2017) and Canadian authorities detained 18 tonnes of eel meat in Vancouver in May 2018 (CITES, 2018b). In relation to that case and further investigations, Environment and Climate Change Canada estimated in June 2018 that at least 50% of the eel in the Canadian retail market is probably European eel that was brought in unlawfully (National Post, 2018). Spanish authorities found 17% positive European eel samples when inspecting 3.25 tonnes of eel meat imported from China via Greece (pers. comm.,

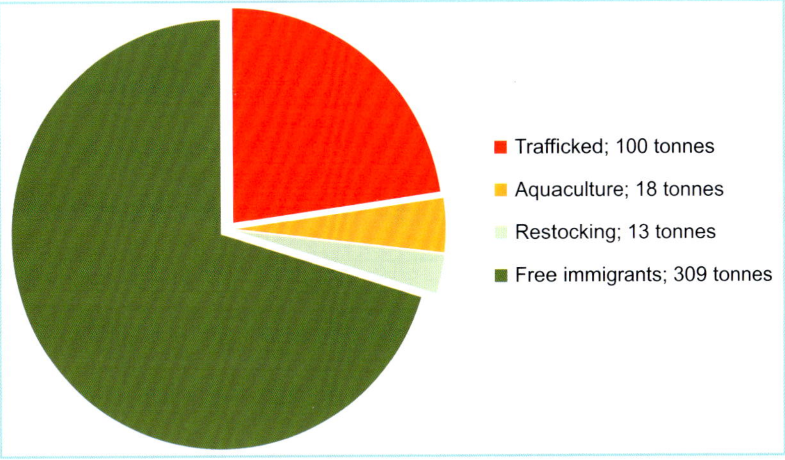

Figure 23.9 The destination of the full (observed and reconstructed) European glass eel catch. Notes: The full pie represents the glass eel recruitment according to Bornarel et al. (2017). 'Free immigrants' represent the proportion of recruitment that is not caught.

José J. A. Alfaro Moreno, Europol, 12 October 2018). Furthermore, European eel was found in genetic samples from UK-based sushi restaurants (Vandammen et al., 2016) that were likely purchased as ready-made kabayaki fillets from Chinese production.

Glass eel trafficking and aquaculture demand in East Asia

Global trafficking in freshwater glass eels is closely linked to the eel farm demand in Asia, since on a global scale, the aquaculture of freshwater eels is primarily concentrated in East Asian countries. From the 1970s to the 1990s, live eels imported by Japan mainly came from Taiwan, PoC (UNIDO, 2013). Owing to the rapid development and associated increase in land prices and parallel opening of China to the global market, the centre of eel culture has gradually moved from Taiwan, PoC, to southern provinces in mainland China where the Fujian province has specialized in European eel (UNIDO, 2013). Nevertheless, European eel has also been farmed in Korea and Taiwan, PoC (Anonymous, 2014a, 2017c). According to UNIDO (2013), European eels farmed in Fujian are mainly processed and exported to the United States, Russian Federation and the EU and used to supply the domestic market.

Contrasts between declared glass eel farm input and Chinese eel farm production

In 2015, the Wildlife Trade Monitoring Network TRAFFIC estimated the eel production of Chinese eel farms. Based on the reported glass eel input into Chinese farms, as reported in a joint statement of mainland China, Japan, South Korea and Taiwan, PoC (Anonymous, 2014a), Shiraishi and Crook (2015) accounted for a total Chinese production of 55,000 tonnes in 2013. For the same year, FAO accounted the *Anguilla* aquaculture production to be 206,000 tonnes. Shiraishi and Crook (2015) analysed the different data sources and concluded that the Joint Statement provides more reliable data and consequently, FAO data might be overestimated.

In 2017, mainland China, Japan, South Korea and Taiwan, PoC, submitted a joint press release for the CITES Animal Committee in Geneva including data describing the 'input of

Table 23.5 Chinese eel farm production and estimated, potential glass eel input demand (in tonnes).

	Year	2010	2011	2012	2013	2014	2015	2016	Mean
	Chinese eel farm production according to FAO ($Prod_{FAO}$) (*Anguilla sp*)	213 811	208 266	212 464	206 026	218 498	232 622	244 777	218 895
	Declared export (*Anguilla sp*)	45 162	40 838	36 378	33 917	34 116	37 089	38 469	44 413
$Prod_{pot1}$ 200 g	Potential production from declared glass eel input	31 188	36 186	29 580	21 600	25 920	73 920	43 008	37 343
	Difference between $Prod_{FAO}$ and $Prod_{pot2}$	182 623	172 080	182 884	184 426	192 578	158 702	201 769	182 152
	Converted into potential glass eel equivalent (*A. anguilla*)	435	410	435	439	459	378	480	434
	Converted into potential glass eel equivalent (other *A.* species)	190	179	191	192	201	165	210	166
$Prod_{pot1}$ 250 g	Potential production from declared glass eel input	38 985	45 233	36 975	27 000	32 400	92 400	53 760	46 679
	Difference between $Prod_{FAO}$ and $Prod_{pot2}$	174 826	163 034	175 489	179 026	186 098	140 222	191 017	172 816
	Converted into potential glass eel equivalent (*A. anguilla*)	333	311	334	341	354	267	364	329
	Converted into potential glass eel equivalent (other *A.* species)	146	136	146	149	155	117	159	126
$Prod_{pot1}$ 300 g	Potential production from declared glass eel input	46 782	54 279	44 370	32 400	38 880	110 880	64 512	56 015
	Difference between $Prod_{FAO}$ and $Prod_{pot2}$	167 029	153 987	168 094	173 626	179 618	121 742	180 265	163 480
	Converted into potential glass eel equivalent (*A. anguilla*)	265	244	267	276	285	193	286	259
	Converted into potential glass eel equivalent (other *A.* species)	116	107	117	121	125	85	125	99
$Prod_{pot2}$ 200 g	Potential production from declared glass eel input	NA	31 188	36 186	29 580	21 600	25 920	73 920	36 399
	Difference between $Prod_{FAO}$ and $Prod_{pot2}$	NA	177 078	176 278	176 446	196 898	206 702	170 857	184 043
	Converted into potential glass eel equivalent (*A. anguilla*)	NA	422	420	420	469	492	407	438
	Converted into potential glass eel equivalent (other *A.* species)		184	184	184	205	215	178	192
$Prod_{pot2}$ 250 g	Potential production from declared glass eel input	NA	38 985	45 233	36 975	27 000	32 400	92 400	45 499
	Difference between $Prod_{FAO}$ and $Prod_{pot2}$	NA	169 281	167 232	169 051	191 498	200 222	152 377	174 943
	Converted into potential glass eel equivalent (*A. anguilla*)	NA	322	319	322	365	381	290	333
	Converted into potential glass eel equivalent (other *A.* species)	NA	141	139	141	160	167	127	146
$Prod_{pot2}$ 300 g	Potential production from declared glass eel input	NA	46 782	54 279	44 370	32 400	38 880	110 880	54 599
	Difference between $Prod_{FAO}$ and $Prod_{pot2}$	NA	161 484	158 185	161 656	186 098	193 742	133 897	165 844
	Converted into potential glass eel equivalent (*A. anguilla*)	NA	256	251	257	295	308	213	263
	Converted into potential glass eel equivalent (other *A.* species)	NA	112	110	112	129	135	93	115

Notes: Data on declared export and glass eel input origin from the joint press release (Anonymous, 2017c). Glass eel equivalent was converted from the difference between $Prod_{FAO}$ and $Prod_{pot1}$ and $Prod_{pot2}$ and the parameters in Tab. 23.3. NA = not available.

other species (Japanese, American and shortfin). The theoretical glass eel equivalent estimated for European eels ranges from 193 tonnes to 480 tonnes considering that the market size is reached in the following season after glass eel input and from 85 tonnes to 210 tonnes for the other species. Considering that market size is reached 2 years after glass eel input, the mean estimates for European eels range from 251 tonnes to 492 tonnes and from 93 tonnes to 215 tonnes for the other three species. According to the China Eel Industrial Association, eels for the domestic market are fattened to weights above 600 g. This has not been considered in the estimation but would significantly lower the theoretical glass eel demand in dependence of the domestic market size.

The exact sources for FAO figure are unclear and therefore we were unable to assess its reliability. Nevertheless, the origin of the glass eels that would theoretically be required to meet the Chinese eel farm production level as reported by FAO, is unknown.

In summary, the exploitation of the European eel has gone through a number of phases. First, small-scale subsistence fisheries all over the continent. From the early 1900s onwards, the fishing companies, and the inter-European trade increased in scale, culminating in the mid-1900s. In the late 1900s, aquaculture developed in Europe, using predominantly glass eels caught in south-western countries for on-growing in central-northern Europe. Finally, in the early 2000s, those glass eels have increasingly been exported to Asia, which export was illegal since 2010. The number of eels illegally exported to Asia is estimated to be at least three times as large as the European legal demand for aquaculture and restocking. The illegal character and scale of these exports constitutes a major challenge for the current policies to protect and restore the stock of the European eel.

American eel (*Anguilla rostrata*): elver fishing in the USA

Nick Walker, C. Andrew Dolloff, Kristen Steele and A. Alonso Aguirre

The American eel historically was the most abundant fish by biomass in North America, found in all watersheds east of the continental divide (Prosek, 2011). American eels are capable of living in more habitats than any other species of native freshwater fish and may be found in coastal marine waters, estuaries, rivers and lakes (Shepard, 2015). American eel leptocephali (larvae) are transported by ocean currents from spawning sites in the Sargasso Sea to the Atlantic coast of the Americas by ocean currents. Eight to 12 months after hatching, leptocephali begin to metamorphose into glass eels as they approach estuaries and the tidal portions of rivers (Facey et al., 1987). Arrivals occur during winter and spring, beginning in the southern parts of the range and proceeding northward (Helfman and Bozeman, 1984; McCleave and Kleckner, 1982). Shortly after arrival in continental waters, American glass eels become pigmented elvers. As with other freshwater eel species, these elvers naturally grow into yellow eels (Lecomte-Finiger, 2003), a stage which can last anywhere from 3 to 25 years or more (Day et al., 2012; Thibault et al., 2007). After a final metamorphosis into silver eels, the American species returns to the Sargasso Sea to complete their life cycle, where they are presumed to die after spawning (Helfman et al., 1987).

Despite considerable research, no one has successfully cultivated American eels from egg to maturity in captivity; hence, all American eels that are harvested and sold, from glass eels to silver eels, are wild-caught and began life in the Sargasso Sea. Most juvenile eels are sold live to Asian markets, where they will be raised in aquaculture and later resold, sometimes returning to the USA, where eel sushi is becoming increasingly popular (Prosek, 2011). Generally purchased by weight, one pound of glass eels includes 3000–7000 individuals (Ricker and Squires, 1974) (6600–15,400 per kg) with the numbers even higher from Caribbean waters (local fishers, pers. comm., 2018). Although yellow eels are also harvested in the USA, they are used primarily as bait for commercial and recreational fishing of other species.

Management of the American eel stocks in the USA is overseen by the Atlantic States Marine Fisheries Commission (ASMFC), which was formed as an interstate compact in 1942 to coordinate the conservation, management and sustainable use of state fisheries resources with representatives from 15 East Coast states (ASMFC, 2018). In 2000, the ASMFC promulgated the first Fisheries Management Plan (FMP) for American eel, with the goal being 'to conserve and protect the American eel resource to ensure its continued role in its ecosystems while providing the opportunity for commercial, recreational, scientific, and educational uses' (ASMFC, 2000a: iv).

In 2012 – and again in 2017 – the ASMFC assessed the American eel stock as depleted (ASMFC, 2012, 2017a). No single cause is responsible, but many contributing factors have been proposed including commercial fishing; habitat loss, pollution, barriers to fish passage, parasitic infections (for example, the invasive nematode *Anguillicoides crassus*), climate change and changes in the Sargasso Sea (Shepard, 2015; Walker et al., 2017). On two occasions, in 2007 and 2015, the outcome of petitions for listing the American eel under the Endangered Species Act (ESA) were deemed not warranted by the US Fish and Wildlife Service (Shepard, 2015), although both assessments relied on sparse coastwide datasets. The IUCN Red List however does list the American eel as Endangered (Jacoby et al., 2017), as does the Canadian province of Ontario (OMNRF, 2018).

Multiple factors have increased international demand for American glass eels in recent years. The European and Japanese species are similar in taste and had previously been preferred for aquaculture. The European eel has been in decline since the 1970s, with a 90–95% decline in recruitment across wide areas of its geographic range. It has been listed on the IUCN Red List as Critically Endangered for the past decade (Jacoby and Gollock, 2014b). In 2010, the EU banned imports and exports of all stages of European eel outside its borders (Ebersole, 2017). The Japanese eel is also in decline and was listed as Endangered by Japan's Environment Ministry in 2013 and appeared on the IUCN Red List the following year. The capture of silver eels has been restricted in some parts of Japan since 2012 (Jacoby and Gollock, 2014a). It has been reported in the US media that the 2011 tsunami wiped out stocks in Japanese eel farms and drove up prices of American glass eels, but this is not the case. While the tsunami reduced inland fisheries by about 15% (Craig, 2016), it occurred in the north of Japan, while the glass eel catch and farms are in the south (Pols, 2014).

In the second half of the 20th century, there were glass eel fisheries in the USA all along the coastal states from Florida to Maine, although never in any of the gulf states (ASMFC, 2000a). Today, Maine has the largest commercial glass eel fishery on the east coast, with a much smaller fishery in South Carolina. Colloquially, 'glass eel' and 'elver' are often used interchangeably, but these terms are not biologically synonymous. The glass eel stage begins when the leptocephali metamorphose to an eel-like but transparent body morphology, 4.8 to 6.5 cm in length

(ASMFC, 2006a). Glass eels are differentiated from elvers, which are pigmented and usually 6.5 to 9.0 cm (Hardy, 1978), ranging up to 15.25 cm in length (ASMFC, 2006a). The variability in size reflects the fact that the term 'elver' may include several age classes, with young-of-the-year (YOY) first-year elvers defined as fish less than 8.5 cm. YOY are sampled annually at various sites because this is the one time when it is possible to determine that all of the eels are from the same cohort (ASMFC, 2000a). Another way to tell them apart is that a glass eel will glow bright green under a UV light, while an elver will not (Kumagai et al., 2013).

In the original FMP, the term 'yellow eel' refers to eels >10 cm in length, although this distinction seems variable among coastal states. In 2006, the length was updated to 15.25 in a joint meeting of the ASMFC American eel technical committee and the stock assessment subcommittee (ASMFC, 2006a). In yellow eel fisheries, gear and size restrictions, such as a 0.5 × 0.5″ mesh for nets and a minimum size of 22.5 cm in eel pots, are implemented specifically to prevent the harvest of glass eels (Walker et al., 2017).

Randall Livingston, the first recorded commercial eel buyer in the USA (Ebersole, 2017), entered the glass eel fishery in 1952 (Thomas, 1985). The market took off two decades later in 1972, in response to demand from Japan (ASMFC, 2012), which remained strong until 1977. That year, minimum size limits were imposed in Virginia (15 cm) and Massachusetts (10 cm), effectively ending glass eel fishing in those states (CBP, 1991). The fishery collapsed the following year due to market conditions (ASMFC, 2000a) and in Maine remained poor until the mid-1990s (DMR, 2018b).

Georgia imposed a 15 cm minimum in the early 1980s. From the mid-1980s to early 1990s, glass eel fisheries were established in Connecticut, Rhode Island, New York, New Jersey, Delaware and South Carolina. According to some fishers, there were very few glass eels in 1988, but they returned the following year. The reasons for this resurgence are not clear, but fishers often attribute such fluctuations to natural cycles in eel reproduction and environmental conditions. The Potomac River fishery closed in 1992, when a 15 cm minimum was imposed. From 1992 to 1995, New York, Rhode Island, Delaware, Maryland and North Carolina also instituted 15 cm minimums (ASMFC, 2000a).

The fisheries began to grow again in 1994. The following year, the ASMFC convened the American Eel Board to address concerns over the growing exploitation of this species (ASMFC, 2018). Virginia issued two permits to fish for glass eels for aquaculture in 1996. New Jersey had an active fishery in the mid-1990s but closed in 1998 (ASMFC, 2000a). Maine once had a silver eel fishery, which ran from late summer to fall using weirs across streams and rivers, but it was closed in 2014 (DMR, 2018b; ASMFC, 2013b). New York has the only silver eel fishery that continues to this day, with nine permit holders on the Delaware River (ASMFC, 2014).

By the late 1990s, four states had glass eel fisheries – Maine, Connecticut, South Carolina and Florida (ASMFC, 2000a). Some have described the glass eel fishery of the 1990s as an 'outlaw fishery' (local fishers, pers. comm., 2018). Regulations varied widely by state and were not well enforced. This was caused by market conditions – although Maine has the most coastline of any state in the conterminous USA, eels arrive sooner to southernmost states because of warmer waters and proximity to the Sargasso Sea.

In 1998, glass eels were selling for US$300/lb (US$660/kg) in Maine. The market declined the following year however, with prices tumbling to US$10–15/lb (US$22–33/kg) and rising only slightly in 2000 to US$25/lb (US$55/kg). As a result of lower prices and the FMP, the Maine fishery was considerably reduced. Gear restrictions went into effect and the duration of the season was shortened, with fishing effort declining by at least 79% (ASMFC, 2002). The only types of gear now authorized for use in Maine are dip nets, fyke nets or Sheldon eel traps (DMR, 2018a). As a result of the low prices (approximately US$25/lb [US$55/kg], for 4 years), hundreds of fishers chose not to renew their licences. New rules stated that if licences were inactive for three years, they were automatically revoked. The Connecticut fishery closed around this time as well.

Four American Indian tribes (the Maliseet, the Mi'kmaq, the Passamaquoddy and the Penobscot) are also authorized by the state of Maine to issue fishing licences because of their ancestral rights under historical treaties. Each of these tribes was once part of the Wabanaki Confederacy, also written as Wabenaki, Wobanaki or Waponahki, which translates to 'People of the First Light' or 'Dawnland' (FDDC, 2018). Tribes have the authority to issue licences, but not to create alternative management plans or other restrictions (Maine DMR, pers. comm., 2018), although the Passamaquoddy tribe has created their own eel management guidelines as part of their tribal fisheries plan (Sipayik, 2013). A discussion of the legislation pertaining to tribal fisheries is beyond the scope of this chapter, but see the Maine Indian Tribal-State Commission for an overview of the relevant statutes at MITSC (2018). There are no tribal fisheries in South Carolina.

The history of tribal fisheries is complex. Native Americans have fished for yellow and silver eels for millennia and eels reportedly were even served at the first Thanksgiving (Prosek, 2011). Historically, the tribes would move throughout the year, often staying closer to the coast in the summer and further inland in winter. During colder months, yellow eels were a reliable source of food. However, prior to when Randall Livingston began exploiting glass eels in the 1950s, there are no records of anyone fishing for them in the USA. Conflicts between tribal and state-licensed fishers have been reported. Points of contention include differences in licensing systems, quota allocations and territorial disputes over good fishing spots. Tribal fishers have invoked their ancestral heritage and sovereign rights as agreed to in historical treaties, while state fishers have felt their own fishing rights are undervalued in comparison. Official relations between these groups of fishers remain strained and how to resolve the conflicts that arise continues to be a legally contentious issue.

As of 2000, South Carolina had ten permittees (a practice that has continued to this day), although only two were active that year (ASMFC, 2000c). All fishing in South Carolina is restricted to the Cooper River, although at various times in the early 1990s and in 2008 there were experiments with fishing on the Santee River. Many of the details regarding glass eel fishing in South Carolina are confidential under the 'rule of three', where any aggregate of data (for example, month, year, area, gear, quantity landed or quantity sold) consisting of fewer than three fishery identities (of all relevant types) is considered confidential and cannot be shared, because in cases of only two identities someone with the data of one identity could determine the data of the other. This is done to protect the proprietary interests of fishers (DNR, 2018).

Florida had no glass eel landings in 2001 or 2003–05 (numbers were not available for 2002) (ASMFC, 2002, 2004, 2005, 2006b). Florida briefly had a glass eel harvest in 2013–14 but stopped again in 2015 (ASMFC, 2015b). South Carolina had no glass eel landings in 2003 or 2004. Glass eel landings in Maine were 14,000 lb (6363.64 kg) in 1998 but dropped to a low of 1282 lb (582.73 kg) in 2004 (ASMFC, 2006c). In 2004 and 2005, because so many had exited Maine's fishery, the Maine Department of Marine Resources (DMR) held a draw to give out more licences. By 2006 prices were still so low that licenses were being given to anyone who requested one, with some leftover.

The FMP was modified in 2006 with the passage of Addendum I, which established a mandatory catch and effort monitoring programme with new reporting requirements (ASMFC, 2018). By that time, all states except Maine, South Carolina and Florida had implemented a minimum size of 15 cm for yellow stage American Eel (which was later changed to 22.5 cm). Glass eel landings continued a general decline throughout the 2000s (ASMFC, 2006c), although reported landings for Maine were 3735 lb (1697.73 kg) in 2007 and 6051 lb (2750.45 kg) in 2008, with an additional 0.1 lb (0.05 kg) from South Carolina (ASMFC, 2009). In 2008, the American Eel Board approved Addendum II, which included measures to improve eel abundance and recruitment by facilitating the escapement of silver eels during or just prior to their spawning migration (ASMFC, 2008).

In the early 2010s, prices increased sharply in response to international demand. By 2011, glass eels were selling for nearly US$900/lb (US$1980/kg) on average, with landings of 8548.9 lb (3885.86 kg) reported that year in Maine, valued at US$7.7 million. The majority (6384.48 lb [2902.04 kg]) was taken with fyke nets, with the next most common type of gear being dip nets (1973.64 lb [897.11 kg]). No gear type was reported for 226.78 lb (103.08 kg) (ASMFC, 2013a). The following year, glass eel prices were already over US$2000/lb (US$4400/kg) (ASMFC, 2012), with landings in Maine totalling 20,764 lb (9438.18 kg) (ASMFC, 2013a). With higher prices however, the incentive for poaching increased and in 2013 the ASMFC's law enforcement committee found a significant amount of glass eels being poached outside of South Carolina and Maine (ASMFC, 2013b).

In 2012, the ASMFC's 2012 Benchmark Stock Assessment found the American Eel stock to be depleted. As a result, Addendum III was ratified in 2013, with the goals of reducing mortality and increasing conservation of the American Eel stock across all life stages (ASMFC, 2013b). The Addendum increased the yellow eel size limit from 15 cm to 22.5 cm and reduced recreational creel limits from 50 to 25 eels per day (ASMFC, 2018).

Until 2013, eels could be bought and sold in Maine for cash, which made it difficult to track buyers and sellers. This meant glass eels were sometimes sold off the books and landings were underreported, with poaching and illegal sale of eels across jurisdictional lines identified as a major problem. In some cases, international eel smugglers have resorted blending multiple eel species together to sell them at the price of the more expensive species. This has happened in different ways, with shipments being intentionally mislabelled, or paperwork changed at border crossings to bring in species like the endangered European eel. DNA sequencing has the potential to reduce mislabelling in the future.

In response to the higher prices and poaching, Maine switched to a quota-based system for the 2014 fishing season. For the first year, the quota was set at 11,749 lb (5340.45 kg) as a 35% reduction from the 2013 harvest of 18,081 lb (8218.64 kg). In part because landings were only 9688 lb (4390 kg), the 2014 quota has been retained for subsequent years (ASMFC, 2018). This state-wide quota is divided up among individual license holders. These 'allocations' are based on previous individual catch records, so quantities vary from person to person.

At the same time that the quota system was implemented, Maine began requiring all transactions use a swipe card, with cash no longer accepted to buy and sell eels. This had been requested by many harvesters and dealers. With the swipe card system, all glass eel sales are tracked in real-time. This resulted in a huge decrease in the number of fishery-related infractions reported by the Maine Marine Patrol, from over 200 in 2013 to under 20 between 2014–16. In 2015, there was only a difference of 120 lb (55 kg) (2%) between what dealers reported purchasing from harvesters and what was exported from Maine dealers, which was attributed to shrinkage (die-off after initial purchase) and did not raise concerns (ASMFC, 2016b, 2018).

Tribes receive 14% of the total quota, with the Maliseet receiving 106.6 lb (48.45 kg), the Mi'kmaq 38.8 lb (17.64 kg), the Passamaquoddy 1356.3 lb (616.5 kg) and the Penobscot 620 lb (281.82 kg) (Resources, 2018). The Passamaquoddy tribe has a derby-style fishery, which means each member can catch as much as they want up to 10 lb (4.55 kg) until the entire tribe reaches their quota. The Mi'kmaq have a total of eight licences with 4.85 lb (2.2 kg) for each fisherman. Discrepancies between the quotas and totals reflect a buffer for slight overages.

The quota system also increased the number of tribal fishers. In 2013, there was a dispute over the number of licences issued by the Passamaquoddy. The tribe was statutorily authorized to issue 200 licences, but instead issued 575, which put the state over the limit of 744 total licences established by the ASMFC's FMP. In 2014, when the ASMFC implemented the quota system, the state of Maine worked with the Passamaquoddy to provide them with the authority to issue an unlimited number of licences through a memorandum of agreement, which is renewed annually (Maine DMR, pers. comm., 2018).

Licences and enforcement work differently in South Carolina. There is no swipe card system, primarily because it would be very expensive to implement such a system with ten or fewer fishers. Fishers are required to report their catches daily; however, these reports are confidential and under the rule of threes the total harvest cannot be published. As such, exact numbers are not available for all years. In 2010 and 2011, the total harvest in South Carolina was under <500 lb (227.28 kg) each year. In 2012, the total amount is unknown, but was under <5000 lb (2272.73 kg). The highest year on record is 2013, with 2270 lb (1031.81 kg). The numbers were lower in 2014 and 2015, with 203 lb (92.27 kg) and 124 lb (56.36 kg). In 2016, the latest year for which numbers are available, the total harvest was <500 lb (227.28).

South Carolina limits fishers to ten fyke nets per licensee, with dip nets also permitted. Until now, fishers have been limited to the Cooper River, but starting in 2019 nearly all of the rivers in the state will be open. The allowed catch limit will be 700 lb (318.18 kg), with each of the ten licence holders receiving 70 lb (31.82 kg). This allowed limit keeps South Carolina landings from exceeding the 750 lb (341 kg) trigger that mandates additional life cycle studies. There were ten licence holders for the 2018 fishing season. One person can only hold a single licence, family members may each hold a licence.

Addendum III had also put restrictions on pigmented eels, allowing a maximum of 25 individuals per lb (55/kg) in glass eel landings, using an 1/8″ mesh screen to keep pigmented eels out. This posed some challenges in South Carolina, where many pigmented eels in the Cooper River pass through the 8″ screens, causing some fishers to exceed the limit. This caused a number of fishers to leave the fishery, because they had been selling pigmented eels before and now their activities were far less profitable, leading to a reduction in landings from 2013 to 2014. Another reason for the reduction was that South Carolina restricted the fishery in 2014 to in-state residents only, in an effort to reduce poaching (South Carolina DNR, pers. comm., 2018).

In October 2014, the ASMFC approved Addendum IV, which aimed to reduce overall mortality and improve conservation of the American eel primarily through management recommendations for the glass eel fishery. Additionally, any state or jurisdiction with a glass eel fishery had to implement a fishery-independent survey of American eel covering glass, yellow and silver stage eels within at least one river system (ASMFC, 2015b).

Under Addendum IV, states could submit aquaculture plans for harvesting up to 200 lb (90.9 kg) of glass eels for use in recirculating aquaculture systems (RAS). In 2015, North Carolina submitted a request to the ASMFC for harvesting glass eels for the purpose of raising them in aquaculture. One permit was granted in 2016, but no glass eels were caught due to logistical issues, including delays in the permitting process and fishing late in the season (ASMFC, 2017).

Addendum V was drafted in October 2017 and has not been finalized at the time of this writing. It proposes new management options, specifically a coastwide cap for yellow eel (with management triggers if the cap is exceeded), as well as potential changes to Maine's glass eel quota and aquaculture (ASMFC, 2017b). Options for glass eel management include the possibility of increasing Maine's quota. In addition, up to three states could combine harvest allowances for aquaculture purposes up to a maximum of 600 lb (272.73 kg) (ASMFC, 2018).

As prices continued to rise, several states approached the US Fish & Wildlife Service (FWS) to investigate poaching activity under the Lacey Act (US FWS Special Agent, pers. comm., 2018), a federal law that prohibits trade in species that have been illegally obtained (USFWS, 2006). In response, FWS launched Operation Broken Glass in collaboration with the US Department of Justice's Environmental Crimes Section, the National Oceanic and Atmospheric Administration (NOAA) and numerous state agencies.

The investigation uncovered poaching in North Carolina, South Carolina, Virginia, Maine, New Jersey, Delaware and Massachusetts (Ebersole, 2017). In some cases, dealers sold eels to undercover agents posing as buyers and recording their interactions. As of June 2018, 19 eel smugglers had been sentenced in the USA (Ebersole, 2018). The cases have been prosecuted in federal district courts in Virginia (DoJ, 2017a), South Carolina (DoJ, 2016b), Maine (DoJ, 2016a, 2017c, 2018a), and New Jersey (DoJ, 2018b). The Lacey Act carries a maximum sentence of 5 years and up to US$100,000 in fines, although no one sentenced under Operation Broken Glass has received more than two years (Ebersole, 2018; DoJ, 2017b).

In early 2018, Japanese eel recruitment plunged, with the volume of Japanese glass eels put into aquaculture at only 10% of the previous year (Desk, 2018). This drove demand for American eel glass eels, with prices surging to record highs of over US$2800/lb (US$6160/kg) in March 2018 (Trotter, 2018) (in fact, glass eels were selling for US$3000/lb [US$6600/kg] on opening night, but that was short-lived) and averaging US$2369/lb (5211.8/kg) by the end of the season (DMR, 2018). The sustained demand from East Asia also continued to support illegal exports of European eels from within the EU (Anonymous, 2018a).

With prices for American eel at record highs, several dealers and fishers managed to circumvent the swipe card system during the 2018 fishing season, which led to the DMR closing the season 2 weeks early. At the time of this writing, the amount of glass eels poached has not been published, but based on the quota total and amount caught prior to the closure, it would have exceeded 500 lb (227.27 kg). There was a slight overage of 51.66 lb (23.48 kg) of glass eels by the Passamaquoddy Tribe that will be deducted from next year's total (DMR, 2018), but this is wholly unrelated to the closure of the fishery.

It is unclear what will happen to the US glass eel fishery in years to come, especially since much of the fishing effort and regulatory response is driven by foreign demand, as well as the supply of other eel species, particularly European and Japanese eels. As the range of the American eel encompasses Canada and the Caribbean, it would be helpful in future studies to examine the dynamics of glass eel fisheries in those regions, as well as the effects of yellow and silver eel fisheries on recruitment. Follow-up studies could also include a discussion of additional ways to improve the conservation and management of eels through opening up more habitat by improving upstream and downstream passage, modifying quotas, providing additional options for aquaculture and increasing enforcement measures to reduce poaching.

Acknowledgement

Some of the information in this section was obtained through conversations with commercial fishers, tribal fishers, eel dealers and state agencies. At the request of the publisher and to protect privacy, individual names have not been included. We thank the anonymous fishers and dealers for their contributions.

Tropical eels I: *Anguilla spp.*

Vincent Nijman and Penthai Siriwat

Tropical eels are found in a large number of countries and territories, including at least 12 in Oceania, 13 in South and Southeast Asia and 15 in Africa and the Indian Ocean Islands (Table 23.6). The countries with the most species of tropical eels are all in the Asia-Pacific, with the Philippines having five species of tropical eel (in addition to the Japanese eel) and Indonesia and Papua New Guinea also having five species (Arai, 2016). Common threats to tropical eel populations are generally trade, habitat degradation and modification (for example, river and forest destruction, dam building), agricultural pollution and climate change. Currently none of the tropical eel species are listed on any of the appendices of CITES, but national governments may have restrictions in place regarding the domestic and international trade in eel, such as restrictions in locally protected species, quotas in export volume, eels with minimum length, or weight restrictions.

Following the CITES listing of the European eel, there have been patterns of follow up exploitation of substitution species, namely, American eel, shortfin eel, as well as other *Anguilla* species (Crook and Nakamura, 2013). A major Japanese eel retailer announced in June 2018, that they aim to gradually replace with the shortfin eel and only sell fully traceable eels, certified by the Marine Stewardship Council, until 2023 (Nagano and Ushio, 2018). The trade of glass eels has also been evident for the African longfin eel from Africa and giant mottled eel from Southeast Asia. Key sources involved in the trade of tropical species of glass eels are the Philippines, Madagascar, Vietnam, Malaysia, Thailand and Indonesia (Crook, 2014).

A large part of the region where tropical eels occur naturally adhere to Islamic tenets and indeed this includes some of the countries with the largest Muslim populations such as Indonesia, Pakistan, India and Iran. According to the Shia Islamic jurisprudence (Fiqh) it is haram (forbidden to eat) to consume eel just as any other fish that does not have scales. According to the

Table 23.6 Tropical eel species.

Species	Distribution	IUCN Red List	Reference
Shortfin eel (*A. bicolor*)	Indian ocean population (A. bicolor bicolor) occurs along eastern African coast, Arabian Peninsula, Indian Ocean islands, South Asia and greater Sundaland Pacific ocean population (A. bicolor pacifica) found in East Asian and Southeast Asian waters	NT	(Jacoby et al., 2014a; Sugeha and Suharti, 2008)
Indian mottled eel (*A. bengalensis*)	The Indian Ocean population, A. bengalensis bengalensis occurs in South Asia and Southeast Asia The Eastern African population, A. bengalensis labiata occurs along east-coast Africa, from Kenya to South Africa	NT	(Jacoby et al., 2014b)
Marbled eel (*A. marmorata*)	Pacific and Indian Oceans From east coast of Africa, India and Sri Lanka, to the Indo-Pacific region, including East Asia, Southeast Asia, the Pacific and South Pacific islands	LC	(Jacoby and Gollock, 2014c)
Indonesian longfinned eel (*A. borneensis*)	Smallest range of all fresh water eels Endemic to Borneo Island, the Celebes Sea and the Makassar Strait region	VU	(Jacoby and Gollock, 2014d)
Philippine mottled eel (*A. luzonensis*)	Two narrow ranged populations in the northernmost and southernmost parts of the Philippines	NT	(Jacoby and Gollock, 2014e)
Highlands longfin eel (*A. interioris*)	Endemic to the northern half of New Guinea with leptocephali recorded in Indonesian archipelago	DD	(Jacoby and Gollock, 2014f)
Celebes longfin eel (*A. celebesensis*)	Western Pacific from Indonesia to southern Philippines	NT	(Jacoby and Gollock, 2014g)
African Longfin eel (*A. mossambica*)	South West Indian Ocean, ranging along the African east coast from Kenya to northern South Africa, and in Indian Ocean islands	LC	(Jacoby and Gollock, 2014h)
Pacific long-finned eel (*A. megastoma*)	Endemic to the Western and Central Pacific islands from Pitcairn to Papua New Guinea	DD	(Jacoby and Gollock, 2014i)
Pacific shortfin eel (*A. obscura*)	Found in western New Guinea, Solomon Islands, Fiji and French Polynesia.	DD	(Jacoby and Gollock, 2014j)

Notes: Range distribution and IUCN Red List status for tropical eel species. IUCN Red List are based on categories, including, Least Concern (LC), Near-Threatened (NT), Vulnerable (VU), Data Deficient (DD).

Sunni Islamic jurisprudence it is 'sea game' that is haram, but exclude fish from this list. There is disagreement whether or not eel should be on the list of prohibited fish. Some argue that for Hanafi Sunni Muslims, eel is halal (permissible to eat). Riaz and Chaudry (2003) added that what matters is not just whether or not sea animals have scales but also whether or not they can only live in water (if so, they are not permissible to eat, if not, they are permissible). Species that are fish-like, live in water permanently and that do not have scales but that can breathe oxygen from air rather from water may be permissible to eat. Eels fall into this category. As such Riaz and Chaudry (2003) argue that eels are acceptable to eat for the majority of Muslim consumers, although some of them may consider them makrooh (disliked or detested).

In practice there is a large debate within the Muslim communities whether or not eel is haram or halal, and practices seem to differ between countries. Thus, Hossain et al. (2007) notes that in Bangladesh, another largely Muslim country, the Indian mottled eel is only consumed by local non-Muslim communities implying it is not permissible for Muslim to consume eels. He noted that apart from the small proportion that is consumed locally, the eel trade in Bangladesh is for the international export market. Purwaty et al. (2018) noted that a local eel farm on the Indonesian island of Java, a predominantly Muslim region, had its eel products certified as being halal, thus making it permissible for local consumption.

Perhaps linked to religion, in many countries within the range of tropical eels, eels are not a common fish on the menu. For instance, in Indonesia, a country with 240 million people, Nijman (2015) found only three specialized local Indonesian restaurants that served eel on the menu. Likewise, in Malaysia, eel does not feature on the local menu, or if it does, it is highly unusual (V. Nijman and P. Siriwat, pers. observ., 2003 to 2015). Traditionally in Thailand, Asian swamp eels can be found in 'forest/wild delicacies' but are not commonly eaten due to availability and price (P. Siriwat, pers. observ.). Owing to the popularity especially in urbanizing cities in the past decade, the term 'eel' referring to the domestically available swamp eel is now becoming replaced by anguillid eels used in Japanese cuisine. In Myanmar, local consumption of eel is very limited with no less than 98% of the eel being exported (Muthmainnah et al., 2016). In India, Pantulu (1958) noted that Bengal eel was rarely eaten, apart from by the poorer classes who occasionally ate them, and furthermore remarked that eel fetched higher prices in the south than in the north (indicating either higher demands or lower supply). Le et al. (2010) mention marbled eel is consumed locally in Vietnam. For Sri Lanka's Northwest Province, but aparently not elsewhere, Murray and Little (2000) noted that outside the dry season, specialist fishers catch small volumes of shortfin and African mottled eels using hook and line; these are typically sold from a roadside vantage, close to the point of harvest by the fishers themselves. In eastern Madagascar eel (African longfin eel or Indian mottled eel) were among the most preferred wild meats listed in two local communities; in fact 70% of over a thousand people that were asked ranked eel in their top five preferred meats (Jenkins et al., 2011). However, wild meat consumption overall was low. In other parts of Madagascar, taboos preclude the consumption of eel, or at least for segments of society (for instance pregnant women) (Jones et al., 2008).

All in all, it can be concluded that local consumption of eels throughout most of the range of tropical eels is uncommon and insignificant. However, with globalization of the food industry, food habits may be changing.

Many cities in the region, especially capital cities now have Japanese restaurants, with many of the five-star hotels having one as standard (Table 23.7). Analysis of online menus of Japanese restaurants in Java and Bali, Indonesia, suggests that some 85% (17/20) of them serve eel dishes, although it was unclear which species were served (Nijman, 2015). A search for online menus in Bangkok revealed a similar pattern, with 70% (7/10) of the restaurants offering eel dishes, and for Singapore it was 65% (11/17). Overall this may suggest that, in Southeast Asia, two-thirds or even three-quarter of the Japanese restaurants serve eel, and this most likely includes domestically reared or caught tropical eel species.

The Indonesian Director General of Aquaculture Fisheries, S. Subiakto, stated that in 2012 the domestic demand for eel in Japanese restaurants in Indonesia was high and that there were difficulties meeting the demand. Just in the capital Jakarta the demand for eel, mostly or exclusively shortfin eel, was 36 tonnes per year (Lukita, 2012). A few years later, in 2017, J. Soetanto, the director of the eel farm and exporting company PT Iroha Sidat Indonesia, based in eastern Java, stated that their annual production of shortfin and marbled eel, approached 250 tonnes a year. While the majority of this (80%) was for export to Japan, the remainder (20% or 50 tonnes) was sold to Japanese restaurants in the cities of Surabaya, Denpasar, Makassar and Jakarta (Sudrajat, 2017).

Table 23.7 gives an overview of the number of Japanese restaurants in selected capitals (or former capitals, excluding urban sprawls), in tropical eel range countries – these were obtained through a search in May 2018 on GoogleMaps using 'Japanese restaurant' and 'name of capital' as a search key, and then manually checking each entry. Some of the smallest cities have a large number of Japanese restaurants whereas some of the largest cities have few – overall there is no positive correlation between the size of the city and the number of Japanese restaurants. There is a clear correlation between the number of Japanese restaurants per million people living in

Table 23.7 Eel consumption and Japanese restaurants.

Capital city	Japanese restaurants	Human population in millions	Restaurants per million people	GDP per capita
Singapore, Singapore	276	5.6	49	90
Philippines, Manila	249	1.8	138	8
Malaysia, Kuala Lumpur	224	7.2	31	29
Indonesia, Jakarta	214	10.4	21	12
Thailand, Bangkok	168	8.2	20	18
Cambodia, Phnom Penh	63	1.5	42	4
Myanmar, Yangon*	58	7.4	8	6
Vietnam, Hanoi	53	7.6	7	7
India, New Delhi	56	17.1	3	7
Bangladesh, Dhaka	23	8.5	3	4
Sri Lanka, Colombo	16	0.8	20	13
Brunei, Bandar Seri Begawan	13	0.1	130	77
Kenya, Nairobi	12	3.1	4	3
Pakistan, Karachi	6	14.9	0	5
Oman, Muscat	6	1.3	5	45
Madagascar, Antanarivo	4	1.4	3	2
Tanzania, Dar es Salaam*	3	4.3	1	3
Mozambique, Maputo	3	1.8	2	1
East Timor, Dili	1	0.3	3	5
New Guinea, Port Moresby	1	0.3	3	4

Notes: Number of Japanese restaurants in capital cities of tropical eel range countries, and how this relates to population numbers and GDP per capita (in US$). *The capitals Naypyitaw and Dodoma have 1 and 0 Japanese restaurants, respectively.

the capital and the GDP per capita, that is, wealthier countries have significantly more Japanese restaurants in their capital cities. It is clear that especially the Southeast Asian capital cities have a large number of Japanese restaurants with Singapore, Manila, Kuala Lumpur and Jakarta each having over 200. The South Asian capitals typically have a few dozen Japanese restaurants each, whereas most of the east African capitals have only a few.

Combined with information presented above, we here present a concise overview in trade in the three most commonly traded tropical eels.

Shortfin eel (A. bicolor)

Shortfin eel are mostly traded as adults and live glass eel, though the volume traded is much less than the mottled eel also found in Indian fresh waters (Talwar and Jhingran, 1991). Overall the trade of shortfin eels has increased in recent years. Shortfin eel is found to be traded online as live and frozen specimens (Jacoby et al., 2014a) predominantly in Indonesia, Malaysia and the Philippines, where shortfin eel were most commonly traded as they are favoured replacement species for Japanese eel due to the similarity in taste and texture (Crook, 2014). Shortfin glass eels have been reported to be traded to supply stock farms in China (Jacoby et al., 2014a). Additionally, despite export bans being put in place, the Philippines has also become a key supplier of glass eels to East Asian countries with at least an average of 25% of the species exported being shortfin. In Sri Lanka, the shortfin eel is also reportedly sold for the aquarium trade (Pethiyagoda, 1991).

Marbled eel (A. marmorata)

Owing to its extensive range, marbled eel is an economically important species. It is locally and commercially fished among the islands of the western Indian Ocean. Equally important is its demand from East Asian markets, from stocking eel farms for food purposes and for East Asian traditional medicine (Jacoby and Gollock, 2014b). The trade of eel fry in Madagascar and Mauritius has increased in the past few years. Marbled eel are relatively easy to buy in wholesale online live as elvers, adults, or frozen. Most of the commodities are supplied from Indonesia, Philippines, Malaysia, Vietnam and Sri Lanka. Between 2003 and 2012, an estimated 35 tonnes of marbled glass eel was exported to Hong Kong from the Philippines alone (Jacoby and Gollock, 2014b).

Indian mottled eel (A. bengalensis)

Indian mottled eel is the most commonly traded fresh water eel species found in Indian Ocean region. In India and Southeast Asia, live elvers and adult eels are traded making the Indian mottled eel commercially valuable. The export of Indian mottled eel from South India is often included with similar species such as marbled eel, accounting for 1700 tonnes of eel from 2009 and 2010 from Bangladesh, India and Pakistan (Jacoby et al., 2014b). Locally in India and Bangladesh, eels are sold as yellow or silver eel. In India, mottled eel is believed to have special nutritional value for general consumption, and medical uses to treat arthritis (Talwar and Jhingran, 1991). It is also considered a pristine rare ornamental species of the Himalayan drainage and preserved in temple ponds for religious purposes (Shrestha, 2013). In Myanmar, mottled eel are one of the few anguillid species found and Myanmar exports around 8000 tonnes of live eels annually (Jacoby et al., 2014b).

Recent trade dynamics

Relatively little information is available on the past or recent trade dynamics (volumes, origin, trade routes, destinations) for individual species (little data of tropical eels introduced into Asian farms can be found in Anonymous (2017c). Fortunately, there is information available on the trade in tropical eel species from individual countries as many report their international trade to the UN ComTrade database (comtrade.org). This database allows for the analysis of eel, both in terms of volume and value, on the basis of Harmonized System (HS) codes, and three of these refer to eels (live, chilled, frozen). It does not allow for the identification of individual species or life stages (glass eel, elver, adults).

Overall trade

Based on data from the UN ComTrade database, the overall trade of eels in Asia show a few key countries which play a large role in the eel trade (Table 23.8). Export trade values for the year 2016 or 2017 were obtained for seven key countries exporting tropical eels covering the region of Southeast Asia, South Asia and East Africa (Figure 23.11). Most countries along the eastern African coast, Middle East and Pacific islands do not have reported export values for the trade of eels. Overall, Southeast Asian countries export large volumes of eels, where Philippines and Myanmar have exported significantly high values of live eels while the majority of exports from Malaysia and Indonesia are chilled and frozen eels. Table 23.1 reports the regional price data; average prices of eel exports per kilogram from Southeast Asian countries range from US$1.13 to US$4.28 for live eels and US$1.66 to US$3.34 for chilled and frozen eels, whereas the average

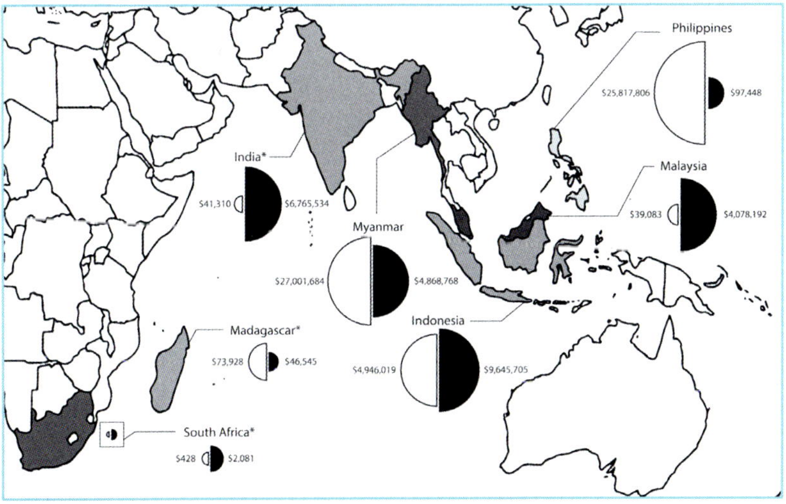

Figure 23.11 Export values (in US$) of live eels (white) and chilled or frozen eels (black) reported in 2017 or 2016 from seven key countries where tropical Anguilla species are found from Africa to Southeast Asia. Note: *represents countries where 2016 data were used.

Table 23.8 Export trade values (in US$) and export volumes (kg) in key range countries where tropical *Anguilla* species spawn or migrate ranging from Africa to Southeast Asia

Country	Export value in 2017 (US$)		Export volume in 2017 (kg)		Average price per kg (US$)	
	Live eels	Chilled/frozen eels	Live eels	Chilled/frozen eels	Live eels	Chilled/frozen eels
Malaysia	39 083	4 078 192	34 562	1 272 634	1.13	3.20
Indonesia	4 946 019	9 645 705	1 154 631	2 888 550	4.28	3.34
Philippines	25 817 806	97 448	8 002 122	58 757	3.23	1.66
Myanmar	27 001 684	4 868 768	8 899 732	1 520 599	3.03	3.20
India*	41 310	6 765 534	4 289	4 445 814	9.63	1.52
Madagascar*	73 928	46 545	11 831	1 710	6.25	27.22
South Africa*	428	2 081	100	321	4.28	6.48

Note: *2016 values used if 2017 values were unavailable.

prices per kilogram reported from Africa and South Asia ranged from US$4.28 to US$9.63 for live eels and US$1.52 to US$27.22 for chilled and frozen eels.

Indonesia and Madagascar

To illustrate some recent trade dynamics, we here focus on the export of tropical eel from two countries, Indonesia and Madagascar. Indonesia is home to five species of tropical eel (shortfin, Indonesian longfinned, Celebes longfin, highlands long-finned and marbled) whereas Madagascar is home to two species (shortfin and marbled).

For the period 1992 to 2016, for both countries we see a gradual or perhaps even exponential increase in the volume of eel that is exported from Madagascar and Indonesia. This increase is especially apparent after the 2007–10 period, when European eel was listed on Appendix II of CITES and when subsequently the commercial export of European eel out of the EU countries was banned. The volume of eel exported out of these two countries varies significantly, with in recent years Indonesia exporting close to 5 million kg of eel and Madagascar exporting around 50,000 kg a year (Figure 23.12). While the volume of trade differs two orders of magnitude, the declared value differs one order of magnitude, given that the declared value of the eel trade out of Madagascar is higher than that from Indonesia. The value of the eel trade out of Madagascar and Indonesia, as declared in the UN ComTrade database is close to US$1 million for Madagascar and close to US$10 million for Indonesia (Figure 23.13).

IUU fishing and illegal trade

The illegal, unreported and unregulated sourcing of glass eels, the most lucrative eel commodity, and to a lesser extent, the illegal trade in live adult eels, chilled and frozen eel complicates traceability within the industry and the development of appropriate management measures (Shiraishi and Crook, 2015). As such, the illegal trade in glass eels is an ongoing concern for many of the species being grown out in mariculture farms in Asia. Customs and other sources of seizures data show that large quantities of glass eel have been exported illegally from Europe, the Philippines,

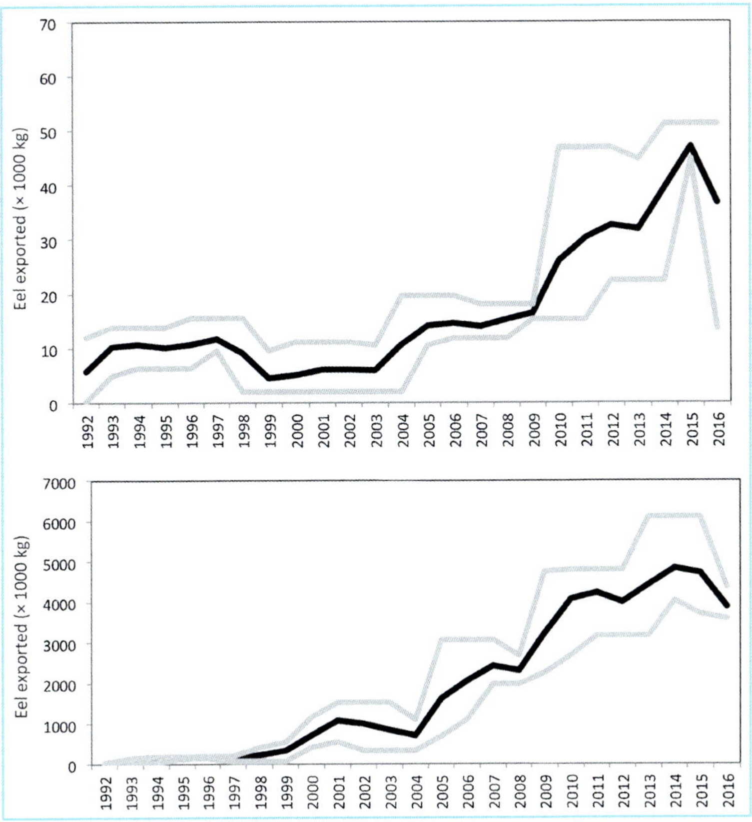

Figure 23.12 Trade in tropical eel from Indonesia (top) and Madagascar (bottom) for the period 1992 to 2016. Notes: Data from the UN ComTrade database. Presented are the three-year running average (black line) and the minimum and maximum for these 3 year periods. While the species is not given, most of this trade refers to *A. bicolor* and *A. marmorata*. Note that the y-axis differs for the two countries.

Indonesia over recent years (Anonymous, 2018b; Europol, 2018b; Nijman, 2015; Shiraishi and Crook, 2015; Stein et al., 2016). For Indonesia, and thus presumably for marbled and shortfin eel, Nijman (2010) noted that it was unclear to what extent the export out of Indonesia abides by the restrictions imposed on it by the Indonesian authorities, adding that it is likely that the real quantities are substantially higher than official figures indicate. For instance, the official records for 2012 for Indonesia indicated than the country exported 4.0 million kg of eel (live, frozen and chilled). For that same year Anonymous (2012a) stated that the province of East Java (one of about 14 to 16 provinces in western Indonesia from where eel is exported) exported an estimated 3.6 million kg. For the same year, based on data from the first 10 months (Anonymous, 2012c), the export to Japan alone was estimated to be 14.9 million kg. These figures indicate that, at least for 2012, the real quantities of eel exported from Indonesia could be easily double or triple that of what the official export figures indicate. Hitherto it is unclear if the two- to three-fold difference is due to underreporting of legal trade, illegal (undocumented) trade or a combination of the two.

Anecdotal reports suggests that smuggling from Indonesia can be rampant. In 2013, Coco Kokarin Soetrisno, Production Director at Indonesia's Ministry of Maritime Affairs and Fisheries stated that smuggling of glass eel was a big problem, and that punishment for smuggling is minimal and the glass eel are usually confiscated, with the smugglers (allegedly mainly from South Korea, China and Taiwan, PoC) charged only with visa misuse (Tjandraningsih, 2013). Smuggling hubs include the international airport in Denpasar on Bali, Surabaya in

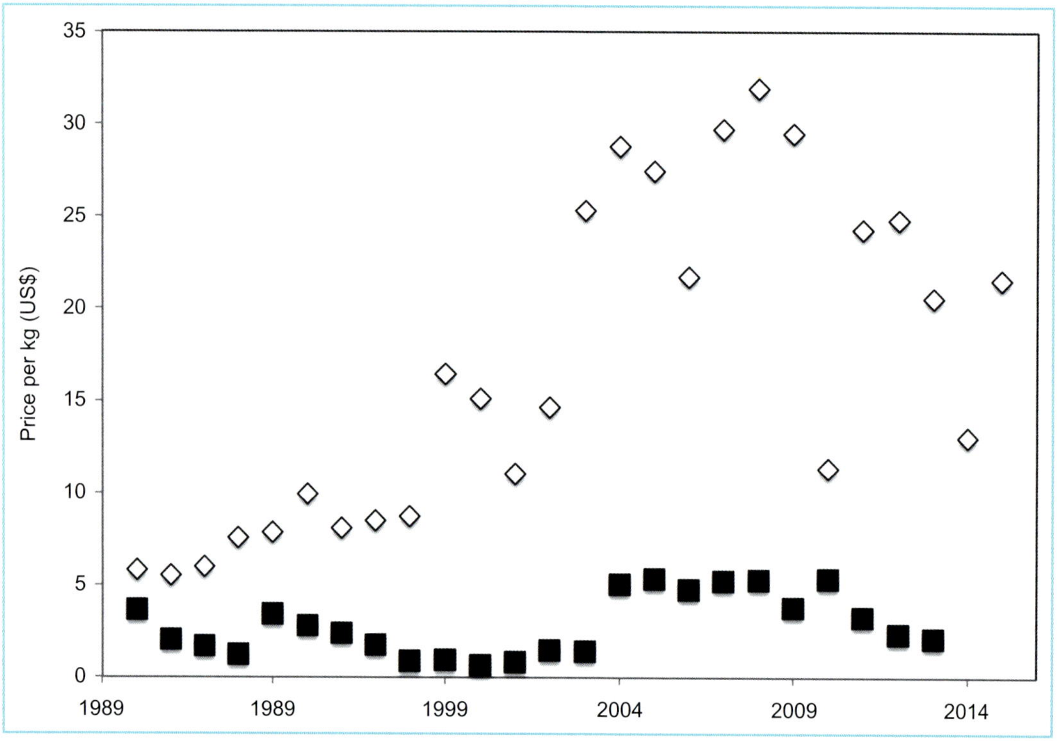

Figure 23.13 Value of tropical eel trade in Madagascar and Indonesia (top) showing that in general declared values of the eel export out of Madagascar (open diamonds) is consistently higher than that declared from Indonesia (black squares). Price difference in eel exported out of Indonesia (bottom). Notes: Live eel are indicated in solid black diamonds and frozen/chilled eel is indicated in open squares. Prices are not corrected for inflation.

East Java and Batam Island, near Singapore. Soetrisno, on the basis of information from Indonesian eel aquaculturists, stated that about 70% of the juvenile eel caught by fishers are smuggled out of Indonesia annually (Tjandraningsih, 2013), but did not provide evidence to back this up.

Seizures in Indonesia, from where the export of glass eel and individual eels with a mass less than 150 g is illegal, indicate that indeed glass eel is exported illegally. However, the number of seizures that have been reported are few and far between. Nurhayat (2015) reported on the successful thwarting of two smuggling operations. The first attempt was on Batam Island, close to Singapore, where 268 glass eels (it is uncertain whether or not this seizure indeed comprised 268 individual glass eels as reported by Nurhayat, or perhaps 268 g of glass eels) were confiscated on a motorboat heading to Singapore and the second involved 20 boxes of glass eel confiscated at a Garuda (Indonesia's national airline) warehouse near Jakarta international airport. The shipment was to be exported to Hong Kong.

In 2012 a trader in Japan reported to have imported about 1 tonne of eel from Madagscar a week (at a price that was about 40% lower than of domestically raised eel) (Real, 2012). If this amount was imported consistently over a one year period this would equate to about 50 tonnes per year; this is equal to the maximum amount of eel Madagscar exported in 2012 (see Figure 23.13). Unless this one trader is the only one to import eel from Madagascar, which is highly unlikely, then the total amount of eel exported by Madagscar is considerably higher than official figures indicate.

Catch and trade statistics available currently offer the most relevant estimation of the use and trade in *Anguilla* species. However, as the reports are rarely species specific, estimations are often based on range locations, where species with broad distributions are more likely to be caught. Differences in custom codes between each export and import country also produces inconsistencies. For example, the unit measured in Korea is weight in grams per eel, Taiwan, PoC, uses number of eel pieces per kilogram, and the EU uses the size in centimetre for each eel. Therefore, current trade values are often limiting, an underestimation and cannot be directly compared (Crook, 2014).

Tropical Eels II: Case study – Some data on the ecology of freshwater eels in French Polynesia

Pierre Sasal

In the Polynesian triangle eels are often considered as sacred and associated with legends and traditions linked with the creation of life. With the legend of Hina and the giant eel, eels are considered the origin of the coconut tree, which is fundamental for isolated people on remote islands (wood for construction, leaves for protection, coconuts for oil and food and water for hydration where freshwater is difficult to find). As a paradox, there is little information concerning the life cycle and biology of eels in French Polynesian waters.

Three eel species are known to occur in French Polynesia: the shortfin eels (*A. obscura*) and two long-finned eels (giant and Pacific). The mottled eel, *A. marmorata*, is the most widely distributed species with an Indo-Pacific distribution from the South African coast in the west to the Far East of French Polynesia. Genetic study suspected the existence of at least four populations of marbled eels, from different nesting sites, two in the North Pacific and one or more in the north-western Pacific and Indian Ocean (Minegishi et al., 2008). However, given the small differences observed, the divergence would be recent with an origin from the West of the Pacific Ocean and an invasion in the Indian Ocean via the central South Pacific. Many grey areas still exist on the lifecycle of this species since only one spawning area would have been identified to date (Kuroki et al., 2006). Data on the others species are by far less abundant. For example, Watanabe and collaborators (Watanabe et al., 2011 along with data from Ege, 1939) showed that there was no evidence of several populations for shortfin eels while, based on the number of vertebrae, longfin eels may be separated into two populations in its distribution area. This last result still remains questionable as it was based on a small number of individuals and probably from several cohorts. Phylogenetic relationships with other species gave some arguments in favour of this hypothesis but the range of vertebrae counts overlap between the studied places making the significance of results very speculative as suggested in the conclusion of the paper (Watanabe et al., 2011). Concerning the biology of these species, Jellyman worked on shortfin eels in the Cook Islands (Jellyman, 1991) suspecting an intermittent recruitment in the lake Te Rotonui and a piscivorous diet of eels supposed to live up to 27 years.

In French Polynesia, apart from works in the 1990s (Marquet, 1992; Marquet and Galzin, 1991), the preliminary study of Budimawan (1997) on otolith microstructure and work showing the presence of eels in the atolls (Lecomte-Finiger et al., 2001), very little information is available on the life cycle of eels (downstream migration period to the spawning ground is still unknown for example and arrival of glass eels from the sea has just been monitored (Helme et al., 2018). Results show a clear seasonality during the wet season for the recruitment of the three species. Interestingly, the three species recruited during the same period with a clear majority of marbled eels and a very low percentage of longfin eels (on more than 15,000 glass eels identified with fin ratios [Ege, 1939], 85% were marbled, 10% shortfin and 5% longfin). After this result, we studied the microstructure of otoliths with an electron microscope in order to estimate the age at recruitment of the three species. Although still preliminary, results revealed very assorted findings with an important age variation within species (Table 23.9). This seems to indicate that there are probably several spawning areas. Only a genetic study on successive cohorts and/or on individuals recruiting early in the season and late in the season may help to answer this question. This work still remains to be done.

In a second step, eels were sampled electrofishing using ponctual abundance sampling (PAS) (Nelva et al., 1979, Robinet et al., 2003). One PAS is a continuous sampling effort on a surface of 10 m^2 for 30 seconds. A minimum of 10 PAS was done on each sampling station. Seven islands belonging to three archipelagos were sampled in French Polynesia: Moorea, Tahiti and Raiatea for the Society Islands, Rurutu, Tubuai for the Australs and Nuku Hiva and Iva Oa for the Marquesas. In each island, a minimum of nine stations were sampled in at least four different rivers. Altogether, 38 rivers have been sampled for a total of 68 stations. Interestingly the results revealed almost the same proportions of each species in rivers as the one obtained in the recruitment data. Marbled eel was by far the most abundant with biomasses over 1000 kg/ha in stations close to estuaries and mean biomasses over 100 kg/ha in all the archipelagos (Table 23.10). In Australs, densities were much lower and because of the presence of more lentic stations (with very low current) we found that shortfin eel was more abundant. It could be interesting to sample glass eels in this archipelago to know (1) if the proportion of glass eel is also different and (2) because of the southern position of this archipelago what is the age at recruitment of the different species.

Finally, despite intensive sampling all year round, we found very little information on the downstream migration of the eels in French Polynesia. Only two small silvering males were found in Moorea and one larger longfin eel in Tahiti during the dry season (June to September). Back calculation with glass eel otoliths allow us to estimate the hatching date between August

Table 23.9 Mean age at recruitment for the three eels species found in French Polynesia.

	A. marmorata	*A. obscura*	*A. megastoma*
Number	51	18	4
Age at recruitment and range (days)	99.4±19.3 65–144	105.3±16.6 81–124	110.3±6.4 102–117

Table 23.10 Mean total lengths (cm ± SE), total weights (g ± SE), densities (in ind./100 m^2 ± SE) and biomasses (in kg/ha ± SE) (range = min-max) of marbled eel in French Polynesia depending on the archipelago.

	Australs	Marquesas	Society
n	23	282	311
Total length	60.2±6.8	39.1±1.5	25.7±1.0
	(6.1–114.5)	(8.3–148.0)	(4.8–117.0)
Total weight	1144±310.4	276.8±35.7	118.9±18.1
	(1.0–5000)	(1–4000)	(1–2500)
Density	2.3±0.9	10.8±2.7	10.3±1.9
	(0–11.0)	(1.3–36.5)	(3.5–28.1)
Biomass	295.0±227.2	136.6±28.4	103.5±31.9
	(0–2550.3)	(6.1–293.4)	(5.8–458.8)

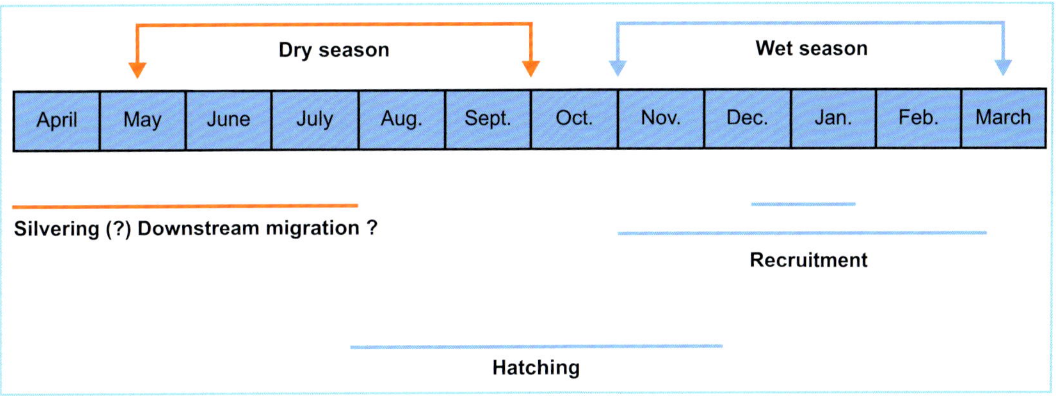

Figure 23.14 Tentative representation of the timing of the life cycle of freshwater eels in French Polynesia. Except recruitment and hatching period based on our personal results, silvering process and downstream migration are based of very limited observations and need to be refined and confirmed.

and December depending on the age and the date of collection. If we assume that silver eels spend as much time as the leptocephalus in the ocean, we can estimate the downstream migration of silver eels between May and September (Figure 23.14). However, this estimation has to be refined because results obtained in New Zealand on other South Pacific eel species seem to be dependent on many factors and may occur at different time depending the sampling site (Boubée et al., 2001; Jellyman et al., 2009).

Acknowledgements

We thank the fishers and dealers who were willing to share their experiences. We thank Kirby Rootes-Murdy and Kristen Anstead at the ASMFC. We thank Bill Post at the South Carolina Department of Natural Resources.

Bibliography

Anonymous (2010) Short Summary of Conclusions of the 54th Meeting of the Scientific Review Group on Trade in Wild Fauna and Flora 3 December 2010. Available at: https://circabc.europa.eu/sd/a/49ab3fc9–646b-4b35-ac42-f0333479ce24/54_summary_srg.pdf (accessed 22 May 2018).

Anonymous (2012a) Industralisasi perikanan: ekspor illegal benih sidat dikeluhkan. *Indonesia Business Daily* 15 May.

Anonymous (2012b) *La Guardia Civil desarticula una red internacional dedicada al comercio ilegal de angulas*. Available at: http://www.guardiacivil.es/en/prensa/noticias/4047.html (accessed 29 June 2018).

Anonymous (2012c) 'Mengatasi kelangkaan dengan impor', *Halo Jepang!*

Anonymous (2014a) Joint Statement of the Bureau of Fisheries of People's Republic of China, the Fisheries Agency of Japan, the Ministry of Oceans and Fisheries of the Republic of Korea and the Fisheries Agency of Chinese Taipei on International Cooperation for Conservation and Management of Japanese Eel Stock and Other Relevant Eel Species. Available at: http://www.jfa.maff.go.jp/j/saibai/pdf/140917jointstatement.pdf. Available at: http://www.jfa.maff.go.jp/j/saibai/pdf/140917unagi_data.pdf (accessed 3 May 2018).

Anonymous (2014b) La Guardia Civil incauta cientos de kilos de angulas valoradas en medio millón de euros que se exportaban de forma ilegal. Available at: http://www.guardiacivil.es/va/prensa/noticias/4975.html (accessed 29 June 2018).

Anonymous (2014c) Le marché de l'anguille européenne. Les Études De FranceAgriMer. Available at: http://www.franceagrimer.fr/content/download/33408/302575/file/ETU-MER-2014-marche%20anguille.pdf (accessed 26 March 2016).

Anonymous (2015) Short summary of conclusions of the 73rd meeting of the scientific review group on trade in wild fauna and flora, 15 September 2015. Available at: https://circabc.europa.eu/sd/a/1e2aa226–4ea3–42f1-a5ff-2b868a7dde12/73_summary_srg.pdf (accessed 22 May 2018).

Anonymous (2016a) EU Action Plan against Wildlife Trafficking (COM(2016) 87 final). Available at: http://ec.europa.eu/environment/cites/pdf/WAP_EN_WEB.PDF (accessed 26 March 2016).

Anonymous (2016b) La Guardia Civil desarticula una red que había exportado ilegalmente más de 2,5 toneladas de angulas. Available at: http://www.guardiacivil.es/en/prensa/noticias/5766.html (accessed 29 June 2018).

Anonymous (2017a) Draft Council conclusions on setting the EU's priorities for the fight against organized and serious international crime between 2018 and 2021. 8654/17. Available at: http://data.consilium.europa.eu/doc/document/ST-8654–2017-INIT/en/pdf (accessed 03 July 2018).

Anonymous (2017b) Guardia Civil: Dismantlement of an organization devoted to the illegal in elvers/baby eels. Available at: http://www.guardiacivil.es/en/prensa/noticias/6121.html (accessed 29 June 2018).

Anonymous (2017c) Joint press release on the occasion of the tenth meeting of the informal consultation on international cooperation for conservation and management of Japanese eel stock and other relevant eel species. Available at: https://cites.org/sites/default/files/eng/com/ac/29/inf/E-AC29-Inf-13.pdf (accessed 03 May 2018).

Anonymous (2018a) Evaluation of eel restocking across Europe and recommendations for improvement. SEG-Report:2018–2-V1. Available at: http://www.sustainableeelgroup.org/wp-content/uploads/2018/04/SEG_Report-2018–2-V1.pdf (accessed 22 May 2018).

Anonymous (2018b) Quantifying the illegal trade in European glass eels (Anguilla anguilla): Evidences and Indicators. SEG-Report:2018–1-V2. Brussels. Available at: http://www.sustainableeelgroup.org/wp-content/uploads/2018/05/SEG-Report-2018–1-V2.pdf (accessed 01 May 2018).

Arai, T. (2016) *Biology and Ecology of Anguillid Eels*. CRC Press, Abingdon.

AMFC (2014) *Draft Addendum IV to the Interstate Fishery Management Plan for American Eel – Written comment summaries*. Atlantic Marine Fisheries Commission, Arlington, VA.

ASMFC (2000a) *Interstate Fishery Management Plan for American Eel*. Atlantic States Marine Fisheries Commission, Washington, DC.

ASMFC (2000b) *Summary minutes – American Eel Technical Committee Meeting January 11, 2000 8:30 a.m.*. Atlantic States Marine Fisheries Commission, Washington, DC.

ASMFC (2000c) *Summary minutes – American Eel Technical Committee Meeting May 18, 2000 2:15 pm*. Atlantic States Marine Fisheries Commission, Washington, DC.

ASMFC (2002) *2002 Review of the Atlantic states marine fisheries commission fishery management plan for American eel (Anguilla rostrata)*. Atlantic States Marine Fisheries Commission, Washington, DC.

ASMFC (2004) *2004 Review of the Atlantic states marine fisheries commission fishery management plan for American eel (Anguilla rostrata)*. Atlantic States Marine Fisheries Commission, Washington, DC.

ASMFC (2005) *2005 Review of the Atlantic states marine fisheries commission fishery management plan for American eel (Anguilla rostrata)*. Atlantic States Marine Fisheries Commission, Washington, DC.

ASMFC (2006a) Joint meeting of the American eel technical committee and the stock assessment sub-comittee. Atlantic States Marine Fisheries Commission, Washington, DC.

ASMFC (2006b) *Review of the Atlantic states marine fisheries commission fishery management plan for American eel (Anguilla rostrata)*. Atlantic States Marine Fisheries Commission, Washington, DC.

ASMFC (2006c) *Stock Assessment Report No. 06–01 of the Atlantic States Marine Fisheries Commission Terms of Reference and Advisory Report to the American Eel Stock Assessment Peer Review January 2006*. Atlantic States Marine Fisheries Commission, Washington, DC.

ASMFC (2008) *ADDENDUM II TO THE FISHERY MANAGEMENT PLAN FOR AMERICAN EEL*. Atlantic States Marine Fisheries Commission, Arlington, VA.

ASMFC (2009) *Review of the Atlantic states marine fisheries commission fishery management plan for American eel (Anguilla rostrata) 2009*. Atlantic States Marine Fisheries Commission, Washington, DC.

ASMFC (2012) *Stock Assessment Report No. 12–01 of the Atlantic States Marine Fisheries Commission American Eel Benchmark Stock Assessment*. Atlantic States Marine Fisheries Commission, Arlington, VA.

ASMFC (2013a) *2012 Review of the Atlantic states marine fisheries commission fishery management plan for American eel (Anguilla rostrata) 2011 FISHING YEAR*. Atlantic States Marine Fisheries Commission, Arlington, VA.

ASMFC (2013b) *Law Enforcement Committee May 20 and 21, 2013 Draft Agenda*. Atlantic States Marine Fisheries Commission, Arlington, VA.

ASMFC (2015a) *2015 Review of the Atlantic States Marine Fisheries Commission Fishery Management Plan for American Eel (Anguilla rostrata)*. Atlantic States Marine Fisheries Commission, Arlington, VA.

ASMFC (2015b) *Atlantic States Marine Fisheries Commission American Eel Management Board November 3, 2015*. Atlantic States Marine Fisheries Commission, Arlington, VA.

ASMFC (2016a) 'Introduction to the commission', *Commissioner Manual*. Atlantic States Marine Fisheries Commission, Arlington, VA.

ASMFC (2016b) *Memorandum Re: Advisory Panel Review of North Carolina's Aquaculture Plan*. Atlantic States Marine Fisheries Commission, Arlington, VA.

ASMFC (2017a) *2017 American Eel Stock Assessment Update*. Atlantic States Marine Fisheries Commission, Arlington, VA.

ASMFC (2017b) *2017 Review of the Atlantic states marine fisheries commission fishery management plan for American eel (Anguilla rostrata) 2016 FISHING YEAR*. Atlantic States Marine Fisheries Commission, Arlington, VA.

ASMFC (2018) *DRAFT ADDENDUM V TO THE AMERICAN EEL FISHERY MANAGEMENT PLAN FOR PUBLIC COMMENT Commercial Yellow and Glass/Elver Eel Allocation and Management*. Atlantic States Marine Fisheries Commission, Arlington, VA.

CBP (1991) *Chesapeake Bay American eel Fishery Management Plan*. U.S. Environmental Protection Agency, Washington, DC.

Bornarel, V., Lambert, P., Briand, C., Antunes, C., Belpaire, C., Ciccotti, E., Diaz, E., Diserud, O., Doherty, D., Domingos, I., Evans, D., de Graaf, M., O'Leary, C., Pedersen, M., Poole, R., Walker, A., Wickström, H., Beaulaton, L. and Drouineau, H. (2017) Modeling the recruitment of European eel (*Anguilla anguilla*) throughout its European range. *ICES Journal of Marine Science* 75(2), 541–552.

Boubée, J.A., Mitchell, C.P., Chisnall, B.L., West, D.W., Bowman, E J. and Haro, A. (2001) Factors regulating the downstream migration of mature eels (A*nguilla* spp.) at Aniwhenua Dam, Bay of Plenty, New Zealand, *New Zealand Journal of Marine and Freshwater Research* 35(1), 121–134.

Briand, C., Bonhommeau, S., Beaulaton, L. and Castelnaud, G. (2008) An appraisal of historical glass eel fisheries and markets: landings, trade routes and future prospect for management. In: Moriarty, C. (ed.) *The Institute of Fisheries Management Annual Conference 2007*. The Institute of Fisheries Management, Westport, CT.

Budimawan (1997) The early life history of the tropical eel Anguilla marmorata (Quoy and Gaimard, 1824) from four Pacific estuaries, as revealed from otolith microstructural analysis. *Journal of Applied Ichthyology* 13(2), 57–62.

CITES (2017) *Species specific matters. Eels (Anguilla spp.). Illegal trade in Anguilla anguilla. Sixty-ninth meeting of the Standing Committee Geneva (Switzerland), 27 November -1 December 2017. SC69 Doc. 47.2*. Geneva. Available at: https://cites.org/sites/default/files/eng/com/sc/69/E-SC69–47–02.pdf (accessed 29 June 2018).

CITES (2018a) *Implementation of the CITES Appendix II listing of European Eel Anguilla Anguilla. CITES AC30 Doc. 18.1 Annex 1*. Available at: https://cites.org/sites/default/files/eng/com/ac/30/E-AC30–18–01-A1.pdf (accessed 27 July 2018).

CITES (2018b) *Month-long transcontinental operation hit wildlife criminals hard*. Available at: https://cites.org/eng/news/month-long-trans-continental-operation-hit-wildlife-criminals-hard_20062018 (accessed 03 July 2018).

Craig, J.F. (2016) *Freshwater Fisheries Ecology*. Wiley, Chichester.

Crook, V. (2014) Slipping away: International *Anguilla* eel trade and the role of the Philippines. TRAFFIC International and ZSL, UK.

Crook, V. and Nakamura, M. (2013) Glass eels: assessing supply chain and market impacts of a CITES listing on *Anguilla* species. *Traffic Bulletin* 25(1).

Day, J.W., Kemp, W.M., Yanez-Arancibia, A. and Crumb, B.C. (2012) *Estuarine Ecology*, 2nd edn.Wiley-Blackwell, Hoboken, NJ.

Dekker, W. (2003a) Did lack of spawners cause the collapse of the European eel, *Anguilla anguilla*? *Fisheries Management and Ecology* 10, 365–376.

Dekker, W. (2003b) On the distribution of the European eel (*Anguilla anguilla*) and its fisheries. *Canadian Journal of Fisheries and Aquatic Sciences* 60, 787–799.

Dekker, W. (2016) Management of the eel is slipping through our hands! Distribute control and orchestrate national protection. *ICES Journal of Marine Science* 73(10), 2442–2452.

Dekker, W. (2019) The history of commercial fisheries for European eel commenced only a century ago. *Fisheries Management and Ecology* https://doi.org/10.1111/fme.12302.

Dekker, W. and Beaulaton, L. (2016a) Faire mieux que la nature – the history of eel restocking in Europe. *Environment and History* 22(2), 255–300.
Dekker, W. and Beaulaton, L. (2016b) Climbing back up what slippery slope? Dynamics of the European eel stock and its management in historical perspective. *ICES Journal of Marine Science: Journal du Conseil* 73(1): 5–13.
Desk, N. (2018) Higher prices expected amid plunge in glass eel catches in Japan, *The Japan News*, 26 February 2018. Available at: http://annx.asianews.network/content/higher-prices-expected-amid-plunge-glass-eel-catches-japan-67725 (accessed 23 Jul 2018).
DMR (2018) The Maine Eel and Elver Fishery. Maine Department of Marine Resources. Available at: https://www.maine.gov/dmr/science-research/species/eel-elver/factsheet.html (accessed 31 July 2018).
DMR (2018a) *Chapter 32 – Eels.* ME. Available at: https://www.maine.gov/dmr/laws-regulations/regulations/documents/Chapter32.pdf (accessed 27 July 2018).
DMR (2018b) *The Maine Eel and Elver Fishery.* ME: Maine Department of Marine Resources. Available at: https://www.maine.gov/dmr/science-research/species/eel-elver/factsheet.html (accessed 28 July 2018 2018).
DNR (2018c) *Confidential Fishery Dependent Data Explained.* South Carolina Department of Natural Resources, Columbia, SC.
DoJ (2016a) Seven Men Plead Guilty for Illegally Harvesting and Selling American Eels. *In:* Affairs, O.o.P. (ed.). Department of Justice, Washington, DC.
DoJ (2016b) Three Additional Men Plead Guilty for Illegally Harvesting and Selling American Eels. *In:* Affairs, O.o.P. (ed.). Department of Justice, Washington, DC.
DoJ (2017a) Brooklyn Seafood Dealer Pleads Guilty for Illegally Trafficking American Eels. *In:* Affairs, O.o.P. (ed.). Department of Justice, Washington, DC.
DoJ (2017b) Maine Fisherman Sentenced for Illegally Trafficking American Eels. *In:* Affairs, O.o.P. (ed.). Department of Justice, Washington, DC.
DoJ (2017c) Three Men Plead Guilty to Illegally Trafficking American Eels. *In:* Affairs, O.o.P. (ed.). Department of Justice, Washington, DC.
DoJ (2018a) Maine Men Sentenced for Illegally Trafficking American Eels. *In:* Affairs, O.o.P. (ed.). Department of Justice, Washington, DC.
DoJ (2018b) Two Men Indicted for Illegally Trafficking American Eels. *In:* Affairs, O.o.P. (ed.). Department of Justice, Washington, DC.
Ebersole, R. (2017) Inside the multimillion-dollar world of eel trafficking. *National Geographic* 13 June.
Ebersole, R. (2018) 19 eel smugglers sentenced, but lucrative trade persists. *National Geographic* 27 June.
Eding, E. (2014) RAS Technologie for Eels in Europe. Available at: https://docplayer.net/50967939-Ras-technologies-for-eels-in-europe.html (last accessed 8 October 2018)
Ege, V. (1939) A revision of the Genus A*nguilla,* shaw: systematic, phylogenetic and geographical study. *Dana Report* 16(3), 8–256.
EU (2007) *Council Regulation (EC) No 1100/2007 of 18 September 2007 establishing measures for the recovery of the stock of European eel.* Official Journal of the European Union. Available at: http://eur-lex.europa.eu/eli/reg/2007/1100/oj (accessed 26 March 2016).
Europol (2018a) *3.4 tonnes of seized glass eels reintroduced into their natural habitat.* Available at: https://www.europol.europa.eu/newsroom/news/34-tonnes-of-seized-glass-eels-reintroduced-their-natural-habitat (accessed 29 June 2018).
Europol (2018b) *Glass eel traffickers earned more than EUR 37 million from illegal exports to Asia.* Available at: https://www.europol.europa.eu/newsroom/news/glass-eel-traffickers-earned-more-eur-37-million-illegal-exports-to-asia (accessed 05 May 2018).

Facey, D.E., Van Den Avyle, M.J., Group, U. S. A. E. W. E. S. C. E. and Center, N. W. R. (1987) *American Eel.* Fish and Wildlife Service, US Department of the Interior, Washington, DC.

FDDC (2018) *Wabanaki Tribes.* Four Directions Development Corporation, Orono, ME. Available at: http://www.fourdirectionsmaine.org/wabanaki-tribes/ (accessed 19 Jul 2018 2018).

Han, Y., Hung, C., Liao, Y. and Tzeng, W. (2010) Population genetic structure of the Japanese eel *Anguilla japonica:* panmixia at spatial and temporal scales. *Marine Ecology Progress Research* 401, 221–232.

Hardy, J.D.J. (1978) *Development of fishes of the Mit-Atlantic Bight: an atlas of egg, larval and juvenile stages. Vol. II Anguillidade through Syngnathidae.* Fish & Wildlife Service, Department of Interior, Washington, DC.

Helfman, G.S. and Bozeman, E.L. (1984) Size, age, and sex of American eels in a Georgia river. *Transactions of the American Fisheries Society* 113(2), 132–141.

Helfman, G.S., Facey, D.E., Hales, L.S., Jr and Bozeman, E.L., Jr (1987) Reproductive ecology of the American eel. American Fisheries Society Symposium 1:42–56. Winston Salem, NC.

Helme, H., Hirschinger, J., Wolff, Y. and Sasal, P. (2018) Temporal dynamics of the recruitment of glass eels in two valleys of French Polynesia (Tahiti and Moorea Islands). *Cybium* 42(4), 341–348.

Hossain, M., Islam, M., Hosain, S., Khan, M. and Khalequzzaman, S. (2007) Status and potential of eel fisheries in Bangladesh. *Journal of Soil and Nature* 1, 46–51.

ICES (2010) Report of the Planning Group on Recreational Fisheries (PGRFS), 7–11 June 2010, Bergen, Norway. ICES CM 2010/ACOM:34.

ICES (2013) Report of the Joint EIFACC/ICES/GFCM Working Group on Eel (WGEEL), 18–22 March 2013 in Sukarietta, Spain, 4–10 September 2013 in Copenhagen, Denmark, ICES CM 2013/ACOM:18.

ICES (2016) Report of the Working Group on Eels (WGEEL), 15–22 September 2016, Cordoba, Spain. ICES CM 2016/ACOM:19.

ICES (2017) Report of the Joint EIFACC/ ICES /GFCM Working Group on Eel (WGEEL), 3–10 October 2017, Kavala, Greece. ICES CM 2017/ACOM:15.

Interpol (2018) Wildlife Crime: global seizures and arrests in transcontinental operation. Available at: https://www.interpol.int/News-and-media/News/2018/N2018–058 (accessed 29 October 2018)

Jacoby, D. and Gollock, M. (2014a) *Anguilla japonica. The IUCN Red List of Threatened Species 2014: e.T166184A1117791* (accessed 20 July 2018).

Jacoby, D. and Gollock, M. (2014b) *Anguilla anguilla. The IUCN Red List of Threatened Species 2014: e.T60344A45833138.* Available at: www.iucnredlist.org (accessed 20 July 2018).

Jacoby, D. and Gollock, M. (2014c) *Anguilla marmorata. The IUCN Red List of Threatened Species 2014: e.T166189A45832585* (accessed 2 July 2018).

Jacoby, D. and Gollock, M. (2014d) *Anguilla borneensis. The IUCN Red List of Threatened Species 2014: e.T198973A2545681* (accessed 09 July 2018).

Jacoby, D. and Gollock, M. (2014e) *Anguilla luzonensis. The IUCN Red List of Threatened Species 2014: e.T18435966A45833133.*

Jacoby, D. and Gollock, M. (2014f) *Anguilla interioris. The IUCN Red List of Threatened Species 2014: e.T198972A2545630* (accessed 02 July 2018).

Jacoby, D. and Gollock, M. (2014g) *Anguilla celebesensis. www.iucnredlist.org* (accessed 02 July 2018).

Jacoby, D. and Gollock, M. (2014h) *Anguilla mossambica. The IUCN Red List of Threatened Species 2014: e.T183155A1730550* (accessed 20 July 2018).

Jacoby, D. and Gollock, M. (2014i) *Anguilla megastoma. The IUCN Red List of Threatened Species 2014: e.T196301A2443332.*

Jacoby, D. and Gollock, M. (2014j) *Anguilla obscura . The IUCN Red List of Threatened Species 2014: e.T196302A2443435* (accessed 20 July 2018).

Jacoby, D., Harrison, I. and Gollock, M. (2014b) *Anguilla bengalensis. The IUCN Red List of Threatened Species 2014: e.T61668607A15501445* (accessed 09 July 2018).
Jacoby, D., Harrison, I. and Gollock, M. (2014a) *Anguilla bicolor. The IUCN Red List of Threatened Species 2014: e.T166894A67015710* (accessed 20 July 2018).
Jacoby, D., Casselmann, J., DeLucia, M., Gollock, M. (2017): *Anguilla rostrata.* The IUCN Red List of Threatened Species 2017: e.T191108A121739077 (accessed 31 July 2018).
Jellyman, D. (1991) Biology of the shortfinned eel Anguilla obscura in Lake Te Rotonui, Mitiaro, Cook Islands. *Pacific Science* 45(4), 362–373.
Jellyman, D.J., Booker, D.J. and Watene, E. (2009) Recruitment of *Anguilla* spp. glass eels in the Waikato River, New Zealand. Evidence of declining migrations? *Journal of Fish Biology* 74, 2014–2033.
Jenkins, R., Keane, A., Rakotoarivelo, A., Rakotomboavonjy, V., Randrianandrianina, F., Razafimanahaka, H., Ralaiarimalala, S. and Jones, J. (2011) Analysis of patterns of bushmeat consumption reveals extensive exploitation of protected species in eastern Madagascar. *PloS One* 6(12).
Jones, J.P., Andriamarovololona, M.M. and Hockley, N. (2008) The importance of taboos and social norms to conservation in Madagascar. *Conservation Biology* 22(4), 976–986.
Kaifu, K., Maeda, H., Yokouchi, K., Sudo, R., Miller, M. J., Aoyama, J., Yoshida, T., Tsukamoto, K. and Washitani, I. (2014) Do Japanese eels recruit into the Japan Sea coast?: A case study in the Hayase River system, Fukui Japan. *Environmental Biology of Fishes* 97(8), 921–928.
Kaifu, K., Itakura, H., Amano, Y., Shirai, K., Yokouchi, K., Wakiya, R., Murakami-Sugihara, N., Washitani, I. and Yada, T. (2017) Discrimination of wild and cultured Japanese eels based on otolith stable isotope ratios. *ICES Journal of Marine Science* 75(2), 719–726.
Kaifu, K., Yokouchi, K., Higuchi, T., Itakura, H. and Shirai, K. (2018) Depletion of naturally recruited wild Japanese eels in Okayama, Japan, revealed by otolith stable isotope ratios and abundance indices. *Fisheries Science* 84(5), 757–763.
Kumagai, A., Ando, R., Miyatake, H., Greimel, P., Kobayashi, T., Hirabayashi, Y., Shimogori, T. and Miyawaki, A. (2013) A bilirubin-inducible fluorescent protein from eel muscle. *Cell* 153(7), 1602–1611.
Kuroki, M., Aoyama, J., Miller, M., Wouthuyzen, S., Arai, T. and Tsukamoto, K. (2006) Contrasting patterns of growth and migration of tropical Anguillid leptocephali in the Western Pacific and Indonesian seas. *Marine Ecology Progress Series* 309, 233–246.
Le, D., Nguyen, D., Harino, H., Kakutani, N., Chino, N. and Arai, T. (2010) Distribution of trace metals and methylmercury in soft tissues of the freshwater eel A*nguilla marmorata* in Vietnam. *Archives of Environmental Contamination and Toxicology* 59, 282–290.
Leander, N.J., Shen, K.-N., Chen, R.-T. and Tzeng, W.-N. (2012) Species composition and seasonal occurrence of recruiting glass eels (A*nguillas*pp.) in the Hsiukuluan River, eastern Taiwan. *Zoological Studies* 51, 59–71.
Lecomte-Finiger, R. (2003) The genus Anguilla Schrank, 1798: current state of knowledge and questions. *Reviews in Fish Biology and Fisheries* 13(3), 265–279.
Lecomte-Finiger, R., Lo-Yat, A. and Yan, L. (2001) First record of Anguilla glass eels from an atoll of French Polynesia: Rangiroa, Tuamotu Archipelago. *Atoll Research Bulletin* 486, 1–6.
Linnaeus, C. (1758) *Systema Naturae per Regna Tria Naturae*, 10 edn. Stockholm.
Lukita, B. (2012) Manisnya budidaya ikan sidat. *Kompas* 2 April 2012.
Marquet, G. (1992) Létude du recrutement et de la physiologie des anguilles de Polynésie Françaize permet-elle de cerner leur aire de ponte. *Bulletin de linstitut océanographique de Monaco* 10, 129–147.
Marquet, G. and Galzin, R. (1991) The eels of French Polynesia: taxonomy, distribution and biomass. *La Mer* 29, 8–17.

McCleave, J.D. and Kleckner, R.C. (1982) Selective tidal stream transport in the esturine migration of glass eels of the American eel (*Anguilla rostrata*). *Journal Du Conseil* 40(3), 262–271.

Minegishi, Y., Aoyama, J. and Tsukamoto, K. (2008) Multiple population structure of the giant mottled eel, *Anguilla marmorata*. *Molecular Ecology* 17, 3109–3122.

Minegishi, Y., Toyama, J., Yoshikawa, N. and Tsukamoto, K. (2012) Lack of genetic heterogeneity in the Japanese eel based on a spatiotemporal sampling. *Coastal Marine Science* 35(1), 269–276.

MITSC (2018) *Section: Statutes*. ME: MITSC. Available at: http://www.mitsc.org/library.php?do=section&name=Statutes (accessed 31 July 2018).

Moriarty, C. and Dekker, W. (1997) *Management of the eel. Irish Fisheries Bulletin 15*. Available at: http://hdl.handle.net/10793/197 (accessed 24 November 2016).

Murray, F. and Little, D. (2000) Fisheries marketing systems and consumer preferences in Puttalam District Sri Lanka. Working Paper, UK Department for International Development, London.

Muthmainnah, D., Honda, S., Suryati, N. and Bi, P. (2016) Understanding the current status of anguillid eel fisheries in Southeast Asia. *Fish for the People* 14, 19–25.

Nagano, T. and Ushio, A. (2018) Retailer Aeon in move to keep eel eating culture alive in Japan. *Asahi Shimbun* 30 June.

National Post (2018) Largest detention ever: Canadians seized eel meat in 92-country wildlife sting. Available at: https://nationalpost.com/pmn/news-pmn/canada-news-pmn/canadian-officials-seize-eel-meat-in-92-country-wildlife-sting-2 (accessed 29 October 2018).

Nelva, A., Persat, H. and Chessel, D. (1979) Une nouvelle méthode détude des peuplements ichtyologiques dans les grands cours deau par échantillonnage ponctuel dabondance. *C.R. Acad. Sci. Paris* 289 (D), 1295–1298.

Nijman, V. (2010) An overview of international wildlife trade from Southeast Asia. *Biodiversity Conservation* 19, 1101–1114.

Nijman, V. (2015) CITES-listings, EU eel trade bans and the increase of export of tropical eels out of Indonesia. *Marine Policy* 58(0), 36–41.

Nurhayat, W. (2015) Dilarang diekspor, benih Ikan Ular sidat diselundupkan. *Detik Finance* 26 June 2015.

OMNRF (2018) American eel. Ontario Ministry of Natural Resource and Forestry. Available at: https://www.ontario.ca/page/american-eel (Accessed 28 July 2018).

Pantulu, V. (1958) Studies on the biology of the Indian fresh-water eel *Anguilla bengalensis* Gray. *Proceedings of the National Academy of Sciences, India* 22, 259–280.

Pethiyagoda, R. (1991) *Freshwater fishes of Sri Lanka*. The Wildlife Heritage Trust of Sri Lanka, Colombo.

Pols, M. (2014) Why should Asia harvest the long-term elver profits? *Portland Press Herald* 10 October 2014. Available at: https://www.pressherald.com/2014/04/13/why_should_asia_harvest_the_long-term_elver_profits__/ (accessed 13 July 2018).

Pramod, G., Pitcher, T.J. and Mantha, G. (2017) Estimates of illegal and unreported seafood imports to Japan. *Marine Policy* 84, 42–51.

Prosek, J. (2011) *Eels: An Exploration, from New Zealand to the Sargasso, of the Worlds Most Amazing and Mysterious Fish*. Harper Perennial, New York.

Purwaty, H., Oktaviana, R. and Suparno, O. (2018) Eel business development strategy in PT Laju Banyu Semesta (Labas). *Indonesian Journal of Business and Entrepreneurship* 4, 99–110.

Real, N. (2012) Importers target eels as domestic prices soar. *Fish Information Services* 26 July.

Resources, S. o. M. D. o. M. (2018) *Elver Landings Reported as of Noon May 24, 2018*. ME. Available at: https://www.maine.gov/dmr/news-details.html?id=792787 (accessed 19 July 2018).

Riaz, M. and Chaudry, M. (2003) *Halal Food Production*: CRC Press, Baca Raton, FL.

Ricker, F.W. and Squires, T. (1974) *Spring Elver Survey – Pilot Project.* Maine Department of Marine Resources, August, ME.

Robinet, T., Lecomte-Finiger, R., Escoubeyrou, K. and Feunteun, E. (2003) Tropical eels *Anguillas* recruiting to Reunion Island in the Indian Ocean: taxonomy, patterns of recruitment and early life histories. *Marine Ecology Progress Series* 259, 263–272.

Shepard, S.L. (2015) *American Eel Biological Species Report.* US Fish & Wildlife Service, Hadley, MA.

Shiraishi, H. and Crook, V. (2015) Eel market dynamics: an analysis of *Anguilla* production, trade and consumption in East Asia. TRAFFIC, Tokyo.

Shrestha, T. (2013) Conservation and Management of Fishes in the Large Himalayan Rivers of Nepal. International Symposium on the Management of Large Rivers for Fisheries: Sustaining Livelihoods and Biodiversity in the New Millennium, Phnom Penh, Nepal.

Sipayik, P.A. (2013) Part 5: American Eel Management, ME: www.wabanaki.com. Available at: http://www.wabanaki.com/wabanaki_new/documents/American%20Eel%20Management%20Plan%20Part%205.pdf (accessed 27 Jul 2018).

South Carolina Department of Natural Resources (2018) *Confidential Fishery Dependent Data Explained.* South Carolina Department of Natural Resources, Columbia, SC.

Stein, F.M., Wong, J.C.Y., Sheng, V., Law, C.S.W., Schröder, B. and Baker, D. M. (2016) First genetic evidence of illegal trade in endangered European eel (*Anguilla anguilla*) from Europe to Asia. *Conservation Genetics Resources* 8(4), 533–537.

Sudo, R., Okamura, A., Fukuda, N., Miller, M.J. and Tsukamoto, K. (2017) Environmental factors affecting the onset of spawning migrations of Japanese eels (*Anguilla japonica*) in Mikawa Bay Japan. *Environmental Biology of Fishes* 100(3), 237–249.

Sudrajat, S. (2017) Sidat Banyuwangi dan Situbondo Merambah Negeri Sakura. *Detik Finance* 18 December 2017.

Sugeha, H. and Suharti, S. (2008) Discrimination and distribution of two tropical short-finned eels (*Anguilla bicolor bicolor* and *Anguilla bicolor pacifica*) in the Indonesian water. *The Nagisa Westpac Congress*: 1–14.

Talwar, P. and Jhingran, A. (1991) *Inland Fishes of India and Adjacent Countries.* A.A. Balkeman, Rotterdam.

Tanaka, E. (2014) Stock assessment of Japanese eels using Japanese abundance indices. *Fisheries Science* 80(6), 1129–1144.

Tanaka, H., Kagawa, H., Ohta, H., Unuma, T. and Nomura, K. (2003) The first production of glass eel in captivity: fish reproductive physiology facilitates great progress in aquaculture. *Fish Physiology and Biochemistry* 28 (1–4), 493–497.

Thibault, I., Dodson, J.J. and Caron, F. (2007) Yellow-stage American eel movements determined by microtagging and acoustic telemetry in the St Jean River watershed, Gaspe, Quebec, Canada. *Journal of Fish Biology* 71(4), 1095–1112.

Thomas, M. (1985) Eel farmer wriggles in -- trouble in the potato patch. *Orlando Sentinel* 9 April. Available at: http://articles.orlandosentinel.com/1985–04–09/news/0290080086_1_eel-whiteside-potato-farmers (accessed 13 July 2018).

Tjandraningsih, C. (2013) Indonesia eel hot item for smugglers. *Japan Times* 29 July 2013.

Trotter, B. (2018) Price offered for Maines baby eels hits record high. *Bangor Daily News* 27 March. Available at: https://bangordailynews.com/2018/03/27/business/fisheries/price-offered-for-maines-baby-eels-hits-record-high/ (accessed 19 July 2018).

UNIDO (2013) Case Study: Chinese Eel Exports. Regional Trade Standards Compliance Report – East Asia 2013. United Nations Industrial Development Organization, Vienna.

USFWS (2006) Lacey Act O. o. L. Enforcement. 18 USC 42–43. US Fish and Wildlife Service, Washington, DC.

Vandammen, S., Griffiths, A., Taxlor, S., Di Muri, C., Hankard, E., Towne, J., Watson, M. and Mariani, S. (2016) Suhsi barcoding in the UK: another kettle of fish. *PeerJ* 4(e1891).

Violi, L., Falcone, G., De Luca, A.I. and Chies, L. (2015) Sustainabillity of European Eel population: a statistical survey on production, conservation status and market trends. *Food Safety Management* 16(148), 83–89.

Walker, N.J., Lee, L., Rootes-Murdy, K., Dolloff, C., Prasad, V., De Mutsert, K. and Aguirre, A.A. (2017) Stock assessment, prioritization of habitat, conservation priorities, and cultural significance of the American eel (*Anguilla rostrata*) in the Chesapeake Bay Watershed. *FISH Magazine* 3(127), 22–25.

Watanabe, S., Miller, M., Aoyama, J. and Tsukamoto, K. (2011) Analysis of vertebral counts of the tropical anguillids, A*nguilla megastoma,* A. *obscura,* and A.*reinhardtii,* in the western South Pacific in relation to their possible population structure and phylogeny. *Environmental Biology of Fishes* 91(3), 353–360.

PART VII

Appendices

Appendix 1 Programme from the 1st International Eel Science Symposium

ZSL
LET'S WORK FOR WILDLIFE

ifm
Institute of Fisheries Management

Environment Agency

The 1st UK International Eel Science Symposium

The Zoological Society of London. June 13th - 15th 2017.

'SCIENCE INTO MANAGEMENT'

Symposium Supporters

Main symposium sponsor

hydrolox®

Symposium dinner sponsor

hydrolox®

Thames Water

Networking events sponsor

Sustainable *Eel* Group

Student attendance sponsor

RESEARCH INSTITUTE
NATURE AND FOREST

Symposium co-conveners

ZSL
LET'S WORK FOR WILDLIFE

Environment Agency

Tuesday June 13th 9:30 – 17:15

	Session 1	
	Setting the Scene	
	Session Chair. (tbc)	
09.30	**Andy Don**, Environment Agency (UK)	Welcome from the Symposium Steering Group
09.40	**Official Symposium Opening** Presented by **Hugh Fearnley-Whittingstall**	
09.50	**Alan Walker**, Cefas (UK) and Chair ICES Anguillid Section	**Key Note** Challenges of management from a scientific perspective.
10.20	**Matthew Gollock**, Zoological Society of London (UK) and IUCN	Anguillid conservation – the Global context.
10.45	**Hiromi Shiraishi**, TRAFFIC (Japan)	The illicit global trade in eels.
11.05	**BREAK**	
	Session 2	
	Species and Regions	
	Session Chair. Andrew Kerr, Sustainable Eel Group (UK)	
11.35	**Tom Kwak**, North Carolina State University (USA)	Ecology and Conservation of the American Eel (*Anguilla rostrata*) in the Caribbean Region.
11.55	**Robert Schabetsberger**, University of Salzburg (Austria)	From myth to reason – search for a spawning area of tropical eels in the western South Pacific.
12.15	**Kenzo Kaifu**, Chuo University (Japan)	How science can contribute to conservation and sustainable use of Japanese eel?
12.35	**Céline Hanzen**, University of KwaZulu Natal (S. Africa)	Changes in historical distribution and genetic diversity of anguillid eels along the east coast of KwaZulu-Natal, South Africa.
12.55	**Peter Phillipsen**, Nature at Work (Holland)	The Rhine Eel Roadmap.
13.00	**Markus Diekmann**, Dept. for Inland Fisheries (Germany)	On the actual recruitment of European eel in the River Ems, Germany.
13.05	**Antoine Legault**, Fishpass (France)	Silver Eel stock Evaluation in France.
13.10	**LUNCH**	

	Session 3	
	Species and Regions	
	Session Chair. Andy Don, Environment Agency (UK)	
14.10	**Willem Dekker**, Swedish University of Agricultural Sciences (Sweden)	What eel did my Grandma eat? Dynamics of the European eel fisheries in historical perspective.
14.30	**Emna Derouiche**, INSTM (Tunisia)	How many downstream silver eels escaped from Ichkeul Lake (Northern Tunisia)? A mark-recapture experiment.
14.50	**Alejandro Belen/Matthew Gollock**, ZSL (Philippines)	Developing an Eel Management Plan for the Philippines – Process and Challenges.
15.10	**BREAK**	
	Session Chair. Charles Crundwell, Environment Agency (UK)	
15.40	**Pierre Sasal**, University of Perpignan (France)	Research and conservation of eels in the South Pacific region.
16.00	**Kazuki Yokouchi**, National Research Institute of Fisheries Science (Japan)	Decadal monitoring of the Japanese eel demography in several model river systems in Japan.
16.20	**Al Amouri Mohamed**, Mohamed V University (Morocco)	Eel stock in Morocco.
16.40	**Hikaru Itakura**, Kobe University (Japan)	Site fidelity, diel and seasonal activity of growth phase Japanese eel in the freshwater habitat inferred from acoustic telemetry.
17.05	**Jonah Tosney**, Norfolk Rivers Trust (UK)	A catchment-based approach to eel conservation on a small Norfolk river: The Glaven Eel Project.
17.10	**Pieterjan Verhelst**, Ghent University (Belgium)	Emigration of European silver eel (*Anguilla anguilla*) from a polder system into the Schelde estuary.
	17.15 Close of Day 1	

19.30 – 23.00 Drinks and Poster Reception *RSA Vaults, the Strand*
Wednesday June 14th 8:50 – 17:20

	Session 4	
	Marine Phase Studies (1)	
	Session Chair. Paul Coulson, Institute of Fisheries Management (UK)	
08.50	**Paul Coulson**, Institute of Fisheries Management (UK)	Opening remarks from the Organising Committee.
08.55	**David Righton**, Cefas (UK)	**Key Note** Extreme swimming: eels on their oceanic spawning migration.
09.25	**Nobuto Fukuda**, National Research Institute of Fisheries Science (Japan)	Oceanic migration of each developmental stage in the Japanese eel.

Session 4		
	Marine Phase Studies (1)	
	Session Chair. Paul Coulson, Institute of Fisheries Management (UK)	
09.45	**Martin Castonguay**, Fisheries and Oceans Canada (Canada)	Behavioural plasticity in American eels migrating across the continental shelf into the Sargasso Sea.
10.05	**Elsa Amilhat**, University of Perpignan (France)	Silver Eel migration behaviour in the Mediterranean Sea.
10.25	**Hakán Wickstrom**, Swedish University of Agricultural Sciences (Sweden)	Growth in stocked eels—promising results from a full scale marking experiment in a marine environment.
10.45	**Yoichi Miyake**, University of Tokyo (Japan)	Recruitment mechanism of *Anguilla japonica* glass eels implied by 50-year catch and oceanographic data and model.
11.05	**BREAK**	
	Session 5	
	Monitoring and stock management	
	Session Chair. Matthew Gollock, Zoological Society of London (UK)	
11.35	**Caroline Durif**, Institute of Marine Research (Norway)	Depth, home range and migratory behaviour of European eels tracked in coastal marine waters.
11.55	**Nick Walker**, George Mason University (USA)	Prioritizing Fish Habitat for American eel (*Anguilla rostrata*) in the Chesapeake Bay Watershed.
12.15	**Derek Evans**, AFBI (UK)	Statistical analysis of the eel population dynamics of Lough Neagh, Northern Ireland.
12.35	**Kohma Arai**, University of Tokyo (Japan)	A study on the natural distribution of Marine Phase.
12.55	**Priit Bernotas**, Estonian University of Life Sciences (Estonia)	First results of using an enclosure fyke net system for estimating European eel abundance in a large and shallow lake.
13.00	**LUNCH**	
	Session 6	
	Monitoring and Stock Management	
	Session Chair (tbc)	
14.00	**Jonathan Bolland**, Hull International Fisheries Institute (UK)	Route choice, passage and survivability of eel at a lowland pumping station.
14.20	**Sarah Mount**, State University of New York (USA)	The Development of a Non-Lethal Maturity Index for American Eels.
14.40	**Jeroen Van Wichelen**, Research Institute for Nature and Forest (Belgium)	Evaluation of adjusted tidal barrier management for European glass eel migration: upstream dispersal and feeding behaviour.

	Session 6	
	Monitoring and Stock Management	
	Session Chair (tbc)	
15.00	**Adam Piper**, ZSL and Environment Agency (UK)	Response of seaward-migrating European eel (*Anguilla anguilla*) to hydrodynamics.
15.20	**Andrew Vowles**, Southampton University (UK)	Summary of passage/screening ecohydraulic studies carried out at flumes.
15.40	**BREAK**	
	Session 7	
	Monitoring and Stock Management (Restocking)	
	Session Chair (tbc)	
16.10	**Thomas Pratt**, Fisheries and Oceans Canada (Canada)	Lessons from a large-scale American eel conservation stocking experiment.
16.30	**Maidu Silm/Mehis Rohtla**, Estonian University of Life Sciences (Estonia)	Can we distinguish strontium chloride marked European eels by country from isotope signals?
16.50	**Shingo Kimura**, University of Tokyo (Japan)	Inhabitation of the European eel in the Japanese rivers.
17.10	**Bjoern Kullmann**, Hamburg University (Germany)	Mass-marking of farmed European eels *Anguilla anguilla* (Linnaeus, 1758) with alizarin red S.
17.15	**Billy Nzau Matondo**, University of Liège (Belgium)	Long-term monitoring of European eels in the Belgian Meuse River basin. From the historical drastic decline to recent outcomes of restocking practices.
	17:20 Close of Day 2	

19:00 – 22:45 - Symposium Dinner on the River Thames

Departing from Tower Pier at 19:00

During the evening we will have a number of presentations as we enjoy a glass of wine and some fine food.

Speakers

Andrew Kerr, Sustainable Eel Group (UK)

Working towards a sustainable future for the eel

Sophie Edwards, Thames Water (UK)

The Thames Water Eel Programme. Screening, science, challenges and opportunities

Karl Bousfield, Hydrolox (UK/USA)

Hyrdolox Eel and Fish Screening Solutions

Steve Colclough, Institute of Fisheries Management (UK)

The history of eels in the Thames

Thursday June 15th 8:50–16:15

Session 8		
Threats		
Session Chair. Iain Turner, Institute of Fisheries Management (UK)		
08.50	**Paul Coulson**, Institute of Fisheries Management (UK)	Opening remarks from the Organising Committee.
09.00	**Olga Haenen**, Wageningen University (Holland) Key Note	**Key Note** The most important diseases of European eel in Europe in the last 30 years.
09.30	**Neil Lewin**, Environment Agency (UK)	Silver Service: A Standard Protocol for Eel Health Examinations.
09.50	**Kuan-Mei Hsiung**, University of Tokyo (Japan)	Effect of ENSO events and global warming on larval and juvenile duration and transport process of the Japanese eel (*Anguilla japonica*).
10.10	**Maria Mateo**, IRSTEA (France)	Assessing the impact of anthropogenic pressures on temperate eels using Genetics & Evolutionary ecology based model for eels (GenEveel).
10.30	**Kieran McCarthy**, National University of Ireland Galway (Ireland)	Parasite assemblages of European eels in Irish freshwater, mixohaline and marine habitats.
10.35	**Ben Griffioen**, Wageningen Marine Research (Holland)	Silver eel migration in the river Meuse 2002 – 2015: mortality estimations and behavioural patterns.
10.40	**Brittany Dixon**, University of New Brunswick (Canada)	Near dam, spatio-temporal distribution of migrating American eel elvers and juvenile eels in the downstream reach of the Mactaquac Generating Station.
10.45	**BREAK**	

Session 9		
Threats		
Session Chair. Alan Walker, Cefas (UK)		
11.15	**Tatsuki Yoshinaga**, Kitasato University (Japan)	CITES-listing and other social forces shrank the commercial utilization of the European eel in the Japanese market between 2011 and 2016.
11.35	**David Buysse**, Research Institute for Nature and Forest (Belgium)	Mortality of European eel after downstream migration through three types of pumping station.
11.55	**Claude Belpaire**, Research Institute for Nature and Forest (Belgium)	Standardisation of eel quality assessments and the effects of contaminants on the eel stock.
12.15	**Estibaliz Diaz**, Marine Research. Itsas Ikerketa (Spain)	European eel (Anguilla anguilla) under climatic change.

	Session 9	
	Threats	
	Session Chair. Alan Walker, Cefas (UK)	
12.35	**Florian Stein**, Sustainable Eel Group (UK)	Combating the illegal trade in European eels.
12.40	**Rob Pilcher**, AECOM (UK)	New guide to eel screening and passage in the UK.
12.45	**Hendrik Doerner**, European Commission (Italy)	Feeding ecology of anguillid eels.
12.50	**LUNCH**	
	Session 10	
	Aquaculture and Genetics	
	Session Chair (tbc)	
13.50	**Oliviero Mordenti**, University of Bologna (Italy)	Captive Breeding of the European Eel. A review of the work in Italy.
14.10	**Anke Zernack**, Tanenuiarangi Manawatu Inc (New Zealand)	PROJECT TUNA - Overcoming the hurdles to establish an environmentally, economically and socially sustainable shortfin eel aquaculture industry in New Zealand.
14.30	**Chrysoula Gubili**, Helenic Agricultural Organisation (Greece)	Panmixia in the tropics: lack of genetic structure in sympatric eels.
14.50	**Andrew Griffiths**, University of Bristol (UK)	Sushi barcoding in the UK: another kettle of fish.
	Matthew Gollock, Zoological Society of London (UK) and IUCN	
15.10	**BREAK**	
	Session 11	
	15.40 – 16.15 Open discussion. The future	
	Facilitated by Matthew Gollock, Zoological Society of London (UK)	
15.40	**Alan Walker**, Cefas (UK) and Chair ICES Anguillid Section	Highlights and further work from the symposium.
16.15	**SYMPOSIUM CLOSE**	

Appendix 2
Pictures from the symposium

USGS

ZSL
LET'S WORK
FOR WILDLIFE

Ocean Tracking Network
ZSL
LET'S WORK
FOR WILDLIFE

ZSL
LET'S WORK
FOR WILDLIFE

ZSL
LET'S WORK FOR WILDLIFE
CONSERVATION PROGRAMMES
hydro
#OneLess
mti

pumping stations
PAUL

ifm
Institute of Fisheries Management
ZSL
LET'S WORK FOR WILDLIFE
hydrolox
Environment Agency
ZSL
LET'S WORK
FOR WILDLIFE

Appendix 3 *Fisheries Management and Ecology*, special issue, Volume 26, Number 1, February 2019

The front cover and table of contents from the special issue are reproduced below with kind permission of John Wiley & Sons Ltd.

The issue is open access and available here: https://onlinelibrary.wiley.com/toc/13652400/2019/26/1.

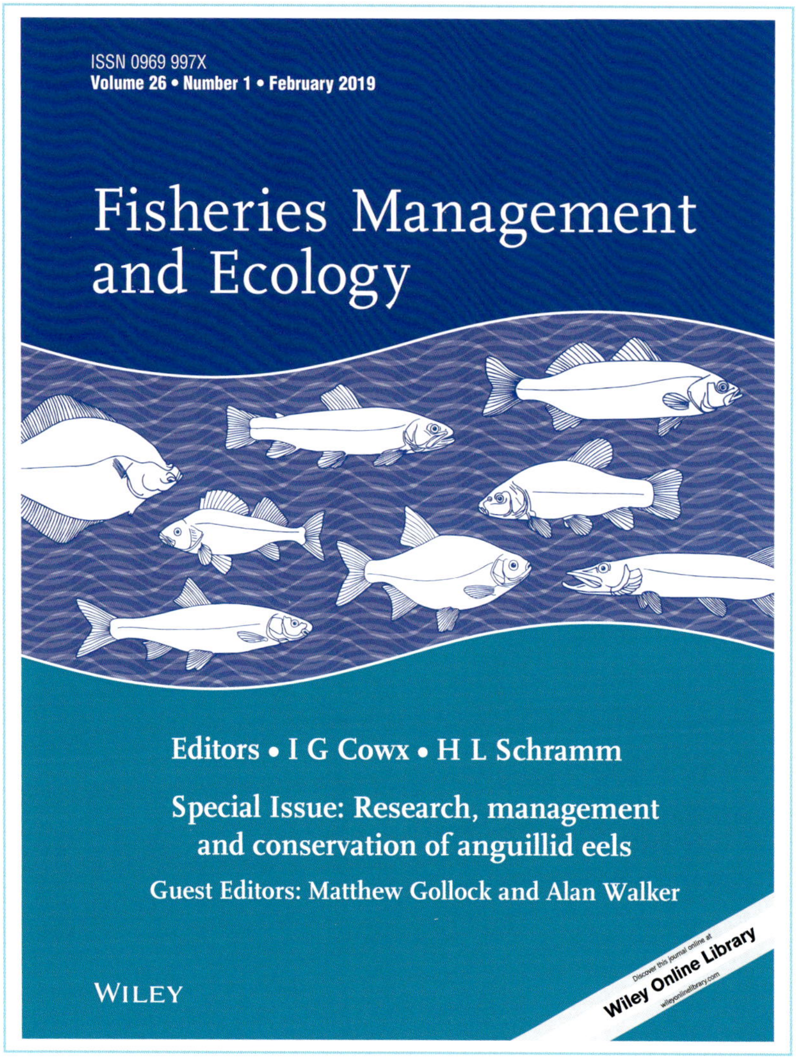

Volume 26 • Number 1 • February 2019

Fisheries Management and Ecology

Contents

Fisheries Management and Ecology is covered by Current Contents/Fisheries, FISHLIT, ISI Alerting Services, SciSearch (Science Citation Index-Expanded) and Wildlife Review

Typeset by SPS Ltd, Chennai, India.
Printed in Singapore.

WILEY

This journal is available online.
Visit wileyonlinelibrary.com/journal/fme
to search the articles and register for table
of contents e-mail alerts.

Index